PROTEIN PURIFICATION

Principles, High Resolution Methods, and Applications

Editors

Jan-Christer Janson
and
Lars Rydén

VCH

Jan-Christer Janson
R & D Department
Pharmacia LKB Biotechnology AB
S-751 82 Uppsala,
Sweden

Lars Rydén
Department of Biochemistry
Uppsala University
Uppsala Biomedical Centre
Box 576
S-751 23 Uppsala, Sweden

Library of Congress Cataloging in Publication Data

Protein purification: principles, high resolution methods, and
  applications/editors, Jan-Christer Janson and Lars Ryden.

     p.   cm.
     Bibliography: p.
     Includes index.
     ISBN 0-89573-122-3
     1. Proteins—Purification. 2. Chromatographic analysis.
  3. Electrophoresis. I. Janson, Jan-Christer. II. Rydén, L. (Lars)
QP551.P69754   1989
547.7′5—dc20   89-9134
                                          CIP

British Library Cataloguing in Publication Data

Protein purification
   1. Proteins. Purification
   I. Janson, Jans—Christer, 1938–
   II. Ryden, Lars, 1940–
   547.7′5

   ISBN 0-89573-122-3

Printed in the United States of America.

ISBN 0-89573-122-3 VCH Publishers
ISBN 3-527-26184-2 VCH Verlagsgesellschaft

Printing History:

10 9 8 7 6 5 4

Published jointly by:

VCH Publishers, Inc.
220 East 23rd Street
Suite 909
New York, NY 10010

VCH Verlagsgesellschaft mbH
P.O. Box 10 11 61
D-6940 Weinheim
Federal Republic of Germany

VCH Publishers (UK) Ltd.
8 Wellington Court
Cambridge CB1 1HW
United Kingdom

Contributors

John Brewer
Scientific and Technical Service
Pharmacia LKB Biotechnology AB
S-751 82 Uppsala
Sweden

Jan Carlsson
Dept. of Explorative Chemistry
Pharmacia Diagnostics AB
S-751 82 Uppsala
Sweden

Kjell-Ove Eriksson
Department of Biochemistry
Uppsala University
Uppsala Biomedical Centre
Box 576
S-751 23 Uppsala
Sweden

Bo Ersson
Biochemical Separation Centre
Uppsala University
Uppsala Biomedical Centre
Box 577
S-751 23 Uppsala
Sweden

Millicent A. Firestone
Center for Separation Science
University of Arizona
Bldg. 20
Tucson, AZ 85721
USA

Lars Hagel
R & D Department
Pharmacia LKB Biotechnology AB
S-751 82 Uppsala
Sweden

Milton T. W. Hearn
Centre for Bioprocess Technology
Monash University
Clayton, Victoria
Australia

T. William Hutchens
Department of Pediatrics
USDA/ARS Children's Nutrition
    Research Center
Baylor College of Medicine
Houston, TX 77030
USA

Jan-Christer Janson
R & D Department
Pharmacia LKB Biotechnology AB
S-751 82 Uppsala
Sweden

Göte Johansson
Department of Biochemistry
University of Lund
Chemical Center
Box 124
S-221 00 Lund
Sweden

Karl-Erik Johansson
The National Veterinary Institute
Box 7073
S-750 07 Uppsala
Sweden

Jan-Åke Jönsson
Department of Analytical Chemistry
University of Lund
Chemical Centre
Box 124
S-221 00 Lund,
Sweden

Lennart Kågedal
R & D Department
Pharmacia LKB Biotechnology AB
S-751 82 Uppsala
Sweden

Evert Karlsson
Department of Biochemistry
Uppsala University
Uppsala Biomedical Centre
Box 576
S-751 23 Uppsala
Sweden

Torgny Låås
R & D Department
Pharmacia LKB Biotechnology AB
S-751 82 Uppsala
Sweden

Jorge A. Lizana
Scientific and Technical Service
Pharmacia LKB Biotechnology AB
S-751 82 Uppsala
Sweden

Lars Rydén
Department of Biochemistry
Uppsala University
Uppsala Biomedical Centre
Box 576
S-751 23 Uppsala
Sweden

Marianne Sparrman
R & D Department
Pharmacia LKB Biotechnology AB
S-751 82 Uppsala
Sweden

Wolfgang Thormann
Dept. of Clinical Pharmacology
University of Bern
Murtenstrasse 35
CH-3010 Bern
Switzerland

Preface

Over the last two decades the scientific community has witnessed an unprecedented expansion within the biosciences and biotechnology. This expansion has been to a large extent driven by advances in several key areas, most notably recombinant DNA technology, hybridoma and cell culture techniques and, finally, in biochemical separation methods. This book is a description of the current status of one of these areas: modern techniques for protein purification and analysis.

The research on which the progress in separation techniques is based has been conducted both in university departments, devoted to basic research, and in industrial laboratories whose main concern is the development of new equipment and tools. In many cases the two communities have cooperated to their mutual benefit. In fact, a great number of the products now available for the separation and purification of proteins, such as chromatographic media with a wide range of selectivities and efficiencies, as well as equipment for electrophoretic separation and analysis, were originally developed in a university setting. This book is also the result of a joint effort between university researchers, in particular at Uppsala University, and the research staff of a company, Pharmacia LKB Biotechnology. Although it is thus a product of this condition of mutual benefit, the ambition has not been to give a selective description of methods or materials from a single commerical source, but rather to give an unbiased account of all key techniques in the field.

Today it is to a great extent possible to base the separation of proteins on knowledge of their molecular properties, structural as well as functional. Suggestions on how to solve a separation problem can best be made if data on protein structure and function, including particular structural details, is available. Conversely, results from the application of a particular separation method can often be interpreted in terms of molecular properties of the protein under study. Throughout the text of this book, separation results are related to protein properties, often in a detailed manner. We are the first generation to be on the verge of rational protein management.

Starting with this general concept, we have aimed at providing students, teachers and research workers in biomedicine, bioscience and biotechnology with a concise and practical treatise covering, in a single volume, all important chromatographic and electrophoretic techniques used in preparative and analytical protein chemistry. The book contains a general introductory chapter on protein preparative work, Chapter 1, where the key concepts are introduced. Similarly, a general introduction to chromatography is given in Chapter 2 and an introduction to analytical electrophoresis in Chapter 12. The major chromatographic and electrophoretic techniques are presented in individual chapters, including one chapter on affinity partitioning in aqueous polymer two-phase systems.

No single person can today be even close to acquiring the amount of experience necessary to describe with confidence the wealth of techniques and methods which makes up the arsenal for protein separations. We have thus chosen to produce a multi-author volume recruiting expertise from the entire field. All chapters have, however, been thoroughly worked through by the editors to achieve a reasonable uniformity of style and organization. Each chapter deals first with the theory and underlying principles of each separation technique, followed by a section on methodology, and ends with a number of representative application examples described in detail.

The preparation of this book has been a matter of several years. We would like to thank the authors for their cooperation, from the first planning stage to the last phase of updating and addition. We would also like to thank our editors at VCH Publishers in New York, in particular Dr. Edmund H. Immergut who took the first initiative and who followed the project up to its realization. The management and staff of Pharmacia LKB Biotechnology are thanked for their cooperation and support which allowed the selling price to be considerably reduced. Many staff members have made invaluable contributions to the final result, which are gratefully acknowledged. We also thank Elizabeth Hill and Ursula Snow for their contributions in the early phase of the project; Inger Galvér, Gull-Maj Hedén, Inga Johansson and Madeleine de Sharengrad for secretarial work; Bengt Westerlund for handling the computer programmes for the chemical structures; Uno Skatt and Lilian Forsberg for producing a number of the illustrations; and David Eaker and John Brewer for keeping our freedom with the English language within limits.

Finally, we would like to add that we are well aware that much of our own efforts, occasional achievements and sometimes hard-won experience, as well as that of several of the other authors of this book, spring from the tree planted long ago by The Svedberg and Arne Tiselius, and later kept alive by Jerker Porath and Stellan Hjertén and many of their colleagues and pupils through fifty years of separation science at Uppsala University. We offer this book as the latest fruit of this tree, hopefully to be enjoyed by many.

Uppsala, Sweden, June 21, 1989

Jan-Christer Janson
Lars Rydén

Foreword

Biosciences have experienced a phenomenal development during the last 50 years. Discoveries and new concepts have moved the frontiers forward, revealing details of the inner structures of living matter and transcending boundaries which many biochemists a generation ago would have thought to be beyond the reach of science for many centuries ahead. The accelerated speed of advancement of knowledge in molecular biology and related fields would have been impossible without the tools provided by separation science.

Most, if not all, of the contributors to this monograph are in some way, directly or indirectly, "lineal decendents" of the pioneers of the 1940s: Arne Tiselius, Archer J. P. Martin and others. Already in the late 1930s Tiselius realized that electrophoresis must be supplemented by other methods if it ever should be possible to isolate pure antibodies from the gamma globulin fraction of human serum. He concentrated his efforts on chromatography, formulating three modes of development: elution, frontal and displacement chromatography.

It is somewhat amusing to compare two expert opinions about the relative importance of electrophoresis and chromatography. "At present chromatography is the premier method for separating organic and biological mixtures" (Barry Karger) and "Electrophoresis is the premier separation method for biological compounds such as proteins and polynucleotides, despite recent applications of HPLC to these biopolymers" (Wolfgang Thormann and M. A. Firestone, this volume). These contradictory statements reflect the present activities in both of these areas. Other fields, such as affinity extraction, are also treated; whereas, for example, ultracentrifugation is not discussed in this monograph.

I remember discussions we had some 15 years ago concerning whether it was desirable to continue research and development in those areas of separation science which are supposed to contribute to the solution of actual problems in biochemistry and molecular biology. Why should we, so went the arguments, hand over the keys necessary to open the doors to the treasures of life sciences? Why not grab the chance to make a scoop in some field of extreme importance? I advocated the idea that most of us could make more important contributions behind or beside the stage on which most exciting discoveries are made. In keeping to current problems in separation science we will always enjoy continuous contact with the exciting fields in biosciences that happen to be in focus. Therefore, a tradition in separation science is likely to be more long-lasting than are traditions in any other field of chemistry.

Apropos, Scheidekunde in old Dutch—literally the art of separation—is equivalent to chemistry, an art or a science that none disclaim the importance of.

It is interesting to note that ninety percent of the references given in this book are in fact concerned with work done after 1970. How many of these references will appear in an imagined new edition of the book in the year 2000? Will they only be found among the ten percent of old references the majority of which still will date back to before 1970? We can be sure that the introduction of new concepts in the field will be remembered. From this, my conviction, we may learn to be careful not to declare all the ideas of our forerunners and teachers outdated and impossible to revive. In a short perspective, however, the present book will serve as a very useful guide to the daily practice in many laboratories throughout the world.

Uppsala, Sweden, May, 1989

Jerker Porath
Department of Biochemistry and Biochemical Separation Centre
Biomedical Center, Uppsala University
Box 576 and Box 577, respectively
S-751 23 Uppsala, Sweden

Contents

Part I   Introduction

Part II   Chromatography

Part III   Electrophoresis

I   Introduction

1 Introduction to Protein Purification

Bo Ersson
Biochemical Separation Centre
Uppsala University
Uppsala Biomedical Centre
Box 577, S-751 23 Uppsala, Sweden

Lars Rydén
Department of Biochemistry
Uppsala University
Uppsala Biomedical Centre
Box 576, S-751 23 Uppsala, Sweden

Jan-Christer Janson
Pharmacia LKB Biotechnology AB
S-751 82 Uppsala, Sweden

1.1   Introduction

The development of techniques and methods for the separation and purification of biological macromolecules such as proteins has been an important prerequisite for many of the advancements made in bioscience and biotechnology over the past three decades. Improvements in materials and utilisation of microprocessor-based instruments have made protein separations more predictable and controllable, although to many they are still more an art than a science. However, gone are the days when an investigator had to spend months in search of an efficient route to purify an enzyme or hormone from a cell extract. This is a consequence of the development of new generations of chromatographic media with increased efficiency and selectivity as well as of new electrophoretic techniques for fast analysis of protein composition and purity.

In the area of chromatography the development of new bonded porous silicas, new porous resin supports and new cross-linked beaded agaroses has enabled a rapid growth of high resolution techniques (HPLC, FPLC) on an analytical and laboratory prepara-tive scale as well as of industrial chromatography in columns with bed volumes of several hundred litres. Another field of increasing importance is micropreparative chromatogra-phy, a consequence of modern methods for amino acid and sequence analysis requiring submicrogram samples only. The data obtained are efficiently exploited by recombinant DNA technology, and biological activities previously not amenable to proper biochemi-cal study can now be ascribed to identifiable proteins and peptides.

A wide variety of chromatographic column packing materials such as gel-filtration media, ion exchangers, reversed phase packings, hydrophobic interaction adsorbents and affinity chromatography adsorbents are today commercially available. These are based on low pressure media (90–100 μm beads), medium pressure media (30–50 μm) and high pressure media (5–10 μm) in order to satisfy different requirements of efficiency, capacity and cost.

However, not all problems in protein purification are solved by the acquisition of sophisticated laboratory equipment and column packings which give high selectivity and efficiency. Difficulties still remain in finding optimum conditions for protein extraction and sample pretreatment as well as in choosing suitable methods for monitoring protein and biological activity. These problems will be discussed in this introductory chapter. There will also be an overview of different protein separation techniques and their principles of operation. In the subsequent chapters each individual technique will be discussed in more detail. Finally, some basic equipment necessary for efficient protein purification work will be described in this chapter.

Several useful books covering protein separation and purification from different points of view are available on the market[1–3]. In Methods of Enzymology, especially volumes 22, 34, and 104 [4–6], a number of very useful reviews and detailed application reports will be found. Also the booklets available from manufacturers of separation equipment and media can be helpful by providing detailed information regarding their products.

1.2   The Protein Extract

1.2.1   Choice of Raw Material

In most cases, interest is focused on one particular biological activity, such as an enzyme, and the origin of this activity is often of little importance. Then great care should be taken in the selection of a suitable source for the enzyme. Among different sources there might be considerable variation with respect to the concentration of the enzyme, the availability and cost of the raw material, the stability of the enzyme, the presence of interfering activities and proteins and difficulties in handling a particular raw material. Very often it is compelling to choose a particular source because it has been described previously in the literature. However, sometimes it is advantageous to consider an alternative choice. One of us had difficulty in obtaining guinea pig brains for the preparation of histamine-methyl transferase. It turned out to be quite easy to prepare the same enzyme from pig kidney as a cheap and easily available source[7]. In an attempt to study the subunit structure of human ceruloplasmin, a major difficulty was its sensitivity to proteolytic degradation. Here is was demonstrated that porcine ceruloplasmin is quite similar, is present in higher concentrations, and is not nearly as sensitive to proteases[8].

The traditional animal or microbial sources may today be replaced by genetically engineered micro-organisms or cultured eucaryotic cells. Protein products of eucaryotic orgin, cloned and expressed in bacteria such as E. coli, may either be located in the cytoplasm or be secreted through the cell membrane. In the latter case they are either collected inside the periplasmic space or they are truly extracellular, secreted to the culture medium. Proteins which accumulate inside the periplasmic space may be selectively released either into the growth medium by changing the growth conditions[9], or following cell harvesting and washing of resuspended cell paste. Already at this stage, a considerable degree of purification is thus achieved by choosing a secreting strain as illustrated in Fig 1-1.

1.2.2   Extraction Methods

Some biological materials constitute themselves a clear or nearly clear protein solution suitable for direct application to chromatography columns after centrifugation or filtration. Some examples are: blood serum, urine, milk, snake venoms, and—perhaps most importantly—the extracellular medium after cultivation of micro-organisms and mammalian cells as mentioned above. It is normally an advantage to choose such a starting material because of the limited number of components and since extracellular proteins are comparatively stable. Some samples, e.g. urine or cell culture supernatants, are normally concentrated before purification begins.

In most cases, however, one has to extract the desired activity from a tissue or a cell paste. This means that a considerable number of contaminating molecular species are set free and that proteolytic activity will make the preparation work more difficult. The extraction of a particular protein from a solid source often involves a compromise

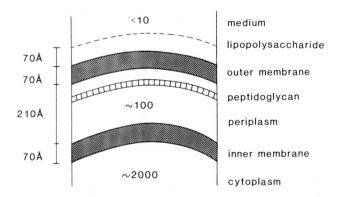

Figure 1-1. Location and approximate numbers of the proteins in *Escherichia coli*. Courtesy M. Uhlén, Royal Institute of Technology, Stockholm, Sweden.

between recovery and purity. Optimization of extraction conditions should favour the release of the desired protein and leave difficult-to-remove contaminants behind. Of particular concern is to find conditions under which the already extracted protein is not degraded or denaturated while more is being released.

Various methods are available for the homogenization of cells or tissues. For further details and discussions the reader is referred to the paper by Kula and Schütte[10]. The extraction conditions are optimized by systematic variation of parameters such as the composition of the extraction medium (see below), time, temperature and operating conditions such as speed of rotation, size and shape of blades, pestles etc.

The proper design of an extraction method thus requires preliminary experiments in which aliquots are taken at various time intervals and analyzed for activity and protein content. Since the number of parameters can be very large this part of the work has to be kept within limits by proper judgement. It is however not recommended to accept a single successful experiment. Further investigations of, in particular, the extraction time required often pays in the long run.

Major problems in preparing a protein are in general denaturation, proteolysis and contamination with pyrogens, nucleic acids, bacteria and viruses. These can be limited by proper choice of extraction medium as we shall show. However, we can already now point out that many of the above problems can be reduced by short preparation times and low temperature. It is thus good biochemical practice to carry out the first preparation steps as fast as possible and at the lowest possible temperature. However, low temperatures are not always necessary and are sometimes inconvenient. The working temperature is therefore one of the parameters which should be optimized carefully, especially if a preparation is to be done routinely in the laboratory.

The extract must be clarified by centrifugation before submission to column chromatography. A preparative laboratory centrifuge is normally sufficient for this step.

A common phenomenon when working with intracellularly expressed recombinant DNA proteins is their tendency to accumulate as insoluble aggregates (inclusion bodies, refractile bodies) which have to be solubilized and refolded to recover their native state. At a first glance, the formation of insoluble aggregates in the cytoplasm might be considered a major problem. However, as the inclusion bodies seem to be fairly well defined with regard to both particle size and density[11], they should provide a unique means for rapid and efficient enrichment of the desired protein simply by low speed fractional centrifugation and washing of resuspended sediment. The critical step is solubilization and refolding, often combined with chromatographic purification under denaturing conditions in the presence of high concentrations of urea or guanidine hydrochloride. This area has been reviewed by Marston[12].

1.2.3 Extraction Medium

To arrive at a suitable composition for the extraction medium one must first study conditions where the protein of interest is stable and, secondly, where it is most efficiently released from the cells or tissue. The final choice is usually a compromise between maximum recovery and maximum purity. The following factors have to be taken into consideration:

(1) *pH*. Normally the pH-value chosen is that of maximum activity of the protein. However it should be noted that this is not always the pH that gives the most efficient extraction, nor is it necessarily the pH of maximum stability. For example, trypsin has an activity optimum at pH 8–9 but is much more stable at pH 3, where autolysis is avoided. The use of extreme pH-values, e.g., for the extraction of yeast enzymes in 0.5 M ammonia, is sometimes very efficient and is acceptable for some proteins without causing too much denaturation.

(2) *Buffer salts*. Most proteins are maximally soluble at moderate ionic strengths, 0.05–0.1, and these values are chosen if the buffer capacity is sufficient. Suitable buffer salts are given in Table 1-1. An acceptable buffer capacity is obtained within one pH unit from the pK values given. The proteins as such also act as buffers and the pH should be checked after addition of large amounts of proteins to a weakly buffered solution. Some extractions do not give rise to acids and bases and thus do not need a high buffer capacity. In other cases this might be necessary and occasional control of the pH value of an extract is recommended.

Table 1-1. Buffer salts used in protein work. Buffer concentration refers to total concentration of buffering species. Buffer pH should be as close as possible to the pK_a value, and not more than one pH unit from the pK_a.

Buffer	pK-values	Properties
Sodium Acetate	4.75	
Sodium Bicarbonate	6.50; 10.25	
Sodium citrate	3.09; 4.75; 5.41	Binds Ca^{2+}
Ammonium acetate	4.75; 9.25	Volatile
Ammonium bicarbonate	6.50; 9.25; 10.25	Volatile
Tris-Chloride	8.21	
Sodium Phosphate	1.5; 7.5; 12.0	
Tris-phosphate	7.5; 8.21	

Table 1-2. Detergents used for solubilization of proteins.

Detergent	Ionic character	Effect on protein	Critical micelle concentration % w/v
Triton X-100	non-ionic	mild non-denaturing	0.02
Nonidet P 40	non-ionic		
Lubrol PX	non-ionic		0.006
Octyl glucoside	non-ionic		0.73
Tween 80	non-ionic		0.002
Sodium deoxycholate	anionic		0.21
Sodium dodecyl sulphate, SDS	anionic	strong denaturing	0.23
CHAPS	zwitter-ionic		1.4

(3) *Detergents, chaotropic agents.* In many extractions the desired protein is bound to membranes or particles, or is aggregated due to its hydrophobic character. In these cases one should reduce the hydrophobic interactions by using either detergents or chaotropic agents (not both!). Some of the commonly used detergents are listed in Table 1-2. Several of them do not denature globular proteins or interfere with their biological activity. Others, such as SDS, will do that. Quite often it is not necessary to continue using a detergent in the buffer after the first step(s) in the purification, so its use is restricted to the extraction medium. In other cases it might be necessary to use a detergent at all times. One is then actually purifying a protein-detergent complex. More information about detergents, including their chemical structures can be found in[13].

Detergents are amphipatic molecules. When their concentration increases they will eventually form micelles at the so called critical micell concentration, CMC. Since micelles often complicate purification procedures, in particular column chromatography, concentrations below the CMC should be used.

Instead of detergents for dissolving aggregates one could try chaotropic agents such as urea or guanidine hydrochloride, or moderately hydrophobic organic compounds such as ethylene glycol. Urea and guanidine hydrochloride have proven particularly useful for the extraction and solubilization of inclusion bodies[12].

(4) *Reducing agents.* The redox potential of the cytosol is lower than that of the surrounding medium where atmospheric oxygen is present. Intracellular proteins often have exposed thiol groups and these might become oxidized in the purification process. Thiol groups can be protected by reducing agents such as DTE, DTT or mercapto-ethanol, see Table 1-3. Normally 10–25 mmolar concentrations are sufficient to protect thiols without reducing internal disulphides. In other cases a higher concentration might be needed[14].

(5) *Chelators or metal ions.* The presence of heavy metal ions can be detrimental to a biologically active protein mainly for two reasons. They can enhance the oxidation of thiols by molecular oxygen and they can form complexes with specific groups, which may cause problems. Heavy metals can be trapped by chelating agents. The most commonly used is ethylenediamine tetraacetic acid, EDTA, in the concentration range 10–25 mM. An alternative is EGTA, which is more specific for calcium. It should be noted that EDTA is a buffer. It is therefore best to add the disodium salt of EDTA before final pH adjustment. The chelating capacity of EDTA increases with increasing pH.

In other cases stabilizing metal ions are needed. Many proteins are stabilized by calcium ions. The divalent ions calcium and magnesium are trapped by EDTA and cannot be used in combination with this chelator.

Table 1-3. Reducing agents.

Agent	Structure
Mercaptoethanol	$HS-CH_2-CH_2-OH$
1,4 Dithioerythritol, DTE	CH_2SH H—OH H—OH CH_2SH
1,4 Dithiotreitol, DTT	CH_2SH H—OH HO—H CH_2SH

(6) *Proteolytic inhibitors.* The most serious threats to protein stability are the omnipresent proteases. The simplest safeguard against proteolytic degradation is normally to work fast in the cold. An alternative or added precaution is to add protease inhibitors (Table 1-4), especially in connection with the extraction step. Often there is a need for a combination of inhibitors, e.g., for both serine proteases and metalloproteases. In general, the protein inhibitors are expensive chemicals which may limit their use in large scale applications. Proteolysis can also be reduced by rapid extraction of the fresh homogenate in an aqueous polymer two phase system[15] or by adsorption of the proteases to hydrophobic interaction adsorbents[16]. Sometimes it is sufficient to adjust the pH to a value at which the proteases are inactive, but where the stability of the protein to be purified is maintained. The classical example is the purification of insulin from pancreas.

(7) *Bacteriostatics.* It is wise to take precautions to avoid bacterial growth in the protein solutions. As was mentioned in the previous section on proteolytic inhibitors, the simplest remedy here is to use sterile filtered buffer solutions as a routine in the

Table 1-4. Proteolytic inhibitors.

Inhibitor	Enzymes inhibited	Working concentration
Diisopropyl fluorophosphate, DFP	Serine proteases	(avoid DFP)
Phenylmethylsulfonyl fluoride, PMSF	Serine proteases	0.5–1 mM
Ethylenediamine tetraacetate, EDTA	Metal-activated proteases	around 5 mM
Cysteine reagents	Cysteine-dependent proteases	
Pepstatin A	Acid proteases	0.1 μm

laboratory. This will also reduce the risk of bacterial growth in columns. A common practice in order to avoid bacterial growth in chromatographic columns is to allow the column to flow at reduced rate even when it is not in operation. Some buffers are more likely than others to support bacterial growth, e.g, phosphate, acetate and carbonate buffers at neutral pH values. Buffers at pH 3 and below or 9 and above usually prevent bacterial growth, but may occasionally allow growth of molds.

Whenever possible it is recommended to add an antimicrobial agent to the buffer solutions. Often used are azide at 0.001 M or Merthiolate at 0.005 % and alcohols such as n-butanol at 1 %. Azide has the drawback that it is a nucleophilic substance and binds metals. In cases where these substances may interfere with activity measurements or the chromatography itself it is always possible to add the substances to solutions of the protein to be stored.

1.3   An Overview of Fractionation Techniques

In early work, complex protein mixtures were fractionated mainly by extraction and precipitation methods. These methods are still used today as preliminary steps for crude fractionation or to obtain more clear or more concentrated sample solutions. Preparative electrophoretic and chromatographic techniques developed during the 1950s and 1960s, made possible rational purification protocols and laid the foundation for the situation we have today. This section gives a short overview of the various techniques normally used in preparative biochemical work. Chapter 2 of the book contains an introduction to chromatography and Chapter 12 gives an introduction to electrophoresis. Each individual chromatographic and electrophoretic separation technique is then treated in detail in subsequent chapters.

1.3.1   Precipitation, Extraction and Centrifugation

Precipitation of a protein in an extract may be achieved by adding salts, organic solvents or organic polymers, or by varying the pH or temperature of the solution. The most commonly used precipitation agents are listed in Table 1-5. The property of a particular salt as a precipitation agent is described by the so-called Hofmeister series

anions: PO_4^{3-}, SO_4^{2-}, CH_3COO^-, Cl^-, Br^-, NO_3^-, ClO_3^-, I^-, SCN^-

cations: NH_4^+, K^+, Na^+, guanidine $C(NH_2)_3^+$

The so-called *antichaotropic salts* to the left are the most efficient salting out agents. They increase the hydrophobic effect in the solution and promote protein aggregation by

Table 1-5. Precipitation agents.

Agent	Type	Properties
Ammonium sulphate	Salt	Easily soluble, stabilizing
Sodium sulphate	Salt	
Ethanol	Solvent	Flammable, risk of denaturation
Acetone	Solvent	Flammable, risk of denaturation
Polyethylene glycol, PEG	Polymer	Uncharged, unflammable

association of hydrophobic surfaces. The chaotropic salts on the right hand side in the series decrease the hydrophobic effect, and thus help to maintain the proteins in solution.

Organic solvents promote the precipitation of proteins due to the decrease in water activity in the solution as the water is replaced by organic solvent. They have been widely used as precipitation agents, especially in the fractionation of serum proteins. The following five variables are usually kept under control: concentration of organic solvent, protein concentration, pH, ionic strength and temperature[17]. Low temperature during the precipitation operations is often necessary to avoid protein denaturation: the addition of an organic solvent decreases the freezing point of the solution and temperatures below 0°C can be used. In reversed phase chromatography proteins are chromatographed in solutions that contain up to about 50% organic solvent often with retention of biological activity.

Organic polymers function in a way similar to that of organic solvents. The most widely used polymer is polyethylene glycol, (PEG), with an average molecular weight of either 6000 or 20,000. The main advantage of PEG over organic solvents is that it is more easily handled. It is inflammable, it is not poisonous, it is uncharged and it is cheap. Rather low concentrations are required (often less than 25%) to precipitate most proteins. One disadvantage is that concentrated solutions of PEG are highly viscous. PEG can also be difficult to remove from protein solutions. However, after dilution with buffer the viscosity decreases and since the substance is uncharged the solution may be applied directly to an ion exchange column to further separate the proteins and to remove the polymer.

pH adjustment has been used as a simple and cheap way to precipitate proteins. Proteins have their lowest solubility at their isoelectric point. This is sometimes used in serum fractionation and also in the purification of insulin.

Besides pH, another parameter which influences precipitation of proteins in salt solutions is temperature (see below). Keeping the salt concentration constant and varying the temperature is another way of fractionating a protein solution.

The salting out of a protein can be described by the equation

$$\log S = B - K \cdot c$$

Where S = the solubility of the protein in g/l of solution
  B = intercept constant
  K = salting out constant
  c = salt concentration in moles/1.

The value of B depends on the salt used, the pH, the temperature and the protein itself. K depends on the salt used and the protein. It should be stressed that addition of a salt or another precipitating agent to a protein solution only decreases the solubility of the proteins. This is why a very dilute protein solution for precipitation may lead to low recovery, since a major part of the protein simply remains in solution. Reproducible results can only be achieved if all parameters mentioned above, including the protein concentration, are kept constant.

Centrifugation is used routinely in the protein purification laboratory to recover precipitates. It can also be used to separate two immiscible liquid phases. Another application is density gradient centrifugation. Today this is used mostly for the fractionation of subcellular particles and nucleic acids. An alternative is the use of liquid-liquid phase extraction which seems to offer several advantages over the more classical methods.

1.3.2 Electrophoresis

Electrophoresis in free solution or in macroporous gels such as 1–2% agarose separates proteins mainly according to their net electric charge. Electrophoresis in gels such as polyacrylamide separates mainly according to the molecular size of the proteins.

Today, analytical gel electrophoresis requiring μgram amounts of proteins is an important tool in bioscience and biotechnology (see Chapter 12). Convenient methods for the extraction of proteins after the electrophoresis have been developed, especially protein blotting (see Chapter 16), making the technique micropreparative. There are also many instances where a very small amount of protein is sufficient for the analysis of size and composition as well as the primary structure. Finally there are cases where the starting material is extremely limited, e.g., protein extracts from small amounts of tissue (biopsies etc). In these cases the protein "extract" might be just large enough for gel-electrophoretic analysis.

Larger scale (milligrams to gram of protein) electrophoresis was an important method for the fractionation of protein extracts during the 1950s and early 1960s. It was carried out using columns packed with, e.g., cellulose powder as convection depressor as in the "Porath-column"[18]. An innate limitation of preparative column electrophoresis is the joule heat developed during the course of the experiment. This means that the column diameter, to allow sufficient cooling, should not exceed approximately 3 cm. Several hundred milligrams of protein can, however, be separated on such columns. Column zone electrophoresis has the advantage of allowing a precise description of the separation parameters involved and is besides gel filtration the mildest separation technique available for proteins. It can be recommended for special situations, but practical aspects and the excessive time required precludes routine use. Methods for large and medium scale preparative electrophoresis have been developed, e.g., the flowing-curtain electrophoresis of Hannig[19] and, more recently, the "Biostream" apparatus of Thomson[20].

Isoelectric focusing, the other main electrophoretic technique, separates proteins according to their isoelectric points. This technique gives very high resolution, but presents major difficulties as a preparative large- or medium scale technique. Special equipment is required to allow cooling during the focusing. Proteins often precipitate at their isoelectric point and this precipitate can contaminate the other bands when a vertical Sephadex bed or column with sucrose gradient is used as anticonvection medium in the focusing experiment. Modern equipment for preparative isoelectric focusing[21] avoids these problems by dividing the separation chamber into compartments. Another solution is to carry out the isoelectric focusing in a horizontal trough of sedimented gel particles such as Sephadex[22]. Here, precipitation in one zone will not distrub the other bands. On the other hand the recovery of proteins is more tedious.

For routine preparative protein fractionation the electrophoretic techniques have become less important than chromatography. Ion exchange chromatography depends on parameters similar to those for electrophoresis. Chromatofocusing fractionates proteins largely according to their isoelectric points and is thus a more convenient alternative to preparative isoelectric focusing. The description of the various electrophoretic techniques in part III of the book thus concentrates on analytical use and microscale preparations.

1.3.3 Chromatography

Separation by chromatography depends on the differential partition of proteins between a stationary phase (the chromatographic medium or the adsorbent) and a mobile phase (the buffer solution). Normally the stationary phase is packed into a vertical column of plastic, glass or stainless steel, whereas the buffer is pumped through this column. An alternative is to stir the protein solution with the adsorbent batch-wise and then pour the slurry onto an appropriate filter and make the washings and desorptions on the filter.

Column chromatography has proved to be an extremely efficient technique for the separation of proteins in biological extracts. Since the development of the first cellulose ion exchangers by Peterson and Sober[23] and of the first practical gel filtration media by Porath and Flodin[24,25] a wide variety of adsorbents has been introduced which exploit various properties of the protein for the fractionation. The more important of these are listed below, together with the chromatographic method where they dominate the separation:

(a) Size and shape—gel filtration
(b) Net charge and distribution of charged groups—ion exchange chromatography
(c) Isoelectric point—chromatofocusing
(d) Hydrophobicity—hydrophobic interaction chromatography and reversed phase chromatography
(e) Metal binding—immoblized metal ion affinity chromatography
(f) Content of exposed thiol groups—covalent chromatography
(g) Biospecific affinities for ligands, inhibitors, receptors, antibodies etc.—affinity chromatography

The methods often have very different requirements with regard to chromatographic conditions. This applies to ionic strength, pH and various additives such as detergents, reducing agents and metals. By proper adjustment of the buffer composition the conditions for adsorption and desorption of the desired protein can be optimized. It should be stressed that the result of a particular chromatographic separation often depends on more than one parameter. In ion exchange chromatography the charge interaction is the dominant parameter but molecular weight and hydrophobic effects can also contribute to some degree depending on the experimental conditions.

Highly specific methods, such as those based on bioaffinity, e.g., antibody-antigen interaction, do in some cases give a highly pure protein in a single step. Normally, however, one has to combine several chromatographic methods to achieve complete purification of a protein from a crude biological extract. With the wide variety of chromatographic media available today, this can normally be done in a short period of time.

All of the chromatographic methods mentioned above are treated in part II, which begins with a general description of the concepts used in protein chromatography.

1.4 Fractionation Strategies

1.4.1 Introductory Comments

Before attempting to design a protein purification protocol, one should collect as much information as possible about the characteristics of the protein and preferably also about the properties of the most important impurities. Useful data involve approximate molecular weight and pI, degree of hydrophobicity, presence of carbohydrate (glycoprotein) or free —SH, etc. Some of this information might be obtained already on a DNA level, if nucleotide sequence data are available, but is otherwise often collected easily by preliminary trials using crude extracts.

One should also establish criteria with regard to the stability of the protein to be purified. Important parameters affecting structure are temperature, pH, organic solvents, oxygen (air), heavy metals and mechanical shear. Special concern should be addressed to the risk of proteolytic degradation.

According to a recent study of 100 published successful protein purification procedures[26] the average number of steps in a purification process is four. Very seldom can a protein be obtained in pure form from a single chromatographic procedure, even when this is based on a unique biospecificity. In addition to the purification steps there is often a need for concentrations and sometimes changes of buffers by dialysis or membrane filtrations.

The preparation scheme can be described as consisting of three stages:

(a) The preliminary or initial fractionation stage
(b) The purification stage
(c) The final polishing stage

The purpose of the initial stage is to obtain a solution suitable for chromatography by clarification, coarse fractionation and concentration of the protein extract. The purpose of the final stage is to remove aggregates and degradation products and to prepare a solution suitable for the final formulation of the purified protein.

Sometimes one or two of these stages coincide. An initial ion exchange adsorption step can thus serve as a preliminary fractionation applied directly to the protein extract, or a gel filtration can give a product that is suitable as a final product. However, as the purposes of the three stages are different it is useful to discuss them separately.

The design of the preparation scheme will be different depending on the material at hand and the purpose. If the starting material is very precious one should favour high yield over speed and convenience. If one wants to extract several different proteins from a starting material, that will of course also influence the planning of the work. Finally, the final step is designed such that the product will be suitable for its purpose, which can vary. These aspects will be discussed below.

1.4.2 Initial Fractionation

There are many methods for the clarification of protein solutions. Extracts of fungal or plant origin often contain phenolic substances or other *pigments*. These can be removed by adsorption to celite, either batch-wise or on a short column.

Similarly, *lipid* material can be removed either by centrifugation, where the lipids will float and one thus needs to extract the protein solution from below, or by a chromatographic procedure. Lipids adsorb to a number of materials. Aerosil, a fused silica, has been used for the adsorption of lipids, but agarose is sometimes a simple choice.

Contamination with nucleic acids can in some cases, especially when preparing proteins from bacteria, constitute a problem due to their high viscosity. The classical way to solve this problem is to precipitate the nucleic acids. Streptomycin sulphate and polyethylenimine have been used as precipitants as have protamine sulphate and manganese salts[27]. Another way to solve the problem is to add nucleases, which cut the nucleic acids into smaller pieces thereby reducing the viscosity.

1.4.2.1 Clarification by Centrifugation and Microfiltration

The clarification of any cell homogenate is usually no problem on a laboratory scale, where refrigerated high speed centrifuges covering operating speeds from 20,000 rpm to 75,000 rpm generating from about 40,000 × g to about 500,000 × g can be used. A useful review of centrifugation and centrifuges in preparative biochemistry is found in reference[28]. As a complement to centrifugation, in recent years tangential or cross-flow microfiltration has received increased attention, especially for large scale applications. For a review of the advantages of cross-flow microfiltration we suggest reference[29].

1.4.2.2 Ultrafiltration

Ultrafiltration has become a widely used technique in preparative biochemistry. Ultrafiltration membranes are available with different cut-off limits for separation of molecules from 1000 Daltons up to 300,000 Daltons. The method is excellent for the separation of salts and other small molecules from a protein fraction with higher molecular weight and at the same time can effect a concentration of the proteins. The process is gentle, fast and inexpensive. For a recent review see reference[30].

1.4.2.3 Precipitation

Crude extracts are seldom suitable for direct application to chromatographic columns. Preparative differential centrifugation seldom results in a sufficiently clear solution. This is one reason why it is often necessary to use other means for clarification which simultaneously concentrate the solution and at the same time remove most of the bulk proteins. Such an initial fractionation step should also result in the removal of proteases and membrane fragments that sometimes bind the protein of interest in the absence of detergents. The classical means is to make a fractional precipitation. Firstly bulk proteins in the solution are precipitated together with residual particulate matter and then the protein of interest can be precipitated from the resulting supernatant solution. Sometimes the protein of interest is allowed to remain in the mother liquor solution for direct application to chromatographic columns, e.g. hydrophobic interaction adsorption of proteins in ammonium sulphate solutions and ion exchange chromatography of proteins in polyethylene glycol mother liquors. The most commonly used precipitating agents are listed in Table 1-5 together with some of their properties. A typical precipitation curve is shown in Fig 1-2.

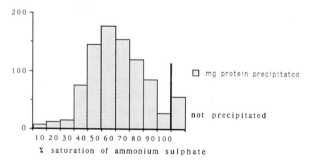

Figure 1-2. Example of a precipitation curve. Amount protein precipitated by stepwise increase of the ammonium sulphate concentration.

Of the various methods available for protein precipitation the classical ammonium sulphate has some disadvantages. The resulting protein solution often needs to be dialysed to obtain an ionic strength that allows ion exchange chromatography. This problem is avoided when using polyethylene glycol, PEG. Organic solvents, in particular ethanol and acetone, often produce extremely fine powder-like precipitates that are difficult to centrifuge and handle. They have also been shown often to cause partial denaturation of proteins, which can, for example, prevent subsequent crystallization. This is why organic solvents are not recommended as first-choice precipitating agents.

1.4.2.4   Liquid-Liquid Phase Extraction

A radically different way of making an initial fractionation is by partitioning in an aqueous polymer liquid-liquid two phase system[31]. These systems often contain polyethylene glycol, PEG, as one phase constituent and another polymer, such as dextran or even salt, as the other. Under favourable conditions it is possible to obtain the protein of interest in the upper, normally the PEG phase. The contaminating bulk protein as well as particles will become collected in the lower phase and can be removed by centrifugation. Particles sometimes stay at the interphase and are thus also removed in the centrifugation step. By covalent attachment of affinity ligands to PEG molecules these can be used for affinity partitioning. This technique is dealt with in detail in Chapter 11.

1.4.3   The Chromatographic Steps

1.4.3.1   Choice of Adsorbent

The first information of the chromatographic behaviour of a protein is often obtained most simply by preliminary analytical scale experiments, e.g., by gel filtration and by ion exchange chromatography using salt and pH gradients. In these runs approximate values of molecular size and ionic properties such as isoelectric points are obtained, information that is fundamental to the further planning of the work. A more thorough survey of the behaviour of the protein on various adsorbents can then be done using a panel of adsorbents. This can be carried out either in a panel of parallel columns or using tandem columns.

The parallel column approach has been developed by Scopes[32] for a panel of dye adsorbents. In this case he used up to 20 small columns containing various dye adsorbents. The columns were equilibrated with a predetermined application or starting buffer. A small volume of the protein extract was applied to each column and the protein content (280-absorption) and the activity in the effluent were measured. Then a predetermined terminating buffer was applied to each column and again the protein and enzyme activity in the effluent was determined. A column where the bulk of the proteins, but not the activity, was adsorbed was chosen as a "minus-column" while an adsorbent where the reverse happened was chosen as a "plus-column". These two columns in combination effected a considerable purification of the desired substance in the actual preparation. In a similar approach, a panel of parallel columns was used earlier by, e.g., Shaltiel[33] for evaluation of hydrophobic adsorbents. The technique can, however, be used for any set up of adsorbents such as different ion exchangers, the same ion exchanger under different conditions, thiol-gels, metal-chelating gels, etc. The elution of the columns can also be performed with more than two elution buffers. The purpose, however, is to get a quick idea of the behaviour of a previously unknown protein and thus the set-up should not be enlarged beyond what can be handled easily in the laboratory.

If the adsorbents used have well defined and continuously increasing adsorption capacities for proteins in general the panel can also be arranged as tandem columns. This approach was used by Porath and co-workers for the immobilized metal ion (IMAC) adsorbents[34]. Here, three columns (Zn, Fe and Cu e.g.) were connected in series and a sample was pumped through all of them. After washing with starting buffer the three columns were disconnected and eluted separately, mostly using gradients. The approach requires that the first column adsorbs few of the proteins present whereas the last adsorbs almost all of them. This technique is not as generally applicable as the use of parallel columns.

1.4.3.2   The Order of the Chromatographic Steps

A priori one would expect that the order in which the different chromatographic steps are applied in a protein purification protocol is of minor importance. The total purification factor should be constant and the product of the factors obtained in each individual step should be independent of the other steps of the protocol. In the ideal case, where each chromatographic technique is utilized optimally with regard to the resolution and recovery, i.e., within the linear regions of the adsorption isotherms (see Chapter 2), with adequate sample volume to column volume ratios and with no adverse viscosity effects, this is probably true. However the real-life situations are always far from ideal or at least such that adaptation to ideality becomes highly impractical. For example, a fractionation gel filtration step can be optimized to give very high resolution (Chapter 3) but only at the cost of time and sample volume. To choose fractionation gel filtration as the first step when the sample volume might be much larger than the total volume of the column, means repetitive injections and excessive and impractical total process times, which would probably also be deleterious to the proteins in the sample solution. Likewise, to choose affinity chromatography on immobilized monoclonal antibodies as the first step would probably result in an extraordinarily high purification factor. However, the high cost of such adsorbents prohibits the use of large columns, which makes repeated injections of sample in smaller columns almost mandatory. This leads to long process times and the risk of product losses and/or modifications due to proteolytic attack. Proteolytic activity can also threaten the stability and life length of the actual immunosorbent. Furthermore, protein-based adsorbents are difficult to maintain to a sufficiently high degree of hygiene. There are limitations with regard to means for regeneration (washing) and sterilization (Chapter 10). This is why they should be saved for the later steps of the purification protocol.

The consequence of these considerations is that there are a number of practical rather than theoretical reasons why one should choose certain chromatographic techniques[26] for the early steps and others for the final steps of a protein purification process. The choice is primarily governed by the following parameters: (1) the sample volume, (2) the protein concentration and viscosity of the sample, (3) the degree of purity of the protein product, (4) the presence of nucleic acids, pyrogens and proteolytic enzymes in the sample and (5) the ease with which different types of adsorbents can be washed free from adsorbed contaminants and denatured protein. The last parameter governs the life length of the adsorbent and, together with its purchasing price, the material cost of the particular purification step.

In the light of what has been said above, the logical sequence of chromatographic steps would be to start with more "robust" techniques which combine a concentration

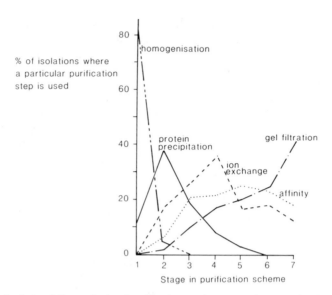

% of isolations where
a particular purification
step is used

Figure 1-3. Analysis of the methods of purification used at successive steps in the purification schemes. The results are expressed as a percentage of the total number of steps at each stage. Adopted from ref. 26 by permission of the authors and publisher.

effect with high chemical and physical resistance and low material cost. The obvious candidates are ion exchange chromatography and to some extent hydrophobic interaction chromatography. As the latter often requires the addition of salt for adequate protein binding it is preferably applied after salt precipitation or after salt displacement from ion exchange chromatography thereby excluding the need for a desalting step. Thereafter, the protein fractions can preferably be applied to a more "specific" and more expensive adsorbent. The protocol is often finished with a gel filtration step (Fig. 1-3).

It is advisable to design the sequence of chromatographic steps in such a way that buffer changes and concentration steps are avoided. The peaks eluted from an ion exchanger can, regardless of the ionic strength, be applied to a gel filtration column. This step also functions as a desalting procedure which means that the buffer used for the gel filtration should be chosen so as to allow direct application of the eluted peaks to the next chromatographic step. The different chromatographies have, in practice, widely different capacities, even though it is possible to adapt several of the methods to a larger scale. However, in the initial stages of a purification scheme it is most convenient to start with the methods that allow the application of large volumes and which have the highest capacities. To this category belong, e.g., ion exchange chromatography and hydrophobic interaction, but any adsorption chromatographic method can be used to concentrate larger volumes, especially in batchwise operations.

1.4.4   The Final Step

The purpose of the final step is to remove possible aggregates or degradation products and to condition the purified protein for its use or storage. The procedure will thus be different depending on the fate of the protein. Aggregates and degradation products are preferably removed by gel filtration and if the protein is to be lyophilized this step is also suitable for transfering the protein to a volatile buffer (see Table 1-1). This can sometimes be done by ion exchange chromatography, but more seldom by the other forms of chromatography. If the protein solution is intended to be frozen, stored as a solution or used immediately the requirements for specific buffer salts might be less stringent.

Several of the adsorption chromatographies might be adapted in such a way that they result in peaks of reasonably high protein concentration. This is an advantage when gel filtration is chosen as a final step. Gel filtration will always result in dilution of the sample and is often followed by a concentration step.

If the protein is to be used for physico-chemical characterization, especially for molecular weight studies, gel filtration has the advantage of giving a protein solution of defined size and also in perfect equilibrium with a particular buffer. Biospecific methods, by definition, give a product that is homogenous with respect to biological activity. This was taken advantage of for papain, where the enzyme eluted from a thiol column was twice as active as any previous preparation—most earlier enzyme batches apparently contained molecules in which the thiol necessary for activity was oxidized (see Chapter 9).

Proteins which after purification and formulation are intended for parenteral use in human beings should not contain endotoxins (lipopolysaccarides, LPS) or nucleic acids. The purification protocols must be designed such that these compounds are efficiently removed and validation studies should be performed to prove this. To prepare sterile protein solutions sterile filtration is used.

1.5 Monitoring the Fractionation

Proper analysis is a prerequisite for successful protein purification. Most important is the establishment of a reliable assay of the biological activity. In addition one needs to determine the protein content in order to be able to assess the efficiency of the different steps. It is beyond the scope of this chapter to go into details of the particular assay methods. This is covered by the special literature dealing with the activity in question— hormone, enzyme, receptor etc.

We recommend that each preparation be continually recorded in a purification table (Table 1-6). In combination with results from, e.g., gel electrophoresis this will serve as a guide for judging the reproducibility and the outcome of each preparation. In addition, each chromatography experiment should be accompanied by a suitable protocol such as the one examplified in Fig 1-4. However, the need for measurements of biological activity and protein concentration—especially the latter—should not be allowed to delay the

Table 1-6. Example of a purification table. The activity (e.g. enzyme activity) is expressed as units (u) and specific activity as units per mg of protein (u/mg).

Material	Volume ml	Protein mg/ml	Total protein mg	Activ. u/ml	Total activ. u	Spec. activ. u/mg	Yield %	Purific. factor x
Extract	500	14	8000	7	3500	0.5	100	1
First purif. step	50	10	500	60	3000	6	85	12

Chromatography Data Sheet

Date...   Chrom. No...

Column
Length..................... cm V_T........... m l   Column filling:
Diameter................. cm V_0........... m l   Adsorbent...
C.s. area.................. cm^2 V_{salt}...... m l   Gel...
Packing etc...   Lot No..
....   Preparation...
    ....

Equilibration of column
Buffer system or solvent......................................   Conc..................mol/l
pH......................... Conductivity..................ohm^{-1}cm^{-1} Temp..................°C
Detergent/Denaturant..   Conc............................
Bacteriostat..   Conc............................
Comments..

Sample
Description.. Designation..................
Amount....................mg Vol.......... m l   Conc................ A_{280}............A................
Buffer or solvent...
Specific acitivty etc..
Pretreatment...

Elution
Only buffer ❐   Gradient system..
....
....
Flowrate...................ml/h Prefraction............... m l   Fraction size ....ml..........min
Comments...

Analyses
280 nm ❐.................nm ❐   pH❐ Conductivity ❐ Volume ❐
Other reaction.. Aliquote volume...........................
Procedure...
Comments...

Fate of fractions
Pools..
....
Comments...

Figure 1-4. Example of a chromatography protocol.

preparation, and in many cases it is sufficient to save aliquots for analysis at one's convenience.

1.5.1 Measurements of Activity

In general, biochemical activities depend on the interaction between molecules. This can be measured in different ways. The classical method of enzyme catalysis is only one of these. In addition, the monitoring of the components can be done in several ways, such as spectrophotometry, measurements of radioactivity and immunological methods. Some of them are:

(a) Enzyme activity by direct spectroscopy
(b) Enzyme activity by secondary measurements on aliquots
(c) Binding of ligand
(d) Binding of antibody

The immunological methods require that the protein studied has already been purified once to allow production of an antibody or an antiserum by immunization. The detection of the antigen-antibody precipitate can be done at almost any sensitivity down to the extreme sensitivity afforded by the use of sandwich techniques and radioactively or enzymatically labelled reagents. (see, e.g., Chapter 14).

1.5.2 Determination of Protein Content

In general, a measure of protein content is obtained upon monitoring the effluent in chromatography by UV-absorption. However, it is not always easy to relate these measurements to the protein content. In fact, the only certain measure of protein content is total amino acid analysis after hydrolysis. Strictly speaking even this latter analysis suffers from some shortcomings since tryptophan and cysteine normally have to be analysed separately.

Large deviations from true protein values sometimes occur in the first steps in a purification scheme. The extract itself often contains substances that interfere with the protein analyses. An overestimation might result, especially if measurements of absorption at 280 nm are used, since the solutions are often turbid and absorbing substances of non-protein origin are present. This in turn will make the calculated values of specific activity erroneous.

Three main procedures for protein determination are used routinely:

(a) Spectrophotometry at 280 nm
(b) Colorimetry by Lowry-Folin-Ciocalteau reagent
(c) Dye binding with Coomassie Brilliant Blue G-250

Each of these methods has its advantages and its disadvantages.

UV-absorption measurements require knowledge of the extinction coefficient of the protein(s) to be measured. These vary widely. In the low end there is, e.g., serum albumin with an OD of 0.6 for a 1 mg/ml solution and the extreme parvalbumin with no absorption at all in the 280 band. At the other extreme there is e.g. lysozyme with an OD of 2.7 for a 1 mg/ml solution. The values are due to a corresponding variation in the content of the aromatic amino acids tryptophan and to a lesser extent tyrosine. As a rule of thumb it is convenient to assume a mean extinction of 1.0 for a 1 mg/ml solution and this is often sufficient for practical purposes. When the protein is purified the extinction coefficient and the wavelength for maximum extinction should however be determined on a solution by spectral and amino acid analysis.

An alternative to measurements at 280 nm is the low wavelength measurements at 225 nm or below. This absorption is due to the peptide bond, which has a maximum at 192 nm but still a considerable absorption at 205 nm (50 % of maximum or an OD of 31 for a 1 mg/ml protein solution) and at 220 nm an OD of 11. These measurements are, of course, even more sensitive to contamination and they also require that buffers which are transparent in the low UV regions are used. The use of sensitive UV monitors at these wavelengths in chromatographic equipment thus allows an extreme sensitivity but their use is not possible unless great care is taken to avoid contaminants and impurities in the buffer salts.

The Lowry methods[35] are less problematic but also less sensitive. Aliquots for analysis should have protein concentrations of 0.1 mg/ml or more. It is often a good alternative in the beginning of a purification where direct UV might be impossible due to turbid solutions. The same applies to the use of Coomassie Brilliant Blue. This method is 5–10 times more sensitive than the colorimetric one but is more cumbersome to use[36,37].

Both of the latter two methods are destructive, whereas the UV method allows the sample to be recovered.

1.5.3 Analytical Gel Electrophoresis

The gel electrophoresis techniques allow the investigator to get an idea of the complexity of the sample and, in particular, what the main contaminating species are. By using sieving electrophoresis, e.g. in the presence of SDS, and isoelectric focusing a considerable amount of information about the sample can be obtained in a couple of runs. The amount of each component, its molecular weight, its isoelectric point and even its titration curve can be obtained (see Chapter 13). If an antiserum directed towards the complete mixture is available one can also see whether some of the components are immunologically related and thus also structurally related by means of crossed immuno-electrophoresis.

The gel electrophoresis techniques are introduced in Chapter 12.

1.6 The Final Product

1.6.1 Buffer Change

The high resolution chromatographic steps for protein fractionation usually result in a product which is not directly suitable for the intended use, storage or distribution. The salt content may be too high, the pH of the protein solution may be unsuitable for long-term storage, the concentration of the protein may be too low or the solution may contain desorption agents from an affinity chromatography step which must be removed before the protein can be used.

The classical way of changing the buffer composition of a protein solution is dialysis. The protein solution is here included in a dialysis bag consisting of cellophane or a similar semipermeable material. Salts and low molecular weight substances (< 10.000) can diffuse through the membrane whereas high molecular weight material remains within the dialysis bag. The bag is placed in a larger stirred vessel containing the desired buffer, which is changed several times.

A faster way of changing the buffer composition of protein solutions is by gel filtration on, e.g., Sephadex® G25 equilibrated in the desired buffer. Proteins and other high molecular weight substances (> 6000) eluate at the void volume whereas substances with lower molecular weights are retarded and are separated from the proteins. The method is fast, depending on the equipment and volume, the cycle time is often less than one hour. As in every type of column chromatography the protein solution must not contain particles or colloidal material.

Ultrafiltration (diafiltration) is a third way of changing the buffer composition of a protein solution. With this technique the protein solution is diluted with the desired buffer, concentrated to the original volume, diluted again and so on. After a number of cycles the original buffer has in practice changed to the dilution buffer. The last concentration cycle may be driven longer so that the protein solution after the buffer change is concentrated.

1.6.2   Concentration

Concentration is another operation often required after the final step in a protein purification procedure. Ultrafiltration is the most frequently applied technique for this purpose. Smaller volumes of protein solutions can alternatively be concentrated by inclusion in a dialysis bag which is covered with a high molecular weight substance that cannot penetrate the dialysis bag but creates an osmotic pressure which drives the liquid out through the dialysis membrane. Often used polymers for this purpose are polyethylene glycol and Ficoll™ [38].

All chromatographic techniques which adsorb protein can also be used for the concentration of protein solutions. Especially suitable is ion exchange chromatography due to its high capacity and easy handling of the ion exchange medium. Other concentration methods that have proved useful also in large scale applications are freeze concentration[39] and concentration using dry Sephadex®[40].

1.6.3   Drying

Most biological processes occur in water solution and one way to stop these is to freeze the protein sample. For minimum risk of inactivation or denaturation a storage temperature of $-70°C$ or below is required. If the protein under study cannot stand repeated freezing and thawing, storage in aliquots is recommended. Another way to stop biological processes is to remove the water. The method used most for biologically active proteins is freeze-drying or lyophilization. In this method the protein solution is frozen below the eutectic point of the solution to ensure that all liquid is frozen. The frozen solution is then placed in a chamber which can be set under high vacuum. In the chamber or connected to the chamber is a condensing surface, a cold trap, with a temperature of $< -40°C$. After the vacuum is applied the protein sample is gently heated such that it does not melt to speed up the sublimation of water to the condenser. Normally all proteins maintain their biological activity and are fully recovered upon adding water. A techique often used for commercially available biochemicals is to lyophilize aliquots of protein solution in ampoules.

1.7 Laboratory Equipment

1.7.1 General Equipment

Laboratories for the preparation and separation of proteins may look very different, from the well-equipped special laboratory serving many research groups to the small lab with few people and limited resources. The large laboratory, in addition, often has dedicated service groups for special analyses, etc.

For the successful preparation and separation of proteins certain basic equipment is needed. Standard lab glassware will not be discussed. To the basic equipment belongs two or three balances, one spectrophotometer, one centrifuge, one pH-meter plus stirrers and micropipettes.

A good combination of balance equipment is two preparative balances; one double-range digital balance 0–1200 g or 0–3000 g with the possibility to weigh with 10 mg resolution and the other range 0–120 g or 0–300 g with 1 mg resolution; and one analytical balance for the interval 0–150 g with 0,1 mg resolution. For the measurement of larger amounts of material a simpler, and therefore cheaper, balance suffices, e.g., a balance for use in the food industry. It is important that the balances are calibrated as well as that they are serviced regularly.

Spectrophotometers for protein work should cover the wavelength interval 190–800 nm. Absorption around 280 nm is used routinely for estimating protein concentration whereas light in the visible region is often needed for measurement of different enzymatic activities. A double beam UV-VIS spectrophotometer with a thermostated cuvette holder is a good choice. It is important that the wavelength setting is correctly calibrated and that the instrument shows low drift. Regular servicing of the instrument is recommended.

A refrigerated floor centrifuge is standard in a preparative protein laboratory (Fig 1-5). The largest rotor should take 6 flasks of 500 ml, whereas the smallest rotor should accommodate 8 tubes of 50 ml. The maximum G-force for those rotors are normally about $13,000 \times g$ and $50,000 \times g$ at the tube bottom, respectively. An additional rotor for 6 flasks of 250 ml will increase the flexibility of the centrifuge. Some centrifuges accommodate zonal-rotors with accessories for continuous operation allowing larger volumes containing relatively low contents of fine particles to be centrifuged at high g-forces (up to $40,000 \times g$). The standard centrifuges normally need no or little service except changing of motor brushes. If the rotors are carefully maintained, including thorough cleaning after each run when liquid is found inside the rotors, they will function for many years. The rotors mentioned above are angle rotors and the flasks are normally filled completely. The special, completely tight lids which are available for most centrifuge flasks are strongly recommended. For small, easily centrifuged samples, a simple desk top centrifuge without cooling often is a good complement to the larger high speed refrigerated centrifuge.

The maximum speed of the refrigerated floor centrifuge is normally 20,000–21,000 rpm. Recently, a new type of centrifuge allowing speeds up to 28,000 rpm has been introduced. At speeds over 20,000 rpm, the rotor compartment can be

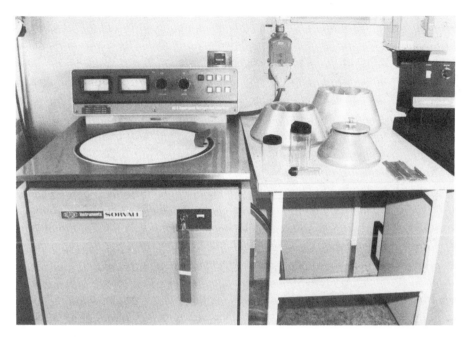

Figure 1-5. A refrigerated high speed centrifuge, a standard apparatus in every laboratory working with protein preparation.

maintained under partial vacuum. Small samples can thus be centrifuged at $100,000 \times g$.

pH determinations are critical in the protein laboratory. Buffers should be checked and titrated routinely. For certain enzyme assays equipment for monitoring the formation or consumption of protons is needed. The pH-meter can then be equipped with an automatic burette that adds acid or base to hold the pH constant, and the consumption of acid or base is recorded. Combination electrodes are generally the most convenient. The linearity of the electrode changes with time and for accurate measurements the electrode has to be calibrated on both sides of the interval within which the pH will be determined.

At least two types of stirrers are integral parts of the basic laboratory equipment (Fig 1-6). Firstly, magnetic stirrers with variable speed, at least one of which is equipped with a thermostat-controlled heating plate, are recommended. A variety of PTFE- and glass coated magnets are required. For extractions, propeller stirrers with interchangeable propellers and with variable speed regulators are convenient. These should preferably be of the electronic type which compensate for variation in load. Stirrers where the speed is controlled only with a variable resistor should be avoided.

Pipetting small volumes, up to 1 ml, is a standard laboratory procedure. Micropipettes of different types, most commonly with adjustable volume and with disposable tips, are used. Two or three pipettes of different sizes are normally sufficient. Check that the length of the pipette allows samples to be collected from the bottom of the longest type of test tube which is used in the laboratory.

For personnel safety the laboratory must have a ventilated hood or fume cupboard where procedures involving handling or preparation of hazardous substances can be safely performed. Even such a simple and common procedure as dissolving sodium

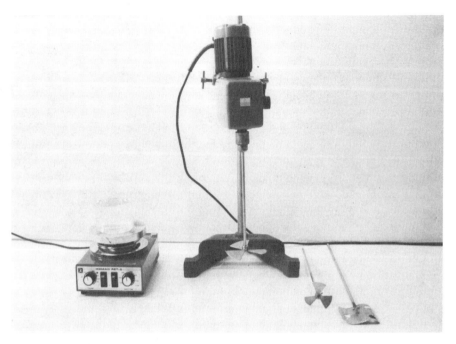

Figure 1-6. Stirrers: left, magnetical stirrer, right stirrer with variable speed for larger volumes or solutions with high viscosity.

hydroxide in water creates aerosols which are very irritating and this procedure should be done in the hood.

1.7.2   Equipment for Homogenization

Starting materials for protein preparation can be obtained from plants, animals, and microorganisms. Plant and animal materials first have to be homogenized to individual cells but then the preparation procedures are in principle similar for the three different kinds of starting material.

Waring Blendors or similar types of equipment and "Turrax"—type apparatus are frequently used for the homogenization of plant and animal tissues (Fig 1-7). The material is first cut into pieces with a knife or a pair of scissors to a size that suits the type of blendor used. At the same time the tissue or organ is trimmed and unwanted material like fat, ligaments and vessels are discarded before grinding the material in the blendor. The blendor can be used both dry and wet. Plant material like seeds, for example, can be ground dry if the time of grinding can be kept short. Dry ice can be added to keep the temperature low. Wet grinding prevents the formation of harmful dust particles and the heat formed in the grinding procedure is dissipated by the liquid. In most homogenization equipment only about 10 % of the energy input is normally used for the disintegration procedure: the rest of the energy is lost as heat. A common household meat grinder is also useful in many instances eventually in combination with a Waring blendor. Further homogenization is provided by the Elvehjem homogenizer (Fig 1-8). There are different sizes and also motor-driven varieties of this homogenizer.

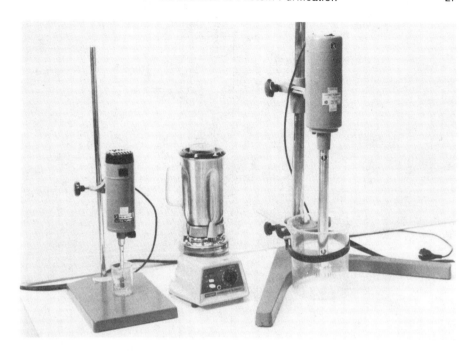

Figure 1-7. Homogenization equipment: middle, Waring Blendor, sides Turrax stirrers of different size.

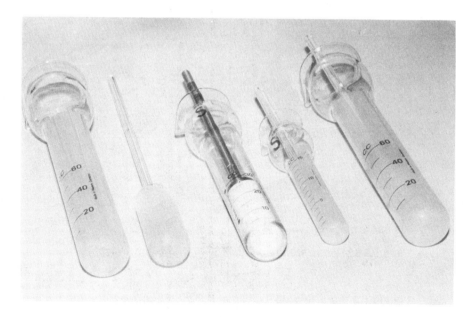

Figure 1-8. Elvehjem homogenizers.

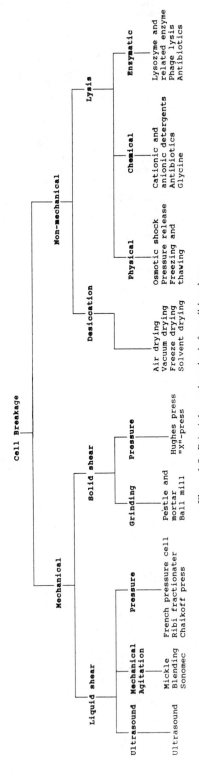

Figure 1-9. Principles and methods for cell breakage.

Micro-organisms are much more difficult to disintegrate. The polysaccaride cell wall withstands most homogenization procedures used for plant and animal cells. The well-cited overview by Wimpenny[41] for breakage of micro-organisms is still valid for small scale preparations (Fig 1-9). For large scale homogenization of micro-organisms, which means kilogram quantities of cell paste, the most frequently used equipment is the Dynomill[42] and the Manton-Gaulin[43] homogenizer. Both are designed for continuous operation and are available in various sizes for the breakage of from 100 g up to several kilograms of microbial cell paste per hour.

1.7.3   Equipment for Chromatography

1.7.3.1   Column Design

In order to utilize the packed chromatographic gel particles optimally, the column should be designed such that it does not significantly contribute to band spreading or peak distortion. Most modern columns are of the closed type with two fixed end pieces or with one fixed end piece and one adjustable adaptor. Sometimes it is convenient to use two adjustable adaptors, one at each end of the column. The end pieces or adaptors are equipped with porous frits made of, e.g., sintered metal, glass or plastic, pore diameters and surface structures of which should be optimized to avoid clogging and abrasion of the chromatographic particles. The frits should be easily exchangeable. For larger diameter columns the frits have to be combined with the flow distribution system. The prime requisites of such a system are that its volume should be negligible compared to the total column volume and that there should be no pressure drop in the system between the eluent inlet and the column wall. In many modern columns for low and medium pressure chromatography the frits have been replaced by a combination of a fine mesh (10 micron) polyamide fabric and a coarse mesh (e.g. 0.93 × 0.61 mm) polypropylene support net. The flow distribution layer will then be only 1 mm in depth and the support net will reduce the dead volume by as much as 32%. There is no indication that this uniform support net affects the sample application. This is the way many very large diameter industrial columns are designed. For the largest columns (>40 cm diameter) the fine mesh nylon fabric is in most cases replaced by a high quality stainless steel mesh. In most large-diameter industrial columns the number of sample- and eluent in- and outlets have been increased to 4–6. The linear flow rate of the eluent in the column feed pipe is very high in large diameter columns and it is recommended to prevent this jet from directly hitting the fine mesh fabric (when using a frit this problem is less pronounced) which could cause flow inhomogeneity and band distortion. The effect is reduced by the application of a small disc covering the support net under each pipe inlet.

The materials used in the parts of the chromatographic system that come in contact with the eluent should preferably be non-corrosive and compatible with all normal protein samples and eluent buffers/solvents used. However, they should also preferably be resistant to the conditions applied in the cleaning and maintenance of the packed gel particles. This includes high concentrations of sodium hydroxide and solvents such as alcohols. The CIP concept (Cleaning In Place) has become an important part of process design in biotechnology. In many instances, sterilization by autoclaving is preferred. Most laboratory column tubes are made of borosilicate glass or plastics with end pieces and adaptors made of polypropylene, nylon or fluorocarbon. Many HPLC columns are made of stainless steel to withstand the mechanical stress.

The column is connected to the pump and to the monitors by small bore tubing made of stainless steel or titanium in HPLC systems or fluorocarbon, polypropylene, polyethene or nylon in medium or low pressure systems. The inner diameters of the tubings should be optimized for each application. They should be as small as possible without too much contribution to the pressure drop of the system. To avoid zone mixing, the tube length should be reduced as much as possible. Ideally, the column should be attached directly to the monitor cell on top of the fraction collector. In normal laboratory systems 1 mm inner diameter fluorocarbon tubing is recommended.

1.7.3.2   Pumps and Fraction Collectors

The traditional laboratory-scale pumps for low-pressure systems (0.1 MPa) are of the peristaltic type, of which several brands are on the market. Care must be taken to keep the rollers and pumping tube clean. Otherwise, the tube life length is severely decreased. Peristaltic pumps are available with flow rates for a few ml per hour to cubic metres per hour. The pumping tube is available in several different materials, but for the protein work silicone tubing is recommended. Pumps with three or more channels can be used for gradient formation. The normal way to form a gradient is to use two connected vessels, with stirring in the output mixing vessel. In medium and high pressure chromatography, gradients are usually formed using two pumps connected to a programmable controller.

For HPLC most pumps are of the piston type and made of stainless steel or titanium, as are the sample injectors. In most cases they function according to the reciprocal displacement principle and are capable of pressues up to 35 MPa or more. Among medium pressure (5 MPa) pumps is the positive displacement type in which a fluorocarbon-sealed piston made of titanium or stainless steel is allowed to move in a thick-walled borosilicate glass tube. When the piston reaches the end of the tube, a valve switches over to a second identical piston-equipped glass tube filled with eluent and which will continue feeding the column with eluent at the same pressure while the first is filled. Unless proper damping is provided, a pressure transient will occur giving rise to spikes in the UV-monitor recordings at each piston change.

Traditional fraction collectors consist of a large plastic or metallic ring with holes bored at the periphery moved by an electric motor controlled by a timer. This type of fraction collector requires a relatively large bench area. Alternatives are compact designs with the holes for the test tubes arranged in a spiral or with the test tubes placed in racks. These collectors are controlled by microprocessors and different tube sizes and fraction collection times can be programmed in advance. In large scale chromatography, when the results normally are predictable from pilot experiments on a small scale, only a few fractions are needed and the fractions are obtained by using magnetic valves connected to the controlling equipment.

1.7.3.3   Monitoring Equipment

The classical way to monitor a chromatographic experiment is to take fractions from the outlet of the column in a fraction collector and analyze each tube manually for the different substances one wishes to determine. This is very time-consuming work and flow monitors connected directly to the column eluate have largely replaced this practice today. The most commonly used parameter in protein work is the absorption at 280 nm and a lot of detectors are available for this purpose. In new diode array monitors even a spectrum can be obtained directly on the eluate. In ion exchange chromatography the ionic strength is of interest and conductometers with flow cells are available. The pH of the eluate can also be monitored in, e.g., chromatofocusing by continuously-working monitors. Even complicated enzyme assays can be made directly on line by the use of autoanalyzers. Here a small part of the process stream is shunted through the autoanalyzer. This part of the sample is normally destroyed during the analysis.

1.7.4   Equipment for Chromatographic and Electrophoretic Analyses

For the characterization of the product at various stages of a protein preparation in principal the same separation techniques that are used for preparation of the protein can be used. The techniques are however scaled down to the μg scale. Even more important are analyses by electrophoretic separation techniques. The equipment for the different electrophoretic techniques are described in connection with the techniques themselves in Chapters 12-16.

References

1. R. Scopes, *Protein Purification Principles and Practice*, Springer-Verlag, New York, Heidelberg, Berlin, 1987.
2. D. Freifelder, *Physical Biochemistry; Applications to Biochemistry and Molecular Biology*, W. H. Freeman, New York, 1982.
3. T. G. Cooper, *The Tools of Biochemistry*, John Wiley & Sons, New York, London, Sydney, Toronto, 1977.
4. *Meth. Enzymol.*, *22*, W. B. Jakoby, ed., Academic Press, New York, San Fransisco, London (1971).
5. *Meth. Enzymol.*, *34*, W. B. Jakoby, and M. Wilchek, eds., Academic Press, New York and London (1974).
6. *Meth. Enzymol.*, *104*, W. B. Jacoby, ed., Academic Press, Orlando (1984).
7. L. Rydén, unpublished.
8. L. Rydén, Studies on the structure of human ceruloplasmin, Thesis, University of Uppsala (1972).
9. T. Moks, L. Abrahamsen, B. Österlöf, S. Josephson, M. Östling, S.-O. Enfors, I. Persson, B. Nilsson, M. Uhlén, *Biotechnology*, *5*, 379-382 (1987).
10. M.-R. Kula, H. Schütte, *Biotechnology Progress*, *3*(1), 31-42 (1987).
11. G. Taylor, M. Hoare, D. R. Gray, F. A. O. Marston, *Biotechnology*, *4*, 553-557 (1986).
12. F. A. O. Marston, *Biochem. J.*, *240*, 1-12 (1986).
13. L. M. Hjelmeland, *Meth. Enzymol.*, *124*, 135-164 (1986).
14. L. Rydén, H. F. Deutsch, *J. Biol. Chem.*, *253*, 519-524 (1978).
15. A. Veide, T. Lindbäck, S.-O. Enfors, *Enzyme Microb. Technol.*, *6*, 325-330 (1984).
16. P. Hedman, J. G. Gustavsson, *Develop. Biol. Standard.*, *59*, 31-39, S. Karger, Basel (1985).
17. E. J. Cohn, L. E. Strong, W. L. Hughes, Jr., D. J. Mulford, J. N. Ashworth, M. Melin, H. L. Taylor, *J. Am. Chem. Soc.*, *68*, 459 (1946).
18. J. Porath, *Science Tools*, *11*, 21-29 (1964).
19. K. Hannig, *Hoppe-Seyler's Zeitschrift für physiologische Chemie*, *338*, 211-227 (1964).
20. P. Mattock, G. F. Aitchison, A. R. Thomson, *Separation and Purification Methods*, *9*, 1-68 (1980).
21. M. Bier, in *Recovery and Purification in Biotechnology*, ACS Symposium Separation Series 314, 185-192, J. A. Asenjo, and J. Hong, eds., American Chemical Society (1986).
22. B. J. Radola, in *Meth. Enzymol.*, *104*, 256-275, W. B. Jacoby, ed., Academic Press (1984).
23. E. A. Peterson, H. A. Sober, *J. Am. Chem. Soc.*, *78*, 751-755 (1956).
24. J. Porath, P. Flodin, *Nature*, *183*, 1657-1658 (1959).
25. J.-C. Janson, *Chromatographia*, *23*, 361-369 (1987).
26. J. Bonnerjea, S. Oh, M. Hoare, P. Dunnill, *Biotechnology*, *4*(11), 954-958 (1986).

27. J. J. Higgins, D. J. Lewis, W. H. Daly, F. G. Mosqueira, P. Dunnill, M. D. Lilly, *Biotechnol. Bioeng.*, *20*, 159–182 (1978).
28. C. M. Ambler, F. W. Keith, in *Separation and Purification* 3rd edition, E. S. Perry, and A. Weissberger, eds., pp. 295–347, J. Wiley & Sons, New York (1978).
29. D. Derise, V. Gekas, *Process Biochemistry*, *23*, 105–116 (1988).
30. H. Strathman, *Trends in Biotechnology*, *3*, 112–118 (1985).
31. P.-Å. Albertsson, *Partition of Cell Particles and Macromolecules*, Wiley, New York, 1971.
32. R. Scopes, *J. Chromatogr.*, *376*, 131–140 (1986).
33. S. Shaltiel, *Meth. Enzymol.*, *34*, 126–140 (1986).
34. J. Porath, B. Ohlin, *Biochemistry*, *22*, 1621–1630 (1983).
35. O. H. Lowry, N. J. Rosebrough, A. L. Farr, R. J. Randall, *J. Biol. Chem.*, *193*, 265–275 (1951).
36. J. J. Sedmak, S. E. Grossberg, *Anal. Biochem.*, *79*, 544–552 (1977).
37. T. Spector, *Anal. Biochem.*, *86*, 142–146 (1978).
38. Ficoll™ is a polysucrose from Pharmacia LKB Biotechnology AB, Uppsala, Sweden.
39. J.-C. Janson, B. Ersson, J. Porath, *Biotechnol. Bioeng. XVI*, 21–39 (1974).
40. P. Flodin, B. Gelotte, J. Porath, *Nature*, *188*, 493 (1960).
41. J. W. T. Wimpenny, *Proc. Biochem.*, *2*(7), 41–44 (1967).
42. Dynou mill: Willy A. Bachofen, Manufacturing Engineers, Basel, Switzerland
43. Manton-Gaulin; APV GAULIN International SA, P.O. Box 58, 1200 AB Hilversum, Holland.

II Chromatography

2 Introduction to Chromatography

Jan-Christer Janson
Pharmacia LKB Biotechnology AB
S-751 82 Uppsala, Sweden

Jan-Åke Jönsson
Department of Analytical Chemistry
University of Lund
Box 124, S-221 00 Lund, Sweden

2.1 Basic Concepts and Versions of Chromatography

The term chromatography refers to a group of separation techniques which are characterized by a distribution of the molecules to be separated between two phases, one stationary and the other mobile. Molecules with a high tendency to stay in the stationary phase will move through the system at a lower velocity than will those which favour the mobile phase.

The most common physical configuration is *column chromatography* in which the

Table 2-1. Versions of protein liquid chromatography

Separation principle	Type of chromatography
Size and shape	Gel filtration
Net charge	Ion exchange chromatography
Isoelectric point	Chromatofocusing
Hydrophobicity	Hydrophobic interaction chromatography
	Reversed phase chromatography
Biological function	Affinity chromatography
Antigenicity	Immunoadsorption
Carbohydrate content	Lectin affinity chromatography
Content of free —SH	Chemisorption ("Covalent chromatography)
Metal binding	Immobilized metal ion affinity chromatography
Miscellaneous	Hydroxyapatite chromatography
	Dye affinity chromatography

stationary phase is packed into a tube, a column, through which the mobile phase, the eluent, is pumped. The sample to be separated is introduced into one end of the column. The various sample components travel with different velocities through the column and are subsequently detected and collected at the other end. Other configurations, such as thin-layer chromatography and paper chromatography are also used.

In the context of this book, the mobile phase is always a liquid, most often an aqueous buffer. Consequently, these techniques are versions of *liquid chromatography*. For the separation of more volatile compounds, gas chromatography, i.e. with a gaseous mobile phase, is an extremely powerful and widely applied technique.

For protein separation, several versions of liquid chromatography are used, differing mainly in the types of stationary phase (Table 2-1). One of these, *gel filtration chromatography* (also called *size-exclusion chromatography*) is based on quite different principles than are other versions of liquid chromatography. Therefore, much of the theoretical description of that technique must be made separately. In this book, the basic principles of gel filtration chromatography will be described in Chapter 3, while other techniques will be dealt with briefly here. A thorough treatment of the theory and principles of chromatography is outside the scope of this book, but can be found in specialized texts[1-3].

2.2   The Stationary Phase

The stationary phase in a chromatographic experiment is composed of a porous matrix and imbibed immobile solvent. Typically, the solvent constitutes most of the stationary phase, often more than 90 percent, and such materials are generally refered to as gels. In protein chromatography the solvents are normally aqueous buffers and the gel-forming materials are usually composed of hydrophilic polymers. Modern stationary phases are almost exclusively bead shaped with average particle diameters ranging from a few to approximately one hundred microns.

In principle, one may distinguish between two types of gels, *xerogels* and *aerogels*. The xerogels are characterized by their ability to shrink and swell in the absence and presence of the solvent used, whereas the volume of aerogels is independent of the solvent. Typical xerogels are cross-linked dextran gels (Sephadex®) and cross-linked polyacrylamide gels. Typical aerogels are porous glass, silica and most gels based on macroreticular organic polymer gels such as polystyrene gels and polymethacrylate gels.

2.2.1   Matrix Properties

In addition to being hydrophilic, an ideal general matrix for protein chromatography should not contain groups which spontaneously bind protein molecules. However, it should contain functional groups which allow the controlled synthesis of a wide variety of protein adsorbents. Furthermore, the matrix should be chemically and physically stable in order to withstand extreme conditions during derivatization and maintenance (regeneration, sterilization etc.) and be rigid enough to allow high linear flow rates (5 cm/min or more) in columns packed with particles with diameters down to a few microns. Finally, the matrix substance should allow the production of gels with a broad range of controllable porosities.

A wide variety of material has been used for the design of protein chromatography matrices. These can be classified as being either inorganic, synthetic organic polymers or polysaccharides. Among all three groups we find traditional standard chromatography media as well as modern high performance chromatography media. Examples of these are listed here:

Inorganic materials
— porous silica
— controlled pore glass
— hydroxyapatite
Synthetic organic polymers
— polyacrylamide
— polymethacrylate
— polystyrene
Polysaccharides
— cellulose
— dextran
— agarose

None of these materials fulfils all criteria for an ideal general matrix for protein chromatography: they are all compromises. Since a combination of hydrophilicity with chemical and physical inertness is best achieved by use of alcohol hydroxyls or amido groups, the most widely used standard chromatographic media for proteins are based on neutral polysaccharides and polyacrylamide.

Among the polysaccharides, *cellulose* (Fig. 2-1 and Fig. 2-2) is still a widely used matrix for the synthesis of protein ion exchangers more than 30 years after its introduction by Peterson and Sober in 1956[4]. Modern cellulose gel media are marketed under the trade names Whatman[TM 5], Cellulofine[TM 6] and Sephacel®[7]. *Cross-linked dextran* (Fig. 2-1 and Fig. 2-3) was introduced in 1959 by Porath and Flodin[8] and is marketed under the trade name Sephadex®[7]. This material is best known as a gel filtration medium[9] but is also widely used as a matrix for ion exchangers[10]. An account of

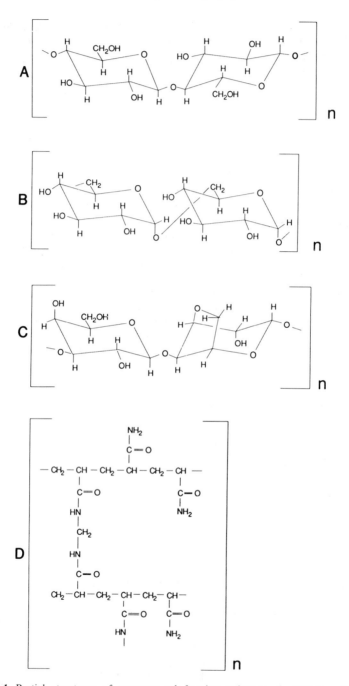

Figure 2-1. Partial structures of common gel forming polymers. A. Cellulose (β-1,4-linked D-glucose), B. Dextran (α-1,6-linked D-glucose; the glycosidic bond is here stretched out for lay-out reasons), C. Agarose (alternating 1,3-linked β-D-galactose and 1,4-linked, 3,6-anhydro-α-L-galactose), D. Polyacrylamide cross-linked with N,N'-methylene bisacrylamide.

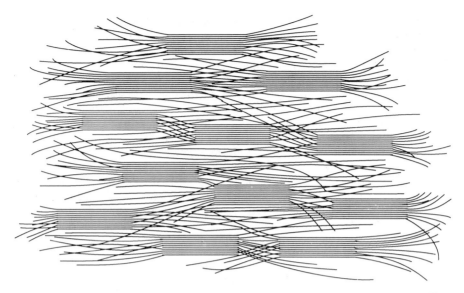

Figure 2-2. Schematic representation of part of a cellulose fibre composed of ordered (crystalline) and disordered (amorphous) regions.

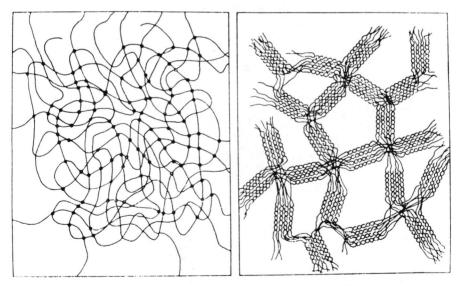

Figure 2-3. Schematic representation of the gel structures of agarose (right) and point cross-linked polymers such as polyacrylamide and dextran (Sephadex) (left). Reproduced by permission of the authors[16] and publishers.

the history of the development of Sephadex is given in Reference 11. The neutral hydrophilicity of dextran gels makes them compatible with most proteins and quantitative recovery is thus the rule. Their most important application area today is in desalting and buffer exchange of protein solutions. Protein fractionation by gel filtration is nowadays performed largely with composite gel matrices (see below).

Agarose (Fig. 2-1 and Fig. 2-3) is a low charge fraction of the sea weed polysaccharide agar. It was introduced as a medium for chromatography by Hjertén in 1962[12] and marketed under trade names such as Sepharose[R] and Superose[R] [7], Ultrogel™ A[13] and BioGel™ A[14]. Bead-shaped gels suitable for chromatography are formed by cooling 2–15 % aqueous solutions of agarose dispersed in an apolar organic solvent in the presence of suitable emulsifiers[15]. The agarose gel structure is an open three-dimensional network of fibres composed of spontaneously aggregated galactan helices[16], (Fig 2-4). In a 90 micron average diameter 4 % agarose gel the total surface area is approximately 5 $m^2 \cdot ml^{-1}$ with an average pore diameter of approximately 30 nm[17]. By chemical cross-linking[18,19] of the spontaneously aggregated galactan polymers the gel rigidity is improved considerably, allowing the manufacture and use of particles down to 10 micron diameter for a 12 % agarose gel[20]. The cross-linking does not change the gel pore structure, but the number of hydroxyl groups is reduced by around 50 % which, however, does not significantly affect the binding capacity for proteins of ion exchangers and affinity chromatography adsorbents prepared from cross-linked agarose gels.

From a functional point of view it is convenient to distinguish between *microporous* and *macroporous* (or macroreticular) gel matrics. The microporous gels are prepared by point cross-linking of linear polymers such as dextran and polyacrylamide (Fig. 2-3). This type of gel lends itself ideally to molecular-sieving separations (size exclusion chroma-

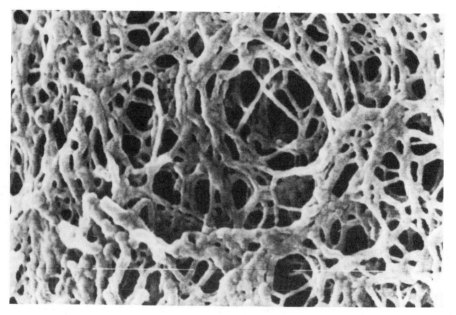

Figure 2-4. Scanning electron micrograph of 2 % agarose gel. The white bar represents 500 nm. (Preparation and photo: A. Medin, Institute of Biochemistry, Uppsala University, Uppsala, Sweden).

tography, gel filtration). At porosities suitable for protein chromatography these gels become impractically soft and this is why they often participate in the design of composite gels (see below).

The macroporous gels are most often obtained from aggregated and physically cross-linked polymers. To this group belong agarose, macroreticular polyacrylamide, silica and several synthetic organic polymers. These gels are best suited for the design of stationary phases intended for ion exchange chromatography, affinity chromatography and other adsorption chromatographic techniques. By introducing microporous gel forming polymers into the pores of macroreticular gels, *composite gels* are formed which combine the strength of and reduce the weakness of each separate gel moiety. Examples of such gels are Sephacryl[®] (poly-N,N'-bisacrylamide-dextran) (Fig 2-5)[7], Superdex[®] (agarose-dextran)[7] and Ultrogel[TM] AcA (agarose-polyacrylamide)[13].

The advent of protein HPLC increased the demand for matrix rigidity. This gave rise to new derivatization procedures for porous silica to make it compatible with proteins, e.g., the introduction of diol silanols[21]. Despite attempts to stabilize the silica surface by using, e.g., zirconium oxide[22], the major weakness of this matrix is still its instability at alkaline pH. This has encouraged the development of rigid matrices based on porous synthetic organic polymers[23] and agarose[24].

Finally, hydroxyapatite should be mentioned because it represents a widely used inorganic matrix with high selectivity for a wide variety of proteins and also nucleic acids[25]. Bead shaped hydroxyapatite particles with diameters of only a few microns are commercially available in prepacked columns from several sources.

2.2.2 Stationary Phase Technology

The shape, rigidity and particle size distribution profile of the gel matrix are important parameters which govern the performance of the stationary phase. The original batch polymerization procedures followed by grinding and sieve classification of the irregular particles have, with few exceptions (e.g. cheap silica variants), largely been replaced by bead polymerization technologies.

Cross-linked dextran gels, Sephadex[®], are thus produced by allowing aqueous solutions of dextran containing an excess of alkali to be dispersed in an emulsifier-containing apolar solvent[26]. The average droplet diameter and size distribution is controlled by the emulsifier concentration and the type and speed of the stirrer. Cross-linking occurs upon addition of a predetermined quantity of epichlorohydrin. The pore diameter and size distribution of the formed gel is governed by the concentration and the molecular weight distribution of the dextran and by the concentration of epichlorohydrin.

Standard polyacrylamide gels (Fig. 2-1 and Fig. 2-3) for chromatography are formed by bead polymerization of aqueous droplets containing acrylamide (monomer) and methylenebisacrylamide (crosslinker) dispersed in an emulsifier-containing apolar solvent[27]. The gel pore dimensions depend on the total concentration of monomer (%T) and the relative concentration of cross-linker (%C), respectively.

One major step forward in stationary phase technology was the bead polymerization technique developed by Ugelstad[28], which is based on controlled swelling of submicron particles in the presence of monomer and which gives rise to practically monodisperse beads (Fig 2-6) of any desired diameter up to approximately 30 microns. Such particles are commercially available with a diameter of 10 microns and with controlled porosity[23].

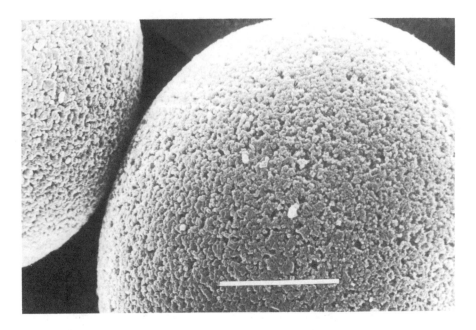

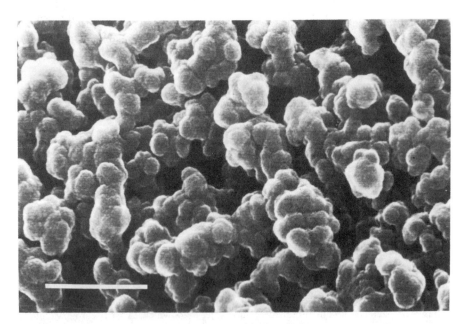

Figure 2-5. Scanning electron micrograph of Sephacryl® S-500[7]. The white bars represent 10 μm (top picture) and 0.5 μm (lower picture). (Preparation and photo: A. Medin, Department of Biochemistry, Uppsala University, Uppsala, Sweden).

Figure 2-6. Scanning electron micrograph of MonoBeads[TM 7].

As mentioned above, the development of new cross-linking procedures for agarose have enabled the use of this polysaccharide for synthesis of stationary phases which fulfil most of the requirements of a modern high performance matrix. There are available today stationary phases based on 10-micron[20,24], 30-micron[29] and 90-micron[30] average diameter beads of cross-linked agarose which cover the demands of protein chromatography from micrograms to several kilograms per cycle and with cycle times from a few minutes to less than two hours in adsorption chromatography applications such as ion exchange chromatography and affinity chromatography.

2.2.3   Introduction of Ligand Groups

The ideal base matrix for the synthesis of stationary phases for adsorption chromatography techniques is identical with the ideal gel filtration medium in all respects (as indicated in the beginning of section 2.2.1.) but one: it should not give rise to excessive mass transport constraints but should have an open, non-sieving and easily accessible pore structure, allowing unhindered diffusion of the protein molecules to their adsorption sites on the gel matrix polymer surface.

The surface should be covered with hydrophilic functional groups which do not themselves participate in protein binding but should, on the other hand, be amenable to derivatization for the synthesis of ion exchangers and other adsorbents. However, all base matrices interact with at least some proteins under certain conditions. Cross-linking and derivatization procedures may also inadvertently introduce non-specific adsorption sites which may complicate the interpretation of the purification data.

By the introduction of new chemical structures on the surface of the matrix it is possible to design stationary phases which interact more or less specifically with a particular protein. Least specific are gels for ion exchange chromatography (IEC), which separates proteins primarily on the basis of their content and distribution of charged groups; and hydrophobic interaction chromatography (HIC) adsorbents, which interact with hydrophobic patches and crevices on the protein surface. Low specificity should, however, not be confused with low selectivity. Both IEC and HIC can be very selective indeed when time can be spent on the optimization of the operating conditions. The stationary phases with the highest specificities are found among the affinity chromatography adsorbents and particularly among immunoadsorbents based on immobilized monoclonal antibodies. Medium specificities are obtained using so-called general ligands, i.e., group-specific ligands such as co-enzyme analogues.

The synthesis of an ion exchanger usually takes place by a one-step chemical reaction in which a charged molecule containing a reactive group (often a halogenide) is allowed to react with the alcohol hydroxyl-group-containing matrix (such as a polysaccharide) under strongly alkaline conditions. Most DEAE- and CM-ion exchangers are produced in this way. The synthesis of an affinity chromatography adsorbent, on the other hand, is normally a three-step procedure. In the first step the matrix is activated by the introduction of reactive groups. In the second step the ligand is covalently attached to the matrix by reaction with the activated group. In the third and last step the excess reactive groups are inactivated by blocking with a low molecular weight substance which displays low adsorptivity when coupled. Different coupling methods are discussed in some detail in Chapter 10. Some of the more general approaches are mentioned below for hydroxyl group containing matrices. For synthetic organic polymer based matrices and for silica gels special techniques are required (see Chapter 10).

As with ion exchangers, the activation reactions normally take place at alkaline pH. The most frequently used reagents for hydroxyl-group-containing matrices are CNBr, bis-epoxides or divinylsulfone, introducing reactive cyanoesters, epoxides and vinylsulfones, respectively. These can then be reacted with nucleophilic groups such as thiols, amines, alcohols, etc. The blocking is often carried out with an excess of ethanolamine. Several alternatives to these coupling methods are available, also for the binding via carboxylates. Amino groups are the most commonly used functional groups for coupling of low molecular ligands as well as proteins. Carbohydrates (sugars, oligosaccharides) can be attached to epoxides and vinylsulfones via their hydroxyls.

2.2.4 Ligand-Protein Interactions

The binding of a dissolved protein to an immobilized ligand is due to one or several of the following interactions:

(1) ion-ion or ion-dipole bonds
(2) hydrogen bonds
(3) dispersion or van der Waals forces
(4) aromatic or π–π interactions
(5) hydrophobic effect or hydrophobic interaction

Sometimes covalent bonds are formed, as in covalent chromatography (Chapter 9), or metal ligand bonding, as in immobilized metal ion affinity chromatography (Chapter 8). The hydrophobic interaction differs from the others in that it has a large entropic component.

The system solution-protein-stationary phase at each moment minimises its free energy and this gives rise to the binding of the protein to the ligand. This interaction is normally the result of many secondary bonds formed and several of the above mentioned factors might contribute. Each one of these might be weak, often around 1 kcal/mole, but several of them add up to a considerable binding constant. A K_D of 10^{-5}, which gives rise to retardation, corresponds to a binding energy of 7 kcal/mole in standard conditions according to

$$\Delta G = RT \ln K_D$$

and thereby 4–8 weak interactions (see e.g. Ref. 31). A general treatment of the subject is given in Ref. 32.

The surface of globular proteins typically has around 45% hydrophobic residues and 55% charged or hydrophilic uncharged residues accessible to surrounding water (Fig 2-7). In the latter groups the positive charges of lysine, arginine and histidine side chains as well as the α-amino group change according to pH (Table 4-2). The negative charge of glutamic acid and aspartic acid side chains and the α-carboxyl group as well are pH dependant. Of these the histidine side chain and the α-amino groups titrate within the pH region that is commonly used in chromatography. The cysteine thiol side chain is not typically available in extracellular proteins, but might be so in intracellular proteins. It does not play a role as an anion although it has a pK of about 8.5. Other hydrophilic groups such as asparagine, glutamine, serine and threonine as well as peptide bonds are mostly involved in hydrogen-bonding interactions.

Although these interactions determine the strength of the binding of proteins, the specificity of binding is to a large extent explained by the fit between the ligand and the protein. This is particularly true in the case of affinity chromatography. But even if other types of chromatography might to a larger extent depend on a single of the interactions discussed it is certainly not a sufficient explanation for protein selectivity. Ion exchange thus depends on ionic forces and hydrophobic interaction chromatography on hydrophobic interactions. Still, net charge or net hydrophobicity cannot alone explain the differences in retardation of different proteins. Here also the distribution of interacting groups on the protein surface and thereby the fit between the protein and the ligand is of major importance as well as secondary interactions, such as H-bonding.

The different interactions are each influenced by the solvent in some particular way. Increased ionic strength thus decreases ionic interactions, whereas hydrophobic interactions are favoured. The different ways to elute adsorbed proteins are discussed in detail for each type of chromatography. It should, however, be kept in mind that since a single type of interaction is seldom responsible for the binding, non-conventional approaches are sometimes worth trying.

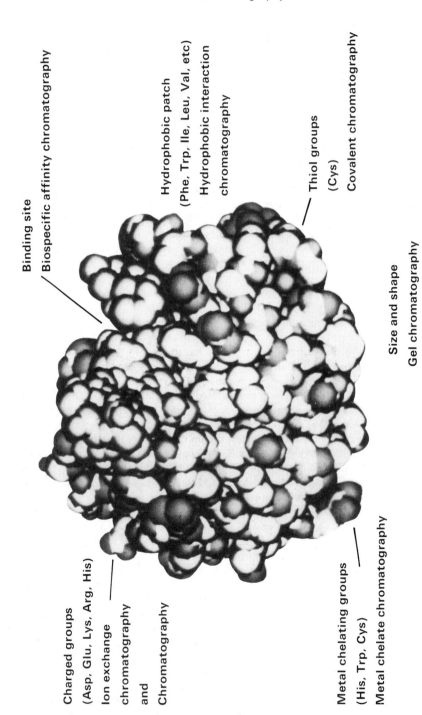

Binding site
Biospecific affinity chromatography

Hydrophobic patch
(Phe, Trp, Ile, Leu, Val, etc)
Hydrophobic interaction
chromatography

Thiol groups
(Cys)
Covalent chromatography

Charged groups
(Asp, Glu, Lys, Arg, His)
Ion exchange
chromatography
and
Chromatography

Metal chelating groups
(His, Trp, Cys)
Metal chelate chromatography

Size and shape
Gel chromatography

Figure 2-7. Separation principles and corresponding chromatographic techniques.

2.3   Chromatographic Theory

2.3.1   Chromatographic Quantities

A large number of quantitative concepts are used to describe a chromatographic experiment. Some of the most significant of these will be discussed here.

2.3.1.1   Retention Parameters

A chromatogram is a plot of concentration at the column exit versus time or eluent volume. The volume that has passed the column from the introduction of the sample until the emergence of a certain component as a peak in the chromatogram is the *retention volume*, V_R, see Fig. 2-8.

The void or *holdup volume*, V_0, is usually taken as the volume of the mobile phase in the column. With liquid chromatographic columns, packed with porous particles or gels this simple correspondence is unfortunately not exact, since small molecules penetrate the pores to a greater extent than large molecules and thereby distribute in a larger liquid volume. Even in the absence of physical or chemical interactions with the stationary phase itself, this effect leads to a separation of molecules of different sizes: molecules of a particular size will emerge at the column outlet after passage of a volume that is equal to the column volume that is available to that size. This is the principle of gel filtration chromatography, where the porosity of the materials used is chosen to maximize this effect. In other types of chromatography this effect should be minimized by using materials with pores that are wide enough to accommodate all molecules of interest. For protein chromatography this is especially important.

The volume of the mobile phase is composed of the *interstitial volume*, i.e., the volume outside (between) the porous particles and the *pore* or *inner volume*. Adding the volume of the solid (matrix) material, we obtain the *total* or *geometrical volume*, i.e., the volume of the empty tube. In gel filtration chromatography, the void volume is equal to the interstitial volume and the inner volume is considered as a part of the stationary phase.

In conclusion, one should not confuse the void volume concept as it is used in gel filtration theory (see Chapter 3) with the void volume concept used in other types of chromatography.

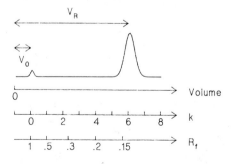

Figure 2-8. Relation between various retention parameters.

Instead of retention volume, *retention time*, t_R, is often used, which may be easier to measure. Obviously, $V_R = F t_R$ where F is the *volumetric flow rate*. The numerical value of the retention volume depends largely on the dimensions of the column used. To obtain a normalized retention quantity, the *capacity factor k* is generally used. It is equal to the number of void volumes needed to elute the compound of interest minus one. Thus:

$$k = \frac{V_R}{V_0} - 1 \tag{2-1a}$$

which also can be written:

$$V_R = V_0(1 + k) \tag{2-1b}$$

In Fig. 2-8 an alternative *x*-scale is graduated in units of k. Often the capacity factor is written k'. This is, however, in variance with official recommendations for the terminology of chromatography. In this chapter, the liquid chromatography recommendations of ASTM[33] are followed where relevant. The term "capacity factor" is generally accepted, but can be misleading: it should not be confused with the adsorbent capacity Q as defined below.

Another normalized retention measure is the *retention ratio*, R_f which is the ratio between the velocity of the actual compound and the velocity of the mobile phase. Consequently:

$$R_f = \frac{V_0}{V_R} = \frac{1}{1 + k} \tag{2-2}$$

R_f is especially used in non-column versions of chromatography (e.g. thin-layer chromatography). When plotted in a chromatogram as in Fig. 2-8, a scale in units of R_f is non-linear.

2.3.1.2 Peak Shape and Width; The Theoretical Plate Concept

A separation of two compounds is possible only when their velocities (R_f-values) and thereby their k and V_R differ by some amount.

As is obvious from Fig. 2-9, additional parameters related to the width of the peaks are needed to specify the resolution of two compounds.

To describe the width of chromatographic peaks, the concept of *theoretical plates, n,* is commonly used[33]. This term is related to fractional distillation and is nowadays obsolete as there are no "plates", either "theoretical" or otherwise in a chromatographic column. The proper definition is based on statistical theory:

$$n = \frac{\mu^2}{\sigma^2} \tag{2-3}$$

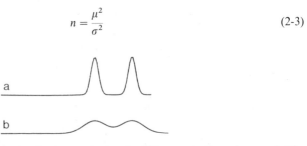

Figure 2-9. (a) A chromatogram showing two resolved peaks. (b) Same retention volumes as in Fig. 2a, but wider peaks.

where μ^2 and σ^2 are the mean and variance of the chromatographic peak, respectively. In practice, one of the following formulae is used to calculate n from a chromatogram:

$$n = 5.55\left(\frac{t_R}{w_{1/2}}\right)^2 \tag{2-4a}$$

$$n = 16\left(\frac{t_R}{w_b}\right)^2 \tag{2-4b}$$

$$n = 6.28\left(\frac{t_R \cdot h_p}{A_p}\right)^2 \tag{2-4c}$$

Here, $w_{1/2}$ and w_b are the width of the peak at half the peak height, h_p, and at the base, respectively (see Fig. 2-10). A_p is the area of the peak.

To apply these formulae, it is not important whether t_R or V_R is used. It is, however, crucial that the widths are measured in the same units at t_R. Typically they are measured in cm directly from the recorder paper using a ruler. Equation (2-4c) is intended to be used with an electronic integrator, which usually supplies values for t_R, h_p and A_p, but no peak widths.

Equations (2-4) are derived from the gaussian (normal) probability curve, which is a fair description of a chromatographic peak in simple cases. It should be emphasized that they are only valid in isocratic chromatography and when the peak is reasonably symmetrical. For tailing peaks, which are often obtained (see below p. 54), it is not meaningful to calculate the number of theoretical plates.

The theoretical-plate concept is also used widely to characterize the performance of a chromatographic column. To a first approximation, all (symmetrical) peaks in a chromatogram show roughly the same plate number. Consequently, this number can be considered as a property of the commmn used. A large plate number means narrow peaks and thus a "good" column.

As the plate number is approximately proportional to the column length, L, column quality can also be expressed in terms of theoretical plates per meter column or *height of a theoretical plate* (HETP), h, which is defined by:

$$h = \frac{L}{n} \tag{2-5}$$

Thus, for a "good" column the value of h is small.

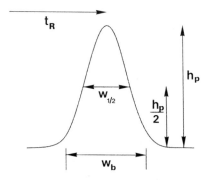

Figure 2-10. Definitions of different widths of a chromatographic peak.

2.3.1.3 The van Deemter Equation

According to the theory originally developed by J. J. van Deemter and co-workers[34] for gas chromatography, the plate height h is a sum of three independent contributions, which depend on the flow rate of the mobile phase according to the equation

$$h = A + \frac{B}{u} + C \cdot u \qquad (2\text{-}6)$$

Here, u is the *linear flow rate* and A, B and C are constants. A plot of the van Deemter equation will show a minimum at a certain flow rate (see Fig. 2-11). This is the "optimal" flow rate at which a maximal plate number is obtained. In practice, especially for liquid chromatography of macromolecules, the "optimal" flow rate is often unpractically low and columns are operated at considerably higher flow rates. The third term is the van Deemter equation then becomes especially important.

A detailed interpretation of the terms is complex but rewarding. The classical work in this field is Giddings' "Dynamics"[1]. Other more recent and fairly complete treatments also exist[3].

Briefly, the first term, A, the 'eddy' term, describes peak broadening resulting from the complicated geometry in a packed column: some molecules will find a relatively straight path through the column, whereas others may follow a longer, more tortuous path. The second term, B/u, results from diffusion of sample molecules along the column. The third term, $C \cdot u$, originates from various kinetic parameters, such as slow transfer of molecules into and out of pores or within the stationary phase, non-instantaneous equilibrium, etc.

Both the A-term and the C-term depend on the diameter d_p of the column packing particles. A is directly proportional to d_p, whereas the relation for C is more complex: the contribution to the C-term of slow mass transfer in pores, etc. is proportional to d_p^2 whereas the non-equilibrium contribution is independent of d_p[1]. Using smaller particles thus leads to a lower plate height and a larger plate number. Observe that the effect might be largest for the C-term, thereby moving the minimum in the van Deemter plot towards higher flow rates and decreasing the plate height, especially for high flow rates, thereby permitting efficient separations in shorter times. The price to be paid is the necessity of using high pressure pumps to force the mobile phase at high flow rates through columns with high resistance. This is the reason why the "P" in "HPLC" can mean "performance", "pressure" and "price".

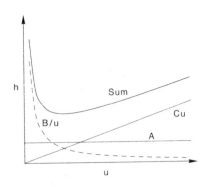

Figure 2-11. Plot of the three terms in the van Deemter equation and their sum.

In order to facilitate comparisons among columns of different types the concepts of *reduced plate height*, h_r, and *reduced flow rate*, v, are often used. They are defined as

$$h_r = \frac{h}{d_p} \quad \text{and} \quad v = \frac{u \cdot d_p}{D} \tag{2-7}$$

where D is the diffusion coefficient to the solute in question in the mobile phase. The adoption of reduced parameters removes the influence of d_p and D from the A and B-terms in the van Deemter equation, whereas the C-term still depends on these parameters.

2.3.1.4 Resolution

In a chromatogram of a complex sample a great number of peaks are to be separated. It is only necessary, however, to consider one pair of peaks at a time. The *chromatographic resolution* of this pair is defined by:

$$R_s = \frac{V_{R2} - V_{R1}}{\frac{1}{2}(w_{b1} + w_{b2})} \tag{2-8}$$

i.e., the distance between the peak maxima divided by the mean peak width (see Fig. 2-12). Just as for Eqs. (2-4), retention times can be used instead of volumes. For $R_s < 1$, the peaks are incompletely separated, for $R_s = 1$, they just touch each other at the base, whereas for $R_s > 1$ there will be a stretch of baseline between the peaks. Optimal separation conditions are reached when $R_s > 1$ for all compound pairs of interest. If R_s is considerably greater than 1 for all pairs, the resolution is unnecessary good and the separation can probably be made faster.

Upon combination with Eq. (2-4b), the resolution can be written:

$$R_s = \frac{\alpha - 1}{\alpha + 1} \cdot \frac{\bar{k}}{1 + \bar{k}} \cdot \frac{\sqrt{n}}{2} \tag{2-9}$$

Here, $\bar{k}$ is the mean of the k-values for the two peaks, $(k_1 + k_2)/2$, or $(V_{R1} + V_{R2})/2V_0$ and α is the relative retention, k_2/k_1 or V_{R2}/V_{R1}.

The resolution depends, according to Eq. (2-9), on three more or less independent factors: the selectivity factor, the retention factor and the plate number factor. To optimize the resolution, each of these factors should be considered.

The selectivity factor, $(\alpha - 1)/(\alpha + 1)$, is usually the most important. For example, an increase of α form 1.01 to 1.02 (i.e., only 1% change) will double the resolution. On the other hand, if $\alpha = 1$, no separation is possible. The retention factor, $\bar{k}/(1 + \bar{k})$ increases with $\bar{k}$, but when $\bar{k}$ is larger than 5 the increase is marginal and will mainly result in slower

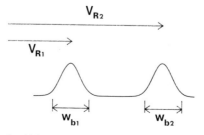

Figure 2-12. Peak widths and retention volumes for definition of resolution.

separation. If $\bar{k} < 1$, the resolution is unnecessarily low. Finally, an increase in the plate number, n, e.g., by using a longer column, will increase resolution. However, a four-fold increase in column length is necessary to double the resolution.

Facing an inadequate resolution, the most effective remedy would probably be to change the chromatographic conditions aiming at a higher relative retention. Changes in pH or in the composition of the eluent are the usual approaches tried to increase α. A mere increase of column length might be impractical, and would considerably increase separation time (but not necessarily the total time spent on the entire task, especially if only a limited number of runs are to be made!).

If the problem is the opposite and more pleasant one, i.e., to speed up an unnecessarily effective separation, the use of a shorter column will be the best solution, offering a considerable time saving on each run. Also here a higher relative retention is beneficial as it permits the use of still shorter and faster columns.

2.3.2   Retention in Adsorption Chromatography

Most versions of chromatography used for protein separation, (except gel filtration chromatography), may be more or less adequately treated together under the term adsorption chromatography. This implies that the sample molecules are adsorbed onto the surface (or in a thin surface layer) of the stationary phase. The precise nature of the adsorption forces varies among the techniques and can for the moment be disregarded.

2.3.2.1   The Adsorption Isotherm

The central concept in adsorption is the *adsorption isotherm*, which is a plot of the sample concentration on the adsorbing surface of the stationary phase versus the sample concentration in the mobile phase. These concentrations are written C_S and C_M, respectively. The concentration C_S may be expressed either in units of moles per surface area or moles per gram of adsorbent. C_M is expressed in moles per litre.

A general shape of such a plot is shown in Fig. 2-13. The curved shape of the isotherm originates from competition among sample molecules for adsorption sites. The most simple description of this competition is due to Langmuir[35].

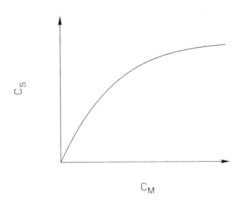

Figure 2-13. Typical adsorption isotherm.

Assume that the adsorbent has a fixed number of equal ligands or adsorption sites, S, to which the sample protein molecules, P, bind one-to-one in a reversible way. The following equilibrium then applies:

$$P + S \rightleftharpoons PS; \quad K = \frac{[PS]}{[P] \cdot [S]} \tag{2-10}$$

where K is an association constant. Clearly $[PS] = C_S$, $[P] = C_M$ and $[S] + [PS] = Q$, the *adsorbent or binding capacity*, i.e. the number of sites per unit surface area (or weight). This has nothing to do with the "capacity factor" k as defined in Eq. (2-1). We can easily solve the equilibrium equation for C_S, which leads to Langmuir's adsorption isotherm equation:

$$C_S = \frac{Q \cdot K \cdot C_M}{1 + K \cdot C_M} \tag{2-11}$$

For most adsorption equilibria we must consider the competition of two different counter ligands, one of which is the sample molecule, P, and one is a component, E, of the eluent.

$$P + ES \rightleftharpoons E + PS; \quad \frac{[PS][E]}{[P][ES]} = K_E^P \tag{2-12}$$

Here, $[PS] = C_S$, $[P] = C_M$ and $[PS] + [ES] = Q$, and K_E^P is a selectivity constant, the quotient of the relevant association constants.

From this, we again obtain Eq. (2-11) with $K = K_E^P/[E]$. For a fixed eluent concentration the Langmuir equation is thus valid also for competition equilibria. The model is also immediately applicable to an ion-exchange equilibrium where P signifies the protein ions, E is an ion in the eluent, and Q is the ion exchanger capacity.

In practical protein purification applications, there may be (and usually are) interfering molecules of several kinds in the sample. Introducing additional molecules of various kinds, A_i, which compete for the adsorption sites and other molecules, B_j, which bind to the sample molecules, we must consider in addition to Eq. (2-12), the relations

$$A_i + ES \rightleftharpoons A_i \cdot S + E; \quad \frac{[A_iS][E]}{[A_i][ES]} = K_E^{A_i} \tag{2-13a}$$

and

$$B_j + P \rightleftharpoons B_jP; \quad \frac{[B_jP]}{[B_j][P]} = K_{B_j} \tag{2-13b}$$

and $[PS] = C_S$, $[P] + [B_jP] = C_M$ and $[PS] + [ES] + [A_iS] = Q$.
This leads again to Eq. (2-11) if K is properly adjusted:

$$K = \frac{K_E^P}{([E] + \sum_i K_E^{A_i}[A_i])(1 + \sum_j K_{B_j}[B_j])} \tag{2-14}$$

The Langmuir model thus still applies, but with a conditional constant which incorporates the influences of the competing molecules.

As the natures and concentrations of interfering sample components are generally unknown or at least incompletely known, the value of measured K_E^P-values for prediction of retention volumes is limited.

The assumptions upon which the derivations above are based may not be entirely applicable in all cases: the adsorption sites may be unequal, there may be multi-site binding (certainly relevant for macromolecules) and there may be interactions (eg. repulsion) between adsorbed molecules. In these cases the Langmuir model is not quantitatively correct, but in practice it often has a reasonable semiquantitative validity.

2.3.2.2 Chromatographic Retention

In one of the first papers on chromatographic theory[36] the relation between the retention volume and the adsorption isotherm was derived:

$$V_R = V_0 + A_S \frac{dC_S}{dC_M} \tag{2-15}$$

Here A_S is the total area of the adsorbent in the column or, alternatively, the weight of the adsorbent, depending on the definition of C_S (see above). Inserting Eq. (2-11), we obtain after differentiation:

$$V_R = V_0 + \frac{A_S \cdot Q \cdot K}{(1 + K C_M)^2} \tag{2-16}$$

which is the proper expression of the retention volume, assuming a Langmuir adsorption isotherm.

If the concentration C_M of sample is small enough, the denominator in the second term of Eq. (2-16) is practically unity, leading to

$$V_R = V_0 + A_S \cdot Q \cdot K = V_0 + n_T \cdot K \tag{2-17}$$

where n_T is the total number of adsorption sites in the column. From Eq. (2-1), we find that:

$$k = \frac{n_T}{V_0} \cdot K \tag{2-18}$$

The retention volume is here independent of sample concentration and solely determined by column parameters and by the equilibrium constant, K. As mentioned in section 2.3.1.1, V_0 may vary slightly with molecular size, which influences the retention.

This case is termed *linear chromatography*. In analytical applications of chromatography this is the preferred case. It is characterized by a simple theoretical description of peak retention and dispersion (broadening). The concepts related to theoretical plates as described above are valid. The term "linear" refers to the fact that it is equivalent to the assumption of a linear adsorption isotherm. In reality the isotherm must be curved (see Fig. 2-13) but at sufficiently low concentrations the curvature becomes negligible.

In many cases, especially when chromatography is used for preparative purposes (which is important in the context of this book) the assumption of linearity is not valid, due to the relatively high concentrations of sample. If this happens, the retention volume varies with sample concentration, C_M, as described by Eq. (2-16). This leads to asymmetric "tailing" peaks as parts of a peak with low concentrations are retarded more than the parts with high concentrations. In Fig. 2-14 tailing peaks of different sizes are shown. The width of a tailing peak depends partly on the dispersion factors as described on p. 50, but these are interrelated with the peak broadening due to the tailing itself in an unknown way. Consequently, theoretical plates and related quantities are not

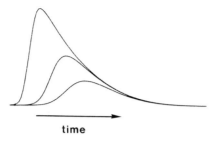

time

Figure 2-14. Tailing peaks of different sizes.

applicable to tailing peaks and should not be calculated. It is even impossible to treat this general problem of *non-linear chromatography* in a mathematical way. Equation (2-15) describes fairly well the retention volume of the maximum of a tailing peak with C_M equal to the concentration at the maximum, but for the rest of the peak, no accurate mathematical description exists.

Observe that although the component of interest may not be present at high concentration, other components may, and often are. All components can then compete for the same adsorption (ion-exchange) sites and, consequently, influence each others' retention, see Eq. (2-14).

The conditions prevailing when a mixture of proteins is chromatographed are very complex and attempts to calculate retention volumes, etc. from batch experiments or chromatography of pure compounds can be very inaccurate.

An erroneous interpretation of retention in non-linear chromatography, usually implicitly expressed as a non-constant capacity factor, is found in several texts. The capacity factor k can be written (see Eq. (2-18) and the definition of K)

$$k = \frac{A_s}{V_0} \cdot \frac{C_S}{C_M}$$
(2-19)

This is correct in the case of linear chromatography and is often stated as an alternative definition of k. However, this is not generally true, and applying Eq. (2-16) to non-linear chromatography is dangerous. It would imply that

$$V_R = V_0 + A_S \cdot \frac{C_S}{C_M}$$
(2-20)

instead of the correct Eq. (2-15). The difference between Eqs. (2-15) and (2-20) is conceptually difficult and leads seemingly to a paradox. The matter was clarified by Helfferich[37].

2.4 Chromatographic Procedures

2.4.1 Sample Introduction

The usual way to perform a chromatographic experiment involves *plug injection*, e.g., the introduction of a small volume of sample at the beginning of the column. The width of this plug obviously influences the width of the resulting peaks. In theoretical discussions, the width of the plug is usually assumed to be negligible. However, in

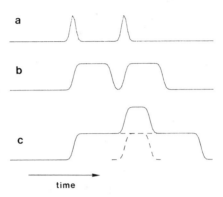

Figure 2-15. The influence of sample size in preparative chromatography. (a) Small (analytical) sample size (b) Sample size optimized with regard to sample throughput. (c) Too large samples size (frontal chromatogram).

preparative chromatography wider plugs may be tolerated, thereby permitting the introduction of a larger volume of sample. See the example in Fig. 2-15.

As the introduction of larger amounts of sample may cause overloading and non-linear conditions (see section 2.3.2.2) the effect is generally more complex than an increase in peak widths.

A special case is *frontal chromatography*, where the pure eluent is exchanged for a solution of sample which is pumped into the column. The sample will appear, after some time, as a more or less steep concentration step in the detector. The midpoint of the step corresponds to the retention time and the height of the fully developed step to the initial sample concentration. With several components, several steps will build up and only the fastest component will be (partly) separated from the other components. If the sample is again exchanged for pure eluent, negative steps will occur and the result is equivalent to a large plug. Figure 2-15c is an example of such a frontal chromatogram. The technique is used mostly for determination of physico-chemical parameters by chromatography, where it has some advantages, and for sampling and preconcentration of dilute samples. It may also unintentionally be encountered in preparative chromatography after introduction of excessively wide plugs. In the latter case, it is important to realise that a chromatogram such as that in Fig. 2-15c contains two components, not three.

2.4.2 Chromatographic Development

2.4.2.1 Isocratic Elution

The most simple mode of chromatographic development is the *isocratic elution* mode, wherein all conditions are held constant throughout the experiment. Each component of the sample will thus travel through the column according to Eqs. (2-15) to (2-17).

The retention times of the various components and thereby the degree of separation are determined by the corresponding values of K. As seen above (Section 2.3.2.1), this constant may be interpreted in several ways, depending on the physical process on which the separation is based. Generally, a necessary condition for a successful isocratic elution is that all sample components elute in a reasonable time and that compounds of interest are not eluted too early (which would destroy resolution, c.f. Eq. (2-9)). This can only be accomplished for a narrow range of sample types, and the application of isocratic elution to a wide range of sample types leads to disturbing trade offs regarding separation power and time. This is sometimes referred to as the *general elution problem*. To solve this problem, it is necessary to employ *gradient elution*, where the composition of the eluent is changed during the development. The change may be continuous, usually linear, or stepwise, and the object is to decrease K successively for each component with time.

2.4.2.2 Gradient Elution

This increase in "elution power" may be accomplished in various ways for different chromatographic techniques: in reversed phase liquid chromatography the polarity of the eluent is decreased, thereby decreasing the partition coefficients: in ion exchange chromatography the concentration of an eluent ion is increased, leading to a decrease in the apparent K (see Eq. (2-14)) and in several other techniques the successive addition of competing compounds will affect the apparent K in the desired direction.

In the beginning of the chromatographic run, a low elution power is chosen. Then, the components which are most loosely bound to the adsorption (etc.) sites will elute under favourable conditions. Other more strongly held compounds will be successively eluted as the elution power is increased. The result is usually an increased resolution over a wide range of sample compounds

The retention times in a gradient elution run cannot be calculated directly from Eq. (2-17) (and similar equations), since the parameter K varies with time. If the variation of K with time is known, the retention time can in principle be calculated by integration, a procedure which is rarely applied.

The plate number concept, as defined on p. 48 is not relevant in the case of gradient elution and the application of Eq. (2-4) to a gradient chromatogram leads to gross overestimates of n. Beware of excessive column performance claims produced in this way!

2.4.2.3 Displacement Chromatography

An extreme case of gradient elution is *displacement chromatography* which is based on competitive binding between the sample components themselves and an additional compound, the terminal displacer or developer, which is more strongly adsorbed than any of the proteins in the sample. During the course of the experiment the continuously added terminal displacer will push the sample components in front of itself, forcing them to displace each other, thereby forming a so-called displacement train in which the different molecules will arrange themselves in the order of their interaction strength with the adsorption sites of the column. In order to improve the resolution between the sample components, spacers, i.e., molecules with intermediate adsorption strengths are usually added. Examples of such compounds are the carrier ampholytes used in isoelectric focusing or carboxymethyl dextrans. A detailed description of displacement chromatography is given in Reference 38. A review of displacement chromatography of proteins is given in Reference 39 and of peptides in Reference 40.

2.4.3 Determination of the Column Capacity and Association Constants

The chromatographic retention depends, as seen above (e.g. in Eq. (2-16)), on essentially two parameters apart from column dimensions: namely, the column binding capacity Q and the (apparent) association constant K.

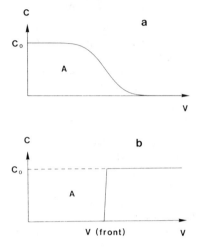

Figure 2-16. Non-linear chromatographic fronts.

The capacity Q for a particular protein is a complex function of several parameters: matrix composition and matrix pore structure, particle diameter and particles size distribution profile, protein molecular weight and solubility, forward and backward rate constants for the binding reaction, bulk-, film- and gel pore-diffusion constants of the protein and finally, the event of possible competitive binding and displacement effects of other proteins present in the sample solution. It is convenient and useful to distinguish between the *nominal* binding capacity of a particular stationary phase such as an ion exchanger and its *dynamic* or *functional* binding capacity. The nominal binding capacity of an ion exchanger can, for example, be determined by an acid-base titration.

This value, however, does not tell how much protein will bind under normal operating conditions in a running column. This can only be measured by chromatographic methods and can for a single column vary considerably for different proteins. For a single protein the dynamic binding capacity varies significantly with the flow-rate used during adsorption and washing.

The determination of the dynamic, or functional, binding capacity is usually performed by frontal chromatography (see section 2.4.1) under isocratic conditions. Figure 2-16 shows typical fronts in non-linear adsorption chromatography.

A positive step-injection, i.e., the exchange of pure eluent for a solution of sample with concentration C_0 will produce a sharp front at the outlet as shown in Fig. 2-16b, whereas a negative step-injection (removal of the sample feed) produces a tailing, "diffuse" concentration step (Fig. 2-16a). It is easiest to treat the second case, as the tailing slope follows Eq. (2-16).

The area A in Fig. 2-16 is found by integration:

$$A = \int_0^{C_0} V_R \cdot dC_M \tag{2-21}$$

If we arrange the experiment to make K very large, we obtain with Eq. (2-16):

$$A = V_0 C_0 + A_S Q \tag{2-22}$$

The area A in Fig 2-16b corresponds to the amount of sample that has been held on the column, whereas the analogous area in Fig. 2-16a corresponds to the same amount which is released when sample feed is removed. Consequently, these areas are equal. The capacity Q can easily be calculated from a sharp frontal chromatogram, as the area A in such a case will be nearly rectangular,

$$Q = \frac{C_0}{A_S}(V_{R,\,front} - V_0) \qquad\qquad (2\text{-}23)$$

provided that K is large. $V_{R,\,front}$ is the retention volume measured from the start of sample introduction to the sharp front. In order to obtain a larger K we should chose a suitable sample component (we intend to measure a property of the column, so there is a choice of sample) and an eluent with low eluting power, free from competing compounds. To obtain valid values of Q it is advisable to repeat the measurement under different conditions. In some cases, the binding equilibrium might be slow. This is often the case in affinity chromatography. The result is that the shape of the front is not sharp but "rounded". The measurements described are still valid, provided that the area A is calculated by integration of the recorder signal.

The determination of the association (partition) coefficients, K, can in some cases be performed by non-chromatographic methods. Hutchens and Yip determined binding constant and capacities of several proteins to IMAC gels by means of Scatchard plots, i.e., equilibrium binding[41]. The values obtained agreed nicely with those derived from frontal chromatography. The chromatographic determination of K is simple: the retention volume of a peak obtained after the injection of a narrow sample plug of low concentration gives directly (Eq. (2-17)) $A_s \cdot Q \cdot K$. If $A_s \cdot Q$ is determined as described above, K is easily calculated.

2.5   Chromatographic Techniques

2.5.1   General Comments

Modern column chromatography utilizes sophisticated equipment to obtain high resolution separations. However, for some applications this might not be necessary. The most simple way to carry out an adsorption experiment is batch-wise, by simply stirring the adsorbent with the protein sample choosing proper conditions for adsorption and subsequent desorption. This has the advantage of being applicable with large volumes of protein solution, which do not necessarily need to be clear, and is sometimes of use at an early stage in a purification procedure or simply to test whether adsorption occurs under the conditions selected.

Column chromatography can be performed as *low pressure* (or standard), *medium pressure*, or *high pressure liquid chromatography*, HPLC. Medium pressure techniques also include fast protein liquid chromatography, for example the Pharmacia chromatography system FPLC®. For desalting experiments, adsorption tests, etc., simple column chromatography equipment is often sufficient. High resolution results can also be obtained with standard, low pressure equipment provided the selectivity of the column packing material is sufficiently high. High resolution fast techniques require small diameter beads and equipment able to withstand the often high pressures necessary to force the buffer or solvent through the column.

In all three techniques the size, and thus the capacity, of the chromatographic column can vary considerably. Medium scale, i.e., from a mg to a single gram of protein, is usually the easiest to handle. Microgram scale requires sensitive analytical techniques, high purity of buffers and solvents—although less volume is consumed—and often special care has to be taken to avoid adsorption to the walls of vessels. Equipment for large scale chromatography runs into other difficulties, in particular the high cost of larger amounts of modern HPLC media.

The most crucial point in column chromatography is to achieve a good column packing. This is therefore treated in some detail in the final section of this chapter. FPLC and HPLC columns are normally delivered prepacked from the manufacturers. However most other gels are delivered in bulk and must be packed in the user's laboratory.

2.5.2 Packing of Columns

In any chromatographic experiment the result obtained can never be better than the quality of the column packing allows. This is why it always pays off in the long run to learn how to pack the most commonly used stationary phases. One should be particularly cautious when packing columns for gel filtration and other isocratic techniques. Detailed packing instructions are usually provided by the manufacturers of a particular gel material. Here only some general principles will be discussed.

The packing techniques used differ depending on the type of stationary phase. The most important discriminating parameter is the rigidity of the gel matrix. It is thus convenient to distinguish between soft, semi-rigid and rigid gel matrices. Particle shape, diameter and size distribution are also important parameters to consider in column packing. The first step is to mount the column with its extension tube on a steady laboratory stand and to ensure that the column tube is perfectly vertical. The stationary phase slurry is degassed to remove all trapped air.

Rigid gel materials such as silica with particle diameters in the range 5–15 microns are preferably slurry packed in dry acetone or chloroform. Slurry concentrations around 10% and packing pressures up to around 300 kg cm^{-2} usually give satisfactory results.

Semirigid gels such as Sephacryl®HR, Sepharose®FF and Superose® are preferably packed in two steps. In the first step, a homogeneous slurry (10–50%) containing all the stationary phase intended for the column is poured into the column fitted with an extension tube. The extension tube is connected to a pump which is adjusted to a medium high flow-rate. For a 45 micron Sephacryl HR this means around 30 cm h^{-1} for columns 40 cm–100 cm in length. When the bed has settled, the second step in the column packing involves a doubling of the flow-rate to, in this case, around 60 cm h^{-1}. The packing of the column should not be considered complete until 4–5 column volumes of packing buffer have been pumped through. The crucial feature of this procedure is that it prevents the formation of a plug of hard packed gel at the bottom frit of the column, which will inevitably occur upon packing in one step at constant pressure with an initial high flow-rate. This plug will block the flow and give rise to badly packed columns.

Every column should be tested for packing quality. A zone of acetone (0.5% V_t) at 30 cm h^{-1} is suitable for this purpose. The reduced plate height should fall in the range 2–3, at the lower end for experienced column packers and at the upper end for beginners.

The packing buffer composition for semi-rigid gels does not seem to be critical. Similar results are obtained with distilled water and with various buffer salt solutions. Most convenient is to use the slurry obtained by shaking the original bottle in which the gel is delivered, and dilute with distilled water to the desired slurry concentration. The column bottom frit or filter mesh should be wetted and all air removed. 20% ethanol is recommended for this purpose. A few cm of this solution can be left in the column before the addition of the gel slurry.

A critical point in the column packing is the application of the adaptor on top of the packed bed. As a general recommendation one should allow the adaptor to compress the upper part of the bed approximately 5 mm. Some workers prefer to pack the columns upside down towards the adaptor. In this way they get the best possible starting conditions for the sample zone, which is especially desirable in gel filtration. The bottom end piece is then allowed to compress the packed bed a few millimeters as described above.

For short columns (5-15 cm in length) normally used in various types of adsorption chromatography (ion exchange chromatography, affinity chromatography, etc) packing quality is less critical. Often one-step procedures based on either constant flow or contant pressure give equally satisfactory results. Larger diameter, semi-rigid stationary phases such as derivatives of Sepharose FF (90 micron) give too little flow resistance to be packed efficiently with normal laboratory pumps. This is why many workers pack this matrix using compressed air or nitrogen as a pressure source and regulate the flow with a needle valve at the column outlet. Sepharose FF is preferably packed in two steps in the way described above. The bed is thus allowed to settle at a linear flow-rate of around 3 cm min^{-1} and the final packing takes place at around 5 cm min^{-1}. One should be aware of the danger of using pressurized vessels and never allow the pressure to exceed the ratings of the equipment used. For normal laboratory columns a pressure drop of 2 kg cm^{-2} is sufficient to give efficient packing of Sepharose FF. As a general precaution, the column and accessories should always be placed behind a protective screen or cover during column packing. Some workers claim that the packing of Sepharose FF is faciliated by the presence of 0.05-0.1% Tween 20.

Soft gel matrices such as cellulose, Sephadex with higher G-numbers, non-cross-linked Sepharose and conventional polyacrylamide gels with low degrees of cross-linking are, in principle, packed in the same way as the semi-rigid gel materials—with the exception that the flow rates used are considerably lower and are never allowed to approach the maximum flow-rate obtainable in a particular column. Neither, of course, should the operating flow-rate exceed the packing flow-rate of the bed.

2.6   References

1. J. C. Giddings, *Dynamics of Chromatography*, Marcel Dekker, New York, 1965.
2. E. Heftmann, (ed.) *Chromatography*, Elsevier, Amsterdam, 1983.
3. J. Å. Jönsson, (ed.), *Chromatographic Theory and Basic Principles*, Marcel Dekker, New York, 1987.
4. E. A. Peterson, H. A. Sober, *J. Am. Chem. Soc.*, 78, 751-755 (1956).
5. Whatman BioSystems Ltd., Springfield Mill, Maidstone, Kent ME14 2LE, England.
6. Cellulofine[R] manufactured by Chisso Corp. Ltd., Kumamoto, Japan and marketed by Amicon Div., W. R. Grace & Co., Danvers, MA 01923, USA.
7. Pharmacia LKB Biotechnology AB, Uppsala, Sweden.
8. J. Porath, P. Flodin, *Nature, 193*, 1657-59 (1959).
9. Gel Filtration, Theory and Practice. Booklet published by Pharmacia LKB Biotechnology AB, Uppsala, Sweden.
10. Ion exchange chromatography. Principles and Methods. Booklet published by Pharmacia LKB Biotechnology AB, Uppsala, Sweden.

11. J. C. Janson, *Chromatographia*, *23*, 361-369 (1987).
12. S. Hjertén, *Arch. Biochem. Biophys*, *99*, 466-475 (1962).
13. IBF Biotechnics, Villeneuve La Garenne, France.
14. BioRad Chemical Division, Richmond, CA 94804, USA.
15. S. Hjertén, *Biochim. Biophys. Acta*, *79*, 393-398 (1964).
16. S. Arnott, A. Fulmer, W. E. Scott, I. C. M. Dea, R. Moorhouse, D. A. Rees, *J. Mol. Biol. 90*, 269-284 (1974).
17. A. Amsterdam, Z. Er-el, S. Shaltiel, *Arch. Biochem. Biophys.*, *17*, 673-678 (1975).
18. J. Porath, J-C. Janson, T. Låås, *J. Chromatogr. 60*, 167-177 (1971).
19. J. Porath, T. Låås, J-C. Janson *J. Chromatogr. 103*, 49-62 (1975).
20. T. Andersson, M. Carlsson, L. Hagel, P-Å. Pernemalm, J-C. Janson, *J. Chromatogr.*, *326*, 33-44 (1984).
21. F. E. Reginer, R. Noel, *J. Chromatogr. Sci.*, *14*, 316-320 (1976).
22. R. W. Stout, J. J. DeStefano, *J. Chromatogr.*, *326*, 63-78 (1985).
23. MonoBead™. Polymer based, 10 μm diameter stationary phases. Pharmacia LKB Biotechnology AB, Uppsala, Sweden.
24. Superose™. Agarose based, 10 μm and 13 μm diameter stationary phases. Pharmacia LKB Biotechnology AB, Uppsala, Sweden. See also reference 20.
25. A. Tiselius, S. Hjertén, Ö. Levin, *Arch. Biochem. Biophys.*, *65*, 132-155 (1956).
26. P. Flodin, Dextran gels and their applications in gel filtration, Dissertation, Pharmacia AB, Uppsala, Sweden, pp. 1-85 1962.
27. S. Hjertén, in *Methods of Immunology and Immunochemistry*, M. W. Chase, C. A. Williams, eds., Vol. 2, 142-148, Academic Press, New York, 1968.
28. J. Ugelstad, L. Söderberg, A. Berge, J. Bergström, *Nature*, *303*, 95-96 (1983).
29. Sepharose High Performance™. Agarose based, 34 μm diameter stationary phases, Pharmacia LKB Biotechnology AB, Uppsala, Sweden.
30. Sepharose Fast Flow™. Agarose based, 90 μm diameter stationary phases. Pharmacia LKB Biotechnology AB, Uppsala, Sweden.
31. A. R. Fersht, J-P. Shi, J. Knill-Jones, D. M. Lowe, A. J. Wilkinson, D. M. Blow, P. Brick, P. Carter, M. M. Y. Waye, G. Winter, *Nature*, *314*, 235-238 (1985).
32. T. E. Creighton, *Proteins, Structures and Molecular Properties*, Freeman, New York, 1983.
33. Liquid Chromatography Terms and Relationships, ASTM E682, American Society for Testing Materials, Philadelphia, 1979.
34. J. J. Van Deemter, F. J. Zuiderweg, A. Klinkenberg, *Chem. Eng. Sci. 5*, 271 (1956).
35. I. Langmuir, *J. Am. Chem. Soc.*, *40*, 1361 (1918).
36. D. de Vault, *J. Am. Chem. Soc.*, *65*, 632 (1943).
37. F. Helfferich, *J. Chem. Educ.*, *41*, 410 (1964).
38. C. Horvath, A. Nahum, J. H. Frenz, *J. Chromatogr.*, *218*, 365-393 (1981).
39. E. A. Peterson, A. R. Torres, in *Methods in Enzymol.* W. B. Jacoby, ed., *104*, 113-133, Academic Press, New York.
40. S. M. Cramer, C. Horvath, *Prep. Chromatogr. 1*, 29-49 (1988).
41. T. W. Hutchens, T.-T. Yip, J. Porath, *Anal. Biochem.*, *170*, 168-182 (1988).

3 Gel Filtration

Lars Hagel

Pharmacia LKB Biotechnology AB

S-751 82 Uppsala, Sweden

3.1. Introduction

The use of column chromatography for the separation of proteins according to size was described as early as in 1955 by Lathe and Ruthven[1]. The separating material used was swollen maize starch which required low flow rates for elution[2]. A material, based on cross-linked dextran, providing more favourable flow properties was presented in 1959 by Porath and Flodin[3]. They named this technique for separating substances of different molecular size gel filtration (GF). Several beaded matrices were subsequently developed for gel filtration[4], and in 1964 the technique was extended to non-aqueous solutions by the introduction of a polystyrene-based matrix by Moore[5], who, however, called the technique gel permeation chromatography (GPC).

Traditionally, the materials used for GPC are small rigid beads composed of styrene-based matrices or coated silica while the materials for GF are typically non-rigid, larger beads prepared by cross-linking a hydrophilic natural polymer. However, the development of separation materials for GF in the last few years has reduced the differences to a matter of matrix and solvent polarities only. This has also, unfortunately, resulted in a more ambiguous nomenclature. Thus, gel permeation chromatography and size exclusion chromatography (SEC) are often used as synonyms for gel filtration in addition to a number of less frequently used designations[6]. It should be noted that permeation chromatography, as defined by IUPAC[7], also includes exclusion effects demonstrated by charged solutes and is therefore not applicable. However, present practice is to use size exclusion chromatography as an universal name for the separation principle while gel filtration and gel permeation chromatography are terms for its applications in aqueous and non-aqueous media respectively. Thus, the designation gel filtration, as introduced by Porath and Flodin, is still widely used in the literature, manufacturers catalogues and standard text books for the description of aqueous size exclusion chromatography[8-11].

Many models have been proposed to explain the separation mechanisms in gel filtration and the validities of these have been thoroughly discussed[12,13]. The separation process is schematically illustrated in Fig. 3-1. The separation may simply be regarded as due to the different amount of time different solutes stay within the liquid phase that is entrapped by the matrix. This time is of course related to the fraction of the pores that is accessible to the solute. The interpretation of this fraction in terms of pore dimensions and gel structure, together with various expressions for solute size, results in slightly different equations for relating the distribution coefficient to the size of the solute. Interestingly enough all these equations propose a linear relationship between the logarithms of the two parameters[13]. A more general approach was depicted by Casassa who used a stochastic model to relate the non-available fraction of the pores and the dimension of the solute[14]. It can be concluded from recent work that regions of inaccessibility of the pore volume result in a loss in entropy of the molecules[15]. This loss in entropy is due to the smaller number of possible conformations of the molecules within the pores as compared to an equal volume segment outside the pore. In this case, a linear relationship between the logarithm of the distribution coefficient and the logarithm of solute size is also expected[15].

Even though gel filtration is an uncomplicated and straightforward technique, there are some points worth consideration before starting the experimental work. The actual sample may require a special pH, solvent, additives or pretreatment to yield a true solution. The next step is to select the gel that will cope with the chosen solvent and pH and that has a suitable separation range. Possible adsorption properties of the gel must also be considered. The nature of the separation and the sample may put demands on such parameters as resolution, separation time and sample load which in turn are partly dependent on the selected gel. These parameters are, however, also affected by the choice of column dimensions and the packing efficiency of the column. Obviously, for different separation problems, such as desalting, preparative purifications or analytical separations different requirements should be stressed. Economic factors and the possibilities of scaling up may also be important.

INSTRUMENTATION

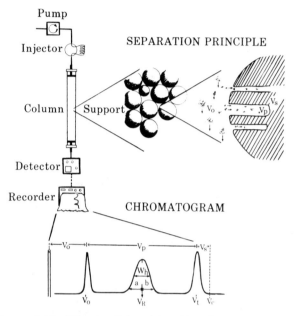

Figure 3-1. Fundamentals of gel filtration. Solutes injected into the column are separated according to decreasing size due to incompatibility between the solute dimension and the pore size of the support. V_0 = void volume between the support particles, V_p = pore volume and V_s = matrix volume of the support. V_R = elution volume of the solute, V_t = total liquid volume of the column and V_c = total geometric volume of the column. Column plate number $N = 5.54 \times (V_R/w_h)^2$ where w_h is the peak width at half peak height and b/a = peak symmetry at 10% peak height.

Please note that the simple pore structure drawn in Fig. 3-1 does not provide a realistic image of the structures of chromatography supports but only serves to illustrate the separation principle in gel filtration (c.f. the authentic pore structure shown in Fig. 3-2).

A thorough description of laboratory techniques for gel filtration has been given by Fischer[6]. An excellent compilation of practical aspects of high performance liquid chromatography has been made by Bristow[16].

The aim of this chapter is to describe some fundamental practical and applied theoretical aspects of experimental work with gel filtration chromatography.

3.2 Selection of the Support

The first constraint on the choice of gel medium is due to the solvent and pH required to provide a good solvent for the sample. Most media withstand the common solvents and pH's used, i.e. pH 2-12, to dissolve proteins, with the exception of silica-based materials. The importance of selecting a gel with a suitable separation range increases with the complexity of the separation problem, i.e. desalting < preparative purification < analytical applications. For difficult separation problems the resolution of the gel may be critical and in these instances properties such as bead size, slope of the selectivity curve and separation volume become important. A third factor to take into account is the effect of the sorption properties of the matrix under the running conditions. Even though virtually no matrix can be expected to be completely free from sorption properties, the nature and degree of these properties varies with the nature of the matrix[17-19]. Sometimes these properties have been used to achieve increased separation of the sample components[19-21]. However it is important to realize that these mixed-mode separations should not be classified as gel filtration!

3.2.1 Characteristics of Available Media

Since the advent of the cross-linked dextran, Sephadex®, in 1959, gels based on agarose, polyacrylamide and combinations of these media have been commercially exploited. These traditional media are characterized by large bead sizes (i.e. typically 100–250 μm), low matrix volumes (e.g. less than 5%) and high deformability. Porous glass may, due to the large bead size, also be included in this category. The chemistry and properties of these gels have been thoroughly described[4,6,19] and compared[17]. Data for some of these supports are complied in Table 3-1.

The gels presently used for high resolution gel filtration of proteins are based on silica, hydrophilized vinyl polymers or highly cross-linked agarose with bead sizes typically between 5 to 50 μm. The smaller particle size yields more efficient columns, resulting in narrower peaks, and may be employed for achieving higher resolution and/or faster separations. However, these supports are more expensive than traditional media and are often only available as pre-packed columns. The chromatographic properties of columns pre-packed with silica-based media have been reviewed by Pfannkoch and co-workers[18]. The gel properties and application of the theoretical aspects of chromatography will be discussed with special reference to the media designed for high resolution gel filtration. Characteristics of some available media are given in Table 3-2. The porous structure of one support as visualized by scanning electron microscopy is shown in Fig. 3-2.

3.2.2 Chemical and Sorptive Properties

The pH resistance of standard gels covers the approximate range pH 2–10. High resolution gels have a somewhat higher pH stability (e.g. pH 1–14) with the exception of silica-based materials, see Tables 3-1 and 3-2. To reduce the dissolution of silica at high pH, a pre-column packed with silica may be placed before the injector. To improve the stability of silica, treatment with zirconium has recently been suggested[22]. However, deterioration of even this material was noted with prolonged use at pH 7[28].

The surface of traditional supports prepared from natural polymers (e.g., agarose and dextran gels) contains predominantly hydroxyl groups and provides a good environment for hydrophilic proteins. Unfortunately the hydrophilicity is reduced somewhat by the introduction of cross-linking reagents (e.g., epichlorohydrin). Matrices composed of styrene-divinylbenzene copolymer must be chemically treated to increase the hydrophilicity of the surface before the material is suitable for gel filtration. The surface of silica and porous glass must also be derivatized or coated to prevent excessive adsorption of proteins.

The presence of charged groups may be due to small amounts of naturally occurring acidic groups in the raw materials (e.g. sulphate groups in agarose or carboxylic acid groups in dextran). Ionic sites may also be introduced by acid or alkaline hydrolyses of the matrix by prolonged exposure to an extreme pH.

The structure of the matrix may yield specific interactions such as aromatic adsorption of, for example, tryptophan-rich peptides, on Sephadex G-25, G-15 and G-10 [29]. The effect is assumed to be caused by interaction between aromatic amino acids and the ether bridges introduced by the crosslinker[30]. Biospecific interactions where the matrix resembles an enzyme substrate or an affinity site may also be noted, e.g., interaction between lectins and glucosidic sites of gels.

Adsorption properties of standard gel filtration media have been utilized advantageously to improve the separation of substances and the basic principles of the phenomena have been fully described (e.g. see Refs. 29–35).

Non specific interactions between sample and high resolution gels are mostly due to ionic interaction with residual silanol groups or carboxylic acid groups, and hydrophobic interactions with the coating or the cross-linking sites. Pfannkoch et al. tested various interactions by measuring the distribution coefficients of citrate, arginine and phenylethanol at ionic strengths from 0.026 to 2.40 in a phosphate buffer at pH 7 [18]. Deviations of K_D from unity indicated anionic-exclusion, cationic-adsorption and hydrophobic effects. Large differences between different coated silica materials were noted.

A general recommendation for all types of materials is to use a buffer with intermediate ionic strength (e.g. 0.15 M) to suppress ionic effects but still not promote hydrophobic interactions. This will also effectively mask the Donnan effect, i.e. the increase of the electrolyte concentration in the pores caused by the presence of macro-ions in the void volume[36]. Buffers commonly used for different materials are listed in Table 3-2 and more detailed information on buffer preparation and additives is found in paragraph 3.5.

3.2.3   Selectivity Curve and Separation Range

The selectivity of a gel filtration medium is, in contrast to other types of media (e.g. media for ion-exchange or reversed-phase chromatography), not adjustable by changing the composition of the mobile phase (as long as this change does not influence the solute shape or pore structure). The selectivity of the gel is thus an inherent property of the material and the separation volume of the column is limited by the total pore volume of the packed bed.

The selectivity curve of the separation material is obtained by plotting the elution volume, or some function thereof, versus an expression of the solute size. Very often, the distribution coefficient, K_D, is related to the logarithm of the molecular weight, M, of the solute. K_D is a column independent variable and is calculated from the elution, or retention, volume, V_R (also denoted V_e), the void volume, V_0, and the pore volume, V_p (sometimes denoted V_i) according to

$$K_D = (V_R - V_0)/V_p = (V_R - V_0)/(V_t - V_0) \tag{3-1}$$

where V_t is the total liquid volume of the bed (see Fig. 3-1). The plot of K_D versus log M will yield a sigmoid selectivity curve which in the middle range may be approximated by

$$K_D = a - b \log M \tag{3-2}$$

Table 3-1. Some traditional media for gel filtration of proteins

Media	Type	Source	Particle size of hydrated beads (μm)	Fractionation range for globular proteins (kDaltons)	pH stability
Sephadex	Dextran	1			2–10
G-50				1.5–30	
G-75			20–100[a]	3– 80	
G-100			100–300[a]	4–100	
G-150				5–300	
G-200				5–600	
Sepharose	Agarose	1			4–9
6B			45–165	10–4,000	
4B			45–165	60–20,000	
2B			60–200	70–40,000	
Sepharose CL	Agarose	1			3–14
6B			45–165	10–4,000	
4B			45–165	60–20,000	
2B			60–200	70–40,000	
Ultrogel	Agarose	2			Not stated
A6				25–2,400	
A4			60–140	55–9,000	
A2				120–23,000	
Bio-Gel	Agarose	3			4–13
A-0.5 m			40– 80	1–500	
A-1.5 m			80–150	1–1,500	
A-5 m			150–300	10–5,000	
A-15 m				40–15,000	
A-50 m				100–50,000	

	Matrix	Manufacturer	Wet bead diameter (µm)	Fractionation range	pH range
Bio-Gel	Polyacrylamide	3			2–10
P-10			<40, 40–80 ⎫	1.5–20	
P-30			⎬ 80–150	2.5–40	
P-60			<80 ⎭	3–60	
P-100			⎫ 150–300	5–100	
P-150			⎬	15–150	
P-200			⎭	30–200	
P-300				60–400	
Trisacryl	Polyacrylamide	2	40–80	10–15,000	Not stated
GF 2000					
Sephacryl	Dextran/	1			2–11
	Bisacrylamide				
S-200			⎫ 40–105	5–250	
S-300			⎬	10–1,500	
S-400 HR			⎭	20–8,000	
S-200 HR			⎫ 25–75	5–250	
S-300 HR			⎬	10–1,500	
S-400 HR			⎭	20–8,000	
Ultrogel	Agarose/	2			3–10
	Acrylamide				
AcA 202				1–15	
AcA 54				5–70	
AcA 44			60–140	10–130	
AcA 34				20–350	
AcA 22				100–1,200	
Glycophase CPG	Silica	4	37–74		<8
100				1–30	
200				2.5–100	
460				11–320	

[a] Calculated from the stated dry bead diameter through:

$$d_{p,\,wet} = d_{p,\,dry}\,(1 + Wr \times d)^{1/3}$$

where Wr = water regain of bead and d = density of matrix.

Sources: Data given as stated by the manufacturers: 1, Pharmacia (Uppsala, Sweden); 2, Réactifs IBF (Villeneuve la Garenne, France); 3, Bio-Rad Laboratories (Richmond, CA, USA); 4, Pierce Eurochemie B V (Rotterdam, Holland). A more extensive tabulation of gel media is found in Ref. 17.

Table 3-2. Characteristics of some media for high resolution gel filtration of proteins

Media	Type	Source	Particle size[a] (μm)	Fractionation range[a] of globular proteins (kDaltons)	Operation range[b] (kDaltons)	Selectivity $-\Delta K_D/\Delta \log M$	Particle porosity $V_p/(V_p + V_s)$	Permeability V_p/V_o	Void fraction V_o/V_c	pH stability[a]	Buffer composition
Superose 12	Agarose	1	10	1-2,000	0.5-600	0.28	0.84	1.84	0.30	1-14	0.05 M phosphate + 0.15 M NaCl, pH 7.0 [24]
Superose 6	Agarose	1	13	5-40,000	5-10,000	0.23	0.93	1.87	0.33	1-14	
TSK SW 2000	Silica	2	10	0.5-60	0.7-100	0.52	0.66	0.95	0.39	2.5-7.5	0.05 M tris-HCl + 0.5 M Na$_2$ClO$_4$, pH 7.5 [25]
TSK SW 3000	Silica	2	10	1-300	2-400	0.43	0.81	1.33	0.38	2.5-7.5	
TSK SW 4000	Silica	2	13	5-1,000	6-10,000	0.28	0.89	1.40	0.39	2.5-7.5	
Synchropak GPC 100	Silica[c]	3	10	3-630		0.44	0.79	1.23	0.39		0.1-0.6 M phosphate pH 7 [18]
Waters I-125	Silica[c]	4	10	0.8-450		0.36	0.55	0.92	0.38		0.08 M phosphate + 0.32 M NaCl in 20% ethanol [26]
Lichrosorb Diol	Silica[c]	5	10	0.8-450		0.36	0.57	0.63	0.47		0.1-0.6 M phosphate pH 7.0 [18]
Shodex OH PAK	Silica[c]	6	10				0.53	0.85	0.38		
Zorbax GF-250	Silica[d]	7	4	12-120	14-100		0.52	0.78	0.40		
Superose 12 Prep Grade	Agarose	1	20-40	1-2,000	1-600	0.32	0.92	1.86	0.33	1-14	0.05 M phosphate + 0.15 M NaCl, pH 7.0 [24]
Superose 6 Prep Grade	Agarose	1	20-40	5-40,000	5-10,000	0.24	0.97	2.23	0.30	1-14	
Fractogel TSK HW-55S[e]	Vinyl polymer	5	25-40	1-1,000	0.2-100	0.18	0.96	2.17	0.31	1-14	25 mM tris-HCl + 0.3 M NaCl, pH 7.5 [27]

[a] Data stated by the supplier: 1, Pharmacia (Uppsala, Sweden); 2, Toyo Soda (Tokyo, Japan); 3, SynChrom (Linden, IN, USA); 4, Waters (Milford, MA, USA); 5, Merck (Darmstadt, F.R.G); 6, Showa Denko (Tokyo, Japan); 7, DuPont (Wilmington, DE, USA).
[b] For globular proteins, calculated from $0.1 \le K_D \le 0.9$
[c] Data from Ref. 18
[d] Data calculated from Ref. 22
[e] Toyopearl HW type gel.
A more extensive tabulation of gel media is found in Ref. 23.

Figure 3-2. Surface pore structure of an agarose based matrix (Superose® 6) as visualized by SEM after coating the surface with a layer of gold. Magnification factor is 100,000X.

The determination of these parameters is discussed in paragraph 3.7.3, p. 96. The slope of the selectivity curve, i.e. the value of b, depends on the width of the pore size distribution of the material and the intercept, a, is a function of the mean pore size. A narrow pore size distribution will result in high selectivity, i.e., a large value of b, but a small separation range for the gel since $0 \leq K_D \leq 1$ for ideal gel filtration. The sigmoid nature of the selectivity curve reduces the practical working range of the gel to approximately $0.1 \leq K_D \leq 0.9$. It is also important to realize that the exclusion limit quoted by many manufacturers is often an extrapolated value of the selectivity curve and will therefore only give an indication of the size of molecules that will be excluded from the matrix and not the ultimate separation range. Data on selectivity and separation range for some materials are found in Table 3-2. The separation range may be extended by the combination of materials with different pore sizes. If the bead sizes of the materials are equivalent they may be mixed before packing but it is more usual to pack the gels separately in the same or different columns. The gels are often placed in order of descending pore size. However, it has been shown that the column arrangement may be optional and is preferentially random[37]. It is very important to use material with overlapping porosity curves or else artefacts may be created[38,39].

3.2.4   Support Pore Volume

The pore volume of a material, V_p, can be calculated by subtracting the void volume, V_0, from the total liquid volume, V_t. The matrix or support volume, V_s, is obtained by subtracting the total liquid volume from the geometric volume of the column, V_c (c.f. Fig. 3-1). The relative pore volume of the gel is expressed by:

$$V_{p,rel} = \frac{V_p}{V_s + V_p} = \frac{V_t - V_0}{V_c - V_0} \tag{3-3}$$

A low pore volume will result in a low separation volume of the column since solutes are eluted between V_0 and $V_0 + V_p$.

3.3   Theoretical Considerations

3.3.1 Estimation of Molecular Size by Gel Filtration

The retention of solutes is, in ideal gel filtration, governed solely by the differences between the solute dimensions and the pore dimensions. The relationship between size and molecular weight of solutes is strongly dependent upon solute shape which may be illustrated by the relationship between the radius of gyration, R_g, of a solute and its molecular weight, M

$$R_g \propto M^a \tag{3-4}$$

where $a = 1$ for rods, $a \sim 1/2$ for flexible coils and $a = 1/3$ for spheres[40]. The influence of the solute shape on the retention in gel filtration is illustrated in Fig. 3-3. It is readily seen that calibration versus molecular weight is only meaningful for solutes of similar shape.

The shapes of solvated proteins vary considerably, i.e. there are spherical proteins, slightly asymmetrical globular proteins, rodshaped fibrous proteins and denatured flexible coil structures. The frictional coefficient, f, of a solvated protein, obtained from

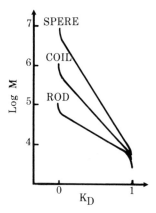

Figure 3-3. Theoretical calibration curves for solutes of various shapes. (Reproduced from Ref. 40 by kind permission of the author and the publisher.)

sedimentation or diffusion experiments, may be compared to the theoretical frictional coefficient, f_0, of a non-solvated sphere of equal volume. It is tempting to interpret deviations of the frictional ratio, f/f_0, from unity as deviations of the protein shape from a sphere. This is facilitated by the theoretical relationship between the frictional ratio of an ellipsoid of rotation and the geometry of the particle (e.g. $f/f_0 = 1.25$ corresponds to a prolate ellipsoid with an axial ratio of $1:5$)[41]. However, the frictional ratio is a function of solute solvation as well as asymmetry (and surface roughness) and therefore interpretation of the shape of, for example, globular proteins as ellipsoidal hydrodynamic particles may not be realistic. It has been concluded that globular proteins (e.g. with $f/f_0 < 1.25$) are neither highly solvated nor highly asymmetric[41]. The situation may be illustrated by reference to apoferritin. This molecule has a perfect spherical shape but yields a frictional ratio of 1.29 due to the large degree of solvation of this hollow sphere shaped molecule[42].

Information on shape may be obtained by comparison of data from hydrodynamic measurements (e.g., intrinsic viscosity and sedimentation or diffusion coefficients) with the radius of gyration from light scattering. A gel filtration column calibrated with such reference substances may be utilised to provide structural information about solutes of known molecular weight (c.f. Fig. 3-3). Conversely, the column may be used for the estimation of molecular weight of compound with shapes similar to the shapes of the reference substances.

Tanford and co-workers proposed calibration of a Sephadex column by plotting K_D versus protein Stokes radius for the accurate (i.e. $\pm 5\%$) estimation of the molecular weight of proteins in a detergent solution[43]. The Stokes radius, R_{St}, is defined as the radius of a sphere that would have the same frictional coefficient as the protein[41]. The molecular weight of a compact globular protein may be inferred from the Stokes radius through

$$\log M \sim 0.147 \pm 0.041 + \log(R_{St}^3/\bar{v}) \qquad (3\text{-}5)$$

where the partial specific volume, $\bar{v}$, if unknown, may be assigned a value of 0.74[42].

The size of solutes may also be estimated from viscosity data yielding an expression of the hydrodynamic volume.

$$V_h = |\eta| M/v N_a \qquad (3\text{-}6)$$

where $|\eta|$ is the intrinsic viscosity, N_a is the Avogadro's number and v is Simhas factor. The value of this factor depends on the shape of the molecule and is 2.5 for spheres and > 2.5 for ellipsoids (e.g. an axial ratio of 1:5 corresponds to $v \sim 5$). For flexible polymers and spherical solutes a viscosity radius, R_{vis} or R_h, may be calculated from the hydrodynamic volume by $R_h = (V_h \times 3/4\pi)^{1/3}$. This estimate is often called the hydrodynamic radius. However, R_{St} is also obtained from hydrodynamic measurements and therefore the term hydrodynamic viscosity radius is preferred for R_h or R_{vis}. The following relationship between the molecular weight of compact globular proteins and viscosity radius (Å) may be calculated from Tanford[41]

$$R_h \approx 0.718 \times M^{1/3} \tag{3-7}$$

A similar relationship was calculated by Squire for globular solvated proteins assuming spherical shape (i.e. $R_h \approx 0.794 \times M^{1/3}$)[44].

As pointed out in the work of Tanford and co-workers, it was not known earlier whether partition in the gel was responsive to Stokes radius or viscosity radius. However, observed differences between the two estimates was smaller than 10% for globular proteins, approximately 15% for random coils and up to 100% for fibrous proteins[43]. In another study they found that denatured proteins of quite different conformations eluted according to identical calibration curves when the solute size was inferred from viscosity data[45]. Benoit and co-workers proposed the hydrodynamic volume as the retention decisive parameter for the solute in SEC[46]. This approach has been shown to be applicable to a wide variety of solute shapes and solute-solvent systems[46-52]. Calibration of a column using hydrodynamic volume (or viscosity radius) has therefore been called universal calibration. This type of calibration curve is illustrated in Fig. 3-4 which shows that globular proteins and rod shaped virus particles elute according to the same curve. A similar result would not have been obtained if retention had been plotted versus R_{St}[43].

Thus, the general applicability of size estimates increases in the order: molecular weight $\ll$ Stokes radius $<$ hydrodynamic volume (or viscosity radius). The radius of gyration is proportional to the viscosity radius of spherical solutes and flexible polymers but not to that of rigid macromolecules[41]. The use of a well defined polymer, such as dextran, to calibrate the column according to the hydrodynamic volume seems to be the most appropriate procedure for the characterization of an unknown solute with the aid of gel filtration.

The solute size is often related to the distribution coefficient in a log-linear plot. Various relationships between size and K_D that have been found useful are given in Table 3-3. Many of the relationships are claimed to yield linear calibration curves which probably reflects that variations in solute properties are sufficiently large to disguise the sigmoid nature of the curve. The largest linear portion of the calibration curve is obtained from a single-pore-size support, but significant deviations from the linear portion are noticed for $K_D > 0.9$ and $K_D < 0.1$[58]. When selecting the mode of calibration it is important to examine critically the validity of the data for the calibration probes. Such parameters as shape and density of the molecule and the experimental conditions for obtaining data on Stokes radius have been used to explain disparities between different sets of experimental results[46,47,59].

The conformation of proteins may be normalized to random coils to avoid the ambiguities sometimes involved with less well-defined solute shapes. Denaturating media promote disruption of non-covalent bonds and the native protein reverts to the state of random polymer coils. Since urea is a poor denaturating agent for many proteins, the use of 6 to 8 M guanidine hydrochloride (GuHCl) is preferred[45,60]. The disulfide bonds are

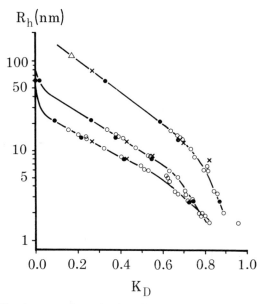

Figure 3-4. Universal calibration curve for evaluation of sample hydrodynamic radius. Columns: TSK 6000 PW (upper curve), TSK 5000 PW (middle curve) and Superose® 6 (lower curve). Sample: Δ = tobacco mosaic virus (TMV) dimers, ● = TMV, spectrin, tropomyosin and ovomucoid, X = DNA and α-actinin, 0 = proteins. (Reproduced from Ref. 28 with kind permission of the author and the publisher.)

broken by reductive cleavage with mercaptoethanol or dithiothreitol and subsequent re-oxidations prevented by carboxylmethylation[60]. Proteins may also be denatured with the aid of low concentrations of detergents, such as sodium dodecyl sulphate (SDS)[45,60,61]. It should be realized that this treatment will yield a protein-detergent complex of a size that is considerably larger than that of the native protein and that charged groups are introduced when using ionic detergents[62]. The hydrodynamic properties of the protein-SDS complex indicates that the complex is a rodlike particle[63]. The shape of the complex will grow more spherical as the molecular weight of the protein decreases. This sets a lower limit, of about 15,000, to molecular weight of proteins that may be estimated by gel

Table 3-3. Relationships used for calibration of analytical gel filtration columns

Relationship	Shape	Reference
$K_D^{1/3} = a - b \times M^{1/2}$	linear	Porath[53]
$K_{av} = a + b \times R_{St}$	linear	Laurent and Killander[54]
$K_D = a + b \times \log M$	partly linear	Granath and Kvist[55]
$\mathrm{erfc}(1/K_D) = a + b \times R_{St}$	linear	Ackers[56]
$K_D = a + b \times \log([\eta]\,M)$	partly linear	Benoit et al.[46]
$\log(1 - K_D) = a + b \times \log M$	linear	Basedow et al.[15]
$1/V_R = a + b \times R_{St}$	linear	Davis[57]

In the formulas are a and b operational constants, erfc is the error function, R_{St} is the Stokes radius of the solute and $[\eta]$ is the limiting viscosity number of the solute.

filtration in SDS[45]. Procedures for fast analytical gel filtration in GuHCl, and SDS (0.1 %) have been presented and were shown to offer attractive alternatives to SDS gel electrophoresis[64,65].

3.3.2   Column Efficiency and Peak Zone Broadening

A separation strategy should, in general, rather be focused on obtaining a high selectivity (i.e. peak-to-peak distance) than a high efficiency (i.e. narrow peaks), unless peak dilution is a critical factor. However, as mentioned earlier, the maximum selectivity of gel filtration is relatively limited and the efficiency is the only variable that is affected by the running conditions. Hence much effort has been spent to find an adequate description of the various phenomena that contribute to peak broadening.

The peak width of a solute can be related to the plate height by[66]

$$H = L/N = L/(V_R/\sigma)^2 = L(w_b/4V_R)^2 \tag{3-8}$$

where H is the height equivalent to a theoretical plate (sometimes denoted HETP) for the solute, N is the number of plates per column length L and w_b is the base width (i.e. 4σ) of a gaussian elution profile. The plate height in Eq. (3-8) is the sum of contributions to peak broadening from different parts of the chromatographic system. The extra-column dispersion can, with the aid of a proper experimental set up, (see section 3.6.5, p. 94), be neglected compared to column effects.

The plate height equation for a chromatography column was early described by van Deemter[67] and later adapted to gel filtration by Giddings and Mallik[68]. Several other plate height equations have been derived starting from slightly different assumptions, but a recent review of the various equations shows that the van Deemter equation is presently the most accurate one for describing zone broadening in SEC[69]. This equation is generally written as:

$$H = A + B/u + Cu \tag{3-9}$$

The first term arises from multiple path dispersion of the solute, the second term describes the effect of axial diffusion of the solute and the third term is due to non-equilibrium conditions in the separation process at the interstitial linear flow velocity u. The terms are given by[66,70];

$$A = \sum_i (1/2\lambda_i d_p + D_m/w_i d_p^2 u)^{-1}$$

$$B = 2(\gamma_m D_m + \gamma_s D_m(1/R - 1))$$

$$C = R(1 - R)d_p^2/30\gamma_s D_m$$

where λ_i and w_i are geometrical factors of order unity, d_p is the average particle diameter of the support, D_m is the diffusion coefficient of the solute in the mobile phase, γ_m and γ_s are obstruction factors to diffusion in the extra-particle space and the pores respectively, and R is the ratio of zone velocity to mobile phase velocity i.e. V_0/V_R. Fortunately, several simplifications of the expressions can be made. Thus, due to the slow diffusion of proteins, the second term of A can be neglected compared to the first term. The value of γ_m was found to be close to 0.6[66]. The plate height equation for macromolecules can thus be approximated by

$$H = 2\lambda d_p + 2(0.6\, D_m + \gamma_s D_m(1/R - 1))/u + R(1 - R)d_p^2 u/30\gamma_s D_m \tag{3-10}$$

The different shapes of the van Deemter plot for small solutes and large macromolecules at the flow rates commonly used in gel filtration are illustrated in Fig. 3-5. The contribution from the B-term is seen to be very small for macromolecules and this term is negligible compared to the C-term at high reduced velocities e.g. $d_p u / \gamma_s D_m \gg 5$. This condition is fulfilled for all the materials listed in Table 3-2 under the conditions commonly used, and the plate height equation for macromolecules may in this case be reduced to

$$H = 2\lambda d_p + R(1 - R)d_p^2 u/30\gamma_s D_m \qquad (3\text{-}11)$$

This equation may be used to predict the zone broadening of a solute, of known molecular weight, under various experimental conditions. Figure 3-5 also illustrates that the minimum value of the plate height is independent of the solute diffusivity.

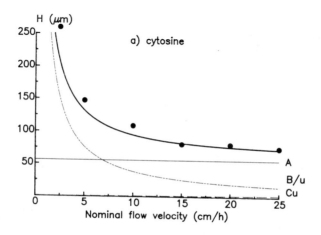

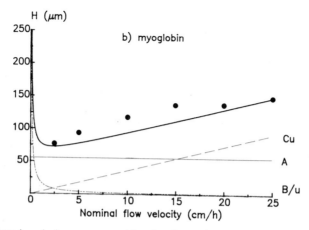

Figure 3-5. Zone broadening as measured by plate height, H, for small and large solutes in gel filtration as a function of flow rate. A, B and C are the terms in the van Deemter equation calculated from Eq. 3-10 with $d_p = 33$ μm and for (a) cytosine and (b) myoglobin. Dots represent experimental data for Superose[R] 6 prep grade[24].

The value of λ can be estimated from the minimum reduced plate height of *any* solute. We have found λ to be close to 1 for many gel filtration columns.

The value of γ_s expresses the restricted diffusion of solutes within the porous network, D_s, as compared to free diffusion in solution, D. The effective pore diffusivity has been calculated from[71]:

$$D_s/D = K_D \times (1 - 2.104(R_h/r) + 2.09(R_h/r)^3 - 0.95(R_h/r)^5)/\tau \qquad (3\text{-}12)$$

where R_h/r is the ratio of solute to pore radius (this is equal to $1 - \sqrt{K_D}$) and τ is the tortuosity factor used to compensate for variations in pore geometry. The value of τ may be obtained from batch experiments as suggesed by Liapis and Arve[72]. However, when no data is available τ is arbitrarily set to $1/V_{p,\text{rel}}$[73]. Equation (3-12) was also used by Ackers and Steere to calculate apparent pore radii of membranes[74]. A similar expression, where K_D was omitted, was recently proposed for predictions of zone broadening in SEC experiments[75]. Unfortunately none of the equations yields data in sufficient agreement with experimentally found obstruction factors[76-78]. The diffusivities of macromolecules in pores was found to be 5–20% of the free diffusion. More research is needed to find appropriate expressions to describe the effective diffusivities of solutes in porous network. From one work, describing restricted diffusion of proteins[77], the simple expression $\gamma_s \sim K_D/4$ for $0.2 \leq K_D \leq 0.8$ may be calculated.

The diffusivity (cm^2/s) of globular proteins may be derived from a formula given by Tanford to[41]:

$$D_{25,H_2O} \sim 2.6 \times 10^{-5} \, M^{-1/3} \qquad (3\text{-}13)$$

which is in good agreement (i.e. better than 6%) with experimentally found data[76,77].

Thus, by using these approximations the zone broadening of any solute of known molecular weight may be estimated with the aid of Eqs. (3-13), (3-11), (3-2) and (3-1). The approach is illustrated in Fig. 3-5 which shows that the experimentally found data are in reasonable good agreement with the predicted plate height. Furthermore, by also using Eq. (3-8), a rough theoretical simulation of the chromatogram may be obtained.

3.3.3 Parameters Affecting the Resolution

The ultimate goal for any separation is the acceptable resolution of a set of sample components. Resolution is greatly affected by the experimental conditions (see section 3.6, p. 89). Equally important, however, is the selection of a gel with optimal properties, e.g. selectivity, for the particular separation problem.

The resolution between two solutes is determined by their elution volumes and peak widths. The difference in elution volumes may, in the linear part of the selectivity curve be expressed in terms of molecular weight from Eqs. (3-1) and (3-2) as

$$\Delta V_R = -V_p \times b \times \Delta \log M \qquad (3\text{-}14)$$

The resolution of two adjacent components, with molecular weights M_1 and M_2, is, with the aid of Eqs. (3-8) and (3-14), expressed in a fundamental equation for resolution in gel filtration;

$$R_s = \frac{V_{R2} - V_{R1}}{(W_{b2} + W_{b1})/2} = \frac{1}{4} \log M_1/M_2 \left(\frac{b}{V_0/V_p + \bar{K}_D}\right)\sqrt{L}/\sqrt{\bar{H}} \qquad (3\text{-}15)$$

The resolution is thus affected by the differences in the sample molecular weights, log M_1/M_2, the selectivity, $b/(V_0/V_p + \bar{K}_D)$, column length, $\sqrt{L}$, and efficiency of the packed column, $\sqrt{\bar{H}}$. Equation (3-15) involves the simplification $(V_0/V_p + K_{D1})\sqrt{H_1} + (V_0/V_p + K_{D2})\sqrt{H_2} \sim 2(V_0/V_p + \bar{K}_D)\sqrt{\bar{H}}$ where $\bar{K}_D = \frac{1}{2}(K_{D1} + K_{D2})$ and $\bar{H} = \frac{1}{2}(H_1 + H_2)$. The equation can, by letting $(V_R - V_t)/V_t = k'$ be shown to be analogous with the resolution equation used in RPC (see Chap. 6).

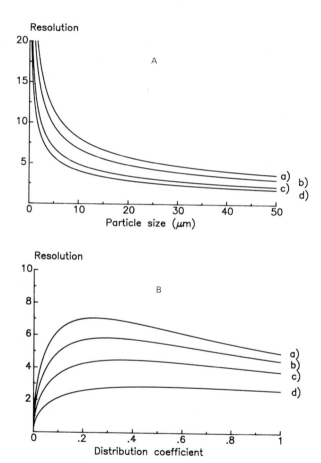

Figure 3-6. Factors affecting the resolution in gel filtration. (A) Resolution as a function of particle size (i.e. efficiency) of the support at various slopes of the selectivity curve, b, and permeabilities, V_p/V_0, of the support.

	(a)	(b)	(c)	(d)
b:	0.4	0.2	0.4	0.2
V_p/V_0	1.0	2.5	0.5	1.0

Calculated from Eq. 3-15 with $K_D = 0.5$, $L = 30$ cm and $H = 2 \times d_p$. (B) Resolution as a function of distribution coefficient at various permeabilities of the support. $V_p/V_0 =$ (a) 2.0; (b) 1.5; (c) 1.0 and (d) 0.5. Calculated from Eqs. 3-10 and 3-15 with $d_p = 10\ \mu m$, $D_m = 7 \times 10^{-7}$ cm^2/s, $L = 30$ cm, $u = 1$ cm/min $M_1/M_2 = 10$ and $b = 0.3$

From Eqs. (3-11) and (3-15) it can be seen that increased resolution is favoured by increasing the slope of the calibration curve, the column length and the permeability, i.e. V_p/V_0, and by decreasing the particle size, and, in most cases, the flow rate. Operating at low K_D-value will theoretically also increase the resolution (the gain in selectivity is larger than the loss in efficiency!). Another conclusion was made by Hjertén who states that "the resolution decreases with increasing R" (and thus decreasing V_R and K_D)[79]. Unfortunately he did not take into account that ΔR varies with variations in R. Substitution of $\Delta V_R \times R^2/V_0$ for ΔR in Eq. 17 in Ref. 79 shows that the resolution increases with increasing R. The effects of these parameters are illustrated in Fig. 3-6 which shows that a large value of the slope of the selectivity curve is an important factor for achieving resolution (Fig. 3-6A a, d) and that a low value may be compensated for by a large value of the permeability (Fig. 3-6A, b, c). The positive effect of operating at a low K_D-value, i.e. at $K_D \sim 0.2$, is illustrated in Fig. 3-6B. However the effect is small compared to the positive effect of a large pore volume.

The resolution increases significantly and the separation time decreases (cf. section 3.6.3, p. 91) by using very small particle sizes (i.e. below 5 μm). However, the trade-off may be a substantial increase in column back-pressure which puts special requirements on the instrumentation used, if the objective is very fast separations. Under such conditions the risk of temperature effects due to frictional heat or shear degradation of elongated solutes, caused by the high flow velocity in the narrow void channels, must be considered. Furthermore, an increase in the pressure resistance of a material is mostly achived by increasing the matrix content and this is likely to have an adverse effect on sorptive properties and resolution. The inherent limitation of particle size for SEC is set by the large pore size necessary for separating macromolecules and the requirement for a large ratio of void channel radius (i.e. particle radius) to pore radius to avoid size exclusion effects in the extra-particle space. Guiochon and Martin calculated the optimal particle size for SEC of macromolecules to 1–2 μm at the modest separation time of 1 h [80]. However, as recently shown by Verzele et al., it is very difficult to pack such small particles efficiently which seems to suggest an optimal particle size around 5 μm with present packing technology[81].

3.4 Packing the Column

3.4.1 Column Materials and Accessories

Whereas most column materials are suitable for the aqueous buffer solutions commonly used in protein chromatography there are often limitations in the use of organic solvents or extremes of pH with different materials. The resistance increases in the order acrylic plastic < polypropylene < glass. However, it is important to realize that the plungers, seals and other components may have a lower solvent resistance than the column tube. When using steel columns special precautions should be taken to reduce problems with corrosion, e.g. extremes of pH and the use of halide ions should be avoided. Recommendations for use should be provided by the manufacturer.

The column of choice should preferably be transparent, pressure and solvent resistant and equipped with an adjustable end adaptor. The dimensions may be chosen according to the application at hand but most laboratory columns have 4–16 mm inner diameter and a length of 25–70 cm. The choice of inner diameter may be based on the desired sample load or considerations of the extra dispersion effects from the wall region which extends 30 particle diameters away from the wall[82]. It is thus safe to use a 8 mm id column for 10 μm beads but not for a 100 μm material. The length of the column is primarily chosen according to the resolution that is required [cf. Eq. (3-15)]. Before packing or repacking the column all parts should be thoroughly cleaned. The bed support filters should be new and have a mesh size compatible with the particle size of the gel. Avoid touching the surface of nets and filters since a fingerprint may cause uneven sample application and a poor separation. It is also wise to inspect the bed support parts for burrs etc to ensure the best possible sample application after the column is packed. The use of screens instead of frits was found advantageous to avoid clogging and extra sample dispersion[83].

3.4.2 Packing Procedure

The gel should be pretreated according to the manufacturers instructions. However, it is often advantageous to decant the diluted slurry to remove fines and to use an excessive amount of gel to let non-dispersed lumps settle, before pouring the gel into the packing reservoir.

An efficiently packed column will have a low, non-separating, void volume and a homogenous bed which prevents channelling and disturbed flow paths. The ways to achieve an efficient packing seem in general to be derived empirically. Thus, small rigid beads are often packed at high pressures (>4 MPa) in solvents that are chosen to prevent inter-particle interactions. Various techniques have been used to pack soft and semi-rigid materials[24,84,85]. This situation probably reflects the many parameters influencing the packing density and bed stability, parameters which will be unique for each type of support[81]. A guideline on how to pack these materials may be given by the pressure-flow curve that is obtained from running a flow gradient through the column[24]. The pressure drop over the bed is proportional to the flow velocity (see section 3.4.4). At sufficiently high flow velocities the beads of semi-rigid materials are compressed (and particles of rigid materials may be partly crushed) by the large cumulative drag force acting on the particles. In the compression region, the flow resistance of the packed bed increases drastically as a result of a decrease in void fraction due to local compression of non-rigid beads. It seems to be most appropriate to pack semi-rigid materials at constant flow rate close to, but before, the compression region. Packing at constant flow rate will generate a column with a uniform packing density due to the constant friction force acting on the particles[86]. It has been suggested that columns should be packed in the compression region to produce a bed with a low non-separating void volume. However, the effect on the average void fraction is small and the increase in back-pressure and irregular flow paths caused by partial blockage of the bed by collapsed particles may be disadvantageous. Furthermore, an impaired stability of the compressed bed may also be expected, resulting in a reduced column life-time.

It is important to stabilise the bed, after the end adaptor has been inserted, by running the column at a somewhat higher flow rate[24,86]. This second step probably predominantly influences the upper, more loosely packed, part of the bed and may be performed using constant flow rate or constant pressure since the bed is already formed.

To obtain a perfect sample application it is essential to obtain a homogenous and well-packed zone at the inlet of the column. This is easily achieved by packing the column towards the inlet adaptor (using two adaptors)[87]. Running a column in the reverse direction to the packing direction is generally not recommended for rigid materials, i.e. silica, since this may cause rearrangement of particles in the bed. This effect is normally not seen with the elastic semi-rigid materials.

3.4.3 Evaluation of Column Packing

Many of the supports for high resolution gel filtration are only available as pre-packed columns, and there is an increasing demand for columns pre-packed with more traditional materials. As shown by Potschka the efficiencies of similar, pre-packed materials vary substantially in the hands of a user[28], and the evaluation of pre-packed columns may be as necessary as it is for user-packed columns to ensure optimal performance.

The column packing may be evaluated by running a low molecular weight solute such as cytidine, azide, glucose, D_2O etc. at a high flow rate (cf. Fig. 3-5) and calculating the reduced plate height from:

$$h = L/5.54\left(\frac{V_R}{w_h}\right)^2 d_p \tag{3-16}$$

where w_h is the peak width at half peak height. For an efficiently packed column, this value should be close to 2[88]. Reduced plate heights of 1.5 have been obtained with very efficiently packed columns[76,77]. The quality of the peak shape may be expressed by the symmetry factor (b/a in Fig. 3-1) which should be close to unity (i.e. 0.9–1.1). The sample application can be evaluated visually by running a coloured sample as depicted in Fig. 3-7.

When the column packing is acceptable it is often useful to run some well defined proteins to measure the actual column performance in the desired separation range. For high resolution columns this is easily done by running a few samples containing mixtures of proteins. Some mixtures which we have found to be useful in the high and middle molecular weight ranges are given in Table 3-4.

The homogeneity of the bed may be checked by comparing the pressure drop over the bed at a certain flow rate to that calculated from the pressure drop equation found below (see Table 3-5). A large deviation (e.g. larger than 50%) indicates a non-uniform packing with a compressed region of beads in the column.

3.4.4 Pressure Drop Over Packed Beds

The back-pressure generated when a liquid is flowing through a packed bed can be calculated from the formula given by Darcy or the more explicit expression derived by Hagen and Poiseuille[89]. Originally, the relationship was utilised to obtain information on particle dimensions from the observed pressure drop[89]. This procedure has also been used in liquid chromatography to estimate an apparent particle size[90]. Many workers, including Blake, Kozeny and Carman, have contributed to refining the equation given by Hagen and Poiseuille to the present formulation[89];

$$\Delta P = F \frac{L}{dp^2} \times \eta \times \frac{(1-\varepsilon)^2}{\varepsilon^3} k \times 36 \tag{3-17}$$

where ΔP is the pressure drop in Pa, F is the nominal flow velocity in cm/min, η is the viscosity of the solvent (1.49×10^{-5} min $\times N/m^2$, H_2O, 25°C) L is the bed height in cm,

Figure 3-7. Visual evaluation of column packing.

Table 3-4. Protein mixtures for evaluation of gel filtration columns

Solute	Source	M (Daltons)	$R_h(\text{Å})^a$	Concentration[b] (mg/ml)
Mixture 1				
Cytidine	A	243		0.1
Bacitracin[c]	B	1,450		10
Cytochrome c	C	12,400	16.6	1
β-lactoglobulin	C	35,000	23.5	2
Human serum albumin	C	67,000	29.2	6
Immunoglobulin G	D	160,000	39.0	2
Thyroglobulin	E	669,000	62.8	4
Mixture 2				
Acetone	A			5
Myoglobin	C	17,600	18.7	1
Ovalbumin	E	43,000	25.2	2.5
Transferrin	C	81,000	31.1	1.5
Aldolase	E	158,000	38.8	6
Ferritin	E	440,000	54.6	1
Blue Dextran 2000	E			2

[a] Calculated from Eq. 3-7
[b] Suggested to be detected at 0.2 AUFS, 280 nm. Adjustment according to instrumentation used and zone broadening will be necessary. Figures include weight of additives.
[c] Freshly prepared.

Sources: A, Merck (Darmstadt, F.R.G); B, Serva (Heidelberg, F.R.G); C, Sigma (St. Louis, MO, USA); D, Kabi (Stockholm, Sweden); E, Pharmacia (Uppsala, Sweden).

d_p is the harmonic mean value of the particle size distribution in cm, ε is the void fraction, i.e. void volume over column volume and k is the aspect factor. For spherical beads the aspect factor is close to 5[89], and the last term in Eq. (3-17) is then reduced to: $k \times 36 \times (1 - \varepsilon)^2/\varepsilon^3 \sim 180 (1 - \varepsilon)^2/\varepsilon^3$. This term is called the column resistance factor and is typically 500–700 for spherical silica materials[91]. This corresponds to a void fraction of 0.42–0.45. It is important to notice the influence of the temperature dependence of the viscosity on the generated pressure drop. The relative viscosity of water is 1.5 at 4°C which yields a twofold increase in the pressure drop in a cold room as compared to ordinary room temperature.

The theoretical pressure drops of columns packed with materials of different particle sizes and void fractions can be inferred from Table 3-5. A large void fraction will lead to a low back-pressure but also to poor resolution due to the decrease in permeability [cf. Eq. (3-15)].

The pressure drop over beds packed with non-solid supports always deviates from the linear relationship predicted by Eq. (3-17) at sufficiently high flow velocities. The steep increase in pressure noticed with the softer materials is due to compression of the beads and a subsequent reduction of the void fraction in a small zone near the column outlet[24,92]. The compression is due to the cumulative drag force caused by the flowing liquid. This force increases with increasing flow velocity, viscosity of the liquid and bed height[93]. Thus, dividing a bed into shorter segments (i.e. stacked columns) will enable the use of relatively higher flow velocities before the beads are compressed. The drag force is counteracted by the supporting force of the column walls. Elastic materials are not destroyed by the compression and the original performance of a column may be restored by repacking it with the material[24].

Comparison of compressibility of different materials, or fragility of e.g. silica, may be obtained by normalising the pressure drop with respect to column length and particle size, as in Fig. 3-8. It is then important that the column diameter is sufficiently large to avoid wall effects (e.g. a ratio of column to particle diameter of 1×10^3 seems to be necessary).

The matrix rigidity may be improved by an increased cross-linking of the polymer[94,95], or by allowing a larger matrix volume of the support[18]. Increased matrix rigidity is thus achieved at the expense of pore volume and resolution. Therefore, high

Table 3-5. Theoretical pressure drop, in KPa*, over packed beds of various void fractions and particle sizes. Calculated, per cm bed height at $F = 1$ cm/min, from Eq. 3-17.

Void fraction V_0/V_c	Particle size (μm)								
	2	5	10	15	20	30	50	75	100
0.50	134	21	5.4	2.4	1.3	0.6	0.2	0.09	0.05
0.45	223	36	8.9	4.0	2.2	1.0	0.4	0.16	0.09
0.40	377	60	15.1	6.7	3.8	1.7	0.6	0.27	0.15
0.35	661	106	26.4	11.7	6.6	2.9	1.1	0.47	0.26
0.33	838	134	33.5	14.9	8.4	3.7	1.3	0.60	0.34
0.30	1217	195	48.7	21.6	12.2	5.4	1.9	0.87	0.49
0.25	2414	386	96.6	42.9	24.1	10.7	3.9	1.72	0.97
0.20	5364	858	214.6	95.4	53.6	23.8	8.6	3.81	2.15

* Conversion factors: 100 KPa = 1 bar = 14.5 psi.

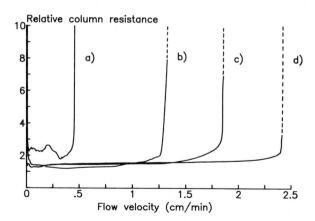

Figure 3-8. Variation in the column resistance factor as a function of flow velocity for materials of different mechanical strength (i.e. degree of cross-linking). Experimental data are normalized with respect to particle size, bed height and initial void fraction of the columns. Supports: (a) Sepharose® 6B, (b) Sepharose® CL 6B, (c) Superose® 6 prep grade and (d) Superose® 6.

matrix rigidity is not an objective per se, but should be related to the back-pressures that are generated by the modest flow velocities that are applicable to the separation of high molecular weight solutes (cf. section 3.6.3).

3.5   Sample and Buffer Preparation

3.5.1   Buffers and Additives

The sample is preferably dissolved in the same medium as that used for the gel filtration step, unless the procedure is a desalting step. It is generally advisable to use aqueous buffer systems with buffering capacities in the pH-range 6–8 since this will produce a good environment for many proteins as well as cope with most gel filtration matrices. The acid dissociation constant, pK_a, of the buffering substance should be close (i.e. ± 0.5 pH) to the desired pH to yield the optimal buffering capacity of the solution. Traditionally, phosphate ($pK_{a2} = 7.2$) and tris(hydroxymethyl)aminomethane ($pK_a = 8.1$) have been used at neutral pH. In cases when the buffering capacity of these substances is too low or the properties of the substance are incompatible with the sample (e.g. phosphate is known to inhibit certain enzymes[96]), the biological buffers proposed by Good et al. may be more suitable[97]. The universal citrate-phosphate-borate buffer described by Teorell and Stenhagen can be used to study the effect of pH on the separation in the pH-range 2–12 with only a slight variation in ionic strength[96]. However, borate buffer may interact with glycopeptides[152]. When running a preparative purification or a desalting step, volatile buffer salts such as ammonium acetate or ammonium bicarbonate may be preferred since these are readily removed by freeze drying. Suggestions for other volatile buffer systems covering the pH-range 2–12 are found in Ref. 96.

To avoid ionic interactions between the solute and the matrix, the ionic strength of the buffer is often increased to 0.05–0.50 M by the addition of a salt. Sodium chloride is used most frequently, but since halides are very corrosive these salts should be replaced with, for instance, sulphates whenever stainless steel is in contact with the liquid.

In some situations the salting-out effect of sulphate may create hydrophobic interactions with the matrix. In those cases, chaotropic ions, such as perchlorate, may be used to increase the ionic strength of the buffer[25]. Interpretation of elution volume in terms of molecular weight from analysis at a high ionic strength must be made with care since the conformation of proteins may vary with variations in ionic strength[41].

Non-ionic interactions have been reported for some types of matrices. These effects may be reduced by increasing the pH, decreasing the ionic strength or by adding small amounts of detergents, ethylene glycol or an organic modifier, such as 1-propanol (1%) or acetonitrile, to the buffer. Gel filtration of lipophilic solutes has been carried out in 45% acetonitrile.[98] A successful separation of membrane proteins on Sepharose[R] CL-4B using a chloroform/methanol mixture as eluent has been reported [99].

Addition of detergents to the buffer is sometimes required to solubilize certain proteins, e.g. membrane proteins. This approach was used for the solubilization of integral membrane proteins in 0.1 M sodium dodecyl sulphate (SDS) followed by high performance gel filtration in 5–50 mM SDS[100].

Denaturing media such as 6 M guanidine hydrochloride is often used to break the hydrogen bonds which stabilise the tertiary structure of proteins and to transform them into random coils for subsequent molecular weight analysis, (see section 3.3.1, p. 72).

The use of 0.02% sodium azide in the buffer has proved to be useful to prevent microbial growth. Other preservatives that have been tested include trichlorobutanol and thiomersal, but they are less effective[6]. Cationic preservatives are not recommended since these may adsorb to residual negatively charged groups on the matrix at low ionic strength.

The choice of buffer, pH, ionic strength and additives must be made with respect for solvent-matrix interactions, the solubility and biochemical properties of the sample and the limitations of the detection system used. Undesired effects may be caused by chelating or solubility properties of the buffer substance (e.g. borate-carbohydrate or phosphate-Ca^{2+} interactions)[96,100], self-association of proteins[101] or high absorbancy of buffer or additives at the wavelength used for the detection of the sample and interferences of these substances with chemical detection steps (cf. section 3.6.4)[96].

Some buffer systems commonly used in high performance gel filtration are given in Table 3-2. An extensive compilation of other buffer systems is found in Ref. 96. It is advisable to prepare the buffer at the running temperature since the pK_a-value of the substance may vary considerably with temperature (e.g. -0.03 pH/$°C$). Finally, filtering the buffer solution through a 0.45 μm filter prior to use will save the column from being blocked by undissolved substances.

3.5.2 Sample Load

The sample load, expressed as the amount of sample applied, depends of course on sample concentration and sample volume. Whereas the constraint on the former is mostly due to the high viscosity of concentrated samples, the latter is affected by the dimensions of the chromatographic bed, and by the pore volume and the particle size of the support.

As a general rule, the viscosity of the sample relative to the viscosity of the eluent should be less than 1.5. This corresponds to a maximal concentration of 70 mg/ml of a globular protein such as human serum albumin. This was experimentally verified in the authors laboratory on a column packed with Superose[TM] 6 prep grade where a sample load of 100 mg/ml gave abnormally broad peaks whereas 50 mg/ml gave normal sharp peaks. If the viscosity of the sample is very high then this may be compensated for by increasing the viscosity of the eluent by adding sucrose or dextran [6]. However, this will require the use of a lower flow rate to maintain high efficiency and a low back pressure.

Viscous fingering effects, caused by the hydrodynamic instability of a relatively viscous sample zone, are more likely to occur for concentrated solutions of high molecular weight solutes. The effect is readily indicated by varying the concentration of the sample as demonstrated in Fig. 3-9. It may be expected that viscosity effects are more pronounced in beds of small particle size since this will result in narrow void channels and less sample dilution.

The influence of sample volume on zone broadening can be estimated by comparing the width of the injected sample plug of volume V_{inj} [102], and the final width of the peak as calculated by the plate height equation.

$$\sigma_{total}^2 = \sigma_{column}^2 + \sigma_{injection}^2 \sim V_R^2(H/L) + V_{inj}^2/K_{inj} \qquad (3\text{-}18)$$

The theoretical value of K_{inj}, i.e. 12, has been achieved with carefully designed injection devices [103], but values of 1 to 5 have been reported for ordinary valves [102,104]. We found a value of 5 to be in accordance with experimental results [105].

The relative contribution to the total variance, $\sigma_{rel}^2 = \sigma_{injection}^2/\sigma_{total}^2$, will then theoretically result from a sample volume of

$$V_{inj} = \sqrt{\sigma_{rel}^2/(1 - \sigma_{rel}^2)} \times A_c \times \sqrt{L} \times (1 + K_D V_p/V_0) \times (V_0/V_c) \times \sqrt{h \times d_p} \times \sqrt{K_{inj}}$$
$$(3\text{-}19)$$

This formula confirms the general expectation that high values of the cross-section area, A_c, the column length, the pore volume and K_D will support a high sample volume. However, it is not always realized that a material of small particle size will only allow applications of small sample volumes, if optimal performance is to be retained. From Eq. (3-19) it is seen that the injection volume needs to be adjusted to a level proportional to the square root of the particle size if σ_{rel}^2 is to be kept constant. Thus, while sample volumes of 1–2% of the bed volume do not impair the performance of traditional gel filtration media with $d_p \sim 100~\mu m$,[6] this is certainly not true for 10 μm materials where 0.3% is a more realistic figure [105]. The variation of peak width, expressed as reduced plate

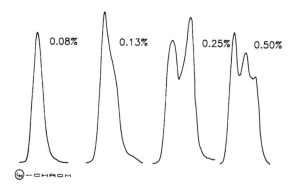

Figure 3-9. Disturbed elution profiles due to viscous fingering effects. Sample: 0.6 ml of native dextran.

Reduced plate height

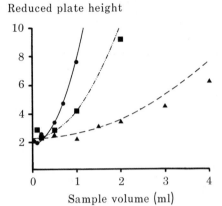

Sample volume (ml)

Figure 3-10. Effect of sample volume of the peak width (i.e. reduced plate height) of cytidine. Lines represents calculated relationship and dots represents experimental data. Columns: (●) Superose[R] 6, 500 × 10 mm; (■) Superose[R] 6 prep grade, 530 × 10 mm; (▲) Superose[R] 6 prep grade, 510 × 16 mm. (Reproduced from Ref 105 by kind permission of the publisher.)

height, with sample volume is shown in Fig. 3-10. The most important implication of this figure is the limited inherent efficiency of small particle sized materials when large sample volumes are applied. The positive effect from increasing the column dimensions to promote large sample volumes is also shown.

The figures for sample load are of course only tentative and due to the nature of the sample, the gel matrix, relative pore volume and the desired preparation, it may be experimentally found that other sample loads are applicable. Thus, small effects on the peak elution volume from the sample concentration have been noted in high precision analytical gel filtration[106]. It is a good practice to study the separation pattern at different sample loads to ascertain that the conditions used are appropriate.

3.5.3 Calibration Substances

The choice of sample substances used for the determination of void volume, V_0, and total liquid volume, V_t, may require special attention. The probe for the void volume must be large enough to be totally excluded from the gel matrix but not so large as to be subjected to secondary exclusion in the void volume or separation due to hydrodynamic chromatography[107]. This will lead to the appearance of a peak before the actual void volume. To avoid these effects the radius of the probe should be less than 0.01 times the particle radius (i.e. c:a 500 Å). The solute must not be charged at the pH used since this can cause ionic adsorption or extra exclusion effects. The most common substance, Blue Dextran 2000, has an assigned molar mass of approximately 2×10^6 which would correspond to a hydrodynamic radius of 350 Å [108]. This substance is not monodisperse and the low molecular weight part of the distribution may therefore permeate the matrix of a sufficiently porous gel. However, the high molecular weight region is useful for the determination of void volumes of matrices with pore sizes of up to at least 1000 Å. In cases where the blue dyestuff of this probe shows adsorption to the gel, it may be replaced by calf thymus DNA [18], or lyophilized Escherichia coli [13].

Sodium azide, cytidine or glycyltyrosine have been used as markers of total liquid volume on highly porous gels with detection by UV absorption. Glucose, sodium chloride or deuterium oxide may be used with detection by refractive index. However special problems arise when measuring the pore volume of a low porosity gel and the ionic strength of the medium has to be kept low. Due to the high surface area and low shielding effect, partitioning of solutes into the hydrated gel layer, adsorption or ionic interactions may be more pronounced. In such cases the total volume is calculated from the geometric dimensions of the bed and the distribution coefficient thus achieved is designated K_{av}[54]. The difficulty of obtaining a correct estimate of the total liquid volume of some types of materials has recently been addressed[109].

If the column is to be calibrated for estimation of molecular size of proteins then the calibration substances should be well defined proteins with similar characteristics, e.g. shape and density, as the sample (see section 3.3.1).

Finally, before applying the sample it is good experimental practice to filter the sample solution through a 0.45 μm filter (e.g. Millipore HA) unless the sample solution is perfectly clear. This will save the inlet filter from being partly blocked with subsequent impaired sample application as a result. The use of filtered buffer and sample solutions may preserve the column life time to over 1000 injections[110].

3.6 The Chromatographic System

A general view of the instrumentation commonly used in a gel filtration experiment is given in Fig. 3-1. The instrumentation may, with respect to pressure resistance, be divided into three categories: high pressure systems capable of overcoming pressure drops of several hundred bars, medium pressure systems working up to fifty bars and low pressure systems used up to one bar. The main difference between the systems is the pumping principle, i.e. reciprocating pump with heads of stainless steel or titanium, syringe pump with wetted parts of glass and tubing pump with inert tubing, respectively.

The choice of column dimensions is closely related to the application, i.e. traditionally, columns for desalting purposes are "short and fat", e.g. 1–10 cm long with 1–50 cm i.d., columns for fractionation purposes are 20–70 cm long with 3–5 cm i.d. and columns for analytical separations are "small and long", e.g. 50–100 cm long with 0.5–1 cm i.d. These figures will change with miniaturisation of support sizes and sample volumes but the relative column sizes for the different application areas are likely to remain the same. The choice of detector, i.e. flow cell, is not as critical as the choice of column dimensions. However, narrow bore flow cells are generally not applicable in large scale fractionations as they seldom tolerate the excessive back-pressure generated over the cell at high flow rates.

3.6.1 Sample Application

There are several ways of applying the sample to the gel bed. On columns that are not equipped with an adaptor (e.g. small columns for desalting purposes) the sample may simply be layered on the bed surface with the aid of a Pasteur pipette or a similar device. More eluent is then carefully added and the elution of the sample started. This technique requires high operator skill but will, if properly performed, result in a very good sample application. A more convenient way is to use an injection device that is coupled to an adaptor through a valve. If the gravity flow of the column is high, the injection device may simply consist of a syringe attached to the valve. However, a more accurate sample application is achieved by using a loop connected to the valve. This is the only technique that is applicable to medium and high pressure systems. When running columns packed with beads of very small particle size (e.g. 3–5 μm) the construction of the injection device becomes critical (see section 3.5.2). If very large sample volumes are to be applied, the use of an ordinary loop, consisting of a long narrow bore tubing, is not recommended. The sample may in these cases be applied through a Superloop[111], or simply by using a peristaltic pump. The valve may be operated manually, electromagnetically or pneumatically. It is important to flush the injection device and the valve extensively between sample applications to avoid carry-over effects from the previous sample. In high resolution gel filtration the tubing lengths between the injection device and the adaptor should be kept as short as possible to minimize zone broadening. In some cases the addition of sucrose to increase the sample density and reduce the broadening of the sample zone between the injector and the column has been used[6], but this is seldom necessary with modern injection devices as demonstrated by Fig. 3-7.

3.6.2 Eluent Delivery

The eluent is contained in a reservoir that may be slightly heated, e.g. with the aid of a heating tape or an ordinary light bulb, to degas the eluent and avoid the creation of air bubbles in the gel bed when using the column for extensive periods of time. We have found the use of a degasser placed before the injection device both convenient and effective[112]. This is not necessary when using a column packing that generates back pressures above 10 bars since introduced air bubbles will be dissolved in the eluent at this pressure.

The eluent is delivered from the reservoir either directly to the column or through a pump. In the first case the hydrostatic pressure from the reservoir to the column outlet must be sufficient to create a flow through the column that exceeds the flow rate used. The actual flow rate is then preferably controlled by a pump placed at the column outlet. This procedure may only be used when the flow rate is low and the bead size is large. A Mariotte bottle is used as a reservoir to maintain a constant operating pressure over the bed (continuous degassing of the eluent is necessary when using a Mariotte bottle). A more convenient, and recommended, way is to place the pump before the injection device. This will also facilitate the collection of fractions after the column. Whatever system is used, measurement of the actual flow rate is of outmost importance due to the possibility of leakage in pump valves etc. and the large effect of an erroneous flow rate on the calculated molecular weights[113]. The most common way to measure the flow rate is by gravimetry but it is also possible to use flow meters for continuous control[114]. Even though high precision pumps are preferable for most work, the long-term flow constancy of peristaltic pumps equipped with Tygon tubing is satisfactory in many cases. An indirect way to control the reliability of the pump is to include an internal standard, e.g., a low molecular weight solute, with each run. This may also provide a way to compensate for any long term drift of the pump.

3.6.3 Optimal Flow Rate

The maximal linear flow rate that is applicable is restricted by both physical and physico-chemical considerations.

The flow of liquid through the column generates a back pressure as described by Eq. (3-17). Significant deviations between the theoretical value of the pressure drop and the experimental one may indicate a non-uniform packing. Abnormally high pressure drops over filters, frits or connection tubings may be caused by clogging from precipitated sample or particulate materials. The column construction and materials, fittings, connections, pump type and the mechanical rigidity of packing material limit the maximal flow rate than can be used.

The useful linear flow rate is mainly restricted by the low diffusion coefficients of macromolecules which may lead to excessive zone broadening when too high flow rates are used[115]. This is eliminated by assuring that equilibrium between the mobile and trapped phases is achieved. This can be done by decreasing the transfer distance of the solute (i.e. reducing the particle size of the support) and/or increasing the available transfer time (i.e. reducing the flow rate of the solvent).

The zone broadening due to non-equilibrium is expressed by the C-term in the van Deemter equation [Eq. (3-9)] and is illustrated in Fig. 3-5. The optimal interstitial flow rate, yielding the minimum zone broadening is found by differentiation of Eq. (3-10) with respect to u which results in Eq. (3-20).

$$u_{\text{opt}} = D_m \gamma_s d_p^{-1} \times (60(2/(3\gamma_s R(1 - R)) + 1/R^2))^{1/2} \qquad (3\text{-}20)$$

showing that the optimal flow rate is proportional to the diffusion coefficient of the solute and inversely proportional to the particle size of the support.

At u_{opt}, the major contribution to the zone broadening is the A-term in Eq. (3-9). It seems practical to allow a flow rate where the A-term is not solely dominating, e.g. allowing an equally large contribution from the C-term. That flow rate is given by

$$u_p = \frac{60 \lambda \gamma_s D_m}{R(1 - R) d_p} \qquad (3\text{-}21)$$

With $\lambda \sim 1$, $\gamma_s = K_D/4$ and $R(1 - R) \sim 0.23$, Eq. (3-21) reduces to: $u_p \sim 65 K_D D_m/d_p$. Practical flow rates for various bead sizes and solutes are indicated in Table 3-6. It can be noted that no single flow rate is optimal for the entire separation range. Thus, at low flow rates high molecular weight solutes will yield sharp peaks due to the large transfer time (low C-term) but small molecular weight solutes will yield broad peaks due to excessive axial zone broadening (large B-term), cf. Fig. 3-5. At high flow rates the conditions are opposite. However, by starting the elution at a low flow rate and increasing the flow rate during the run a more optimal separation is achieved. This approach is illustrated in Fig. 3-11. The effect of flow velocity on peak broadening is, for this 10 μm support, not as large as earlier noticed for 30 μm beads[24]. However, the increased resolution in the high molecular weight region is evident. The separation time can be decreased by a factor of two by increasing the flow velocity during the run, as expected from Table 3-6. The flow gradient may be calculated from Eq. (3-20).

3.6.4 Sample Detection

The most popular method of monitoring the effluent for proteins is continuous measurement of the absorption in the ultraviolet range. Variable wavelength detectors are often expensive and the use of a single or dual wavelength detector is in most cases adequate. Detectors where the wavelength may be selected at 546, 280, 254, 214 or 206 nm or other wavelengths are available. It is thus possible to select the proper

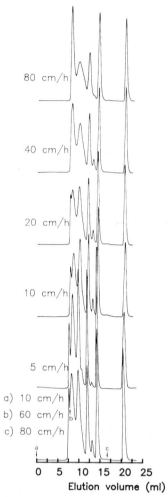

Figure 3-11. Separation of a protein mixture at various constant flow velocities as compared to a stepwise increasing velocity. Column: Superose® 12. Sample: thyroglobulin, ferritin, bovine serum albumin, myoglobin and cytidine.

wavelength for the specific detection problem and for proteins, absorption at 280 nm is often utilized. For the detection of peptides which do not contain aromatic amino acids, lower wavelengths, i.e. below 220 nm, must be used. Care should be taken to avoid buffer solutions and additives which show appreciable absorption at the wavelength used for monitoring. Thus 0.03 M acetate or 0.2 M phosphate buffers, azide, merthiolate and hibitan are not compatible with detection at 214 or 206 nm[116]. Proteins also show different molar absorptivities at different wavelengths and chromatograms at 280 and 254 nm cannot be expected to coincide. This can be used to increase the detection selectivity with the response ratio method[117]. Differences in spectroscopic properties of solutes may be fully utilized with the aid of a fast scanning detector where a total spectrum is recorded in parts of a second[118]. When running unknown samples, the risk of

Table 3-6. Practical flow rates for optimal zone width in gel filtration

Solute molecular weight (Daltons)	D (cm²/s) × 10⁷ [a]	Nominal flow velocity[b] (cm/h) at a support particle size (μm) of						
		2	5	10	15	30	50	100
1,000	26	507	203	101	68	34	20	10
5,000	15	293	117	59	39	20	12	6
10,000	12	234	94	47	31	16	9	5
15,000	10	195	78	39	26	13	8	4
50,000	7.0	137	55	27	18	9	6	3
75,000	6.2	121	48	24	16	8	5	2
100,000	5.6	109	44	22	15	7	4	2
200,000	4.4	86	34	17	11	6	3	2
300,000	3.9	76	30	15	10	5	3	2

[a] Calculated from Eq. 3-13 assuming solutes being globular proteins.
[b] Calculated from $F u_p/3$ and $u_p \sim 65\, K_D D_m/d_p$ with $K_D = 0.5$ and $D_m \sim D$.

losing peaks may be reduced by feeding the signal to a two-channel recorder set at two different sensitivity ranges.

The use of fluorescence monitoring is mainly applicable to compounds containing tryptophan or after chemical derivatisation but, when applicable, the sensitivity and selectivity is substantially higher than with UV-detection.

Refractive index detectors are frequently used for the detection of non-UV absorbing species such as dextran, PEG, saccharides etc. This detector is quite general and will not only respond to variations in the solute concentration, but also to changes in the composition of the mobile phase, e.g. salt peaks or salt depletion, and to temperature and pressure fluctuations. The sensitivity of the RI-detector is in most cases inferior to that obtained by UV-detection. The combination of a low angle laser light scattering detector and an RI-detector can be used to determine the molecular weight of the eluting solutes provided that the refractive index increment of the solute is known[119]. The technique, as applied to proteins, yields data in good agreement with results from sequence determinations[120].

Recently, electrochemical detection has been applied to column effluents and detection of thiol-containing proteins after gel filtration has been described[121]. The use of a bromine detector, where bromine is allowed to react with the protein and the amount of bromine consumed is determined, may provide an interesting general technique for protein determination[122].

In desalting procedures the registration of ionic strength is often of interest and in these cases a conductivity detector equipped with a flow cell can be used. It is, of course, also possible to use an ion-selective electrode equipped with a flow cell for this purpose.

The use of flow-through detectors is mostly straightforward, although certain applications may reveal limitations in instrumental performance including slow electronic response to very fast analysis, refractive index effects in UV-detection when changing mobile phase composition and unbalanced photomultipliers resulting from large differences in absorption of buffers in sample and reference cell. To avoid the formation of air bubbles in the cell the pressure inside the cell may be increased with the aid of a narrow bore tubing, e.g. 3 m × 0.02 mm, or a constant back-pressure device operating at 50 psi connected to the outlet of the cell[16,117].

In the cases where a more specific assay is required the effluent may be collected in fractions and a chemical reaction carried out subsequently. It is also possible in most cases to carry out this reaction on line after the column to obtain a continuous assay[123]. The large potential of such methods may be exemplified by a recently reported method for the rapid diagnosis of myocardial infarction. The main components of serum were separated by fast gel filtration and the content of myoglobin (down to 10 μg/ml) selectively assayed by a post-column chemiluminescence reaction with luminol[124].

3.6.5 System Dispersion

The theoretical contribution to extra-column zone broadening due to the chromatography system may be calculated from[16,102,104]

$$\sigma^2_{system} = V^2_{inj}/K_{inj} + \frac{d^4 F_c l}{122 D_m} + 2V^2_{cell} \tag{3-22}$$

where the first term is related to the injection of the sample volume, the second term is the contribution from tubings or connections with length l and inner diameter d at the volumetric flow-rate F_c, D_m being the diffusion coefficient of the solute and the last term expresses the contribution from the flow cell volume, of V_{cell}.

If the system is allowed to contribute 5 % of the total zone broadening, a maximum of 50 μl sample, 3.5 cm tubing with 0.5 mm i.d and a cell volume of 15 μl can be tolerated if a reduced plate height of 4 is to be maintained for a macromolecule eluting at $K_D = 0.2$ from a 10 × 300 mm column packed with a 10 μm material. For a 30 μm material packed in a 16 × 500 mm column, the corresponding figures are 300 μl sample, 100 cm tubing and 100 μl cell volume. Thus, it is very important to minimize extra-column broadening if the separation power of smaller particle media is to be fully utilized. It should also be noted that a minimum of connections should be used and tubings with different diameters should be avoided in order to prevent turbulent mixing of the sample zones. The effect of sample volume on peak width in high resolution gel filtration is treated more extensively in section 3.5.2.

The risk of shear degradation of macromolecules when reducing the diameters of tubings or particle size while maintaining a high flow rate need to be considered[125,126]. In order to avoid shear degradation the use of tubings with inner diameter of 0.5 mm or larger and keeping the flow velocity below 8 cm/min for 10 μm packing materials is recommended[126].

3.7 Applications

3.7.1 Desalting. Desalting of Large Sample Volume of Hemoglobin

Gel filtration has been used for desalting of protein solutions for almost thirty years[84]. Since the separation is rather simple, the requirements on the gel and the chromatographic system are modest. The porosity of the gel is selected to exclude the solute to be desalted, the particle size is chosen to give low surface area (to minimize adsorptive properties) and low back pressure rather than to yield columns of high efficiency. The buffer salts and additives should be volatile or the desalting should be carried out in distilled water. Desalting may be carried out at small scale as for the

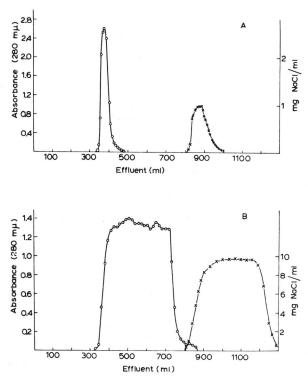

Figure 3-12. Desalting of large sample volumes by gel filtration. Column: 4 × 85 cm packed with Sephadex® G-25. Sample: 0 = hemoglobin, x = NaCl, sample volume; A = 10 ml, B = 400 ml. (Reproduced from Ref. 84 by kind permission of the author and the publisher.)

desalting of radioactive labelled DNA in 1 ml Sephadex G-100 in a Pasteur pipette or on a large scale in stack columns containing 500 l gel[6,127]. Since the total pore volume is employed for the separation, very large volumes of sample may be desalted in one step as illustrated by Fig. 3-12. In this application, 400 ml of a hemoglobin solution, corresponding to a sample load of 37 % of the total bed volume, was desalted in one step on a column packed with Sephadex G-25. The procedure used for desalting may of course also be used for transferring the solute into another buffer system.

3.7.2 Fractionation of Protein Mixtures. Purification of Anti-IgE-β-Galactosidase Conjugate

The most favourable situation is when it is possible to use a gel that will exclude the protein of interest and include the contaminants or vice versa (i.e. a filtration process). In the normal case the separation problem is more difficult and may require the use of a gel with optimal properties, i.e. a high pore volume, a narrow pore size distribution and a suitable pore size to elute the protein of interest at $K_D \sim 0.2$–0.4 (c.f. Fig. 3-6). Since it is often desired to avoid additives the adsorption properties of the gel should be small. Contaminants that show strong affinity for the gel matrix may be removed by filtering the sample solution through a small bed of the gel before applying the sample on the column.

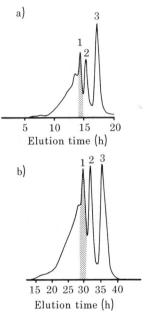

Figure 3-13. Purification of the 1-1 conjugate of anti-IgE-β-galactosidase by gel filtration. (a) Column 16 × 950 mm packed with Superose 6® prep grade, sample volume 500 μl. (b) Column 26 × 1000 mm packed with Superose 6® prep grade, sample volume 6 ml. Peaks are 1, anti-IgE-β-galactosidase; 2, β-galactosidase and 3, anti-IgE. (Reproduced from Ref. 128 by kind permission of the publisher.)

To favour high sample loadings, the porosity and selectivity of the matrix should be high and the cross section of the column large enough to cope with the desired sample volume. It is also advantageous to use gels designed to facilitate the scale up procedure from analytical to preparative runs. This requires that the medium is available in bulk quantities and that the packing procedure to obtain efficient columns is simple. An example is given by the preparative purification of the conjugate of IgE-β-galactosidase in analytical and small-production scale shown in Fig. 3-13 yielding amounts sufficient for diagnostic purposes. The throughput (i.e. amount of processed material per unit time) may be increased by applying a new sample at each time interval corresponding to V_t-V_0 (e.g. after 35 h in Fig. 3-13) since no material from the new sample is eluted during a time period corresponding to V_0. The gain in production rate of consecutive runs equals the void fraction which for many materials is roughly 35% (see Tables 3-1 and 3-2).

With the use of high precision pumps and controllers to govern motorized valves it is possible to use an automatic half-scale system (i.e. 50 × 700 mm column) with repetitive runs as as alternative to traditional large-scale systems.

3.7.3   Analytical Gel Filtration. Monitoring the Purification of Staphylococcal Enterotoxin B. Determination of the Molecular Weight of Chick Interferon

In analytical applications the resolution is of utmost interest. This is affected by the selectivity (i.e. peak-to-peak distance) and the zone broadening as discussed in paragraph

3.3.3. Equation (3-15) may be used to estimate the impact of various parameters on the resolution. It is advisable to check the column for efficiency and symmetry of a well defined solute and these values should be close to $2 \times d_p$ for the "plate height", at optimal flow rate, and 1.0 ± 0.1 for the symmetry at 10% peak height[88].

The risk of sample overload must be considered when high resolution is desired. Some general aspects of sample volume and concentrations were given in section 3.5.2. The applicability of these may be experimentally verified by first decreasing the concentration and then decreasing the sample volume and noting the width of the eluted zone. If this also decreases then the parameter(s) should be adjusted to the region where no effect on the zone is noted[105]. Overload effects are most often noticed for samples eluting in or near the void volume and viscous fingering may result in double or triple peaks as illustrated in Fig. 3-9. This effect is readily precluded by decreasing the sample concentration. To eliminate the influence of sample concentration on peak elution volume a procedure has been suggested where the sample is chromatographed at several concentrations and the elution volume taken as the intercept of a plot of concentration versus elution volume[106]. However, this method is tedious and not necessary in most applications.

Gel filtration may be used in the analytical mode to monitor a purification process selectively as illustrated in Fig. 3-14. The purification of staphylococcal enterotoxin B was controlled by fast gel filtration using Superose™ 12[129].

Interactions between the matrix and the solute should be absent in analytical gel filtration. Such interactions may be influenced by parameters which do not ordinarily affect the gel filtration process and thus can not be expected to have been controlled by the manufacturer. Interactions may also vary unpredictably from lot to lot. Interactions may be indicated by variations in elution volume or peak shape with concentration, temperature, ionic strength, organic modifier or chaotropic salts. However, the same effects may be caused by solute-solvent interactions (cf. section 3.2.2, p. 66). A general procedure for test of matrix properties has been given by Pfannkoch et al.[18].

Analytical gel filtration is frequently used for the assay of molecular weights or molecular weight distributions (MWD) of hydrophilic macromolecules or polymers. Granath and Flodin were the first to describe the use of gel filtration for the analysis of MWD's of clinical dextrans[130]. This application requires the calibration of the column with well-characterised solutes of shapes identical to the sample (e.g. narrow dextran fractions). A similar approach may be used for the estimation of protein molecular weight as exemplified by the characterisation of chick interferon by Phillips and Wood in 1964. They were able to assign a molecular weight of approximately 40,000 to the interferon by eluting the sample on a column of Sephadex G-10, calibrated with standard proteins[131]. The obtained molecular weight was confirmed using an analytical ultracentrifuge. However, calibrating the column with standard proteins is only valid as long as the geometric properties of the sample and standards are similar. In many cases, different shapes of proteins must be normalised by the denaturation of the protein with SDS or quanidine hydrochloride[45]. The size of an intact protein may be inferred from the hydrodynamic volume obtained using the concept of universal calibration of the column. These procedures are discussed in section 3.3.1.

Calibration of the column using the column-independent distribution coefficient, K_D, is recommended if the column is going to be used over a long period of time or the performance is to be compared with other materials. Calculation of K_D requires the determination of the void volume, V_0, and the total liquid volume of the bed, V_t, as well as the sample elution volume. If a proper probe for V_0 is used then the correct value for the

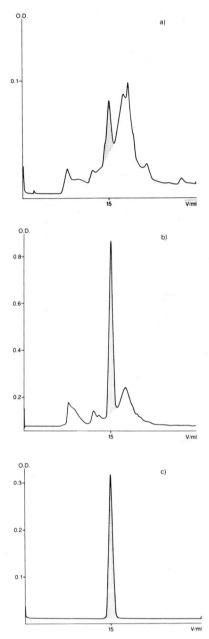

Figure 3-14. Monitoring of a large scale purification process by analytical gel filtration. Column: Superose™ 12. Sample: staphylococcal enterotoxin B in various stages of the purification process[129]. (a) after initial purification of 400 1 cell supernatant on S Sepharose[R] FF, (b) after intermediate purification by gradient elution on S Sepharose[®] FF and (c) after final gel filtration on Sephacryl[R] S-200 HR (By courtesy of H. O. J. Johansson, N. T. Pettersson and J. H. Berglöf, Pharmacia.)

void volume is obtained from the commencement of the sample application to the inflexion point of the ascending part of the peak. This will however result in a value of K_D for the peak apex being close to but not equal to 0. The void volume is therefore usually determined from the injection point to the peak apex. If the probe is too large, then too low a value of V_0 may be expected due to secondary exclusion effects and if the probe partially penetrates the gel too high a value of V_0 can be obtained. Suitable probes for the determination of void volume and total liquid volume are discussed in paragraph 3.5.3. The total liquid volume may be determined from the peak apex of a small molecule anticipated to penetrate the entire pore volume. For gels with very small pores it may be difficult to find a suitable substance that will not interact with the liquid gel phase. In such cases the total volume is calculated from the geometric column volume, V_c, and the distribution coefficient is then designated K_{av}. The two distribution coefficients can be related to each other with the formula

$$K_D = K_{av} (1 + V_s/V_p) \qquad (3\text{-}23)$$

where V_s is the volume of the support matrix and V_p is the pore volume. K_{av} of a totally permeating solute is thus a measure of the relative pore volume of the support.

If there is no need to calibrate the column with the aid of K_D or K_{av} (e.g. for the purpose of comparisons) then the elution volume, expressed by the first statistical moment, may serve as the dependent variable.

Thermostatting the column is preferred to ensure minimal variation in the calibration curve if the column is to be used over a long time period. It is also a good practice to check the efficiency and symmetry by elution of the test probe at regular intervals. When the column is not in use the flow rate may be reduced without impairing the calibration curve but if the flow is switched off then the column must often be re-calibrated. In this case precautions to avoid bacterial growth will be necessary. If it is properly handled an analytical column can be expected to have a lifetime of 1 to 5 years.

The injection point should be set when half the sample volume has been applied. However, when the sample volume is constant, the injection point may be set arbitrarily at the start of the sample application.

3.7.4 Determination of Pore Size. Estimation of the Apparent Pore Diameter of Superose® 6

In gel filtration the elution volume of a solute will be affected by the relationship between the solute size and the pore size of the matrix. Thus, by eluting molecules of well known sizes and applying established theory for gel filtration, approximations of the pore size can readily be achieved[13].

By establishing the entire calibration curve with well characterised standards, estimations of the pore size, pore volume and surface area may be obtained[132,133]. In the method of Freeman et al. the data are obtained from the inflexion point of the plot of K_{av} versus the hydrodynamic radius of the solute[132]. The method of Halász et al. included an initial empirical calibration of the procedure to yield data in accordance with established methods[133]. It has been reported that both methods give data in good agreement with absolute methods[134]. However, the method of Halász was found to yield a broader pore size distribution, PSD, than that obtained by classical methods[135]. The shortcomings of the method are due to an incorrect assumption of the relationship between available pore volume and solute size as recently explained by Knox and Scott[136]. These authors mathematically derived the relationship between the pore size distribution curve and the SEC calibration curve to

$$F(r) = \frac{1}{2}\left[\frac{d^2 K_D}{d(\ln R)^2}\right]_{R=r} - \frac{3}{2}\left[\frac{dK_D}{d(\ln R)}\right]_{R=r} + K_D \qquad (3\text{-}24)$$

where $F(r)$ is the cumulative pore size distribution from cylindrical pores of radius 0 to r and K_D is the distribution coefficient for a solute of radius R. The equation was found to yield data in excellent agreement with mercury penetration of rigid materials[136]. However, the numerical calculation requires a very smooth calibration curve and may show instability. Thus, the authors recently recommended the calculation of pore size distribution from a simulation procedure[137]. An alternative direct calculation procedure based on pores being built up by random spheres has been suggested by Schou et al.[138]. However, these calculation procedures are tedious and a rough estimate of the pore size may be calculated from twice the solute size at the inflexion point of the calibration curve[58].

These methods are of course also applicable to aqueous media for obtaining operational information on the pore size distribution provided that the reference data for the probe molecules are carefully selected to reduce disparities as revealed for data on proteins and dextrans[47]. By calibrating a column packed with Superose[®] 6 with the aid of dextran fractions and using the relationship $R_{h, \text{Dextran}} = 0.271 \times M^{0.498}$ Å, an apparent modal pore diameter of approximately 350 Å could be assigned to this support[58]. A review of methods for the characterisation of pore size as well as a detailed method for obtaining apparent pore size distribution data of gel filtration media was recently given by Hagel[58].

3.7.5 Mixed Mode Separations. Initial Purification of Yeast Cytochrome Oxidase

In mixed mode separations the elution of a solute is not due to size parameters only but is affected by adsorption or affinity to or exclusion from the matrix or layers of solvents or solutes adsorbed to the surface of the matrix. These types of solute-matrix interactions provide the basis for many of the separation principles dealt with in detail in the subsequent chapters. Since mixed mode separations are sometimes noticed on gels designed for gel filtration the subject will be discussed briefly here.

Due to its chemical nature and origin, the matrix may contain residual amounts of anionic groups, as in dextran and agarose gels[19,6], or exposed anionic groups as for silica based materials[18]. The cross-linking agent may introduce hydrophobic sites as will the bounded organic layer of silica materials. The matrix may in rare cases also provide sites for biospecific interactions[13]. These adsorptive properties of the matrices are highly undesirable in gel filtration and may be reduced by using a low pH to suppress ionization or a high ionic strength to decrease ionic interaction, adding detergents or modifiers to prevent hydrophobic interactions or simply by saturating the active sites with, for example, ovalbumin[17,13], phospholipids[61], or basic peptides[139].

Interactions between the solute and the matrix may be utilised advantageously as recently shown for the initial purification of yeast cytochrome oxidase[140]. More than 500 ml of an extract of submitochondrial particles from yeast was applied at high ionic

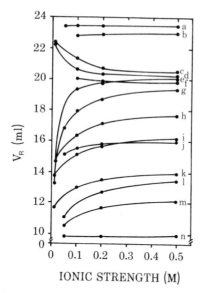

Figure 3-15. The effect of ionic strength on the retention volumes of native proteins. Buffer: 0.01 M sodium phosphate at pH 7.0 containing 0 to 0.5 M ammonium acetate. Solutes: (isoelectric point): (a) L-tyrosine, (b) vitamin B_{12}, (c) cytochrome c (10.3), (d) ribonuclease (9.3), (e) α-lactalbumin, (f) chymotrypsinogen (9.2–9.6), (g) soybean trypsin inhibitor (4.2–4.5), (h) ovalbumin (4.7), (i) serum albumin (4.7–4.9), (j) transferrin (5.5), (k) serum albumin dimer (4.7–4.9), (l) apoferritin (4.6–5.0), (m) β-galactosidase, (n) thyroglobulin (4.5). (Reproduced from Ref. 47 by kind permission of the author and the publisher.)

strength (i.e. 30%, w/w, of ammonium sulphate) to a column packed with Toyoperl (also called Fractogel). A highly purified and concentrated fraction of active enzyme was eluted by a gradient of decreasing salt content. The authors concluded that the very hydrophobic protein was probably adsorbed due to hydrogen bonding and that the interaction provides a favourable alternative to salt fractionation for the initial purification of yeast cytochrome oxidase.

Separations that are influenced by solute-matrix interactions should of course not be referred to as *gel filtration*. The retention volumes of solutes may in a non-gel-filtration mode vary substantially with changes in the mobile phase composition as shown in Fig. 3-15. Mixed mode separations may lead to unexpected results such as the concentration of a solute zone upon passage through the column[19], or the separation of proteins according to their isoelectric points[141]. A separation of ribonuclease *A* and cytochrome *c* was accomplished by utilising cationic effects at low ionic strength[142]. Purification of cytochrome *C* was achieved by adsorption of cytochrome *c* at high ionic strength with subsequent elution with a decreasing salt gradient[143]. Several publications where mixed mode separation was utilized or may be suspected to influence the separation have recently been published[141–146]. It has also been pointed out that such effects may result in poor resolution compared to gel filtration alone[147].

Naturally, this information is very valuable when selecting materials and conditions for which separations based on gel filtration alone are desired.

3.8   Equations and Symbols Used in Gel Filtration

As a result of the many different approaches taken to develop liquid chromatography as a separation technique and also to characterize chromatography supports, the nomenclature in liquid chromatography has, for some parameters, turned out to be

Table 3-7. Symbols used in gel filtration[a]

Symbol Parameter		Proposed by	Alternative symbol
K_D	distribution coefficient	IUPAC	K, ASTM EC
V_R	retention (elution) volume	ASTM, IUPAC	V_e
V_0	void volume	ASTM EC	
V_p	pore volume		V_i, ASTM EC
V_p/V_0	permeability		
V_s	matrix, support volume		
V_t	total liquid volume	ASTM EC	V_M, ASTM
V_c	geometric column volume		
V_0/V_c	void fraction		
ε	void fraction $= V_0/V_c$	ASTM	
R	zone velocity/mobile phase velocity $= V_0/V_R$	ASTM	
F_c	volumetric flow-rate	IUPAC	
F	nominal flow velocity $= F_c/A_c$	IUPAC	
u	interstitial flow velocity, $F \times V_c/V_0$	ASTM, IUPAC	
H	height equivalent to a theoretical plate (HETP)	ASTM EC	h, ASTM, IUPAC
h	reduced plate height	ASTM EC	h_r, ASTM
N	number of theoretical plates per column	ASTM EC	n, ASTM, IUPAC
w_b	base width (4σ)	ASTM	
w_h	peak width at half peak height	ASTM	
R_s	resolution	ASTM	
b/a	symmetry factor		
M	relative molar mass of solute	ASTM EC	
D	diffusion coefficient of solute in free solution		
D_m	diffusion coefficient in mobile phase	ASTM	
D_s	diffusion coefficient in liquid phase within pores	ASTM	
γ_m	obstruction factor in extra particle space		
γ_s	obstruction factor in pores $= D_s/D_m$		
R_h	hydrodynamic viscosity radius of solute (R_{vis})		
R_{St}	Stokes radius of solute		
V_h	hydrodynamic volume of solute	ASTM	
L	column length, bed height	ASTM	
A_c	column cross section area	ASTM	
d_p	average particle diameter of support	ASTM	
ΔP	pressure drop over a packed bed	ASTM	
η	viscosity	ASTM	

[a] Notes: IUPAC   symbols proposed by the International Union of Pure and Applied Chemistry.[7]
ASTM EC   symbols proposed by the American Society for Testing and Materials, Committee for Exclusion Chromatography Nomenclature.[149]
ASTM   symbols proposed by the American Society for Testing and Materials, Committee for Liquid Chromatography Nomenclature.[150]

inconsistent, e.g. in reversed-phase chromatography the total liquid volume is not denoted V_t but V_0. This is unfortunate, and to avoid ambiguities the designation V_m for mobile phase volume has been suggested instead of V_0.[148] A review including a detailed discussion of the, sometimes contradictory, proposals for nomenclature in liquid chromatography issued by IUPAC and ASTM has been given by Ettre.[148] The proposal "Steric Exclusion Chromatography" for SEC made by one ASTM Committee[149], has not yet found broad acceptance by workers in this field of chromatography, as judged from the vast literature on the subject. The symbols given in Table 3-7 are believed to represent an acceptable compilation of designations used today in size exclusion chromatography.[151,70,18]

Some useful formulas for calculation of parameters in gel filtration are, for sake of convenience, listed below. For the interpretation of non-standardised symbols the reader is referred to the original description of the formula in the text.

$$K_D = (V_R - V_0)/V_p \tag{3-1}$$

$$K_D = K_{av}(1 + V_s/V_p) \tag{3-23}$$

$$R_s = \frac{1}{4} \log M_1/M_2 \left(\frac{b}{V_0/V_p + \bar{K}_D} \right) \sqrt{L}/\sqrt{\bar{H}} \tag{3-15}$$

$$H \sim 2\lambda d_p + R(1-R)d_p^2 u/30\gamma_s D_m \tag{3-11}$$

$$h = H/d_p = L \times (w_h/V_R)^2/5.54 \times d_p \tag{3-16}$$

$$u_p \sim 65 \times K_D \times D_m/d_p$$

$$V_{inj} = \sqrt{\sigma_{rel}^2/(1 - \sigma_{rel}^2)} \times A_c \times \sqrt{L} \times (1 + K_D \times V_p/V_0)$$
$$\times (V_0/V_c) \times \sqrt{h \sim d_p} \times \sqrt{K_{inj}} \tag{3-19}$$

$$\Delta P \sim F \times L \times \eta \times d_p^{-2} \times 180 \times (1 - \varepsilon)^2/\varepsilon^3 \tag{3-17}$$

3.9 References

1. G. H. Lathe, C. R. Ruthven, *J. Biochem. J.*, *60*, xxxiv (1955).
2. G. H. Lathe, C. R. Ruthven, *J. Biochem. J.*, *62*, 665 (1956).
3. J. Porath, P. Flodin, *Nature*, *183*, 1657 (1959).
4. H. Determann, J. E. Brewer, in "*Chromatography a Laboratory Handbook of Chromatographic and Electrophoretic Methods*, E. Heftmann, ed., Van Nostrand Reinhold Company, New York, 1975, p. 363.
5. J. C. J. Moore, *Polym. Sci.*, Part A, *2*, 835 (1964).
6. L. Fischer, *Gel Filtration Chromatography*. Elsevier, Amsterdam 1980.
7. D. Amborse, E. Bayer, O. Samuelson, *Pure and Appl. Chem. 37*, 437 (1974).
8. Gel Filtration, Theory and Practice, Pharmacia Fine Chemicals, Uppsala, 1979.
9. TSK-Gel, Toyo Soda Manufacturing Co. Ltd., Tokyo, 1978, p. 3.
10. Chromatography Notes, Waters Associates, Milford, October 1983, p. 6.
11. W. W. Yau, J. J. Kirkland, D. D. Bly, *Modern Size-Exclusion Liquid Chromatography Practice of Gel Permeation and Gel Filtration Chromatography*, Wiley-Interscience, New York, 1979, p. 3.
12. H. Determann, in *Advances in Chromatography*, J. C. Giddings, R. A. Keller, eds., Marcel Dekker Inc., New York, 1969, Vol. 8, pp. 35–42.

13. P. Andrews, in *Methods of Biochemical Analysis*, D. Glick, ed., Wiley-Interscience, New York, 1970, Vol. 18, pp. 2-53.
14. E. Casassa, *J. Poly. Sci., Polym. Lett. 5*, 773 (1967).
15. A. M. Basedow, K. H. Ebert, J. H. Ederer, E. Fosshag, *J. Chromatogr., 192*, 259 (1980).
16. P. A. Bristow, LC in Practice, hetp, Wilmslow, England 1976.
17. H. G. Barth, *J. Chromatogr. Sci., 18*, 409 (1980).
18. E. Pfannkoch, K. C. Lu, F. E. Regnier, H. G. Barth, *J. Chromatogr. Sci. 18*, 430 (1980).
19. P. Flodin, *Dextran Gels and Their Applications in Gel Filtration*, Meijels Bokindustri, Halmstad, Sweden 1963, p. 74.
20. S. Aoyagi, K. Hirayanagi, T. Yoshimura, T. Ishikawa, *J. Chromatogr., 253*, 133 (1982).
21. Y. Kato, T. Hashimoto, *J. High Res. Chromatogr., Chromatogr. Comm., 6*, 45 (1983).
22. R. W. Stout, J. J. DeStefano, *J. Chromatogr., 326*, 63 (1985).
23. T. W. Hearn, F. E. Regnier, C. T. Wehr, *Am. Lab., 14*, 18-39 (1982).
24. L. Hagel, T. Andersson, *J. Chromatogr., 285*, 295 (1984).
25. Y. Kato, T. Hashimoto, *J. High Res. Chromatogr., Chromatogr. Comm., 8*, 79 (1985).
26. F. Hefti, *Anal. Biochem., 121*, 378 (1982).
27. Fractogel TSK, E. Merck, Darmstadt, 1984, p. 9.
28. M. Potschka, *Anal. Biochem., 162*, 47 (1987).
29. J. C. Janson, *J. Chromatogr., 28*, 12 (1967).
30. H. Determan, I. Walter, *Nature, 219*, 604 (1968).
31. B. Gelotte, *J. Chromatogr., 3*, 330 (1960).
32. R. P. Bywater, N. V. B. Marsden, in *Chromatography, Fundamentals and Applications of Chromatographic and Electrophoresis Methods, Part A*, E. Heftmann, ed., Elsevier, Amsterdam, 1983, pp. A297-A305.
33. M. Belew, J. Porath, J. Fohlman, J. C. Janson, *J. Chromatogr., 147*, 205 (1978).
34. L. Politi, M. Moriggi, R. Nicoletti, R. Scandurra, *J. Chromatogr., 267*, 403 (1983).
35. C. L. de Ligny, W. J. Gelsema, A. M. P. Roozen, *J. Chromatogr., Sci., 21*, 174 (1983).
36. L. W. Nichol, W. H. Sawyer, D. J. Winzor, *Biochem J., 112*, 259 (1969).
37. P. M. James, A. C. Ouano, *J. Appl. Polym. Sci., 17*, 1455 (1973).
38. P. C. Christopher, *J. Appl. Polym. Sci., 20*, 2989 (1976).
39. M. R. Ambler, L. J. Fetters, Y. Kesten, *J. Appl. Polym. Sci., 21*, 2439 (1977).
40. W. W. Yau, D. D. Bly, in *Size Exclusion Chromatography (GPC)*, T. Provder, ed., ACS Symposium Series, American Chemical Society Washington, D.C., 1980, pp. 197.
41. C. Tanford, *Physical Chemistry of Macromolecules*, Wiley, New York, 1961, Chapter 6.
42. C. de Haen, *Anal. Biochem., 166*, 235 (1987).
43. C. Tanford, Y. Nozaki, J. A. Reynolds, S. Mikano, *Biochem., 13*, 2369 (1974).
44. P. G. Squire, *J. Chromatogr., 210*, 443 (1981).
45. W. W. Fish, J. A. Reynolds, C. Tanford, *J. Biol. Chem., 245*, 5166 (1970).
46. H. Benoit, Z. Grubisic, P. Rempp, D. Decker, J. G. Zilliox, *J. Chim. Phys., 63*, 1507 (1966).
47. R. P. Frigon, J. K. Leypoldt, S. Uyejl, L. W. Hendersson, *Anal. Chem., 55*, 1349 (1983).
48. Z. Grubisic, P. Rempp, H. Benoit, *Polym. Lett., 5*, 753 (1967).
49. D. J. Harmon, in *Chromatography of Synthetic and Biological Polymers*, Vol. 1, Column Packings, GPC, GF and Gradient Elution, R. Epton, ed., Ellis Horwood, Chichester, 1978, pp. 122-145.
50. D. Berek, I. Novák, Z. Grubisic-Gallot, H. Benoit, *J. Chromatogr., 53*, 55-61 (1970).
51. A. M. Basedow, K. H. Ebert, H. Ederer, H. Hunger, *Macromol. Chem., 177*, 1501-1524 (1976).
52. P. G. Squire, *J. Chromatogr., 210*, 433-442 (1981).
53. J. Porath, *Pure Appl. Chem., 6*, 233 (1963).
54. T. C. Laurent, J. Killander, *J. Chromatogr., 14*, 317 (1964).
55. K. A. Granath, B. E. Kvist, *J. Chromatogr., 28*, 69 (1967).
56. G. K. Ackers, *J. Biol. Chem., 242*, 3237 (1967).
57. L. C. Davis, *J. Chromatogr. Sci., 21*, 214 (1983).
58. L. Hagel, in *Aqueous Size-Exclusion Chromatography*, P. Dubin ed., Elsevier, Amsterdam, 1988, pp. 119-155.
59. S. C. Meredith, G. R. Nathans, *Anal. Biochem., 121*, 234 (1982).
60. C. Tanford, in *Advances in Protein Chemistry*, C. B. Anfinsen, Jr., M. L. Anson, J. T. Edsall, F. M. Richards, eds., Academic Press, New York, 1968, Vol. 23, p. 211.
61. E. Mascher, P. Lundahl, *Biochim. Biophys. Acta, 856*, 505 (1986).
62. C. Tanford, J. A. Reynolds, *Biochim. Biophys. Acta, 457*, 133 (1976).
63. J. A. Reynolds, C. Tanford, *J. Biol. Chem., 245*, 5161 (1970).

64. N. Ui, *Anal. Biochem.*, *97*, 65 (1979).
65. T. Takagi, *J. Chromatogr.*, *219*, 123 (1981).
66. J. C. Giddings, in *Chromatography a Laboratory Handbook of Chromatographic and Electrophoretic Methods*, E. Heftmann, ed., Van Nostrand Reinhold Company, New York, 1975, pp. 27–45.
67. J. J. van Deemter, F. J. Zuiderweg, A. Klinkenberg, *Chem. Eng. Sci.*, *5*, 271 (1956).
68. J. C. Giddings, K. L. Mallik, *Anal. Chem.*, *38*, 997 (1966).
69. E. Katz, K. L. Ogan, P. W. Scott, *J. Chromatogr.*, *270*, 51 (1983).
70. J. H. Knox, P. H. Scott, *J. Chromatogr.*, *282*, 297 (1983).
71. E. M. Renkin, *J. Gen. Physiol.*, *38*, 225 (1954).
72. B. H. Arve, A. I. Liapis, *AICHE Journal*, *33*, 179 (1987a).
73. B. H. Arve, Pharmacia AB, Uppsala, personal communication, 1987.
74. G. K. Ackers, R. L. Steere, *Biochim. Biophys. Acta*, *59*, 137 (1962).
75. B. F. D. Ghrist, M. A. Stadalius, L. R. Snyder, *J. Chromatogr.*, *387*, 1 (1987).
76. J. H. Knox, F. McLennan, *J. Chromatogr.*, *185*, 289 (1979).
77. J. K. Leypoldt, R. P. Frigon, L. W. Henderson, *J. Appl. Polym. Sci.*, *29*, 3533 (1984).
78. M. E. van Kreveld, N. vand den Hoed, *J. Chromatogr.*, *149*, 71 (1978).
79. S. Hjertén, in *New Techniques in Amino Acid, Peptide and Protein Analyses*, A. Niederwiesser, G. Pataki, eds., Ann Arbor Science Publishers, Ann Arbor 1971, pp 227–247.
80. G. Guiochon, M. Martin, *J. Chromatogr.*, *326*, 3 (1985).
81. M. Verzele, C. Dewaele, D. Duquet, *J. Chromatogr.*, *391*, 111 (1987).
82. J. H. Knox, G. E. Laird, P. A. Raven, *J. Chromatogr.*, *122*, 129 (1976).
83. J. L. Glajch, D. C. Warren, M. A. Kaiser, L. B. Rogers, *Anal. Chem.*, *50*, 1962 (1978).
84. P. Flodin, *J. Chromatogr.*, *5*, 103 (1961).
85. Y. Kato, K. Komiyta, T. Iwaeda, H. Sasaki, T. Hashimoto, *J. Chromatogr.*, *208*, 71 (1981).
86. S. A. Karapetyan, L. M. Yakushina, G. G. Vasijarov, V. V. Brazhinkov, *J. High Res. Chromatogr. Comm.*, *8*, 148 (1985).
87. L. Hagel, H. Lundström, T. Andersson, H. Lindblom. *J. Chromatogr.*, *476*, 329 (1989).
88. P. A. Bristow, J. H. Knox, *Chromatographia*, *10*, 279 (1977).
89. T. Allen, *Particle Size Measurement*, Chapman and Hall, London, 1981, p. 432–436.
90. P. A. Bristow, *J. Chromatogr.*, *149*, 13 (1978).
91. J. H. Knox, in *Practical High Performance Liquid Chromatography*, C. F. Simpson, ed., Heyden and Son Ltd, London, 1976, p. 25.
92. M. R. Ladisch, G. T. Tsao, *J. Chromatogr.*, *166*, 85 (1978).
93. S. Katoh, *TIBTECH*, *5*, 328 (1987).
94. J. Porath, T. Låås, J.-C. Janson, *J. Chromatogr.*, *103*, 49 (1975).
95. T. Andersson, M. Carlsson, L. Hagel, P.-Å. Pernemalm, J.-C. Janson, *J. Chromatogr.*, *326*, 33 (1985).
96. D. D. Perrin, B. Dempsey, *Buffers for pH and Metal Ion Control*, Chapman and Hall, London, 1979, pp. 24–61.
97. N. E. Good, G. D. Winget, W. Winter, T. N. Connoly, S. Izawa, R. M. M. Singh, *Biochemistry*, *5*, 467 (1966).
98. G. D. Swergold, C. S. Rubin, *Anal. Biochem.*, *131*, 295 (1983).
99. S. C. Meredith, *J. Biol. Chem.*, *259*, 11682 (1984).
100. E. A. Lance, C. W. Rhodes III, R. Nakon, *Anal. Biochem.*, *133*, 492 (1983).
101. E. T. Adams, Jr., L.-H. Tang, J. L. Sarquis, G. H. Barlow, W. M. Norman, in *Physical Aspects of Protein Interactions*, N. Catsimpoolas, ed., Elsevier, New York, 1978, pp. 29–47.
102. J. F. K. Huber, *Instrumentation for High-Performance Liquid Chromatography*, Elsevier, Amsterdam, 1978, pp. 1–9.
103. R. W. Stout, J. J. De Stefano, L. R. Snyder, *J. Chromatogr.*, *261*, 189 (1983).
104. J. L. DiCesare, M. W. Dong, J. G. Atwood, in *Advances in Chromatography*, A. Zlatkis, ed., Elsevier, Amsterdam, 1981, p. 366.
105. L. Hagel, *J. Chromatogr.*, *324*, 422 (1985).
106. S. Mori, *J. Appl. Polym. Sci.*, *20*, 2157 (1976).
107. P. G. Squire, A. Magnus, M. E. Himmel, *J. Chromatogr.*, *242*, 255 (1982).
108. K. Granath, *J. Colloid Sci.*, *13*, 308 (1958).
109. H. Engelhardt, H. Müller, B. Dreyer, *Chromatographia*, *19*, 240 (1985).
110. B.-L. Johansson, C. Ellström, *J. Chromatogr.*, *330*, 360 (1985).
111. Superloop 10 ml, Pharmacia AB, Uppsala, Sweden.
112. Degasser ERC-3510, Erma Optical Works Ltd., Tokyo, Japan.

114. K. Asai, Y.-I. Kanno, A. Nakamoto, T. Hara, *J. Chromatogr.*, *126*, 369 (1976).
115. J. J. Hermans, *J. Appl. Polym. Sci.*, *Part A-2*, *6*, 1217 (1968).
116. R. Bishop, H. Lundin, *Application Note 315*, LKB, Bromma, 1978.
117. R. L. Stevenson, in *Liquid Chromatography Detectors*, T. M. Vickrey, ed., Marcel Dekker, Inc., New York, 1983, pp. 44, 65–72.
118. H. Elgass, A. Maute, R. Martin, S. George, *Int. Lab.*, *13*, 72 (1983).
119. A. C. Ouano, W. Kay, *J. Polym. Sci. Part A-1*, *12*, 1151 (1974).
120. T. Takagi, *J. Biochemistry*, *89*, 363 (1981).
121. M. L. Hitchman, F. W. M. Nyasulu, A. Aziz, D. D. K. Chingakule, *Anal. Chim. Acta.*, *155*, 219 (1983).
122. K. Isaksson, J. Lindquist, K. Lundström, *J. Chromatogr.*, *324*, 333 (1985).
123. R. S. Deelder, M. G. F. Kroll, A. J. B. Beeren, J. H. M. Van den Berg, *J. Chromatogr.*, *149*, 669 (1978).
124. V. G. Maltsev, T. M. Zimina, A. B. Khvatov, B. G. Belenkii, *J. Chromatogr.*, *416*, 45 (1987).
125. C. N. Trumbore, R. D. Tremblay, J. T. Penrose, M. Mercer, F. M. Kelleher, *J. Chromatogr.*, *280*, 43 (1983).
126. H. G. Barth, F. J. Carlin, *J. Liquid Chromatogr.*, *7*, 1717 (1984).
127. T. Maniatis, E. F. Fritsch, J. Sambrook, in *Molecular Cloning, A Laboratory Manual*, Cold Spring Harbor Laboratory, 1982, p. 464.
128. T. Andersson, L. Hagel, *Anal. Biochem.*, *141*, 461 (1984).
129. H. O. J. Johansson, N. T. Pettersson, J. H. Berglöf, Abstract of Papers 4th European Congress on Biotechnology, Amsterdam, June 1984.
130. K. Granath, P. Flodin, *Macromol. Chem.*, *48*, 160 (1961).
131. A. W. Phillips, R. D. Wood, *Nature*, *201*, 819 (1964).
132. D. H. Freeman, I. C. Poinescu, *Anal. Chem.*, *49*, 1183 (1977).
133. I. Halász, K. Martin, *Angew. Chem. Int. Ed. Engl.*, *17*, 901 (1978).
134. O. Chiantore, M. Guaita, *J. Chromatogr.*, *260*, 41 (1983).
135. W. Werner, I. Hálasz, *J. Chromatogr. Sci.*, *18*, 277 (1980).
136. J. H. Knox, H. P. Scott, *J. Chromatogr.*, *316*, 311 (1984).
137. J. H. Knox, H. Ritchie, *J. Chromatogr.*, *387*, 65 (1987).
138. O. Schou, Novo, Bagsvaerd, Denmark, personal communication, 1987.
139. G. W. Link, Jr., P. L. Keller, R. W. Stout, A. J. Banes, *J. Chromatogr. 331*, 253 (1985).
140. A. V. Galkin, I. E. Kovaleva, S. L. Kalnov, *Anal. Biochem.*, *142*, 252 (1984).
141. W. Kopaciewicz, F. E. Regnier, *Anal. Biochem.*, *126*, 8 (1982).
142. R. A. Jenik, J. W. Porter, *Anal. Biochem.*, *111*, 184 (1981).
143. M. Gurkin, V. Patel, *Amer. Lab.*, January, 64 (1982).
144. B. Renck, R. Einarsson, *J. Chromatogr.*, *197*, 278 (1980).
145. T. Hashimoto, H. Sasaki, M. Aiura, Y. Kato, *J. Chromatogr.*, *160*, 301 (1978).
146. D. E. Schmidt, Jr., R. W. Giese, D. Conron, B. L. Karger, *Anal. Chem.*, *52*, 177 (1980).
147. P. L. Dubin, I. J. Levy, *J. Chromatogr.*, *235*, 377 (1982).
148. L. S. Ettre, *J. Chromatogr.*, *220*, 29 (1981).
149. ASTM D 3016-78; Annual Book of ASTM Standards; American Society for Testing and Materials, Philadelphia, 1978.
150. ASTM E 682-79; Annual Book of ASTM Standards; American Society for Testing and Materials, Philadelphia, 1979, p 541–549.
151. J. H. Knox, R. Kaliszan, *J. Chromatogr.*, *349*, 211 (1985).
152. R. J. Rothman, L. Warren, *Biochim. Biophys. Acta*, *955*, 143 (1988).

4 Ion Exchange Chromatography

Evert Karlsson and **Lars Rydén**
Department of Biochemistry
Uppsala University
Uppsala Biomedical Center
Box 576, S-751 23 Uppsala, Sweden

John Brewer
Scientific and Technical Service
Pharmacia LKB Biotechnology AB
S-751 82 Uppsala, Sweden

4.1 Introduction

Adsorption chromatography of protein depends on several types of protein ligand interaction. The first of these to be successfully employed were the ionic interactions used in ion exchange chromatography (IEC). Today we have long experience of this type of chromatography. It is by far the most utilized chromatographic technique (Table 4-1), included in about 75% of purification protocols[1], followed by affinity chromatography (60%) and gel filtration (50%). The reason for the popularity of IEC is its versatility, its high resolving power, its high capacity and its straightforward basic principle.

Table 4-1. Methods used for protein purification and analysis in 100 articles in five major journals (*Biochemistry, Biochem. J., Biochim. Biophys. Acta, Eur. J. Biochem., J. Biol. Chem.*) during 1987 and 1988. A total of 334 preparative steps were reported (3.34 steps per preparation) and 131 analytical runs. Homogenizations, extractions and precipitation were not counted although they were frequently used as initial steps.

Method	Percentage used	
	Preparative	Analytical
Ion exchange	37	11
Chromatofocusing	3	
Electrophoresis	1	70
Gel filtration	25	11
Hydrophobic interaction chromatography	8 (HIC)	8 (RPC)
Affinity chromatography (incl. dye ligands)	19	
Hydroxyapatite chromatography	7	
Sum	100	100

Protein charge properties are used for fractionation purposes in several techniques. Thus, electrophoresis depends on electrophoretic mobility, which in turn is a function of charge density, while isoelectric focusing separates proteins according to their isoelectric points, i.e. the pH of zero net charge. One might say that the corresponding techniques in liquid chromatography are ion exchange chromatography and chromatofocusing. These methods have larger resolving power when compared to their electrophoretic counterparts. This is due to the fact that IEC not only depends on charge density, but also on the distribution of charges on the protein surface, i.e. charge anisotropy. Similarly separation in chromatofocusing reflects not only the pI of a protein, but also on the shape of its titration curve in the vicinity of the pI. Chromatofocusing, which in fact is a special form of ion exchange chromatography, is treated in Chapter 5.

Historically, IEC of proteins was introduced with the cellulose ion exchangers in the middle 1950's by H. A. Sober and E. A. Peterson[2–4]. They synthesised the DiEthylAminoEthyl, DEAE, and CarboxyMethyl, CM, derivatives of cellulose, which are still in use today. Ion exchangers based on polystyrene had been developed during the Second World War and of course other chromatographic media even earlier. However, none of these were well suited to protein chromatography. Polymethacrylate-based ion exchangers were used for the first successful chromatographies of a number of proteins, mostly basic, including ribonuclease[5–9].

Since then a number of chromatographic media, in particular beaded ones, have been derivatized with ion exchanging groups. These include gels based on cross-linked dextrans, cross-linked agarose, synthetic hydrophilic polymers and the small, rigid beads used in modern high performance chromatography. IEC has thus remained a key technique.

This chapter is devoted mainly to the practical aspects of ion exchange chromatography. Special emphasis will be given to the selection of a suitable ion exchanger and methods for obtaining the optimal result. Hydroxyapatite chromatography will be treated in this chapter since it may be regarded as a specific variant of ion exchange chromatography. The chapter will conclude with some practical applications.

4.2 The Ion Exchange Process

4.2.1 Fundamental Concepts

The basis for the ion exchange process is the competitive binding of ions of one kind, in our case proteins, for ions of another kind, for example other proteins or salt ions of the same charge, to an oppositely charged chromatographic medium, the ion exchanger (Fig. 4-1). The interaction between the proteins and the ion exchanger depends on several factors: net charge and surface charge distribution of the protein; the ionic strength and the nature of the particular ions in the solvent; pH, or strictly speaking the proton activity; and other additives to the solvent, such as organic solvents etc.

The energy gained by the formation of an ionic bond between a protein and a charge on the stationary phase is expressed by the coulombic law:

$$\Delta E = Z_A Z_B e^2 / r_{AB} \tag{4-1}$$

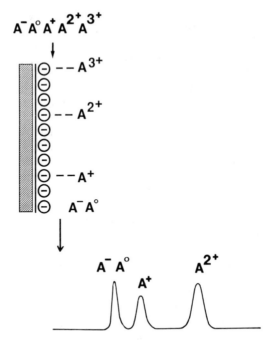

Figure 4-1. Principle of ion exchange chromatography. Species with several positive charges (A^{3+}) are adsorbed to the matrix; those with few charges move slowly, while those with no net charge or a net charge of the same signs the adsorbent pass through the column unretained. The resulting chromatogram is shown below.

where ΔE is the change in energy as two charges A and B with Z_A and Z_B number of unit charges are brought within a distance of r_{AB}. D is the dielectric constant of the medium. If the two charges are of opposite sign there is a decrease in energy and if they are of the same sign there is an increase.

The equation is valid if the two charges are far enough apart. It is important to point out that for the value of D for bulk solution to be valid, there must be a sufficiently thick layer of solvent molecules between the two charged groups. In fact, if there is direct interaction between the two charged groups, as between say an active site residue of a protein and its inhibitor or substrate, the bond is in general stronger than would be predicted by Eq. (4-1) since it is partly covalent. This is also borne out by the specific effects used in affinity elution of proteins from ion exchangers.

It is clear that the more highly charged a protein is, the more strongly it will bind to a given, oppositely charged ion exchanger. Similarly, more highly charged ion exchangers, i.e. those with a higher degree of substitution with charged groups, bind proteins more effectively than weakly charged ones. Conditions, for example pH, which alter the effective charge on either the protein or the ion exchanger will affect their interaction and may be used to influence the ion exchange process.

It should be noted that the terms strong and weak ion exchangers derive from the pK_a's of their charged groups (by analogy with strong and weak acids or bases) and do not say anything about the strength with which they bind proteins. At pH's far from the pK_a, binding will be equally strong to either a weak or a strong ion exchanger.

4.2.2   The pH Parameter

The pH is one of the most important parameters which determine protein binding as it determines the effective charge on both the protein and the ion exchanger. Control of pH by the addition of buffer salt is thus essential if reproducible results are to be obtained. Although proteins have charges of both signs over a wide pH range, as a rule, binding to an ion exchanger only occurs when there is a net charge of opposite sign to that on the ion exchanger. At pH values far away from the pI, proteins bind strongly and in practice do not desorb at all at low ionic strength. Near to its pI, the net charge of a protein is less and consequently it binds less strongly. Since the charges on the protein surface are distributed unevenly, weak binding can also occur close to the pI even when the overall charge is the same as that on the ion exchanger[10], as exemplified by phosphoglycerate kinase[11] and horse heart cytochrome oxidase[12]. Both proteins have clusters of positive charges that explain their binding to a cation exchanger at pH values when their net charges are slightly negative.

The ion exchanger should normally only be used in conditions where its charges will not be significantly titrated by small shifts in pH. Strong ion exchangers have pK_a values outside the pH range in which it is usual to work with proteins and for this reason may be preferred to weak ion exchangers unless experimental studies indicate otherwise (See further section 4.6.2).

4.2.3   Influence of the Concentration of Ions

The binding of proteins to charged groups on the stationary phase competes with the binding of other ions in the solvent. At a low enough concentration of competing ions, binding of proteins occurs through multiple charge interactions between several groups on the protein and a corresponding number, the Z-value, of charges on the ion exchanger, as is often the case when macromolecules interact. At higher concentrations of competing salt ions, the proteins will start to be displaced from the ion exchanger in the order of their binding strength, the least strongly binding being displaced and eluted from the column first. There is no general rule as to what salt concentration is needed to displace a protein with a certain net charge. Most proteins are eluted at salt concentrations in the approximate range 0.01 M and 0.5 M. As will be clear later, the type of ion is also an important factor in the process, and ions which interact specifically with charged groups on the protein will be unusually effective in elution.

4.2.4   Other Factors—Hydrogen-Bonding, Hydrophobic Interaction

Although coulombic forces are mainly responsible for the binding between the stationary phase and proteins, other interactions may also occur. Hydrophobic interaction between, in particular, resin-based ion exchangers and hydrophobic residues of proteins sometimes plays an important role. Similarly, in some cases hydrogen bonds are apparently formed between a protein and the ion exchanger (see below). These additional effects are sometimes crucial for obtaining a separation of two similar proteins. However, the benefits are mostly fortuitous. The effects are difficult to predict and thus difficult to exploit rationally, and might of course just as well constitute a disadvantage.

4.3   Charge Properties of Proteins

4.3.1   Charged Groups in Proteins

Five out of the twenty amino acid side chains of proteins contain groups that are weak acids or bases (Table 4-2). Several of them belong to the most common amino acid residues of proteins. Thus nearly all proteins have both positive and negative charges at pH values used in IEC. It should be mentioned that the amides of asparagine and glutamine for all practical purposes are neither basic nor acidic (although they do accept a proton at about pH −0.5).

The charged groups nearly always reside on the protein surface. The main exceptions are when internal metal ions in metalloproteins use histidines and cysteines (e.g. in Cu, Zn and Fe proteins) or glutamates and aspartates (mainly in Ca and Fe proteins) as ligands to the metal[14]. In these cases, the site of the proton is occupied by the metal, which can be displaced at a low enough pH. It should be added that metal proteins do not form a special case. About one third of all proteins described belong to this category. Only rarely have internal salt bridges been observed. Other prosthetic groups may also bind to the peptide chain via a charged amino acid side chain; e.g. pyridoxal phosphate forms covalent bonds with lysine residues.

Further charged groups (Table 4-2) are provided by the N-terminal α-amino and the C-terminal α-carboxyl groups. The various post-translational modifications that many proteins undergo may also provide new titratable groups. These are most often acidic such as phosphates in the phosphoproteins, γ-carboxylates in the blood clotting factors or sialic acid residues on oligosaccharides in serum glycoproteins[13]. Basic groups are introduced less often, but methylation of lysine and histidine in, for example, histones should be mentioned since it turns these side chains into much stronger bases[13]. Finally a modification may also eliminate the acid/base properties of a group, as in the acylation of the N-terminus by acetic or formic acid, the cyclization of an N-terminal glutamate to pyroglutamate, or the amidation or esterification of the C-terminal α-carboxylate.

Table 4-2. Charged amino acid side chains and other groups in proteins. Data were compiled from Ref. 13.

Group	Structure	pK_a	Average occurrence in proteins (%)
Arginine	guanido	12	4.7
Aspartic acid	carboxylate	4.5	5.5
Cysteine	thiol	9.1 to 9.5	2.8
Glutamic acid	carboxylate	4.6	6.2
Histidine	imidazole	6.2	2.1
Lysine	ε−amino	10.4	7.0
Tyrosine	phenol	9.7	3.5
α−amino	amino	6.8 to 7.9	
α−carboxyl	carboxylate	3.5 to 4.3	
sialic acid	carboxylate		sialoglycoproteins
γ−carboxyglutamate	carboxylate		blood coagulation factors
phosphoserine etc	phosphate		phosphoproteins

The various post-translational modifications often occur in only some of the protein molecules as reflected by a heterogeneity seen in ion exchange chromatography. This is exemplified by the ribonucleases A, B, C, and D from pancreas with different degrees of glycosylation[15] or ceruloplasmin forms I and II[16]. A further source of charge heterogeneity is covalent modification occurring during handling of the proteins and affecting only part of the protein. Nicking (hydrolysis of a single peptide bond) of the peptide chain, and loss of amide groups are especially often encountered.

Experimental chemical modifications also give rise to heterogeneity in proteins. Even if the different isomers formed have the same net charge it may still be possible to separate them by IEC. Thus when a cobra neurotoxin with six amino groups (one alpha and five epsilon) was acetylated, the six monoacetyl derivatives could be separated on the cation exchanger Bio-Rex 70[17] (see also application 4.9.4), a dramatic reminder of the importance of protein charge distribution for binding to the ion exchanger.

4.3.2   Protein Titration Properties

Most proteins have their isoelectric point at a pH below 7^{18-19}. They thus have a slight surplus of acidic groups. This accounts for the fact that in many applications of IEC a first and often successful approach is to use an anion exchanger at a slightly basic pH, i.e. around pH 8.

Proteins which have a preponderance of basic groups and thus a high pI are called basic proteins. Pancreatic ribonuclease is a moderately basic protein. More strongly basic proteins are not so often encountered, but examples are chicken egg white lysozyme, bovine mitochondrial cytochrome c (pI = 10.5)[19] and the histones. Cytochrome c is attached to the reductase and oxidase in the mitochondrial inner membrane by ionic forces, while the histones bind to DNA phosphates to form the nucleosomes in chromatin.

Extremely acidic proteins with a surplus of acidic groups and a pI below 5 are also found. Their acidity may be due to aspartates as in for, example, the acid protease pepsin, but often added groups explain the acidity as in many sialoglycoproteins.

Titration curves of proteins (Fig. 4-2) not only provide information on the position of the point of zero net charge, the pI, but also on how net charge varies with pH. The weak acids and bases in proteins normally titrate over a very wide range. It is especially interesting to study the assembly of titration curves for all the proteins in a sample as a guide to possible separation strategies. Figure 4-2 shows an example of such a collection. It is clear from these curves that the order in which the proteins in a sample may be arranged according to net charge varies with pH. It can, therefore, be expected that the proteins will elute from an IEC column in different orders at different pH values. It is also clear that even if a less successful result is obtained at one pH it is worth trying either a different pH with the same ion exchanger or an ion exchanger of the opposite charge.

4.3.3   Factors Influencing Protein Charge

Protein surface charge depends on the dissociation constants, pK, of the weak acids and bases on its surface. The pK of an acid is defined as the pH of 50% titration or protonation. An acidic group provides a net charge of -0.5 at $pH = pK$, while a basic group provides a net charge of $+0.5$ at its pK value. Only so called acid dissociation constants, pK_a, are used. At a distance of one pH unit from its pK a charged group is

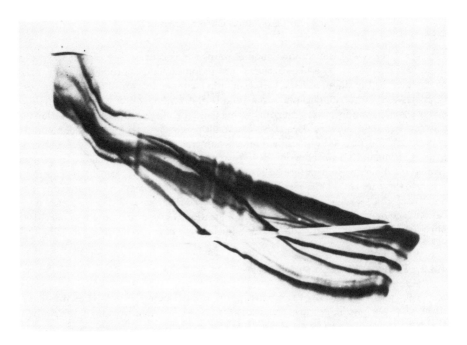

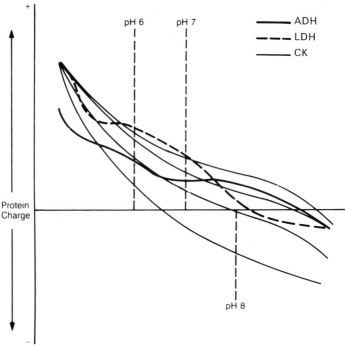

Figure 4-2. Titration curves of proteins showing how protein net charge varies with pH. Above are the actual titration curves of proteins in a chicken skeltal muscle extract as obtained by the technique described in Chapter 12 and visualized by protein staining. The diagram below shows the result of staining with zymogram techniques to identify alcohol dehydrogenase, lactate dehydrogenase and creatine kinase. Reproduced with permission from Ref. 20.

90% titrated as calculated from the Henderson-Hasselbach equation. Thus acidic groups and basic groups provide 0.1 charges one pH unit below and above their pK_a's respectively.

Amino acid side chains have pK values that depend on the neighbouring charges. When the main chain amino and carboxyl groups are part of peptide bonds this influences the pK of the side chain. The effect is to move the pK values closer to neutrality when compared to the free amino acid. However, in proteins many more factors contribute to modifying the pK of the side chains. In particular the position of the side chains in the tertiary structure of a protein is of decisive influence. More extreme deviations from standard pK values are seen for side chains that fulfill special roles in active sites of enzymes. The aspartic and glutamic acid residues in the active site of chicken egg white lysozyme have pK's of 3.5 and 6.3. The elevated pK value of the glutamate 52 is due to its hydrophobic environment[13].

As a corollary it is evident that protein surface charge properties change as the protein's conformation changes. The changes can be brought about by the addition of structure-modifying agents such as chaotropic ions, solvents, e.g. ethylene glycol, or detergents; or they can be a part of the biological function of the protein and influenced accordingly, for example by the use of allosteric effectors. Very specific effects can sometimes be obtained by using small molecules that bind to specific sites on proteins. Enzyme substrates which are charged of course interact specifically and often quite strongly, a fact that is exploited in the so called affinity elution of ion exchange columns.

4.3.4 Importance of Protein Size

The size of a protein is important for the strength of binding to the charged matrix. A larger protein provides more points of contact. Boardman and Partridge[21] showed that the retardation of proteins on Amberlite IRC-50 increased rapidly with the number of contact points, the so-called Z-value. This can occur when a protein aggregates. When cobra neurotoxin, with molecular weight 7820, is chromatographed on IRC-50 at pH 6.5, the monomer elutes at 0.15 M ammonium acetate and the dimers in a long extended peak between 0.5 and 0.7 M. On a cellulose exchanger, CM 52, at pH 5.2 the monomer elutes at 0.19 M ammonium acetate and the dimers at 0.28 M in a less diluted peak[22].

4.4 The Stationary Phase—The Ion Exchangers

Ion exchangers consist of a matrix substituted with either basic or acidic groups. The basic ion exchangers containing positive groups are called anion exchangers, while the acidic ones containing negative groups are called cation exchangers. The charge of the ion exchanger depends on pH as will be explained below.

The matrices are either (1) hydrophobic polystyrene-based or partly hydrophobic polymethacrylate-based polymers, which we will here call resins, cross-linked with divinylbenzene, (2) hydrophilic synthetic and naturally occuring polymers such as cellulose, dextran, agarose and others, (3) various synthetic hydrophilic polymers which make hard or moderately hard beads for high pressure applications, or (4) silica gels. Since the matrix is decisive for how the particular ion exchanger can be used, the description below will be given according to matrix type after an explanation of basic concepts.

4.4.1 Functional Groups and Acid-Base Properties

A number of different functional groups are used (Table 4-3). Most common are the amines in the anion exchangers and carboxylic acids in the cation exchangers. Strong anion exchangers contain quaternary amines and strong cation exchangers sulfonates. Practically all ion exchangers have groups with a single positive or negative charge with the notable exception of phosphate.

Approximate pK values of the groups are given in the table. Precise values are, however, difficult to determine. Titration curves of some ion exchangers show that the groups do not have a single pK_a-value (Fig. 4-3). DEAE-ion exchangers with a high degree of substitution contain significant numbers, up to 30% (Fig. 4.3), of so-called tandem groups formed by the further derivatisation of an already coupled DEAE-group. This leads to the formation of two additional kinds of charged groups, one a quaternary amine, the other a DEAE-group whose pK_a has been lowered to about pH 6 under the influence of the nearby quaternary nitrogen. Note that pK depends on salt concentration[23] and could be 0.2–0.5 units higher for CM cellulose and 1.5 units higher for Bio-Rex 70 in water as compared to 1 M KCl. Correspondingly, DEAE cellulose has pK values about 1.5 unit lower in water as compared to 1 M KCl.

All cation exchangers have a limiting pH below which they cannot be used. It is not good practice to use ion exchangers at pH values where a significant part of the groups have lost their charge. As a rule of thumb, the pK is suggested as a lower limit. Strong cation exchangers such as sulphopropyl can be used at a lower pH than carboxymethyl ion exchangers, but our experience is that some proteins, e.g phospholipase A2 and cellulase (Karlsson, unpublished), adsorb irreversibly at pH 3–3.5 on cation exchangers. In the same way weak anion exchangers have an upper pH limit for their practical use.

Table 4-3. Functional groups used in ion exchangers. Data have been compiled from manufacturers literature. The pK values given mostly refer to an ionic strength of 1. See also text.

Name	Designation	pK	Structure
Anion exchangers			
Diethyl aminoethyl	DEAE	9.0 to 9.5	$-OCH_2CH_2NH^+(C_2H_5)_2$
Trimethyl hydroxypropyl	QA		$-OCH_2CH(OH)N^+(CH_3)_3$
Quaternary aminoethyl, diethyl-(2-hydroxypropyl) -aminoethyl	QAE		$-OCH_2CH_2N^+(C_2H_5)_2CH_2CH(OH)CH_3$
Quaternary aminomethyl	Q		$-OCH_2N^+(CH_3)_3$
Triethyl aminomethyl	TEAE	9.5[a]	$-OCH_2N^+(C_2H_5)_3$
Triethylaminopropyl	TEAP		$-OCH_2CH_2CH_2N^+(C_2H_5)_3$
Polyethyleneimine	PEI		polymerized $CH_3-CH=NH$
Cation exchangers			
Methacrylate		6.5	polymerized $CH_2=C(CH_3)COOH$
Carboxymethyl	CM	3.5–4	$-OCH_2COOH$
Orthophosphate	P	3 and 6	$-OPO_3H_2$
Sulfonate	S	2	$-OCH_2SO_3H$
Sulfoethyl	SE	2	$-OCH_2CH_2SO_3H$
Sulfopropyl	SP	2–2.5	$-OCH_2CH_2CH_2SO_3H$

[a] The pK value apparently does not refer to quaternary groups

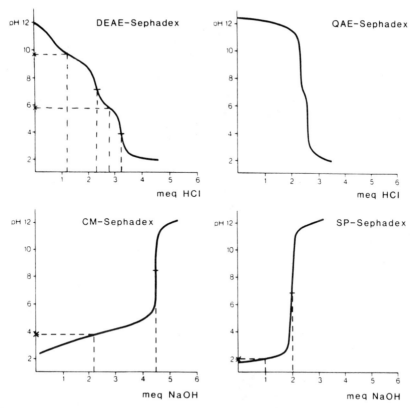

Figure 4-3. Titration curves for ion exchangers used in protein chromatography. The four curves show the titration of Pharmacia LKB DEAE-Sephadex A-50 (pK 9.5 and 5.8), CM-Sephadex C-50 (pK 4.0), QAE-Sephadex A-50 (the irregularity indicates that the ion exchanger is not fully quaternerized), and SP-Sephadex C-50 (pK 2.0).

For the quaternary amines of course, there is no upper limit since they will not deprotonate whatever the pH.

Ionic exchangers with dipolar ligands, such as β-alanine, sulfanilic acid and arginine, have been prepared by Porath and co-workers[24–25]. A protein molecule in the vicinity of a charged dipole should orient itself differently from one close to a single charge and therefore adsorb differently as compared to an ordinary ion exchanger. The electric field potential decreases faster around dipoles than around unit charges which should facilitate desorption. Scopes has suggested that dipolar ion exchangers should be useful for affinity elution[26].

4.4.2   Capacity and Porosity

In general ion exchangers are more densely substituted than any other adsorbent used in protein chromatography. Common degrees of substitution as determined by titration (total ionic capacity) are 100 to 500 micromoles/ml of bed (Table 4-4). This figure corresponds to a concentration of ion exchanging groups in the column of 0.1 to 0.5 M.

Table 4-4. Illustrative examples of ion exchangers for protein chromatography. Data have been compiled from the manufacturers literature. BSA = Bovine Serum Albumin; Lys = Lysozyme; Hb = Hemoglobin. The table is not comprehensive. New products are marketed continuously by several companies. Binding conditions are discussed in the text.

Name	Matrix	Functional group	Degree of substitution μmole/ml	Available capacity mg/ml	Company
Conventional media					
DE 23	fibrous cellulose	DEAE	150	60 BSA	Whatman
CM 23	fibrous cellulose	CM	80	85 Lys	Whatman
DE 52	microgranular cellulose	DEAE	190	130 BSA	Whatman
CM 52	microgranular cellulose	CM	190	210 Lys	Whatman
DE 53	microgranular cellulose	DEAE	400	150 BSA	Whatman
CM 32	microgranular cellulose	CM	180	200 Lys	Whatman
DEAE-Sephacel	beaded cellulose	DEAE	170	160 BSA	Pharmacia
DEAE-Sephadex A-25	Dextran, Sephadex G-25	DEAE	500	70 Hb	Pharmacia
QAE-Sephadex A-25	Dextran, Sephadex G-25	QAE	500	50 Hb	Pharmacia
CM-Sephadex C-25	Dextran, Sephadex G-25	CM	560	50 Hb	Pharmacia
SP-Sephadex C-25	Dextran, Sephadex G-25	SP	300	30 Hb	Pharmacia
DEAE-Sephadex A-50	Dextran, Sephadex G-50	DEAE	175	250 Hb	Pharmacia
QAE-Sephadex A-50	Dextran, Sephadex G-50	QAE	100	200 Hb	Pharmacia
CM-Sephadex C-50	Dextran, Sephadex G-50	CM	170	350 Hb	Pharmacia
SP-Sephadex C-50	Dextran, Sephadex G-50	SP	90	270 Hb	Pharmacia

Name	Matrix	Group			Supplier
DEAE-Sepharose CL-6B	Agarose, 6% crosslinked	DEAE	150	100 Hb	Pharmacia
CM-Sepharose CL-6B	Agarose, 6% crosslinked	CM	120	100 Hb	Pharmacia
DEAE Bio-Gel A	Agarose	DEAE	20	45 Hb	BioRad
CM Bio-Gel A	Agarose	CM	20	45 Hb	BioRad
Q-Sepharose Fast Flow	Crosslinked agarose	Q	150	100 Hb	Pharmacia
S-Sepharose Fast Flow	Crosslinked agarose	S	150	100 Hb	Pharmacia
DEAE-Trisacryl M	Trisacrylate polymer	DEAE	300	90 Hb	IBF
CM-Trisacryl M	Trisacrylate polymer	CM	200	100 Hb	IBF
SP-Trisacryl M	Trisacrylate polymer	SP	230	150 BSA	IBF
Medium Pressure media					
Fractogel TSK DEAE-650	Synthetic organic polymer	DEAE	100	25 Hb	Merck
Fractogel TSK CM-650	Synthetic organic polymer	CM	100	25 Hb	Merck
High performance media					
Mono Q	Synthetic organic polymer	Q	210	25	Pharmacia
Mono S	Synthetic organic polymer	S	210	25	Pharmacia
DEAE-5-PW	Synthetic organic polymer	DEAE	—	1.5–3	BioRad
SP-5-PW	Synthetic organic polymer	SP	—	1.5–3	BioRad
HRLC MA7P	Non-porous synthetic polymer	PEI	—	0.6–2	BioRad
HRLC MA7C	Non-porous synthetic polymer	CM	—	0.6–2	BioRad
DEAE Si500	Silica	DEAE	—	10	Serva
TEAP Si100	Silica	TEAP	—	—	Serva
CM Si300	Silica	CM	—	12	Serva
SP Si100	Silica	SP	—	—	Serva

The available capacity, or protein-binding capacity, of an ion exchanger is defined as the amount of protein that can be bound. This capacity thus depends on the operating conditions and the particular protein. In Table 4-4 the capacity is given for serum albumin for the anion exchangers and lysozyme or hemoglobin for the cation exchangers. Commonly an ionic strength of 0.01 is used and a pH where all groups on the adsorbent are charged. Only part of the ionic groups are available to protein. CM-cellulose with a degree of substitution of 190 micromole/ml binds 130 mg (2 micromoles) of BSA per ml ion exchanger. For Mono Q the corresponding figures are 300 micromoles and 65 mg BSA (1 micromole)/ml. Here the Z-value has been determined to be 9–12. Thus about 4% of the groups are involved in protein binding.

It is thus clear that only a fraction of the group is used for protein binding. This fraction becomes even smaller in actual chromatographic conditions. This is due to the slow kinetics of binding of most of the sites. Thus DEAE-Sephacel has an available capacity of 160 mg serum albumin in 0.01 Tris HCl buffer pH 8.0, but it will take 15–20 min before that value is attained. Adsorption/desorption of about 20% of this amount occurs within 30 sec[27]. This is called the dynamic capacity and is the capacity which can be used in normal chromatographic conditions. Thus an amount of ion exchanger corresponding to 5–10 times the given protein binding capacity should be used in chromatography.

The capacity depends critically on the availability of charged groups and thus on pore size and protein molecular weight. Thus DEAE-Sephadex A-50 binds 250 mg hemoglobin per ml of bed while the corresponding gel with smaller pores, DEAE-Sephadex A-25, only binds 70 mg of the same protein. This is why the more recently developed ion exchangers are based on matrices with large pores. The relationship between capacity and molecular weight is illustrated by the fact that DEAE-Sephacel binds 160 mg/ml of albumin (M 67,000), but only 10 mg/ml of thyroglobulin (M 670,000). Apparently thyroglobulin does not penetrate the matrix and only binds to sites near the surface of the particles. The variation of capacity with protein molecular weight is illustrated by Table 4-5 for a cellulose ion exchanger[28]. It should be noted that the capacity might be higher when a mixture of proteins is used since they might complement each other on the matrix, smaller proteins occupying sites not available for the larger ones. Exclusion limits of ion exchangers are given by the manufacturers.

Table 4-5. Capacity of microgranular CM-Cellulose (Whatman CM-52) for rabbit muscle glycolytic enzymes and some other proteins at pH 6.0, at ionic strength 0.01 (from Ref. 28).

Protein	Mol weight	Capacity mg/ml	Capacity μmole/ml
Chicken egg white lysozyme	14,300	130	9.0
Phosphoglycerate kinase	45,000	70	1.55
Phosphoglycerate mutase	60,000	40	0.67
Creatine kinase	82,000	35	0.43
Enolase	88,000	48	0.55
Lactate dehydrogenase	140,000	21	0.15
Glyceraldehydephosphate dehydrogenase	145,000	26	0.18
Aldolase	160,000	22	0.14
Pyruvate kinase	228,000	12.5	0.055

4.4.3   The Ion Exchangers

A summary of the properties of some of the better-known commerically available ion exchange media is given in Table 4-4.

4.4.3.1   Resin Ion Exchangers

A classical group of ion exchangers are made by polymerization of styrene with divinylbenzene (DVB) as crosslinker in the presence of a catalyst such as benzoylperoxide (Fig. 4-4). Best known are the strongly acidic sulphonated cation exchangers such as Dowex 50-X8 or AG 50-X8. X8 denotes the percentage of crosslinker in the polymerization mixture which is not necessarily the same as that present in the final product. The corresponding strong anion exchangers containing quaternary amines are called Dowex 50 1-X4 and AG 1-X4.

Polystyrene ion exchangers have a low capacity for proteins (1 or a few mg per gram of ion exchanger) due to their small pore size[23,29]. They also bind proteins almost irreversibly by hydrophobic interactions. They have consequently found limited use for protein chromatography, but have for many years been of extreme importance for separation of peptides and amino acids, as for example in the standard Moore and Stein amino acid analysis.

Resins obtained by copolymerization of methacrylic acid and divinylbenzene are available as Amberlite IRC 50 (5% DVB in the polymerization mixture) (Rohm and Haas) and Bio-Rex 70 (5% DVB) (Bio-Rad Laboratories); these two are practically identical. For protein chromatography, a resin with minus 400 mesh size should be used. Polymerization of methacrylic acid alone gives 11.6 mmoles of carboxyl groups per gram. Since the actual capacity is 10 mmoles/g, the aromatic constituent by difference accounts for about 14% of the resin. In the packed chromatographic bed this capacity corresponds

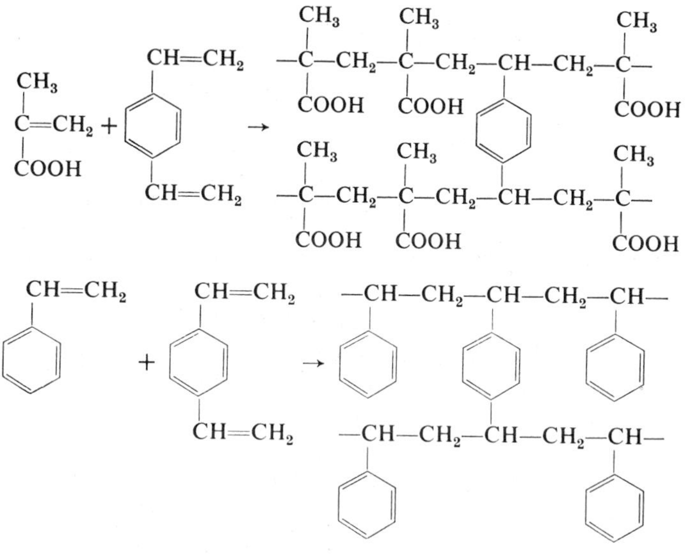

Figure 4-4. Covalent structures of the methacrylate ion exchangers IRC-50 and Bio-Rex 70 (above) and polystyrene-based ion exchangers (below).

to 3.5 mmoles per ml or a 3.5 M concentration of carboxylates. The resin itself is thus a strong buffer in the column and is able to titrate a starting buffer of a deviant pH.

The pK of IRC-50 and Bio-Rex 70 is about 6.0 in 1 M NaCl, 6.5 in 0.1 M NaCl and increases to 7.5 in 0.01 M salt. When the pH is lowered to approximately 5.7, the particles aggregate.

The acrylic acid resins are particularly well suited for chromatography of small basic proteins. They were used for the first reported purifications of proteins by IEC—the purification of lysozyme by Paleus and Neilands[5] and of pancreatic ribonuclease by Stein and Moore[6-9]. They have a resolving power for basic proteins so far hardly surpassed by any other ion exchanger. They have proved of considerable value for the fractionation of the collection of basic proteins that make up the snake neurotoxins as will be illustrated below. Hydrogen-bonding to the carboxylic acid groups and hydrophobic interactions with the matrix may contribute to the adsorption to the stationary phase during chromatography.

4.4.3.2 Cellulose Ion Exchangers

The cellulose ion exchangers were introduced in 1954 by Sober and Peterson[2-4]. Their work was preceeded by a number of reports on the use of cellulose for protein adsorption, but at the time it constituted a major breakthrough in protein chromatography. They are still in wide use today in particular for large scale preparation of neutral and acidic proteins, the group to which most proteins belong.

Cellulose has a macroporous structure and is thus accessible to large macromolecules. Microfibrous cellulose typically binds four times its own weight of water. Its capacity for protein binding is high and its interaction with proteins due to hydrophobic or other non-ionic interactions is low.

The cellulose ion exchangers are prepared from strongly alkaline (mercerized) cellulose by derivatization. Carboxymethyl groups are introduced by reaction with chloroacetic acid and diethylaminoethyl groups by reaction with 2-chlorotriethylamine. Typical degrees of substitution are 1 mmoles/gram of cellulose which corresponds to 200 micromoles/ml bed volume. It is mostly amorphous regions in the otherwise fibrous structure which are derivatized by this treatment. In alkaline conditions, however, the bundles of polyglucan chains (Figure 2-2) open up.

The cellulose ion exchangers exist in several different physical forms. The fibrous celluloses are those that are obtained directly from the derivatization process. They contain fibres of polyglucan chains mixed with amorphous regions. The microgranular ion exchangers are obtained by partial acid hydrolysis of these, a treatment that removes a large part of the amorphous regions and creates large void spaces. The ordered structures are stabilized by crosslinking and derivatization. Microgranular adsorbents have a more uniform distribution of charges. Cellulose ion exchangers in beaded form have been prepared as described by Determan and co-workers[27]. They are available from Pharmacia LKB by the name of Sephacel and from Serva as Servacel.

4.4.3.3 Dextran and Agarose Ion Exchangers

The beaded forms of dextran and agarose gels originally prepared for gel filtration have been derivatized to produce ion exchangers for protein chromatography.

The dextran-based ion exchangers are derived from Sephadex. The designations A and C denote anion and cation exchanger respectively, while the figures 25 and 50 refer to the starting gel which is Sephadex G-25 or G-50. Both of these have a considerably enlarged pore size as compared to the underivatized gels since repulsion of charged groups expands the gel. This swelling depends both on the pH and on the concentration of counter-ions in the solution as well as on the degree of crosslinking. It is more important for the less crosslinked A-50 and C-50; it decreases when the ion exchanging groups are neutralized and is largest at low ionic strength. In fact a column of DEAE-Sephadex A-50 or CM-Sephadex C-50 will shrink to half its volume when the buffer strength is increased from about 0.01 M to 0.5 M. These effects do not seem to have any adverse consequences, at least not with continuous gradients.

Sepharose ion exchangers (Pharmacia LKB) are derived from crosslinked 6% agarose (Sepharose CL-6B). These gels are more porous than the dextran gels and volume changes with changes in ionic strength or pH are insignificant. The most recently developed of these media, Q-Sepharose and S-Sepharose are produced by a different cross-linking procedure. The beads are of the same size as conventional agarose but are much more rigid and particularly suitable for large-scale work and other applications in which high flow rates are desirable. DEAE and CM Bio-Gel A ion exchangers (Bio-Rad Laboratories) also consist of crosslinked agarose beads. Agarose-based ion exchangers are useful for chromatography of larger proteins as they are more macroporous. Their degree of substitution is less than that of the dextran-based ion exchangers, but the available capacity for proteins might not be less if the pore size is the limiting factor.

Several other matrixes are also used for ion exchangers. Trisacryl M (Reactifs IBF-Societé Chimique, France) and Fractogel TSK (Merck) are based on hydrophilic, synthetic organic polymers. These have a more limited use at present.

4.4.3.4   Ion Exchangers for High Performance Applications

In high or medium pressure chromatography ion exchange hard beads of small and uniform size are used. These are based on silica matrices, for example CM-Si300 and DEAE-Si100 (Serva Feinbiochimica), coated with a hydrophilic layer of ion exchange groups. A number of media are based on crosslinked organic non-carbohydrate polymers, for example Mono Q and Mono S (Pharmacia LKB). TSK resins, including DEAE 5 PW and SP 5PW, are produced by Tosoh and distributed by several suppliers including Waters, USA, Merck, Germany and Bio-Rad Laboratories, USA. They consist of a hydrophilic polymer substituted with ion exchanging groups. In general the particle size is 5–10 μm. The matrix of Mono Q and Mono S is practically speaking mono-disperse with respect to bead size (10.0 micrometer) and is stable to aqueous solutions and pH's in the range 2–12. This stability enables them to be cleaned with NaOH solutions (2M) to remove lipids and aggregated proteins which otherwise shorten the life of high performance columns. As explained in Section 2.2.1, the silica derived gels have the drawback that they are unstable at alkaline pH values. It has been shown that organic and silica based stationary phases with the same ionic groups, porosity and particle size have the same chromatographic behaviour[30–32].

In general, the high performance gels are porous with more than 95% of the surface area inside the particles. Their large pore size (> 30 nm) allows high molecular weight proteins to enter easily[33]. A pore diameter of 30 to 50 nm is optimal for proteins in the molecular weight range 30,000 to 100,000. For even larger proteins pore size should increase to 80 to 100 nm[30,34]. Non-porous ion exchangers, where all binding occurs on the surface of the particles, have been developed based on polymethacrylate[35–37] and even crosslinked agarose[38]. In these, the partitioning process is speeded up due to the small diffusion distances, but an inevitable drawback is the substantial decrease in capacity. The recommended protein loading is decreased to about 0.5 mg/ml bed volume for the Bio-Rad HRLC MA 7 ion exchanger. Their use is thus rather fast analytical applications than preparative work.

The high performance gels allow considerably increased resolution as compared to conventional gels. Drawbacks include a higher cost and smaller scale. In practice the HPLC gels are not used for crude extracts both because of their limited capacity and the risk of fouling the column. Removal of bulk protein is thus recommended before HPLC is introduced as a later step in a purification scheme. The porous HPLC ion exchangers allow a practical loading up to 25 mg/ml bed volume. Normal column total volumes are in the range 1-10 ml. Sample volume can be large if the binding (k', i.e. retardation) is sufficiently strong. Of course if only minor quantities of protein are at hand the small scale is rather an advantage.

4.5   The Mobile Phase—Buffers and Salt

Since the charge of the protein is so sensitive to the pH, it is very important that the buffering properties of the mobile phase are given the proper attention. In many cases a minor change in buffer pH or concentration can spell the difference between success and failure[20]. The specific effects of buffer ion species are also decisive in some cases.

4.5.1   Buffer pH and Concentration

Normally the concentration of the buffer salt during protein adsorption is low, around 0.01 M-0.05 M. The buffer capacity needs thus to be reasonable to avoid pH disturbances in the initial part of the chromatogram. A buffer has a maximal buffering capacity at its pK_a. One pH unit away from the pK, the buffering capacity has decreased five-fold since 90% of the protolytic system is in either the base or the acid form. The buffer salt is thus chosen to have its pK_a near to the pH at which the chromatography is to be carried out[26,39].

To avoid disturbances in the pH caused by the buffering species itself participating in ion exchange, the buffering ion should not interact with or bind to the ion exchanger. For an anion exchanger a positive buffering ion, such as Tris, pK_a 8.2, is therefore preferred with Cl^- as counterion. For a cation exchanger, the buffering ion should be negatively charged, for example phosphate, carbonate, acetate or MES (morpholino-ethane-sulphonate), and the counter ion K^+ or Na^+.

The standard buffers are, however, not always the best choice and are not compulsory. It is for example common to use phosphate or acetate buffers with DEAE-ion exchangers, even though the buffering ions bind to the ion exchanger. It is often advantageous to use a volatile buffer salt (Table 4-6) which allows direct lyophilization of the pooled fractions after chromatography. In particular, small amounts of protein are likely to give bad recoveries in desalting or dialysis and the use of volatile salts allows these techniques to be avoided. A further consideration when choosing buffers is the possibility to exploit the special effects of some buffer salts (see below). The possibility that a particular buffer species will interfere with subsequent studies should also be taken into account. For example, low concentrations of Tris, which is non-volatile, will esterify aspartic and glutamic acid during hydrolysis for amino acid analysis (D. Eaker, unpublished).

Table 4-6. Volatile buffer salts used in ion exchange chromatography.

Buffer system (base/acid)	pK-values for buffering ions	Special properties
Formic acid	3.75	
Pyridine/formic acid	3.75; 5.25	smell $\varepsilon_{280} = 3.5$
Trimethylamine/formic acid	3.75; 9.25	
Pyridine/acetic acid	4.76; 5.25	smell $\varepsilon_{280} = 3.5$
Trimethylamine/HCl	9.25	
N-Ethylmorpholine/acetic acid	4.76; 7.8	
Ammonia/formic acid	3.75; 9.25	
Ammonia/acetic acid	4.76; 9.25	
Trimethylamine/carbonate	6.50; 9.25	
Ammonium bicarbonate	6.50; 9.25	
Ethanolamine/HCl	9.5	

4.5.2   Non-Buffering Salt; Specific Buffer and Salt Effects

A non-buffering salt, such as NaCl, is usually added to the medium to elute proteins from an ion exchange column by increased ionic strength. The properties of these non-buffering ions also influence the process. A change in ionic composition is not likely to affect all proteins in the same way. It is therefore also possible to influence both resolution and selectivity of the ion exchanger by a change in the non-buffering salts. However, except for the use of affinity elution (see below), the outcome is difficult to predict and one is left to trial and error to arrive at a proper result.

Polyvalent anions, such as sulphates and phosphates, are more efficient on a molar basis as displacers on an anion exchanger than those with single charges, such as chloride.

The chromatographic behaviour of proteins in the presence of different cations and anions has been studied by several workers[10,30,31,40-49]. Compilations of results are found in[10,43,44]. Proteins were, for example, eluted from the anion exchanger Mono Q with gradients of NaCl, NaBr, NaI and Na acetate. Both the resolution and the order of the eluted proteins changed. For instance alpha-lactalbumin eluted ahead of ovalbumin in NaCl and NaBr gradients, the two proteins eluted together with NaI, and in the reverse order with Na acetate[49]. Yao and Hjertén[45] used QAE-agarose in Tris buffers pH 8.5 for the separation of various protein mixtures. The best resolution was obtained with carboxylate, in particular acetate, anions.

An example of the effect of cations is the elution of amino acids from a sulphonated polystyrene resin, which is decisively different with sodium, potassium and lithium as cations, lithium giving significant additional separation power. Corresponding effects on the elution of proteins have been demonstrated[42,43,49].

Choice of buffering ion also influences the separation. In several instances acetate buffers have proven to be the eluents of choice for chromatography, especially on cation exchangers. Ammonium acetate has proven to be very useful in this respect, not only increasing the resolution, but having the additional advantage of being a volatile buffer that can be easily lyophilized. A fresh solution of ammonium acetate has a pH of 7.0 and a very low buffer capacity. When it is used with an ion exchanger that by itself has a high buffer capacity, such as the methacrylate based resins, the pH will be determined by the equilibrium conditions of the resin and ammonium acetate buffers can very well be used. (See application example 4.9.4). There are several examples of the successful use of ammonium acetate for obtaining separations that could not otherwise be achieved. Two isotoxins from the sea snake *Enhydrina schistosa* differing only by a Pro/Ser mutation could be separated on the cation exchanger Bio-Rex 70 using ammonium acetate, but not using phosphate buffer[50, 51]. In ammonium acetate the Ser 46 mutant toxin was retarded, probably due to hydrogen bonding to the resin. A similar experience was encountered with two isotoxins from *Naja naja* differing in a Ile/Ser mutation. Here too the Ser mutant protein was more retarded using ammonium acetate buffer[52].

Phosphate has also been essential for some separations on anion exchangers. This might be due to the specific binding to the proteins, used in affinity elution (see below). As an example ribonuclease binds tightly to the cation exchanger Amberlite IRC-50 and the chromatography was most easily carried out using phosphate buffers which specifically displaced the protein[53].

The differences in water-structuring, or inversely chaotropic, properties of ions (see Chapter 7) might be one explanantion for some of the specific effects.

4.5.4   Influence of Additives

Additional components in the medium also influence the ion exchange process in an ion exchange chromatography experiment. The additives used are mostly those needed to increase protein solubility. They include in particular those that diminish hydrophobic interactions, such as ethylene glycol or ethanol, urea and sometimes detergents. Note that detergents must be either neutral or of the same charge as the ion exchanger.

In the cases where hydrophobic interactions between the matrix and the proteins in the sample are important it is evident that the solvent modifiers will decrease that binding. However organic modifiers will also influence the dielectric constant of the medium and thereby change the strength of the coulombic interaction. In general the dielectric constant will decrease from a maximum of about 80 in pure water to a lower value. This will increase the ionic forces as is evident from Eq. 4-1. The strong adsorption of proteins to polystyrene ion exchangers might be partly due to a decreased dielectric constant in the hydrophobic microenvironment of the charged groups on the matrix.

The occasional importance of hydrogen bonds for adsorption to the matrix is evident from the chromatography of snake toxins described above where the retardation of the Ser mutants was certainly due to hydrogen bonding.

4.6   Experimental Planning and Preparations

4.6.1 Choice of mode

Ion exchange chromatography may be carried out in a variety of ways according to the properties of the sample and the objective of the separation. Its high capacity and ability to adsorb and concentrate proteins from dilute solution make it particularly useful in early stages of a separation scheme when both the sample volume and the protein mass are large. Batch adsorption in combination with step-wise elution is particularly appropriate as a preliminary step to reduce the sample volume and amount of bulk protein to quantities which are more easily handled in column chromatography.

Column chromatographic techniques are required if advantage is to be taken of the high resolution capabilities of ion exchange chromatography. Special small particle size media, as used in high performance chromatography, are necessary for the highest resolution, usually after a number of earlier fractionation steps. It is often inadvisable to apply crude extracts directly to high performance columns since the risk of column blockage is then relatively high. Excellent results are obtained with modern standard media if proper thought is applied to experimental design, although in some situations HPLC ion exchangers have definite advantages. These include analytical applications and cases where speed is essential, as when the protein of interest is rapidly degraded.

4.6.2   Choice of Ion Exchanger and Buffer

The outcome of IEC depends on the relative charges of the proteins in the sample at the pH at which the separation is performed. Since the sign of the charge on a protein, positive or negative, depends on the pH, it is clear that the choice of both the ion exchanger, cation or anion exchanger, and the buffer will be governed by the pH. The consequences of this choice can be seen from titration curves, e.g. Fig. 4-2.

If the sample is very heterogenous and of large volume, it is in general best to choose an ion exchanger where the protein of interest is adsorbed at the desired pH. The choice of cation or anion exchanger and buffer pH can then be determined by simple test-tube experiments to see which ion exchanger adsorbs the protein of interest at the chosen pH. 1–1.5 ml of ion exchanger equilibrated with starting buffer and the sample in the same buffer are mixed for 1–2 minute. After centrifugation, the supernatant is assayed for the protein. If the protein is only partially adsorbed the chosen conditions can be used for isocratic chromatography, and in fact the retention coefficient can be calculated from the distribution coefficient (see p. 57). Desorption conditions can be examined in the same way.

Alternatively, conditions can be selected to remove a minor contaminant from a partially purified protein by binding the contaminant. This 'subtractive' procedure allows the best use of the capacity of the column which can then be dimensioned according to the small amount of contaminant rather the total quantity of protein.

It is of course possible to use both anion and cation exchangers. The use of a mixture of the two, or two columns in series, is an established technique for deionizing water. This technique has found limited application in protein purification, but for example when bovine serum was adsorbed to a mixed bed ion exchanger at 10 mM phosphate buffer pH 7.5 the IgG_1 (pI = 7.5) was eluted 95 % pure, while the other proteins followed in a linear salt gradient[54,55]. Normally one chooses to run two consecutive chromatographies instead.

The choice of pH value and thus cation or anion exchange may be made more carefully to find the conditions which give the best separation from critical contaminants. Electrophoretic titration curves (Fig. 4-2) provide an excellent guide to both the ion exchanger and the pH of elution. In general however one performs a first chromatography with a linear standard salt gradient, preferably at the pH at which the protein of interest is most stable, and then finds the best resolution by modifying the slope of the gradient.

There is generally little to be gained from chosing a weak as opposed to a strong ion exchanger. As mentioned above, the significant difference between them is the pH range over which they retain their full charge. Possible specific effects of choosing, for example, a DEAE-anion exchanger as opposed to a strong quaternary ion exchanger are serendipitous and can hardly be predicted beforehand, although cases where weak ion exchangers were advantageous have been reported[30,41]. If extreme pH values are to be used, the strong ion exchangers are essential. Thus at pH values below 5 cation exchangers have to be sulphonated or phosphorylated, and at pH values above 9 anion exchangers containing quaternary amines should be used.

The porosity of the matrix influences the choice of ion exchanger. Larger protein molecules require media with larger pore sizes. The largest pores are available in the cellulose and agarose (Sepharose) ion exchangers. Exclusion limits are normally given by the manufacturer. Ion exchangers based on Sephadex have smaller pore sizes and may be advantageous for smaller proteins. When fractionating low molecular weight proteins from snake venom changing from SP-Sephadex C-25 (exclusion limit 30,000) to SP-Sephadex C-50 (exclusion limit 200,000) gave practically identical results. If the porosity is large enough, a further increase will not make a difference. The final choice between adsorbents with similar porosity, capacity and hydrophilicity is probably often arbitrary both with conventional and HPLC ion exchangers[30,32].

In some cases, mechanisms other than ion exchange will influence the separation. It has already been mentioned that methacrylate ion-exchangers, Bio-Rex and Amberlite, have a large resolving power for separations of mixtures of small basic proteins such as ribonuclease and toxins (see above and ref. 17, 57–59). Here hydrogen bonding and possibly also hydrophobic effects will contribute to protein adsorption. Phosphocellulose shows specific interactions with enzymes binding phosphate-containing substrates. Successful purifications of tRNA synthetases[60], glucose-6-phosphate dehydrogenase[61] hexokinase[61] and phosphoglucose isomerase[62] by adsorption to phosphocellulose and affinity elution have been reported.

If the sample size is large and fouling of the chromatographic column is expected at an early stage of the separation, a simple and inexpensive ion exchange material which can be discarded after use is recommended. This will not allow the very fast and high resolution separations that can be obtained with HPLC material, but these can be used later in the process.

The choice of buffer is discussed in detail in section 4.5.

4.6.3 Column Dimensions

The amount of the ion exchanger to be used is considerably larger, about 5–10 times, than that calculated from the protein binding capacity as explained on p. 120. The influence of the protein molecular weight (Section 4.4.2 and Table 4-5) should also be kept in mind. In practice, exact calculations are rarely made, especially since the amount of protein to be adsorbed is often not well known. However, the protein binding capacities of ion exchangers are high and columns are probably in general all too large. But this is hardly a serious disadvantage, except for economy. Thus phospholipase A2 from cobra venom adsorbed to CM cellulose CM 52 and desorbed by gradient elution produced identical chromatograms in a 2 cm and a 10 cm column.

If the ion exchange is carried out as a simple adsorption/desorption experiment, no more stationary phase is required than is needed to bind the protein(s) of interest. The sample will then not need travel through a long column after desorption. Bed lengths which are not more than four times the bed diameter are recommended. Short columns are also to be preferred in gradient elution. Columns to be used for isocratic elution should be longer. Here several proteins migrate on the column simultaneously and their resolution increases as the square root of the column length (Section 2.3.1.4).

4.6.4 Equilibration of Ion Exchanger

An ion exchanger will always have counter-ions at pH values where it is charged and thus ready to be used. The purpose of the equilibration procedure is to ensure that the ion exchanger is in equilibrium with the counter-ions which will be used during elution.

Most of the washing procedures for bulk ion exchangers are best carried out on a filter paper in a Buchner funnel. Pre-packed high performance ion exchange columns are readily equilibrated in the column. If washing protocols are not provided by the manufacturers, and assuming that the ion exchanger medium is not contaminated with lipids or precipitated proteins, the following general equilibration protocol may be used for a bulk ion exchanger:

(a) Rinse with water.
(b) Titrate all groups with 0.1 N NaOH in case of an anion exchanger and 0.1 M HCl in case of a cation exchanger. This step serves to eliminate all previous ions, and possibly proteins, by making the ion exchanger neutral.
(c) Rinse with water.
(d) Transfer to the correct counter-ion by means of the appropriate acid or base. Thus an anion exchanger is transferred to the chloride form by treatment with 0.1 M HCl and to the acetate form by 0.1 M or 1 M acetic acid while a cation exchanger is transferred to sodium form by treatment with 0.1 M NaOH. Note that DEAE-ion exchangers are unstable in the free base form and should be converted to the salt form after washing with strong base.
(e) Rinse with water.
(f) Transfer to 0.1–0.5 M starting buffer. Equilibrium with the pH of the starting buffer is reached much quicker if this buffer is concentrated. If this step is done batch-wise titrate to final pH with dilute acid or base. Check the pH of the eluent.
(g) Transfer to starting buffer of the final concentration. Check the pH and conductivity of the eluent until it is the same as for the starting buffer.

The methacrylate ion exchangers are preferably equilibrated batch-wise because of their high buffer capacity. If the ion exchanger is to be used with ammonium acetate the resin is washed with ammonium hydroxide, and then neutralized with acetic acid. After settling and decantation the resin is suspended in ammonium acetate of the desired concentration, and the pH adjusted with acetic acid or ammonium hydroxide. The procedure is repeated, usually once or twice, until the resin is equilibrated, i.e. until the pH does not change on the addition of a new portion of buffer.

It is important that the ion exchanger is equilibrated to the pH and ionic composition of the starting buffer if changes in the composition of the mobile phase during the chromatography are to be avoided. As the ionic strength of the medium increases, the counter-ions of the exchanger will be released and will influence the pH accordingly if they have significant buffer capacity at the pH of the experiment. These effects are especially pronounced with methacrylate resins.

4.6.5 Preparation of Sample

To ensure controlled binding, the sample should preferably be equilibrated with the starting buffer. This step might sometimes be omitted if the pH and ionic strength of the sample are such that proper protein binding is allowed. This can often be achieved simply by dilution and titration. For a small sample this offers no problem but for samples large enough to affect the equilibration of the column possible effects of sample application have to be considered.

As in all chromatography it is important that the sample is clarified as is described in Chapter 1.

4.6.6   Large Scale Ion Exchange

Ion exchange chromatography is eminently well suited to purification of either large quantities of proteins, because of its high loading capacity, or purification from large volumes of sample, because of its ability to concentrate proteins from dilute solution[63]. Although process scale ion exchange chromatography is routinely carried out with columns as large as 170 litres, even a standard laboratory column with a volume of 500 ml is usually sufficient to concentrate and purify 5 grams of protein from as much as 10 litres of a dilute sample. In practical terms, large scale ion exchange chromatography differs mainly in the increased use of step, as opposed to continuous gradient, elution procedures. Otherwise exactly the same general procedures and principles are used as in a smaller scale.

For obvious reasons, it is clearly desirable to be able to plan a large scale procedure at a smaller scale. This is done by carrying out a series of trial runs under carefully controlled and systematically varied conditions of sample loading and elution to find the set of experimental conditions which gives the desired result in terms of resolution, yield and time. Once this is done at a small scale, the conditions can be accurately extrapolated to the larger scale by application of a few simple scaling factors[64].

When the conditions of binding and elution, i.e. sample composition, ion exchanger, eluent compositions, pH, and loading factor (mg protein/ml ion exchanger), are kept constant, then the column dimensions, eluent volumes and flow rate are scaled up as follows.

Column volume is increased as the increase in sample volume. It is advisable to obtain this increase in volume by choosing a column with a larger diameter to avoid excessively long separation times.

Eluent volumes are increased as the increase in sample volume. This applies equally to step-wise elution schemes and to continuous gradient elution, and to all steps in the elution sequence.

The linear flow rate is kept constant. The volume flow rate will thus be increased by the same factor as the increase in the cross-sectional area of the column. If the large column is longer or shorter than the small one then the elution times will be increased or decreased accordingly.

It is important to note that these factors only apply if the sample loading, elution conditions and the ion exchanger are the same in the two scales. If the sample loading as mg protein/ml ion exchanger is changed in the scale up then the results will be more difficult to predict, particularly at the high loading capacities which are frequently used in large scale work. Changes in the elution scheme will have the same effects on the separation as they would have had without the change of scale. The need to use the same ion exchanger means that it is usually inadvisable to optimise the separation using high performance ion exchangers if subsequent large scale separations will use standard ion exchangers.

The choice of ion exchanger for large scale work follows the same principles as for small scale separations, with the added proviso that care should be taken to choose a gel matrix which will withstand high linear flow rates in large volume columns. The ion exchangers which have been produced for process scale ion exchange chromatography, for example Q-Sepharose and S-Sepharose, are also very suitable for column chromatography in a more usual laboratory scale. However, when batch adsorption is used to concentrate and simplify the sample (see below) there is no need for high physical stability. a simpler medium, for example DEAE-cellulose, will suffice.

Columns for volumes up to ca 1 litre can be of the conventional design. Because it is generally speaking desirable to use short, wide columns for ion exchange, very large columns should have end-pieces and which are specially designed to spread the incoming liquid flow evenly over a relatively large bed area. The multiple inlet design[65] is suitable.

Removal of particulate matter is specially important when large sample volumes are to be applied. Polypropylene depth filters (1 micron) which give a negligible pressure drop over the filter are available for this purpose. In the case of large sample volumes, it may be difficult or impossible to ensure that the sample is exactly equilibrated to the starting conditions for elution. In this case the sample ionic strength should be adjusted by dilution if necessary and the pH adjusted to a value which will give good binding. The column is then equilibrated with the starting buffer and eluted in the usual way. The choice of buffer systems for large scale work will naturally be influenced by cost considerations, but the needs for adequate buffering capacity remain and should not be sacrificed for small cost savings.

4.6.7 Batch-wise Procedures

Although batch-wise procedures are not strictly speaking chromatographic they are nonetheless well worth considering as an alternative to column operations in particular in connection with large scale applications. Batch-wise adsorption and desorption are also very useful for crude or viscous samples independent of scale.

The sample is mixed with gentle stirring with the ion exchanger in conditions which promote binding of the components of interest. The time taken for binding depends on the sample and on the ratio of the ion exchanger volume to the sample volume, and should be determined by activity measurements. As a rule of thumb, an hour is sufficient for protein solutions which behave normally. Unbound components are washed away on a filter funnel under the same conditions; and the bound proteins are then desorbed by a step-wise increase in salt concentration or by a change in pH. This can be done either on the funnel or after the ion exchanger has been packed in a column. If the volume of eluting buffer is kept small a considerable increase in concentration of the protein can be achieved, in particular if a column is used.

After recovering the protein, the ion exchanger may be regenerated, washed with distilled water and re-dried for further use providing it is not heavily contaminated with particulate or lipid material. If this is the case it is often best to simply discard it. However, if one wants to recover the ion exchanger procedures for washing with alkali and ethanol are given below.

The choice of cation or anion exchanger is made so as to achieve binding of the desired components, but it is not necessary to use the most sophisticated matrices. Indeed, inexpensive bulk ion exchangers which can be added in the dry form, for example those based on cellulose or Sephadex, are often an excellent choice. The volume changes associated with ion exchangers based on Sephadex G-50 are even an advantage since the shrinkage of the ion exchanger at high ionic strength improves recovery. Note that some ion exchangers will generate fines if they are stirred too vigorously and this will increase the time it takes for filtration.

4.7   Chromatographic Techniques

4.7.1   Sample Application

In column procedures the sample solution is applied to the column by pumping. The proper flow rate for adsorption depends on several factors. In general it can be quite high if the protein of interest is far away from desorption conditions in the starting buffer. However for samples that are viscous it is wise to reduce the flow rate during the application or to dilute the sample. This generally applies to protein solutions more concentrated than about 30 mg/ml, although only experience of the particular sample and experimental set-up can give definite answers.

In general flow rates can be expressed in the dimension length per time, most commonly cm/h. Consequently the maximal volume flow rate increases as the square of the column diameter, given that other factors are constant. Dextran ion exchangers can be run at about 15 cm/h, whereas the high performance media are typically run at 1 ml/ min corresponding to about 300 cm/h. The rigid agarose and cellulose ion exchangers stand very high flow rates. For adsorption and elution flowrates up to 50 cm/h are used. For equilibration much higher rates can be used. Flow rates have to be decreased by about 50% when changing to cold room temperature due to increased viscosity. In general flow rate is a parameter that should be checked if maximum resolution is of importance.

At this stage it is sometimes possible to see protein bands on the column with the naked eye. The banding is due to protein-protein displacement (see p. 57). The proteins that bind most strongly to the ion exchanger are found closest to the top of the column.

After application the column is washed with starting buffer. The proteins that do not adsorb in starting buffer are recovered in a break-through peak characterized by its sharp front side and tailing end. Normally washing continues until the UV absorption in the eluate decreases to a low value. However, no more than two column volumes should be necessary and in the best case a single column volume suffices. The proteins that elute in the breakthrough peak have a retention of $k' = 1$.

4.7.2   Elution Techniques

The proteins that adsorb to the ion exchanger are eluted from the column either by an increase in ionic strength, by inclusion of new ionic species or by a proper change in pH. It is common practice to increase the concentration of a non-buffering salt, such as NaCl, for elution. Increasing the buffer concentration, might have an undesirable effect on pH. However, as already mentioned exceptions to these rules occur, as when proteins are eluted from either anion or cation exchangers by an increase in the concentration of ammonium acetate or other volatile buffer, or an anion exchanger is eluted by increasing phosphate concentration.

If a pH change is used, anion exchangers should be eluted by a decrease in pH, to make the adsorbed proteins less negative, while cation exchangers are eluted by an increase in pH to make the proteins less positively charged. pH gradients are less often used, but are not necessarily inferior to salt gradients. However, it might be a better alternative to run two consecutive ion exchange chromatographies at different pH values.

The changes can be brought about either step by step, step-wise elution, or gradually, gradient elution. If the chromatography is being run with constant solvent composition and the sample proteins leave the column more or less retarded, it is called isocratic elution. If a protein is being displaced due to its specific interaction with an added ionic species in the elution buffer, this is called affinity elution. Finally it is possible to add a general eluting agent to the buffer that displaces all proteins, which will move ahead of the displacer in the column. This is called displacement chromatography.

To make the best use of the resolving power of ion exchange, a protein should be submitted to chromatography in such a way that it elutes retarded from the column. This normally occurs only in gradient elution. Properly speaking it is only then that the protein chromatographs as opposed to just being adsorbed and then desorbed.

4.7.2.1   Isocratic and Step-Wise Elution

Isocratic elution is used in cases where the sample and its properties are well known and the same kind of sample is run repeatedly, i.e. mostly for analytical separations. To achieve proper reproducibility great care has to be taken to ensure that the column is properly equilibrated and the sample ion composition does not vary too much. A well known example of isocratic elution in an analytical context is the Moore and Stein method of amino acid analysis.

In isocratic elution all components of the sample move simultaneously. Only sample volumes much smaller than the total bed volume can be applied. Separation power increases with the square root of the column length (see p. 51). The peaks widen later in the chromatogram and finally become so wide that they cannot be detected. After several bed volumes of elution buffer—the practical upper limit is 3 to 10 bed volumes—the chromatography has to be interrupted.

Step-wise elution is in fact the serial application of several isocratic chromatographies. However, at each step or buffer concentration only about a single total bed volume of eluent is passed through the column and the breakthrough peak is collected. Thus only protein adsorption/desorption occurs in this case. The step-wise elution mode is mainly used for recovering a specific component in a concentrated peak. Here is it also best if the elution conditions for the desired component in the sample are known. The application conditions are then chosen such that this component just stays bound to the column while contaminant proteins, often bulk proteins, pass through. After washing, the salt concentration or pH value is changed so that the protein of interest is just eluted in a break-through peak with displacing buffer (see application 4.9.6).

The step-wise elution protocol can give a few disagreeable surprises, some of them illustrated in Fig. 4-5. Since a retarded peak is eluted with one buffer, the break-through peak eluted after the following buffer change might well contain the same protein. In fact if the capacity of the column was inadequate the second component might also be found in the first break-through peak. Secondly, as each step produces a new break-through peak the resolving power of the chromatography might not be used optimally with several components eluting together. In general there is a problem to master when choosing the right conditions for step-wise elution. If the buffer is too weak the protein might elute in a broad retarded peak. On the other hand if it is too strong it might elute together with several other components. In general one should avoid stepwise elution of an unknown protein sample and instead use gradients in order to obtain a better picture of its composition.

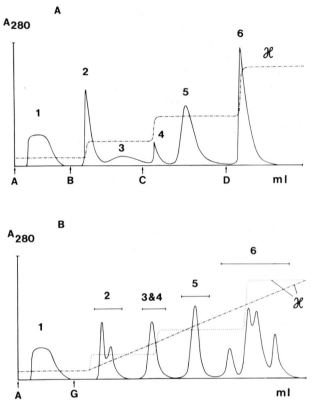

Figure 4-5. Elution of an ion exchange column (A) stepwise and four buffers A-D with increasing ionic strength (broken line), and (B) with a gradient. The gradient starts at G. The conductivity curve from A is inserted. The figure, which summarizes the experience from several experiments, illustrates possible drawbacks of the step-wise elution protocol: poor resolution of several components (2 and 6) and the false separation of one component in two peaks (3 and 4).

4.7.2.2 Gradient Elution

The gradient elution mode has the advantage of the maximal separation obtained in isocratic elution, but because of the continuously increasing elution power of the medium the peaks do not become much broader as the gradient develops. Tailing due to non-linear adsorption isoterms is also diminished. Gradients are the classical answer to the "general elution problem" (see p. 56). There is in principle under a gradient no possibility to recover a single substance artefactually in several peaks, and a maximal use of the resolving capacity of the column is achieved.

Gradients are obtained by mixing starting and final buffer so that the percentage final buffer pumped into the column is gradually increased. Modern chromatographic pumps are equipped with gradient devices that can be programmed to give any conceivable gradient. Simpler gradient mixers can easily be constructed by connecting two cylinders by a tubing. One cylinder, equipped with a stirrer, contains the starting buffer of concentration C_s and the other cylinder final buffer of concentration C_f. The eluting buffer is continuously withdrawn from the stirred chamber. The concentration of the eluent C_v after v ml from a total of V ml is described by the equation[66]

$$C_v = C_s - (C_s - C_f) \cdot (1 - v/V)^{A_f/A_s} \qquad (4\text{-}2)$$

Here A_f and A_s are the cross-sectional areas of the two cylinders. A linear gradient is obtained for $A_f/A_s = 1$, a concave if this measure is larger than 1 and a convex if it is smaller than 1 (Fig. 4-6). The total volume of a gradient might vary but as a rule of thumb they can be made ten times the column volume as a first try.

Linear gradients are recommended as a first choice. When more experience of a particular protein sample is at hand it might be judged advantageous to make a gradient more shallow in areas where peaks are less well resolved. Similarly concentration ranges without proteins eluting can be passed by a steep part of the gradient. However protein resolution problems should not normally be solved by gradient tinkering, but rather by changing elution conditions such as pH (Fig. 4-2 and section 2.3).

4.7.2.3   Affinity Elution

In affinity elution[11,28,60–62,67–70] adsorption of a protein is non-specific but desorption is specific. The specific eluent is normally a substrate ion and the name

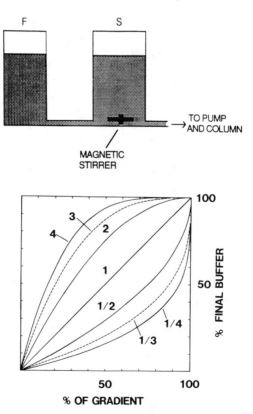

Figure 4-6. *Top.* Simple gradient mixer of two connected cylindric vessels with cross sectional areas A_s and A_f, containing starting (S) and final (F) buffer respectively. *Bottom.* Shape of gradient obtained for different values of A_f/A_s.

substrate elution is sometimes also used for this elution mode. The eluent has a charge of the same sign as the ion exchanger and binds specifically and strongly to charges, often in the active site, of opposite sign on the protein. An early example of affinity elution is the desorption of fructose 1,6-diphosphatase from a column of CM-cellulose with the negatively charged substrate 1,6-disphosphate[11,28]. The enzyme consisting of four subunits could bind four substrate molecules and thus decreased its net charge by 16 units, enough to elute it from the cation exchanger.

Scopes studied affinity elution of a number of enzymes[11,26,28,62,69,70]. He found that on CM-cellulose the inclusion of a charged ligand had the same elution power as an increase in pH of 0.5—0.8 units at constant ionic strength. To avoid non-specific elution a "dummy ligand", with similar ionic properties, was included in the buffer before the affinity ligand. Thus EDTA served as a dummy ligand for phosphoenolpyruvate. For a successful result he found that the number of charges should be at least 4 per 100,000 of molecular weight of protein unless conformational changes facilitate desorption, and that the ligand-protein dissociation constant should be below millimolar.

4.7.2.4 Displacement Chromatography

In displacement chromatography the column is eluted with a substance, the displacer, with a high affinity for the adsorbent while other substances move at the same velocity through the column ahead of the displacer forming rectangular zones that follow each other in a train. Displacement elution is not identical to isocratic elution, since there the components do not displace each other, but move with different rates through the column. A difference between displacement chromatography and step-wise elution is that in the latter case one often searches for conditions where the protein of interest is barely eluted while other components stay behind. Displacement analysis has rarely been used in ion exchange due to difficulties of finding suitable displacers. A few examples will however be mentioned.

In polyion displacement chromatography a polyion with high affinity for the ion exchanger is used as displacer. With carboxymethyldextrans[71] and chondroitin sulfates[72] as polyanion displacers, a 100 mg sample of β-lactoglobulin was fractionated into two rectangular zones of the A and B forms after adsorption to a 3.5 ml column of TSK DEAE equilibrated in sodium phosphate pH 7.0. This method had a two-fold higher capacity as compared to conventional ion exchange.

In ampholyte displacement chromatography, ampholytes intended for isoelectric focusing are used as displacers. Thus two forms of a β-N-acetyl-D-hexoseaminidase were separated by elution with a 4% solution of ampholytes pH 8-10 from a column of CM-cellulose equilibrated in 0.02 M Tris HCl pH 7.6[73]. These forms of the enzymes were not resolved by either isoelectric focusing or ion exchange chromatography. As a pH gradient was obtained in the beginning of the chromatogram it is likely that the separation depended both on both chromatofocusing and displacement chromatography[71].

4.7.3 Collection and Treatment of Sample

Pooling of fractions from a chromatogram is best guided by a specific measurement of activity or an equivalent property of the protein of interest. If only absorbances or another general protein content measurements are available, pooling is more difficult. Figure 4-7 illustrates the cross-contamination of two overlapping peaks of both equal

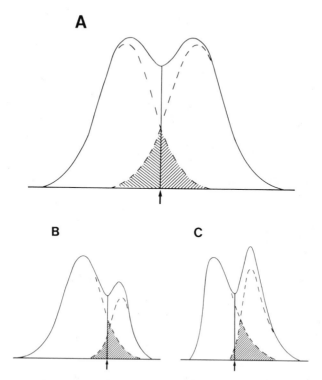

Figure 4-7. Estimation of cross-contamination in two overlapping peaks of equal or unequal size and shape. (A). Two equally large gaussian profiles. If pooled as indicated by the arrow, the crosscontamination is about 10%, much less than judged by the eye. The crosscontamination, C, can in general be found from a measurement of the resolution R_s (see Chapter 2). Thus $R_s = 0.9$ gives $C = 4\%$; $R_s = 0.8$, $C = 6\%$; $R_s = 0.7$, $C = 9\%$; $R_s = 0.6$ $C = 13\%$ and $R_s = 0.5$, $C = 19\%$. (B) A smaller peak is more contaminated than a larger one. (C) A trailing peak (here the first one) increases the contamination of the neighbouring peak.

and unequal size and shape. The underlying assumption here is that the absorbancies reflect the amount available. This might be a reasonable approximation if absorption in the low UV region is used since then the peptide bond is the main chromophore. At the 280-band absorption coefficients vary considerably.

Concentration and drying of pooled fractions is most easily done by lyophilization. This requires that a volatile salt has been used in the buffer or that a change is made to such a buffer (Table 4-6). Lyophilization from concentrated buffers or acetic acid is not recommended. Ammonium acetate is an excellent buffer for lyophilization, but dilution to 0.2 M or less and adjustment of pH to 7–8 is recommended. It is often best to lyophilize twice, dissolving the sample in 0.01 to 0.02 M buffer for the second time. Sonication is recommended to bring proteins adsorbed to walls of vessels into solution.

It is often worthwhile re-running a protein that has presented an asymmetric peak or that has eluted close to another component in the chromatogram. If a contaminant is present in smaller amounts it is much easier to remove from the main component. A re-run also confirms that a component is homogenous with regard to the separation parameter used.

It is more tricky when an originally homogenous protein is presenting a false heterogeneity in a chromatogram. This is not unusual in stepwise elution as described above. However it can happen also in the most perfectly conducted gradient chromatography for various reasons, in particular when a protein complex dissociates on the column and the two components elute in different positions. This can be seen as a complete loss of biological activity. Several presynaptic neurotoxins consist of subunits with very different charge properties held together by non-covalent forces. These dissociate on cation exchange chromatography[74–78]. If one component in a complex is not seen in the monitoring (e.g. by not having 280 absorption), the search for a possible separated component, possibly even of low molecular weight, might be very difficult and only possible by systematic pooling experiments.

Finally some observations suggest that it is possible to obtain multiple peaks also of homogeneous proteins (Karlsson, unpublished), especially when overloading the column. In such situations re-chromatography should be performed. Experience also suggests that binding a protein to the column far below its desorption concentration might give rise to artefacts and several peaks arising from protein denaturation.

4.7.4 Regeneration, Cleaning and Storage of Adsorbent

After each use the ion exchangers usually have at least some material bound, often denatured protein. Depending on the samples there might also be lipids. It is important to remove this material before the ion exchanger is used again. Remaining material will block the adsorption to the column, impair its performance and might even contaminate the sample.

Regeneration procedures vary depending on the stability of the matrix and the functional group. Manufacturers generally describe how their products should be handled, and only a few comments will be given here. The agarose ion exchangers are washed with high salt and then NaOH up to 0.5 M for maximum overnight. The cellulose ion exchangers are washed with only 0.1 M NaOH. Exposure of dextran and agarose to pH below 3 should be avoided and acid is titrated to this pH before washing. Trisacryl adsorbents are however stable in pH 1–13. Lipoproteins and membrane lipids can be removed by non-ionic detergents or ethanol. The HPLC media are regenerated in the column. Mono Q and Mono S can withstand high concentrations of alkali (2 M). Silica based adsorpbents are unstable to alkali and pH above 8 should be avoided. Recommendations for the care of HPLC media have been published[79,80].

Adsorbents that are to be stored for a long time are best transferred to 25 % ethanol containing bacteriostats. HPLC media can likewise be stored in alcohol up to 100 %.

4.8 Hydroxyapatite Chromatography

Calcium phosphate had been used in protein adsorption applications with varying success for many decades when Tiselius and co-workers[81] showed that a much more useful adsorbent was obtained when the brushite, $Ca_3(PO_4)_2$, was changed into hydroxyapatite, $Ca_5(PO_4)_3OH$, by boiling in 1 % NaOH for one hour. The hydroxyapatite crystals generated fines more easily, but these were removed by decantation before column packing.

In hydroxyapatite chromatography the column is normally equilibrated and the sample applied in a low concentration of phosphate buffer, pH 6.8, and the proteins eluted in a concentration gradient of phosphate buffer. Sometimes quite shallow gradients (e.g. 0.02 to 0.05 M) give excellent results[82], while in other instances concentrations up to 0.35 M phosphate are used for elution[83] (application 4.9.7). The presence of other salts such as NaCl and $(NH_4)_2SO_4$ seems not to affect the elution. This can be taken advantage of by applying a sample from an ion exchange column eluted by a NaCl gradient directly to the apatite column. It had also been used for the purification of halophilic enzymes in the presence of 3.4 M NaCl[84]. Some other anions, e.g. citrate, can be used for the elution of apatite columns.

Hydroxyapatite seems to be particularly useful for the purification of medium and high molecular weight proteins and nucleic acids[85]. Low molecular weight solutes show in general very low affinity for hydroxyapatite.

The high selectivity of hydroxyapatite appears to be due to the competition for calcium ions on the crystal surface. This agrees with the finding that the content and distribution of acidic groups, in particular carboxylates, on the surface of proteins is a major determinant of protein binding to hydroxyapatite[86]. Even if in general proteins that chromatograph well on anion exchangers also bind to hydroxyapatite, the two methods do not have the same selectivity. Thus isoforms of human ceruloplasmin were readily separated on apatite but not by conventional ion exchange chromatography[16]. A special possibility is to use the technique for purification of phosphoproteins.

A drawback of hydroxyapatite is its mechanical and chemical instability. This has recently been partly overcome by the introduction of ceramic fluoroapatite. This is available as 10 micron average diameter spherical particles in prepacked HPLC columns[87,88]. Apparently the fluoroapatite has the same chromatographic properties as the hydroxyapatite.

4.9   Applications

4.9.1 Isolation of a Fungal Cellulase by Chromatography on DEAE-Sepharose[89]

The culture filtrate of the fungus *Trichoderma reesei* constitutes an excellent source for the preparation of cellulose-degrading enzymes. After desalting on Sephadex G-25 equilibrated with 0.01 M ammonium acetate, pH 5.0. (the ion exchange starting buffer) the filtrate was applied to a column (2.6 × 30 cm) of DEAE-Sepharose CL-6B, equilibrated with a 0.01 M ammonium acetate, pH 5.0. A 1000 ml linear gradient of ammonium acetate at pH 5.0 from 0.01 to 0.5 M and a flow rate of 40 ml/h was used to develop the column. Three different activities, called A, B and C, were separated (Fig. 4-8).

Component C, the most acidic one, was collected, lyophilized and dissolved in 0.05 M ammonium acetate pH 3.7 and applied on a column (2.6 × 30 cm) of DEAE-Sepharose CL-6B in the same buffer. This time the column was eluted with a 700 ml linear gradient of ammonium acetate at pH 3.7 from 0.05 to 0.3 M. In this chromatographic component C was further separated into two cellobiohydrolases, both pure (Fig. 4-8). Even though the two proteins coeluted at pH 5.0, they separated at the lower pH.

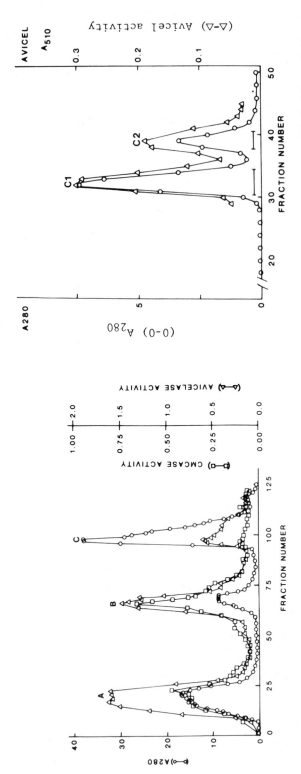

Figure 4-8. Separation of cellolytic enzymes form the fungus *Trichoderma reesei* by ion exchange chromatography on DEAE-Sepharose CL-6B (left) at pH 5.0 and (right) rechromatography of component C at pH 3.5. Details are given in the text. Reproduced with permission from Ref. 89.

4.9.2   Isolation of Immunoglobulins from Egg Yolk[90]

Immunoglobulins can be obtained from immunized hens directly from the egg yolk in a high concentration. Drawbacks are the lower stability of chicken IgG, as compared to rabbit IgG, and the fact that they do not bind to Protein A or G.

Egg yolk was collected and suspended in two volumes of 3.75% polyethyleneglycol (PEG) in 10 mM sodium phosphate buffer, pH 7.5 containing 0.1 M NaCl. After centrifugation the IgG in the supernatant was precipitated by increasing the PEG concentration to 12%. The pellet was dissolved in 10 mM Tris HCl buffer, pH 8.0 containing 0.05 M NaCl and applied to a column (2 × 15 cm) of DEAE-Sephacel in the same buffer. The column was washed with ten column volumes of starting buffer containing 0.5% Triton X-100 to elute lipid and nonadsorbed material, followed by starting buffer without Triton X-100. It was then developed with a 150 ml linear gradient of 0.05 to 0.25 M NaCl in the Tris buffer. The single main peak contained the chicken IgG in 95% or better purity.

4.9.3   Isolation of α-Bungarotoxin From Snake Venom by Chromatography on CM-Sephadex C-50[91]

The venom of the Taiwanese snake krait, *Bungarus multicinctus*, contains a neurotoxin, α-bungarotoxin, widely used as a marker of the nicotinic acetylcholine receptor.

About 1 gram of lyophilized venom was dissolved in 15 ml of 0.05 M ammonium acetate, pH 5.0 and applied to a column (2.5 × 85 cm) of CM-Sephadex C-50 equilibrated in the same buffer. The column was developed with 1) 500 ml of starting buffer 2) a 1400 ml combined pH and ionic strength linear gradient of ammonium acetate, using 1 M buffer pH 7.0 as the final buffer 3) 800 ml of final buffer.

At least fourteen components were resolved in the chromatography (Fig. 4-9). α-Bungarotoxin eluted as component 3 in the first part of the gradient. It was obtained in a pure form after subsequent chromatography on CM-cellulose where it eluted in a gradient of ammonium acetate 0.05 M, pH 5.0 to 1 M, pH 7.0.

The volume of the CM-Sephadex C-50 decreased by about 50% during the elution but apparently without any adverse effects. It is likely that a far shorter column would have been enough for the separation. It is also likely that rechromatography on the same column would have been as successful as the use of the cellulose ion exchanger for the second run. The chromatogram gives an indication of the complexity of snake venoms.

4.9.4   Separation of Six Monoacetyl Derivatives of a Cobra Neurotoxin by Chromatography on the Polymethacrylate Ion Exchanger Bio-Rex 70[17]

The functions of amino acid side chains in proteins can be studied by chemical modifications. In the chemical reactions used several different side chains are normally derivatized especially if they do not differ considerably. Meaningful interpretations of the results in such cases requires that the protein derivatives are purified and assayed separately. A case in point is the acetylation of a neurotoxin from the cobra *Naja naja siamensis*. This toxin is a small basic protein with 71 amino acid residues in a single peptide chain crosslinked by five disulfide bridges. It contains five lysine residues and a free N-terminus, all of which can react with acetic anhydride. An amount of reagent

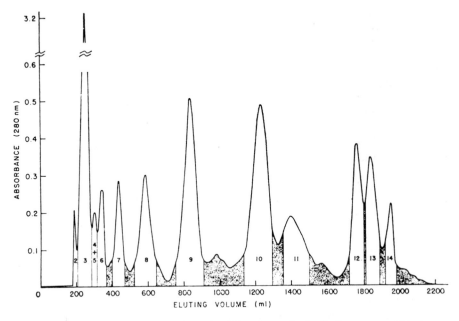

Figure 4-9. Fractionation of venom from the Taiwanese snake krait, *Bungarus multicinctus*, on Sephadex C-50. The chromatography is described in application 4.9.3. Reproduced with permission from Ref. 91.

corresponding to 1/12 of total amino groups produced a mixture of monoderivatives that could be fractionated on the polymethacrylate ion exchanger Bio-Rex 70 in spite of the fact that they had identical net charge.

The derivatized protein was dissolved in 0.02 M ammonium acetate pH 6.7, and applied to a column (2 × 30.5 cm) of Bio-Rex 70 equilibrated in 0.20 M ammonium acetate, pH 6.50. The column was first eluted with 0.05 M ammonium acetate, then with a concave gradient of 0.05 M to 1.4 M ammonium acetate.

In the chromatography (Fig. 4-10) underivatized toxin eluted last (peak 1), as expected since it has all its positive charges intact, preceeded by six peaks of the six

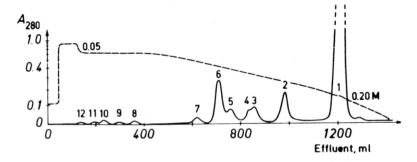

Figure 4-10. Fractionation of acetyl derivatives of a neurotoxin from the cobra *Naja naja siamensis* on a column of Bio-Rex 70 using a gradient of ammonium acetate. The experiment is described in application 4.9.4. Reproduced with permission from Ref. 17.

possible different monoacetyl derivatives (peaks 2–7). These were later identified by peptide mapping[92]. Minor amounts of di- and tri-acetyl derivatives (peaks 8–12) are seen in the beginning of the chromatogram.

4.9.5   Fractionation of Glycolytic Enzymes From Chicken Muscle by Consecutive Chromatographies on the HPLC Ion Exchangers Mono Q and Mono S[20]

Complex samples from natural sources often require the systematic use of several different selectivities to obtain pure proteins. The use of sequential anion and cation exchange chromatography for the purification of creatine kinase from chicken muscle is a case in point. The sample was initially studied by electrophoretic titration curves (Fig. 4-2).

A low ionic strength extract of chicken muscle was transferred to the starting buffer for the first, cation exchange, step, by buffer exchange on Sephadex G-25. The sample was filtered (0.22 micron) and 800 microL was applied to a Mono S HR10/10 column. The column was eluted with 0.05 M MES, pH 6.0 and a gradient from 0–0.5 M NaCl. Creatine kinase and phosphoglyceromutase eluted in the break-through peak with several other enzymes being resolved in the gradient (Fig. 4-11). The break-through peak was collected and transferred to the starting buffer for the anion exchange step by gel filtration on Sephadex G-25. 1 ml of the sample was applied to a Mono Q HR5/5 and eluted with 0.02M Tris, pH 8.0 and a gradient from 0–0.5 M NaCl (Fig. 4-11). Creatine kinase was obtained in good purity, well separated from phosophoglyceromutase.

This application demonstrates the ability to make an informed choice of conditions for ion exchange chromatography by studies of the electrophoretic titration curves as well as the power of modern high performance ion exchange chromatography.

4.9.6   Large Scale Preparative Purification of Human Serum Albumin and IgG by a Step-Wise Elution Protocol[93]

Highly purified human blood plasma proteins are needed for a variety of scientific and therapeutic purposes. The usual procedure for large scale fractionation of human blood plasma proteins is the Cohn cold ethanol procedure. An alternative method uses a combination of anion and cation exchange chromatography starting from cryoprecipitated and Factor IX-depleted plasma.

After centrifugation and buffer exchange to 25 mM sodium acetate buffer, the pH is adjusted to pH 5.2 with acetic acid to allow the euglobins to precipitate. The plasma is then centrifuged and filtered before application to a column of DEAE-Sepharose equilibrated in this buffer at a loading of approximately 35 g protein per liter ion exchanger. After eluting IgG under starting conditions, albumin is displaced by step-wise elution with the same buffer at pH 4.5 (Fig. 4-12) and applied directly to a column of CM-Sepharose equilibrated in this buffer. Highly purified albumin (>99% cellulose acetate electrophoresis) is obtained by elution with 0.11 M sodium acetate buffer, pH 5.5.

4.9.7   Separation of Human Ceruloplasmin Isoforms by Hydroxyapatite Chromatography[83]

Human ceruloplasmin, a blue copper glycoprotein, occurs in serum in two forms that differ only by their carbohydrate content[16]. These forms could not be separated by

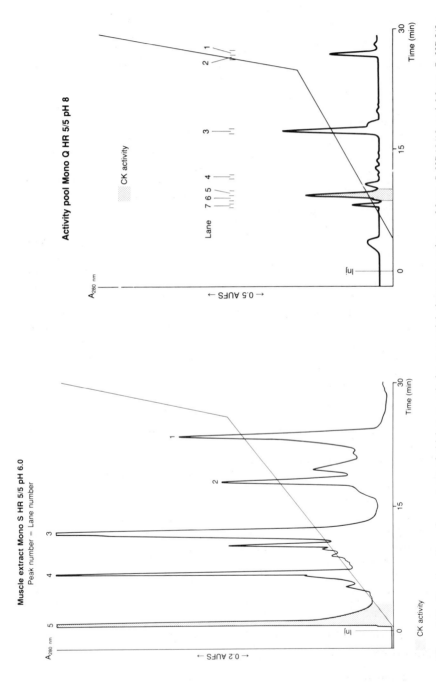

Figure 4-11. Fractionation of an extract of chicken muscle by sequential chromatography on Mono S HR10/10 and Mono Q HR5/5. Details are given in the text. Reproduced with permission from Ref. 20. Electrophoretic titration curves for the sample are shown in Fig 4-2.

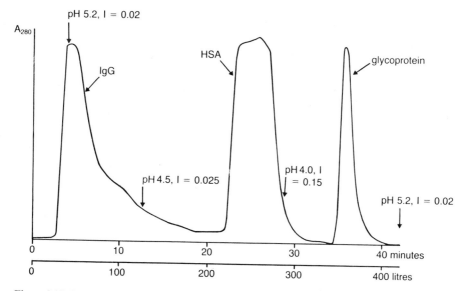

Figure 4-12. Large scale fractionation of human plasma proteins by step-gradient elution from DEAE-Sepharose CL-6B. Details are given in the text.

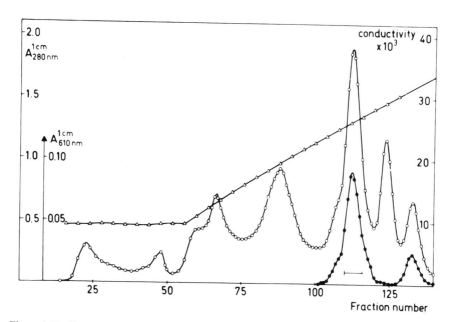

Figure 4-13. Chromatography on hydroxyapatite of a ceruloplasmin-enriched fraction obtained from ion exchange on DEAE cellulose of human serum. For details see text. The cerulopasmin isoforms are monitored by absorption at 610 nm (filled circles). Reproduced with permission from Ref. 83.

ion exchange chromatography, but hydroxyapatite was shown to be a perfect adsorbent for this purpose.

Ceruloplasmin was extracted from serum by batch-wise adsorption of DEAE cellulose. After washing the gel with starting buffer on a funnel the protein was eluted by a fourfold increase in ionic strength. After dilution the sample was applied on a column of DEAE cellulose equilibrated in 0.075 M potassium phosphate buffer pH 6.8, and eluted by a NaCl gradient to 0.45 M. The pooled ceruloplasmin peak, containing about 0.25 M NaCl, was applied directly (on a column (2 × 75 cm) of hydroxyapatite prepared as described by Tiselius and co-workers[81], and equilibrated in 0.075 M potassium phosphate pH 6.8. The column was then developed with a 1000 ml linear gradient of potassium phosphate up to 0.5 M at a flowrate of 12 ml/h (Fig. 4-13). The proteins that coelute on DEAE cellulose separate. The two ceruloplasmin forms contain about 11 % (major form) and 7 % (minor form) total carbohydrate.

4.10   References

1. J. Bonnerjea, S. Oh, M. Hoare, P. Dunhill, *Biotechnology*, 4, 954–958 (1986).
2. H. A. Sober, E. A. Peterson, *J. Am. Chem. Soc.* 76, 1711–1712 (1954).
3. E. A. Peterson, H. A. Sober, *J. Am. Chem. Soc.* 78, 751–755 (1956)
4. H. A. Sober, F. J. Gutter, M. M. Wycoff, E. A. Peterson, *J. Am. Chem. Soc.*, 78, 756–763 (1956).
5. S. Paléus, J. B. Neilands, *Acta Chem. Scand.*, 4, 1024–1030 (1950).
6. W. H. Stein, S. Moore, *J. Am. Chem. Soc.*, 73, 1893 (1951).
7. C. H. W. Hirs, W. H. Stein, S. Moore, *J. Biol. Chem.*, 200, 493–506 (1954).
8. H. H. Tallan, W. H. Stein, *J. Biol. Chem.*, 200, 507–514 (1954).
9. S. Moore, W. H. Stein, *Adv. Protein Chem.*, 11, 191–236 (1956).
10. W. Kopaciewiecz, M. A. Rounds, J. Fausnaugh, F. E. Regnier, *J. Chromatogr.*, 266, 3–21 (1983).
11. R. K. Scopes, E. Algar, *FEBS Lett.*, 106, 239–242 (1979).
12. D. L. Brautigan, S. Ferguson-Miller, E. Margoliash, *J. Biol. Chem.*, 253, 130–139 (1978).
13. R. E. Creighton, *Proteins. Structures and Molecular Principles*; Freeman: New York, 1983, Chapters 1 and 2.
14. L. Rydén, in *Protein Recognition of Immobilized Ligands*; W. T. Hutchins, ed., UCLA Symposium 1987, Alan R. Liss Inc: New York, 1989. in print.
15. T. H. Plummer, C. H. W. Hirs, *J. Biol . Chem.*, 238, 1396 (1963).
16. L. Rydén, in *Copper Proteins and Copper Enzymes*; R. Lontie, ed., Vol III; CRC Publishers, Boca Raton, Fl., 1984. pp. 37–100.
17. E. Karlsson, D. Eaker, G. Ponterius, *Biochim. Biophys. Acta*, 257, 235–248 (1972).
18. P. G. Righetti, T. Caravaggio, *J. Chromatogr.*, 127, 1–28 (1976).
19. P. G. Righetti, G. Tudoe, K. Ek, *J. Chromatogr.*, 220, 115–194 (1981).
20. L. Fägerstam, L. Söderberg, L. Wahlström, U-B. Fredriksson, K. Plith, E. Walldén, in *Protides of the Biological Fluids*, H. Peeters, ed., Vol. 30, Pergamon Press, Oxford, 1982, pp 621–628. and L. Fägerstam, unpublished.
21. N. K. Boardman, S. M. Partridge, *Biochem, J.*, 59, 543–552 (1955).
22. E. Karlsson, unpublished observations.
23. C. J. O. R. Morris, P. Morris, *Separation Methods in Biochemistry*, 2nd ed., Pitman, London, 1976, Chapter 5.
24. J. Porath, L. Fryklund, *Nature*, 226, 1169–1170 (1970).
25. J. Porath, N. Fornstedt, *J. Chromatogr.*, 51, 479–489 (1970).
26. R. K. Scopes, *Protein Purification. Principles and Practice*, 2nd ed., Springer-Verlag: New York, 1987.
27. H. Determan, N. Meyer, Th. Wieland, *Nature*, 223, 499–500 (1969).
28. R. K. Scopes, *Anal. Biochem.*, 114, 8–18 (1981).
29. H. A. Sober, G. Kegeles, F. J. Gutter, *J. Am. Chem. Soc.*, 74, 2734–2740 (1952).
30. F. E. Regnier, *Chromatographia*, 24, 241–251 (1987).
31. F. E. Regnier, *Meth, Enzymol.*, 104, 170–189 (1984).
32. M. A. Rounds, W. D. Rounds, F. E. Regnier, *J Chromatogr.*, 397, 25–38 (1987).

33. F. E. Regnier, *Anal. Biochem.*, *126*, 1–7 (1982).
34. M. A. Rounds, W. Kopaciewicz, F. E. Regnier, *J. Chromatogr.*, *362*, 187–196 (1986).
35. D. J. Burke, J. Keith Duncan, L. C. Dunn, L. Cummings, C. J. Siebert, G. S. Ott, *J. Chromatogr.*, *353*, 425–437 (1986).
36. D. J. Burke, J. Keith Duncan, C. J. Siebert, G. S. Ott, *J. Chromatogr.*, *359*, 533–540 (1987).
37. J. Keith Duncan, A. J. C. Chen, C. J. Siebert, *J. Chromatogr.*, *397*, 3–12 (1987).
38. J. L. Liao, S. Hjertén, *J. Chromatogr.*, *457*, 175–182 (1988).
39. E. A. Peterson, *Meth. Enzymol.*, *5*, 3–22 (1988).
40. M. L. Heinitz, L. Kennedy, W. Kopaciewicz, F. E. Regnier, *J. Chromatogr.*, *443*, 173–182 (1988).
41. W. Kopaciewicz, F. E. Regnier, *Anal. Biochem.*, *133*, 251–259 (1983).
42. K. M. Gooding, M. N. Schmuck, *J. Chromatogr.*, *266*, 633–642 (1983).
43. K. M. Gooding, M. N. Schmuck, *J. Chromatogr.*, *296*, 321–328 (1984).
44. M. A. Rounds, F. E. Regnier, *J. Chromatogr.*, *283*, 37–45 (1984).
45. K. Yao, S. Hjertén, *J. Chromatogr.*, *385*, 87–98 (1987).
46. M. T. W. Hearn, A. N. Hodder, M. I. Aquilar, *J. Chromatogr.*, *443*, 97–118 (1988).
47. *FPLC Ion Exchange and Chromatofocusing. Principles and Methods*, Pharmacia, Uppsala, 1985.
48. F. B. Rudolph, B. F. Cooper, J. Greenhut, in *Progress in HPLC*, H. Parvez, Y. Kato, S. Parvez, eds., Vol. 1, VNU Science Press, Utrecht, 1985, pp. 133–147.
49. L. Söderberg *et al.* in *Protides of the Biological Fluids*, H. Peeters, ed., Vol. 30, Pergamon Press, Oxford, 1982, pp. 629–634.
50. E. Kaolsson, D. Eaker, L. Fryklund, S. Kadin, *Biochemistry*, *11*, 4628–4633 (1972).
51. L. Fryklund, D. Eaker, E. Karlsson, *Biochemistry*, *11*, 4633–4640 (1972).
52. E. Karlsson, H. Arnberg, D. Eaker, *Eur. J. Biochem.*, *21*, 1–16 (1971).
53. D. Eaker, Personal communication.
54. Z. El Rassi, Cs. Horváth, *J. Chromatogr.*, *359*, 255–264 (1986).
55. R. W. Stringham, F. E. Regnier, *J. Chromatogr.*, *409*, 305–314 (1987).
56. D. Josić, W. Hofman, W. Reuter, *J. Chromatogr.*, *371*, 43–54 (1986).
57. H. Hori, N. Tamiya, *Biochem. J.*, *153*, 217–222 (1976).
58. A. S. Arseniev, T. A. Balashova, Yu. N. Utkin, V. I. Tsetlin, V. F. Bystrov, V. T. Ivanov, Yu. A. Ohvcinnikov, *Eur. J. Biochem.*, *71*, 595–606 (1976).
59. A. Rousselet, G. Faure, J. C. Boulin, A. Ménez, *Eur. J. Biochem.*, *140*, 31–37 (1984).
60. F. von der Haar, *Eur. J. Biochem.*, *34*, 84–90 (1973).
61. S. S. Quadri, J. S. Easterby, *Anal. Biochem.*, *105*, 299–303 (1980).
62. R. K. Scopes, *Biochem. J.*, *161*, 253–263 (1977).
63. J. M. Cooney, *Biotechnology*, *2*, 41–55 (1984).
64. E. H. Cooper, R. Turner, J. R. Webb, H. Lindblom, L. Fägerstam, *J. Chromatogr.*, *327*, 269–277 (1985).
65. J-C. Janson, *J. Agr. Food Sci.*, *19*, 581–588 (1971).
66. R. M. Bock, S. N. Ling, *Anal. Chem.*, *26*, 1543–1546 (1954).
67. B. M. Pogell, *Biochem. Biophys. Res. Comm.*, *7*, 225–230 (1962).
68. B. M. Pogell, *Meth. Enzymol.*, *9*, 9–15 (1966).
69. R. K. Scopes, *Biochem. J.*, *161*, 265–277 (1977).
70. R. D. Davies, R. K. Scopes, *Anal. Biochem.*, *114*, 19–27 (1981).
71. E. A. Peterson, A. R. Torres, *Meth. Enzymol.*, *104*, 113–133 (1984).
72. A. W. Liao, Z. El Rassi, D. M. LeMaster, Cs. Horváth, *Chromatographia*, *24*, 881–885 (1987).
73. D. H. Leaback, H. K. Robinson, *Biochem. Biophys. Res. Comm.*, *67*, 248–254 (1975).
74. W. P. Neuman, E. Haberman, *Biochem. Z.*, *327*, 274–288 (1955).
75. H. Breithaupt, K. Rübsamen, P. Walsh, E. Haberman, *Naunyn-Schmiedebergs Arch. Pharmacol.*, *269*, 85–100 (1971).
76. K. Rübsamen, H. Breithaupt, E. Haberman, *Naunyn-Schmiedebergs Arch. Pharmacol.* *270*, 274–288 (1981).
77. R. A. Hendon, H. Fraenkel-Conrat, *Proc. Natl. Acad. Sci. U.S.A.*, *68*, 1560–1563 (1971).
78. J. Fohlman, D. Eaker, E. Karlsson, S. Thesleff, *Eur. J. Biochem.*, *68*, 457–469 (1976).
79. C. T. Wehr, *Meth. Enzymol.*, *104*, 133–154 (1984).
80. C. T. Wehr, *J. Chromatogr.*, *418*, 3–26 (1987).
81. A. Tiselius, S. Hjertén, Ö. Levin, *Arch. Biochem. Biophys.*, *65*, 132–155 (1956).
82. J.-C. Janson, J. Porath, *Meth. Enzymol.*, *8*, 615–621 (1966).
83. L. Rydén, *FEBS Lett.*, *18*, 321–325 (1971).
84. P. Norberg, B. v. Hofsten, *Biochim. Biophys. Acta*, *220*, 132–133 (1970).
85. G. Bernardi, *Nature*, *206*, 779–783 (1965).

86. G. Bernardi, T. Kawasaki, *Biochim. Biophys. Acta, 160*, 301 310 (1968).
87. T. Sato, T. Okuyama, S. Fuginuma, T. Ogawa, *Chromatography (Japan), 9*(1), 51 54 (1988).
88. T. Sato, K. Ohuchi, T. Okuyama, S. Fuginuma, T. Ogawa, *Chromatography (Japan), 9*(2), 171-172 (1988).
89. R. Bhikhabhai, G. Johansson, G. Pettersson, *J. Appl. Biochem., 6*, 336-345 (1984).
90. S. Johansson, unpublished.
91. V. A. Eterovic, M. S. Herbert, M. R. Hanley, E. L. Bennett, *Toxicon, 13*, 37-48 (1975).
92. K. Balasubramaniam, D. Eaker, E. Karlsson, *Toxicon, 21*, 219 229 (1983).
93. J. H. Berglöf, S. Eriksson, H. Suomela, J. M. Curling, in *Separation of Plasma Proteins*, J. M. Curling, ed., Pharmacia Fine Chemicals AB, Uppsala, Sweden, 1983, pp. 51-58.

5 Chromatofocusing

T. William Hutchens
Department of Pediatrics
USDA/ARS Children's Nutrition Research Center
Baylor College of Medicine
Houston, TX 77030, USA

5.1 Introduction and Background

Whether the goal is high resolution separations at the analytical scale or the preparative scale isolation of biologically active macromolecules, often the most definitive separation techniques include exploitation of molecular charge properties. Thus, ion-exchange chromatography, because of its accommodating versatility and simplicity, is a frequently used separation technique in biochemistry. Similarly, electrophoretic focusing procedures, although generally more labor intensive and cumbersome with larger sample volumes, allow macromolecular separations based upon differences in net surface charge. Indeed, isoelectric focusing, as it is often referred to, is perhaps the most discriminating, high-resolution protein separation method available. Recently, during the period from 1977 to 1981, Sluyterman and his colleagues offered a theoretical basis and practical means of realizing the most favorable attributes of both ion-exchange chromatography and isoelectric focusing in a single chromatographic focusing procedure[1-5].

During chromatofocusing, proteins are separated as a result of the isocratic formation of "internal" pH gradients on ion-exchange columns (Fig. 5-1). Because proteins of similar net surface charge may still vary in surface charge distribution, it is even possible to resolve proteins during chromatofocusing which are not well separated by isoelectric focusing.

One of the great benefits of chromatofocusing is the ease of operation. No gradient forming devices or mixers are required. The pH gradient is formed with a single eluent, isocratically. In the case of descending pH gradients formed using an anion-exchange stationary phase, the upper limit or starting pH (e.g. pH 8) is determined by the stationary phase equilibration buffer. The lower or limit pH is determined by titration of

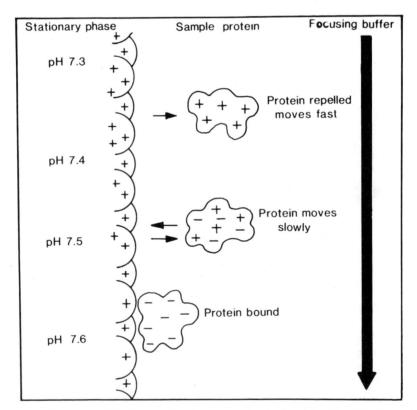

Figure 5-1. Illustration of the chromatofocusing process showing isocratic pH gradient formation, protein adsorption behavior, and migration of the focusing zone.

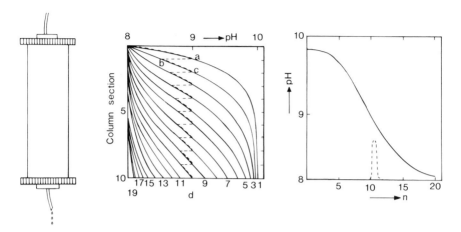

Figure 5-2. Formation of pH gradient in the chromatofocusing column. Each curve in the left diagram shows the gradient in the column (initial pH 10.0) after addition of a given number of unit volumes of focusing buffer (pH 8.0). The pH at the column outlet is given at the bottom of the left diagram and used to construct the final chromatofocusing profile to the right. A sample protein with a pI of 9 in the column and the effluent is indicated by hatched lines. Adapted from an illustration by Sluyterman and Wijdenes (Ref. 2).

the mobile phase or chromatofocusing buffer to the desired end point (e.g. pH 4). The pH gradient slope is a function of the column bed volume, buffer capacity of the matrix and chromatofocusing buffer concentration as well as the initial and final adjusted pH of the stationary and mobile phases, respectively. The protein focusing zone moves more slowly down the length of the column than does the focusing buffer or eluent (Fig. 5-2). Repetitive protein binding events driven by buffer flow against the resistance of the stationary phase to change pH causes the focusing phenomena to occur.

As a powerful analytical and preparative separation technique, chromatofocusing is in its infancy. The purpose of this presentation is to provide enough theoretical and practical guidelines to enable the successful use and further development of chromatofocusing as a high-resolution protein separation technique.

5.2   Chemistry of Protein Surface Charge

The 3-dimensional structure and overall surface molecular architecture of each different protein is unique. Most charged groups on proteins reside at the surface/aqueous interface. Locally, the immediate environment or milieu surrounding individual groups on the protein surface (e.g. His) may specifically effect the reactivity (e.g. pK_a) or chemical potential of that group. These surface reactive properties are influenced by both the mobile phase and stationary phase during chromatofocusing.

The isoelectric point of a protein is defined as that pH in which the net charge is zero. Even though the majority of proteins (ca 70%) examined to date have isoelectric points in the range pH 5–8, considerable heterogeneity exists within this region. Several well-organized and useful compilations of protein molecular weights and subunit compositions[6] as well as apparent isoelectric points[7,8] have been published. It is important to emphasize that even though a protein with a net zero charge will not necessarily migrate in an electric field it nevertheless possesses a given charge distribution which may be operative during chromatofocusing as the pH of the mobile phase alters the pH of the protein and the stationary phase.

5.2.1 Factors Influencing Protein Surface Charge

Proteins and peptides are amino acid polymers (often containing carbohydrate oligomers) with titratable functional groups varying in pK from approximately 3 to 13 (Table 4-2, p. 112). It is the collective contribution of different buffering groups on the protein that allow it to be workably defined as relatively basic (pI > 7) or acidic (pI < 7). In buffers with pH values above its isoelectric point, the protein will have a net negative charge and interact with a positively charged matrix (e.g. anion-exchange resin) (cf. Fig. 5-1). Below its isoelectric point an overall or *net* positive charge will result and the protein will of course not interact with positively charged structures. Most of the individual, titratable groups are located at the protein's surface, and a gradual change in pH will result in a continuously changing surface charge distribution. These changes can occur rapidly, in either direction, and are reversible. At pH values near the protein's isoelectric point, *local* charge distribution effects are important.

There are several factors, both chemical and physical, which can influence the protein surface charge. Intentional perturbations of this kind may include temperature, type and concentration of ions (counter-ions, co-ions and zwitter-ions) present, water structure (activity) modifying reagents such as urea or ethylene glycol, bound metal ions or other specific ligand, as well as detergents. In fact, the accurate determination of pH (see 5.6.2) is significantly influenced by solvent composition and temperature. Gelsema[9–10] and Ui[11] provide a more detailed consideration of these factors relative to the measurement of pH and pI. Unintentional influences on protein charge as a function of pH may include changes in protein quaternary organization (e.g. subunit dissociation or aggregation) and protein interaction with the stationary phase or polymeric ampholytes in the chromatofocusing buffer[12] or other macromolecules present. At a given pH the mobile phase ionic strength can influence protein surface charge and thereby its behavior, including interaction with other proteins and the stationary phase. See further section 4.3, p. 112.

5.2.2 Ionic Strength Effects and Protein Solubility

The solubility of a protein is partially a function of surface charge and thus can vary significantly with pH. Most proteins are least soluble at their isoelectric pH values due to minimized electrostatic repulsion. Thus, low ionic strengths also influence solubility at this pH (pI). The ionic strength of chromatofocusing buffers is best kept low to favor stationary phase - solute (i.e., protein) interactions. However, relative to isoelectric focusing in an electric field, chromatofocusing is to a certain extent independent of small changes in ionic strength.

Protein precipitation or aggregation at the isoelectric point is for many proteins a practical problem in chromatofocusing, and has hampered the wide use of the technique. It is, however, not at all a compulsory reason to avoid the technique. Isoelectric precipitation is counteracted by the fact that proteins do not elute exactly at their pI. They need to have a slight net charge to adsorb to the ion exchanger. Increased ionic strength thus counteracts precipitation both because of the increased net charge needed for adsorption and because of the salt itself. A drawback is, however, that salt in the mobile phase adversely affects the pH gradient. This effect is considerably influenced by the choice and design of the stationary phase. Decreased protein load also diminishes the risk of precipitation. In fact, it should be noted that many proteins of considerably physiological interest occur in very low amounts and the concentrations needed for precipitation are never reached during chromatofocusing. Finally, if precipitation occurs the protein might still be recovered. For example, β-lactoglobulin is only retarded on the column. The protein precipitates but is dissolved continuously as the pH decreases below its isoelectric point.

5.3 Stationary Phase Design

The choice of suitable stationary phases for chromatofocusing has increased with the availability of high-capacity weak anion-exchange materials which are compatible with large biopolymers such as proteins. Those commercially available products listed in Table 5-1 are examples of both conventional and high-performance stationary phases which have been successfully employed for chromatofocusing.

The appropriate stationary phase for chromatofocusing depends very much on the intended application. In general, for linear pH gradient formation, it is a necessary and probably sufficient requirement for the stationary phase to have a significant and *even* buffering capacity over the pH separation range of interest. Depending upon sample loading volumes and column capacity requirements, a percentage of permanent or fixed (non-titratable) charges can aid in separations over pH ranges out-side the area in which the weaker groups titrate. Thus anion exchangers contain an amount of quaternary amines and cation exchangers sulphonates. A fundamental property of the stationary phase should be the lack of non-specific attractions for the proteins or other solutes of interest. Such interactions vary among the several types of columns currently utilized for chromatofocusing due to basic differences in their chemical and physical nature. High porosity and rigidity are also important and have been achieved with several matrices (Table 5-1).

The choice of polymer- or silica-based stationary phases for chromatofocusing depends on, among other things, the pH range of interest. The use of silica-based stationary phases is usually discouraged for applications requiring pH values above pH 8. However, there are exceptions depending upon bonded phase chemistry. Other deciding factors include available equipment, cost, useful lifetime of the material and preference.

5.3.1 Charge, Charge Density, and Capacity

pH-Gradients can be formed on the ion exchangers over the range in which they titrate. To achieve a chromatofocusing effect a descending pH gradient is formed internally on anion-exchange (cationic) stationary phases and, conversely, an ascending pH gradient is formed internally using suitable cation-exchange (anionic) matrices. Most practical experience, and this chapter, involves the use of weak anion-exchange resins (primarily mixtures of secondary and tertiary amines) for the development of descending pH gradients. The charge density (meq/m^2) of various anion-exchange matrices most often reflects the chemistry and capacity of the matrix to be derivatized.

Table 5-1. Anion-Exchange Stationary Phases Used For Chromatofocusing.

	Immobilized ligand or bonded phase	Stationary phase composition	Particle size (μm)	Average pore diameter	Reported charge or loading capacity	pH stability
Bakerbond WP-PEI (J. T. Baker, Inc.)	PEI	Silica	5, 15 and 40	250-300 Å	150-200 mg protein per g silica (40 μm)	pH 2-10
SynChroPak AX-300	PEI	Silica	6.5 and 7	300 Å	96 mg protein per g silica	<pH 8.5
AX-500	PEI	Silica	7	500 Å		<pH 8.5
AX-1000 (SynChrom, Inc.)	PEI	Silica	10	1000 Å		<pH 8.5
Polybuffer Exchanger 94 (Pharmacia)	3° and 4° amines	Polymer Sepharose 6B	40-120	porous	2-4 meq per 100 ml gel	pH 3-12
Polybuffer Exchanger 118 (Pharmacia)	3° and 4° amines	Polymer Sepharose 6B	40-120	porous	0.3-5 meq per 100 ml gel	pH 3-12
Mono P (Pharmacia)	3° and 4° amines	cross-linked hydrophilic	10	porous	15-21 meq per 100 ml gel	pH 2-12

The overall charge capacity (microeq/ml gel) of a given stationary phase per unit bed volume is a property of particle size and porosity (i.e., surface area). Typical figures are 15–30 microeq/ml, half of which are strong non-titratable groups. Not all of the stated charge (binding) capacity is necessarily available for protein due to geometrical constraints. In practical terms, the determination of the stationary phase working capacity is influenced by the choice of protein and interaction conditions (i.e., pH, ionic strength, temperature) as well as non-specific adsorption phenomena. Thus on Mono Q and Mono S (Pharmacia) 80 mg/ml bed volume of human serum albumin and 65 mg/ml of α-lactalbumin is adsorbed. For the much large thyroglobulin the figure decreases to 25–30 mg/ml.

Finally, in addition to those stationary phase charge groups with buffering capacity which are participating in the generation of the internal pH gradient, additional permanent or fixed charges (e.g., quaternary amines) have been found useful to maintain an even protein binding capacity over a broad pH range.

5.3.2   Ionic Adsorption Properties

The ability of weak anion-exchange matrices to participate in electrostatic interactions with biopolymers is affected by pH as well as the type (i.e., pK_a), charge and concentration of competing counterions. Thus, the operable pH range is best kept as narrow as is consistent with good separation. Generally, the initial adjusted pH of the anion-exchange column must be above (e.g., 0.5–1 pH unit) the pI of proteins to be adsorbed, so the buffering capacity of the stationary phase ligand must be adequate at this pH. Similarly, the ionic strength of the chromatofocusing buffer must be less than that required to prevent the interaction of designated protein(s) at the loading pH. These effects are variable and depend upon the actual stationary phase/mobile phase system utilized. Table 5-2 gives recommended column equilibration buffers and elution buffers for developing pH gradients over several different ranges for use with Pharmacia

Table 5-2. Buffer Substances For Focusing in Different pH-Ranges. Based On Instructions Given In Pharmacia Booklet (Ref. 13). Polybuffers Are Oligomeric Ampholyte Buffers Titrating Between pH 9 And 6 And pH 7 And 4 (PB 74).

pH	Starting buffer	Eluent	Dilution factor	Approximate volume (in column volumes)	
				Pre-gradient	Gradient
10.5–7	0.025 M triethylamine pH 11, HCl	Pharmalyte 8–10.5 pH 7.0, HCl	1:45	2.0	11.5
9–6	0.025 M ethanolamine pH 9.4, HCl	Polybuffer 96 pH 6.0, CH₃COOH	1:10	1.5	10.5
8–5	0.025 M Tris pH 8.3, CH₃COOH	30% Polybuffer 96 70% Polybuffer 74 pH 5.0, CH₃COOH	1:10	2.0	10.5
7–4	0.025 M imidazole pH 7.4, HCl	Polybuffer 74 pH 4.0, HCl	1:8	2.5	11.5
5–4	0.025 M piperazine pH 5.5, HCl	Polybuffer 74 pH 4.0, HCl	1:10	3.0	9.0

Polybuffer Exchanger (PBE) and Mono P gels. In practice, both anionic and cationic as well as zwitterionic buffer constituents have been used successfully. Sometimes one good experimental effort provides acceptable results independent of accepted theory.

5.3.3 Buffering Capacity of the Stationary Phase

Anion-exchange columns and packing materials with suitable and even buffering capacity over a given pH range may be used effectively for chromatofocusing in that same pH range. The column or material in question can be titrated with acid (e.g., 0.5–2 N HCl) or base (e.g., 0.5–2 N NaOH) in the presence of elevated ionic strengths to calculate the buffering capacity (β) as originally outlined by Van Slyke[14]. Stationary phases with high and even buffering capacities over a fairly wide pH range have been developed specifically for chromatofocusing and are available commercially (Table 5-1). Also other cationic stationary phases developed for high-performance ion-exchange chromatography have been found to have excellent buffering capacities and pH gradient generating qualities (Fig. 5-3).

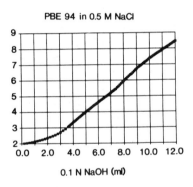

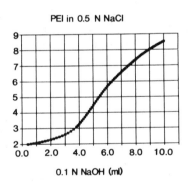

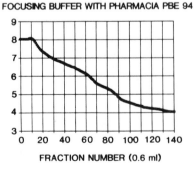

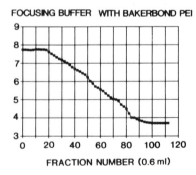

Figure 5-3. Titration curves of (1) a stationary phase specifically designed for chromatofocusing (Pharmacia Polybuffer Exchanger PBE 94) and (2) a silica-based high-performance anion-exchange stationary phase (J. T. Baker, Inc., Bakerbond WP-PEI). The characteristics of the pH gradients formed internally with a chemically-defined focusing buffer are also illustrated for the two stationary phases.

Certain columns have been carefully developed for a wide range of chromatofocusing applications, that is, pH ranges from pH 8 or 9 to pH 4. Yet, it is difficult to maintain wide-range pH gradients which are linear over more than about 3 pH units. Narrow pH gradients (less than 2–3 pH units) may also be formed on columns with a buffering capacity limited to that same pH region. Thus, several other column types, as yet unexploited for chromatofocusing, may be found suitable for certain applications. It also needs to be emphasized that the stationary phase buffering capacity and the pH gradient formed isocratically during separation need not necessarily be linear as long as a satisfactory or optimal separation of the proteins is obtained without it.

5.3.4 Column Volume and Geometry

The smallest possible chromatofocusing bed volume consistent with sample load requirements and resolution should be employed to minimize focusing buffer volumes and the time required for analysis. In general terms, the amount of buffering groups on the stationary phase and column volume define the buffering capacity of the column. This determines the concentration and/or volume of focusing buffer required to generate a pH gradient of defined slope. Excessive column bed volumes also prolong column regeneration times. Good quality analytical separations have been achieved using Pasteur pipettes or other narrow (e.g., 0.4–0.5 cm) diameter open columns packed with 2–4 ml bed volumes of Pharmacia Polybuffer Exchanger 94. An average chromatofocusing colum is, however, 15–30 cm long. The pH gradient becomes less steep as it travels through the column, but after a certain distance further lengthening of the column does not change the result.

5.4 Mobile Phase Design

The single column eluent or mobile phase necessary to develop a particular pH gradient during chromatofocusing, like the mobile phase operative during isoelectric focusing, is actually a carefully designed set of buffering constituents. Collectively, these constituents make up one of the more complex buffer systems used in chromatography today. Relatively simple in concept, it is instructive to understand the mobile phase as an important variable contributing greatly to the efficiency and outcome of the desired separation.

5.4.1 Buffering Capacity of the Mobile Phase

It was Van Slyke[14] who first described the concept of buffering capacity (β) and showed it to be an additive property of weak acids and bases in solution:

$$\beta = \frac{dB}{dpH}$$

where dB is the increment of base (B) added to the buffer solution and dpH is the incremental change in pH which results.

In the context of protein focusing techniques, Svensson first provided the theoretical calculations for using the buffering capacity of "carrier" ampholytes to create natural pH gradients in an electric field[15–16]. The lack of simple ampholytes suitable for this purpose limited the use of isoelectric fractionation until Vesterberg[17] reported the synthesis of polyampholytes with the necessary isoelectric properties. The similarities of commercially available polyampholytes used for isoelectric focusing and chromatofocusing are not coincidental and the reader is urged to consult one of the many reviews on the subject (e.g., Ref. 18) for a more thorough appreciation of the synthesis and properties of polyampholytes.

Polyampholytes are designed to provide an even buffering capacity (β) over the pH range in which separation is to take place. Figure 5-4 illustrates the titration curve of Pharmacia Polybuffers 96 and 74. This curve represents the buffering action of titratable constituents with numerous successive pK_a values.

A generic alternative to oligomeric ampholyte focusing buffers is the use of chemically-defined, non-polymeric (so-called "simple") focusing buffers. Such universal or wide range buffer systems have existed for many years[19–21]. There have been several attempts to extend this philosophy and formulate useful, "simple", focusing buffer systems for chromatofocusing[22–26]. Figure 5-4 also shows the titration curve of one such focusing buffer. Some examples of the utility of these focusing buffer systems are provided in application 5.7.1 (see Fig. 5-5 also). While earlier attempts to prepare simple, chemically-defined, focusing buffers for electrophoretic focusing procedures[27–30] have been met with only partial acceptance, perhaps due to perceptions of electrofocusing theory (e.g., Ref. 18), the development of simple chromatofocusing buffer systems is expected to continue. There are presently no theoretical or practical limitations to the use and further improvement of non-polymeric (i.e., very low molecular mass), chemically-defined chromatofocusing buffer systems, especially for large scale applications. For buffer use in general, Perrin and Dempsey[31] have published a useful book that provides a simplified overview of key considerations in buffer preparation and use.

In practice, pH gradient engineering for chromatofocusing is complicated by the contribution of stationary phase effects.

5.4.2 Physical Properties of Chromatofocusing Buffer Constituents

A desirable chromatofocusing buffer will have sufficient and even buffering capacity over the pH range of interest. The ionic strengths have to be very low in order to give a

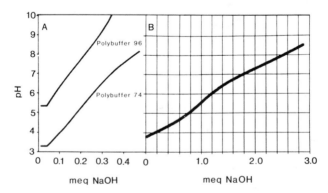

Figure 5-4. Titration curve of polymeric (polyampholyte) buffers (left) as well as of a simple non-polymeric chromatofocusing buffer (right). Polybuffer 96 titrates between pH 9 and 6; polybuffer 74 between pH 7 and 4.

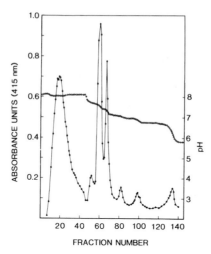

Figure 5-5. Fractionation of a commercial preparation of haemoglobin (15 mg) using the non-polymeric focusing buffer described in Refs. 23, 24 with Pharmacia Polybuffer Exchanger 94 as the stationary phase (1.0 × 5 cm). Two-minute fractions were collected at a flow rate of 0.5–0.6 ml/min. The average pH gradient slope (Δ pH/ml) from pH 7.75 to 6.75 was 0.01–0.02.

separation according to isoelectric point, and thus to favor protein or solute interaction with the charged stationary phase during focusing. The light (UV) absorbing properties should be minimal to permit an adequate signal to noise ratio while monitoring protein or solute elution (e.g., at 280 nm). The most successful chromatofocusing buffer should be composed of constituents that (1) are easily separated from most proteins and (2) do not interact with high affinity or irreversibly with the protein(s) being separated. In particular, the structure and/or function of the protein should not be practically or irreversibly compromised. Separation of proteins from polyampholyte chromatofocusing buffers is treated in more detail in section 5.6.6, p. 165.

The precise physical and chemical properties of oligoampholyte (oligomeric) focusing buffers are variable and poorly defined. The degree to which they interact with proteins remains the subject of considerable debate[12,18]. Even chemically-defined, low molecular weight buffer constituents may interact with proteins[28–29]. Presumably, however, with defined chromatofocusing buffers these constituents can be identified and deleted or replaced if necessary. The degree to which these considerations pose a practical barrier must be determined on an individual base. There are strong arguments for the use and continued development of small, "noninteractive" focusing buffer systems[22–30].

5.5   Mechanism(s) of Separation

In this section we will briefly evaluate existing theories of internal pH gradient formation on stationary phases, as well as the postulated protein separation mechanisms. The aim is to understand better the utility of given experiment variables for the solution of a given separation or analytical problem.

5.5.1   Theory of pH Gradient Formation

The initial assumptions (all-or-none model) and first approximation equations as originally outlined by Sluyterman and his colleagues[2-5] provide a logical approach to the underlying principles generally utilized to explain the generation of internally developed pH gradients during chromatofocusing. They derived the following equation for final eluted pH (pH_f) when the two phases are mixed. The buffering capacity per defined unit (e.g., column length) of stationary phase is β_s and mobile phase is β_m.

$$pH_f = \frac{(\beta_s)pH_s + (\beta_m)pH_m}{(\beta_s) + (\beta_m)} \tag{5-1}$$

In the simple case when β_s and β_m are assumed to be equal (Eq. 5-1) becomes

$$pH_f = \frac{pH_s + pH_m}{2} \quad (\beta_s = \beta_m) \tag{5-2}$$

That is, if the buffering capacities of the mobile phase and stationary phase are equal and *additive*, upon mixing, the resulting pH will reflect the simple average. To best illustrate the consequences of this assumption, Fig. 5-6 shows the theoretical pH profiles

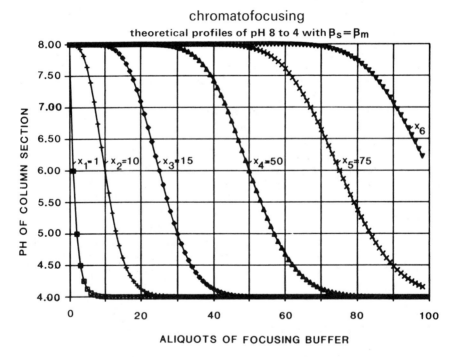

Figure 5-6. Theoretical pH profiles generated from pH 8 to pH 4 as a function of focusing buffer (volume) for each of 100 aliquots as described in the text. The pH gradient of the final column eluate is partially shown (far right). An equal buffering capacity of focusing buffer and stationary phase is assumed per unit volume. x_n = number of column section alternatively column length.

generated using this equation. To begin, the pH of the stationary phase (pH_s) was adjusted to pH 8. The column (e.g., 100 individual units or sections) pH is indicated during the sequential addition of 100 aliquots of focusing buffer at pH 4, where each aliquot of focusing buffer is equal in buffering capacity to one unit or section of column. The internal pH values of column sections 1, 10, 25, 50 and 75 are shown as a function of the number of focusing buffer aliquots added. Except for the extremes at either end, a series of linear pH gradients are developed.

These data reveal an important, experimentally verified, point. The slope of the pH gradient depends upon the number of column sections for a given amount of focusing buffer. Also, if the buffering capacity of the mobile phase is increased relative to that of the stationary phase the pH gradient will be greater in slope. Another important assumption for the model predicted by Eq. 5-2, aside from equal and additive buffering capacities for the stationary and mobile phases, is that of *even* (linear) and parallel buffering capacities over the pH ranges of interest. However, upon physical (electrostatic or hydrophobic) interactions between mobile phase buffer constituents (e.g., polyampholytes) and the stationary phase the functional buffering capacities (pK_a values) may be altered. Indeed, an unequal distribution of buffer constituents between the mobile and stationary phases can result in uneven variations in column buffering capacity as a function of pH. So, while the buffering capacities of soluble buffering agents are additive, the buffering capacities of interacting mobile and stationary phase components may not always be additive. Nevertheless, it has been shown that the contribution of stationary phase and mobile phase buffering capacities alone can indeed be additive and account for the pH gradient formation predicted by Eq. 5-1. This was demonstrated using chemically identical (or similar), non-amphoteric buffering groups in both the mobile and stationary phases[25].

Despite the theoretical uncertainties involved, it has been shown that chemically-defined mixtures of simple (non-polymeric) buffer reagents[22,23], non-amphoteric buffers[25], and heterogeneous, polymeric ampholytes (e.g., Polybuffers 96 and 74) can be utilized to internally generate similar pH gradients.

5.5.2 Control of Separation Variables and Resolution

In practical terms the most efficient means of controlling solute or protein separation efficiency (i.e., resolution) during chromatofocusing is by altering the shape and/or slope of the pH gradient generated. This may be accomplished for a given pH range by altering the stationary phase bed volume (e.g., colum length) or changing the relative concentration of focusing buffer. Thus an increase in column length will only initially make the gradient more shallow, as explained above. The possibility of increasing the concentration of focusing buffer is also limited. A too high buffer strength which does not match the stationary phase will not give an even gradient. Its steepness will increase towards the end of the run.

Alternatively, the initial column equilibration pH (starting pH for gradient) and/or the final adjusted pH of the focusing buffer (limiting pH for gradient) may be adjusted to provide the desired pH gradient slope (Δ pH/ml). Protein resolution can be excellent with chromatofocusing. Protein peaks separated by 0.01 pH units have been reported[26] (see applications).

The specific elution pH of a given protein can also be affected by the inclusion of mobile phase modifiers such as urea, metals, and detergents. These influence the way proteins in the sample behave in the separation.

5.5.3 Ampholyte Displacement Effects

Before the development of chromatofocusing, polyampholytes had been used to displace proteins from conventional anion-exchangers. Ampholyte displacement chromatography, apparently first reported by Leaback and Robinson[32], can be distinguished from chromatofocusing under specifically defined conditions. In particular, the formation of a pH gradient does not always appear to be necessary for the displacement of proteins when high concentrations of competing polyampholytes are used. Discussions have been presented to distinguish, theoretically and practically, chromatofocusing from ampholyte displacement chromatography[33–38].

However, uncertainties remain as to the precise separation mechanism(s) operable in either case. Under the conditions thus far utilized for "chromatofocusing", the distinction appears to be more apparent than real since these two protein separation processes do not necessarily have to be mutually exclusive. For example, as the pH approaches that point at which the protein interaction with the stationary phase becomes weaker, competitive binding of polyampholytes will act more effectively to displace the protein. Stated another way, as pH effects begin to decrease the affinity of protein for the stationary phase, competitive binding of ampholytes or other displacers may become a significant factor. The most basic polyampholytes (high pI values) elute first and others breakthrough in order of their relative basicity. In practice, the separation mechanism may be irrelevant if purification is the desired endpoint. If the investigation is analytical or comparative in nature, unexpected elution behavior must be reconciled with the dominant elution mechanism.

5.5.4 Comparison with Isoelectric Focusing

It is important to emphasize that proteins may elute during chromatofocusing at pH values close to or well away from (e.g., 1–2 pH units) their isoelectric points. The several possible explanations for this phenomena are based upon one of the major mechanistic differences between chromatographic focusing and electrophoretic focusing techniques, namely, the stationary phase.

Individual proteins may vary in their distribution and density of surface charge even if similar in their isoelectric points. Thus, differential interaction with the charged stationary phase is still possible. Since the slope of the titration curve for individual proteins (Δ surface charge/Δ pH) can vary significantly for a given difference between pI and pH, proteins may have markedly different affinities for the stationary phase. The interaction of a protein with a charged column matrix may itself alter the conformational response of that protein to changes in environmental pH. These processes must occur repeatedly during chromatofocusing because the focusing effect is the result of proteins repeatedly adsorbing to and desorbing from the charged matrix. It is during these interactions (at or near the pH of adsorption/desorption) that both concentration and composition of polyampholytes or salts may affect the elution of proteins differentially.

It has been postulated[2,3] that a Donnan potential between the interior and exterior of the stationary phase particles might decrease the elution pH of the proteins from chromatofocusing columns. During chromatofocusing on highly charged (positive), porous matrices, due to electrostatic repulsion, positively charged species in solution (including hydrogen ions, basic ampholytes, etc.) may be excluded and increase the pH of the buffer immediately surrounding the surface of the porous matrix.

To summarize, in contrast to electrophoretic focusing techniques, chromatofocusing relies on interfacial (solid/solution phase) molecular interactions which are influenced, in a collective manner, by such factors as pH, ionic strenth, charge attraction and repulsion, competitive binding reagents (displacers) as well as any "non-specific" adsorption properties the stationary phase may have. It is therefore inappropriate to refer to elution pH as pI. However, it is not trivial to measure the pH on an isoelectric focusing gel, especially in the presence of various additives. Thus, also here it might be questionable to equal the measured pH value and the actual pI.

5.6 Analytical and Preparative Chromatofocusing Techniques

Chromatofocusing is a separation procedure useful for the analysis of protein surface charge heterogeneity as well as alterations in the surface charge properties of individual proteins at different stages of purification and in unfractionated biological fluids and extracts. However, in addition to the small scale analysis and separation of proteins or protein isoforms, chromatofocusing can be useful on a larger scale for preparative protein purifications. Chromatofocusing, as a sample concentrating procedure, is particularly useful at early or intermediate stages in the protein isolation scheme. Since proteins are eluted from anion-exchange columns during the formation of descending internal pH gradients, the protein is not subjected to pH levels below that just required for elution.

It is also possible to collect fractions or individual protein peaks into pH neutralizing buffer solutions to preserve bioactivity (see application 5.7.1).

The pH gradient formed during chromatofocusing can be monitored continuously along with sample elution during the separation experiment. Thus, gradient slope and shape may be modified as needed to alter separation efficiency (see 5.7.1). It is not always desirable to use linear pH gradients. In fact, non-linear gradients can be predesigned to maximize the separation of given components.

5.6.1 Analytical to Preparative Scale Applications

Substantial increases in sample volume, sample protein concentrations and/or column bed volume can often bring changes (e.g., pH gradient slope) not easily anticipated from analytical scale operations. Depending upon sample volumes or quantities of protein needed, utilizing available high-performance stationary phases with rapid analysis times to make replicate "analytical" scale separations under familiar conditions may be more advantageous than altering column dimensions to achieve the desired end product. There are, however, no theoretical or practical reasons why chromatofocusing cannot be utilized on a large scale to address difficult separation problems.

Since chromatofocusing is indeed a focusing procedure, unlike zonal elution chromatography, a large part of the bed volume can be utilized during sample loading. Isocratic pH gradient formation greatly simplifies equipment requirements (i.e., controllers and mixers) for large volume applications. The continued development and commerical availability of suitable, inexpensive focusing buffers may further reduce the cost of large scale chromatofocusing separations.

Large sample volumes can bring unexpected changes in protein elution behavior if sample contaminants or buffer constituents, even if present at the same concentrations as for analytical scale operations, build up in the sample loading zone. EDTA is an example of one common buffer constituent that causes no problems when present in minor quantities (small sample volumes) but can create serious problems with larger sample volumes. EDTA binds to most anion-exchange stationary phases used for chromatofocusing and elutes in a pH-dependent manner with absorbance at 280 nm.

5.6.2 pH Measurements

Particularly for analytical or comparative experiments by chromatofocusing, accurate and reproducible pH measurements are essential. Aside from pH electrode type and stability, the most immediate concerns are temperature of separation and pH measurement (they should be the same), dissolved CO_2 and mobile phase additives such as urea, glycerol or anything else that affects the H^+ activity of aqueous eluents (see Refs. 9-11).

Flow-through pH electrodes are convenient but may not be as accurate or reproducible as needed for sensitive analytical measurements. Of course, if your primary concern is resolution and separation, pH measurements are of less concern.

5.6.3 Sample Preparation and Column Loading

Proper sample preparation and column loading procedures will help increase resolution, sample recovery and useful column lifetime. It is necessary to remove all sample components which are insoluble (e.g., precipitates) or immiscible (e.g., lipid layers or micelles) in the chosen sample and column equilibration buffer. Brief centrifugation at high speeds ($10,000-15,000 \times g$; 10-15 min) or filtration through inert, porous (0.22 or 0.45 μm) membranes (e.g., Gelman or Millipore) will normally suffice to ensure the application of fully soluble sample components if these procedures are performed at the temperature of chromatography. After equilibration of sample into column equilibration buffer, difficulties with protein solubility, when encountered, may be dealt with in several different ways. It is also possible to load the sample in focusing eluent buffer and *after* that the gradient has developed on the column. Very diluted samples can of course also be used.

5.6.4 Solubility of Proteins

Sample protein solubility during chromatofocusing, depending upon the protein, may become a necessary consideration. Protein solubility during chromatofocusing is most probably influenced by pH as well as the concentration and type of salts (i.e., buffer and counter ions) present (see section 5.2.2, p. 152). Several mobile phase (focusing buffer) modifiers are helpful to minimize solubility problems, and also increase resolution during protein separation. The low ionic strength column and sample equilibration buffers utilized during chromatofocusing sometimes induce protein aggregation even at the start pH which may be well above (e.g., 0.5-2 or more pH units) the isoelectric point of the protein of interest. In this case, addition of urea or non-ionic detergent to the sample can sometimes prevent aggregation until protein adsorption to the stationary phase and/or gradient development begins and aggregation may no longer be a problem.

In fact, concentrations of urea from 1 M to 6 M in the sample preparation buffer and both the column equilibration and focusing buffer is an effective, probably under-utilized, means of optimizing sample solubility and resolution during chromatofocusing. Many proteins retain their tertiary and/or quaternary structure as well as their activity in relatively high concentrations of urea, especially at 4°C. Low concentrations of detergents may also be utilized to maintain protein solubility without compromising biological activity. Triton X-100, Nonidet P-40 and Tween 80 are all examples of nonionic detergents (available from Sigma) which have been used effectively during chromatofocusing. CHAPS (Cal Biochem) is a zwitterionic detergent prepared from cholic acid which has also been found useful to prevent aggregation and maintain high recoveries during chromatofocusing. Low concentrations (4–10% w:v) of zwitterions (e.g., betaine or taurine) and glycerol have been found to be effective solubilizing/stabilizing agents for some proteins during chromatofocusing.

5.6.5   Inclusion of Mobile Phase Modifying Reagents

The mobile phase includes the chromatofocusing column equilibration buffer and the focusing buffer used to generate the internal pH gradient. Since the protein migration zone or focusing zone proceeds slowly down the length of the chromatofocusing column relative to the flow of buffer, the sample protein may be exposed to 10–20 column volumes of buffer before finally eluting. Thus, specific buffer constituents required to maintain protein structure (or solubility) and/or function must be included in the focusing buffer. Fortunately, most of them can be included. In fact, in contrast to focusing in an electric field where current (conductivity) is important, even relatively high concentrations of charged components can be included and maintained during chromatofocusing. A good example is the inclusion of 10 mM sodium molybdate in steroid receptor chromatofocusing buffers. The effect of molybdate on the stability of receptor structure and function are easily reversible but could be maintained during chromatofocusing. Other salts can be included (up to 20–30 mM) to influence protein solubility or prevent aggregation.

Urea, detergents and glycerol have been especially useful in chromatofocusing buffers with "sticky" proteins.

There is a problem with mobile phase modifiers when used to analyze surface charge heterogeneity or when comparisons of surface charge are being made. These additives can cause significant error in pH measurements. They can also cause major shifts in protein elution pH which vary in a protein-dependent manner. Again, if purification is the only objective, elution pH may not be a major concern.

5.6.6   Sample Recovery

Some proteins can be encountered that are surprisingly resistant to pH-dependent elution during chromatofocusing. This depends very much on the properties of the individual protein and, just as much, on the properties of the stationary phase chosen for chromatofocusing. The number of different types of protein present (e.g., crude mixtures vs. purified samples) often influences recovery simply due to the mass action of adsorption. Highly purified proteins can often be further purified or analyzed in the presence of carrier proteins to saturate so-called non-specific adsorption sites. Alternatively, mobile phase additives (e.g., detergents) have also been effectively utilized to prevent adsorption.

If the protein being analyzed or purified is a receptor or other ligand-binding protein whose activity is determined by association with ligand (e.g., steroid-receptor complex or metal-protein complex), it is important that the stationary phase have little or no affinity for the ligand in question. For example, ^{3}H-estradiol was found to bind to SynChropak AX-300, AX-500 and AX-1000 silica-based anion-exchanged columns and elute in a pH-dependent manner[39,42]. However, no steroid interaction was evident using Bakerbond PEI (J. T. Baker) as the stationary phase.

The separation of proteins from focusing buffer constituents (most commonly polyampholyte buffers such as Pharmacia Polybuffers 74 and 96) can normally be achieved by size-exclusion chromatography (e.g., Sephadex G-50 or G-75 Fine), precipitation with ammonium sulfate, hydrophobic interaction chromatography, or other types of affinity chromatography. However, certain polyampholyte properties[12,18] and reported difficulties in protein-polyampholyte separations[12,23] indicate that some caution should be used in defining adequate removal of focusing buffer constituents from the specific protein of interest.

5.6.7 Column Regeneration

In general, sample recovery and column maintenance require the routine use of high salt concentrations after completion of the pH gradient to help ensure quantitative elution of adsorbed material. A gradual rise in back pressure is an alert to a growing column contamination problem. Post-analysis column washing and/or column regeneration procedures are best performed with reversed flow if possible. Typically, 1-2 M solutions of neutral salt (e.g., NaCl) should be adequate although more chaotropic salts can be used, especially if hydrophobic adsorption is suspected. A useful practice is to utilize 1-2 M concentrations of equilibrating buffer adjusted to the upper or starting pH to facilitate column regeneration. The removal of more difficult proteins can often be accomplished with 8 M urea or 6 M guanidine-HCl. Lipoproteins and lipids can be removed with one or more of several different organic solvents. Ethylene glycol (e.g., 50%), diethyl ether. dimethylformamide or dimethylsulfoxide (e.g., 10-50%), acetonitrile (up to 100% $\pm$ 0.1% trifluoroacetate), ethanol, methanol and other solvents can be used regularly. Care should be taken to ensure that all salts are washed away (Milli-Q water) and as many proteins as possible are removed before using organic solvents to avoid precipitation. Multiple injections of 0.1 M HCl and/or 0.1 M NaOH is also an effective cleaning procedure. Columns may be stored in 20% methanol as a preservative. Sodium azide should be avoided. The use of clean buffers (especially containers) will reduce the risk of bacterial contaminations of "clean" columns.

5.7 Applications

5.7.1 Analysis of Surface Charge Heterogeneity and Purification of a Labile DNA Regulatory Protein—The Estrogen Receptor

The utilities of chromatofocusing are revealed by the examples provided below illustrating the characterization and purification of various steroid receptor protein isoforms. Since chromatofocusing is preparative even at the analytical scale and because the procedure is amenable to the inclusion of various structure-modifying reagents it has been extremely useful in efforts to better define receptor structures and their functional relationships.

The intracellular receptor protein(s) for the female sex hormone estradiol has been evaluated for surface charge heterogeneity and partially purified by chromatofocusing using several different types of stationary phases including conventional open columns of polymeric anion-exchange gels[39,40] and high-performance polymeric[41] and silica-based[23–26,39,42–44] anion-exchange matrices. During these investigations both polyampholytes and non-polymeric, chemically defined focusing buffers have been utilized to generate similar pH gradients and receptor elution profiles[23,24,26]. One goal of the original investigations[42] was to develop a rapid, high-resolution protein separation technique to better evaluate and purify the different forms of steroid receptor proteins known to exist in the presence and absence of (1) natural steroid ligand[43], (2) receptor structure-stabilizing metal ions[42,43], and (3) receptor structure modifying reagents known to induce exposure of the receptor's DNA-binding site[44].

The steroid receptor proteins are identified by their ability to bind specific steroid ligands with high affinity. To extract and radiolabel the estrogen receptor proteins, endocrine target tissues (e.g., uterus or mammary glands) are homogenized in 2–4 volumes of buffer and centrifuged ($> 100,000 \times$ g, 60 min) to obtain particulate free cytosol. The cytosol is then labeled with [^{3}H]-estradiol-17β or [^{125}I]-iodestradiol-17β in the presence and absence of a radioinert, estrogen receptorspecific competitor (diethylstilbestrol) to identify receptor from non-specific estrogen binding proteins[42]. Cytosol preparations (10–20 mg protein/ml) containing what are considered to be biologically relevant concentrations of receptor protein (typically from as low as 5–10 femtomoles up to several hundred femtomoles receptor per mg cytosol protein) are equilibrated into chromatofocusing column equilibration buffer using minicolumns of Sephadex G-25. The chromatofocusing column equilibration buffer usually consists of 25 mM Tris-HCl, pH 8.0 (4°C), containing dithiothreitol (1 mM) and glycerol (10–20%, v/v). Where noted, 10 mM sodium molybdate, 1–6 M urea, detergent (CHAPS) and other receptor structure modifying reagents are included in the sample buffer, stationary phase equilibration buffer and focusing buffers. All procedures, including equilibration buffer and focusing pH determinations, are carried out at 0–4°C. Buffers and samples are filtered to remove particulate contaminants using Millipore 0.45 μm HAWP filters. Buffers are degassed before final pH adjustments and use. Sample aliquots of 0.5–2.0 ml (5–20 mg total protein) are usually loaded onto the column prior to initiation of the pH gradient. The formation of the internal pH gradient is initiated using either polyampholytes (Pharmacia Polybuffers 96 and 74) or other non-polymeric, chemically-defined focusing buffers[22–24] at flow rates of 0.5 to 1.0 ml/min. The pH of collected fractions (1.0 min) is determined immediately upon elution using a Corning Model 125 pH meter equipped with a microcombination calomel electrode. Flow-through pH electrodes (Phoenix Electroes, Houston, TX) have also been used. The protein elution profile is monitored at 254 or 280 nm using a Beckman model 153 or Model 166 analytical flow-through UV detector (8-μl flow cell) or other suitable spectrophotometer. The elution profile of labeled estrogen receptor proteins was monitored by liquid scintillation counting (30–38% efficiency) or gamma counting (60–80% efficiency).

The estrogen receptor elution profile shown in Fig. 5-7 is typical of that obtained for the human uterine estrogen receptor chromatofocused in the absence of structure-stabilizing or structure-destabilizing mobile phase reagents. Open columns (0.7 cm i.d. × 9 cm) of PBE 94 were utilized with Polybuffers 96 and 74. Relatively small bed volumes of chromatofocusing gel were utilized to minimize separation times. The rather labile steroid receptor proteins are stabilized (reversibly) by the presence of relatively high concentrations of molybdate. Figure 5-8 illustrates an important property of chromatofocusing discussed earlier, namely, the ability to maintain continuous concentrations of mobile phase modifies or stabilizers in both column equilibration and focusing buffers. It shows the elution behavior of human uterine estrogen receptor proteins chromatofocused using SynChropak AX-500 high-performance anion-exchange columns in the absence and presence of 10 mM sodium molybdate.

The practical use of simple, chemically-defined focusing buffers is shown in Fig. 5-9 for the separation of calf uterine estrogen receptor forms showing surface charge heterogeneity. Polyethyleneimine-derivatized high-performance silica-based columns (Bakerbond WP-PEI) from J. T. Baker Inc. were utilized for the experiments shown here with the defined focusing buffer. Since the estrogen receptor proteins have high-affinity binding sites from polyanions and are also relatively hydrophobic, one purpose of these experiments was to determine if polymeric ampholytes in the Polybuffers previously utilized to evaluate receptor surface charge were themselves influencing receptor heterogeneity. These profiles also illustrate nicely how increased column buffering capacity can be used with a single focusing buffer to decrease the pH gradient slope and increase resolution. The average gradient slope (Δ pH/Δ ml) was decreased from approximately

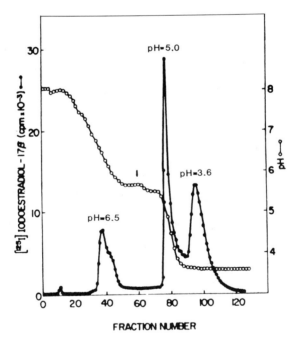

Figure 5-7. Open column chromatofocusing of human uterine estrogen binding proteins on PBE 94. Cytosol was prepared from a fresh specimen of postmenopausal human uterus and labelled with 10 nM [³H]estradiol-17β. Termination of the labelling reaction and details of chromatofocusing on PBE 94 are described elsewhere[39]. Arrows mark initiation of the primary pH gradient (Polybuffers 96 and 74 (30:70) diluted 1:15, pH 4 at 0°C) and secondary eluent (Polybuffer 74 diluted 1:15, pH 3 at 0°C), respectively.

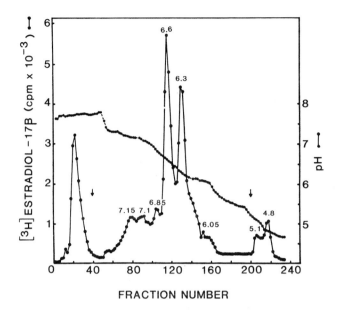

Figure 5-8. Identification of molybdate-stabilized receptor species by HPLC chromatofocusing on AX-500 in the continued presence of molybdate. Human uterine cytosol was prepared and labelled with $[^{125}I]$iodoestradiol-17β in the presence of 10 mM sodium molybdate. Receptor preparations (2–4 mg) were eluted from the AX-500 column with a biphasic pH gradient. The primary eluent was a 30:70 mixture of Polybuffers 96 and 74, diluted 1:10 with 20% glycerol containing 10 mM sodium molybdate and adjusted to pH 5.0. The secondary eluent (initiated at arrow) was Polybuffer 74, diluted 1:10 with 20% glycerol (no molybdate) and adjusted to pH 3.5. The recovery of activity in this representative experiment was 91%.

0.1 to 0.03 over the separation pH range by doubling the column length. The use of this same non-polymeric focusing buffer can be used to generate a narrow pH gradient on a polymer-based chromatofocusing column (Pharmacia PBE 94) to fractionate a commercial source of haemoglobin (Fig. 5-5).

The unliganded form of any steroid receptor protein is even more labile than the steroid-bound form. The separation and recovery of unliganded estrogen receptor (aporeceptor) forms with apparently unaltered biological activity (i.e., steroid binding) was possible by collecting receptor fractions, directly into pH-neutralizing buffer solutions containing radiolabelled steroid.

Figure 5-10 shows how the inclusion of 6 M urea in both column equilibration buffers and focusing buffers was utilized to differentially affect the chromatofocusing elution properties of unfractionated cytosolic versus DNA-affinity purified calf uterine estrogen receptor forms. In this manner, by differential chromatofocusing, two previously uncharacterized receptor forms were identified, separated and partially purified for further analysis.

pH gradient engineering can also be effectively utilized to maximally separate and purify protein isoforms. Affinity-purified estrogen receptors, indistinguishable by sedimentation or size-exclusion analyses, were resolved into two major subfractions using a biphasic descending pH gradient.

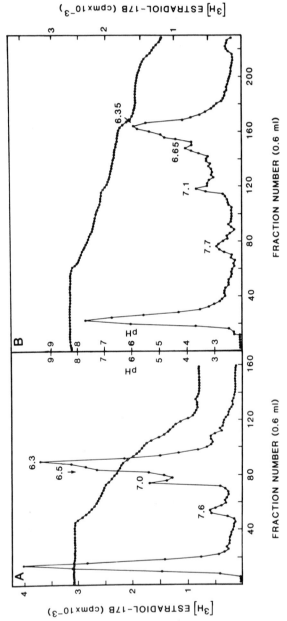

Figure 5-9. Surface charge heterogeneity of calf uterine estrogen receptor proteins evaluated by high-performance chromato-focusing. The internal pH gradients (○) were developed isocratically on Bakerbond WP-PEI columns (J. T. Baker, Inc.) using a nonpolymeric focusing buffer adjusted to pH 4 (see Refs. 23, 24). The wide-range pH gradient (A) was generated using a single 25 cm PEI column. The narrow-range pH gradient (B) was generated using the same focusing buffer with two of the 25 cm PEI columns connected in series. In each case a 500 μl aliquot of labelled cytosol (22 mg protein/ml) was eluted at a flow rate of 0.6 ml/min and 1.0 minute fractions were collected.

5.7.2 Analysis and Preparative Scale Isolation of Heterogeneous Forms of Cyanobacterial Ferredoxin-NADP$^+$ Oxidoreductase

Serrano[45] has achieved nice results using chromatofocusing as both an analytical and preparative tool to evaluate and separate molecular heterogeneous forms of ferredoxin-NADP oxidoreductase (FNR) (EC 1.18.1.2) from the nitrogen-fixing cyanobacterium Anabaena sp. strain 7119. The cells were grown in culture, harvested, washed and stored at −20°C before use. The cells were then thawed and disrupted by exposure to buffered solutions containing a 2% solution of the non-ionic detergent Triton X-100. Duplicate batches of cells were processed in the presence and absence of the serine protease inhibitor phenylmethylsulfonyl fluoride (PMSF). All of the FNR enzyme activity was recovered in the supernatant liquid after centrifugation at 40,000 g for 2 min. The FNR (Mr 33,000–38,000) was purified by chromatography on DEAE-cellulose (Whatman DE-52) and 2',5'ADP-Sepharose 4B (Pharmacia).

Chromatofocusing was utilized to analyze molecular heterogeneity of the FNR purified in the presence and absence of PMSF. A column (1 × 25 cm) of Pharmacia PBE 94 was equilibrated with 25 mM piperazine-HCl buffer, pH 5.3. Samples of the purified FNR (3–13 mg in 5 ml) were dialyzed against the PBE column equilibration buffer before application. The focusing pH gradient was initiated by elution (12 ml/h) with 10-fold diluted Polybuffer 74 adjusted to pH 4 with HCl. As shown in Fig. 5-11 a narrow (shallow slope) pH gradient was quite effective in the separation of (nicked) FNR enzyme

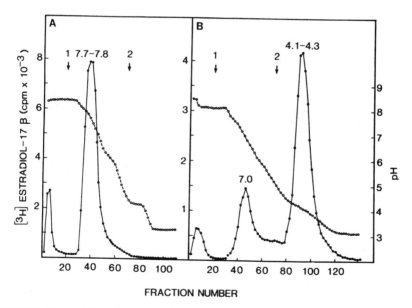

Figure 5-10. Chromatofocusing to evaluate urea effects on the surface charge properties of estrogen receptor purified by DNA-affinity chromatography. Purified estrogen-receptor complexes were analyzed by chromatofocusing on PBE 94 in both the absence (A) and presence (B) of 6 M urea. Non-specific estrogen-binding components were absent after receptor purification by DNA-affinity chromatography in the presence of 6 M urea. The specific elution pH of estrogen-receptor forms is indicated.

generated by purification in the absence of protease inhibitor PMSF. The one minor and three major peaks of FNR resolved (elution pH 4.3–4.6) were recovered in mg quantities for subsequent characterization by SDS-PAGE, isoelectric focusing and amino acid analysis. The high-resolution and preparative nature of chromatofocusing was clearly utilized in this example.

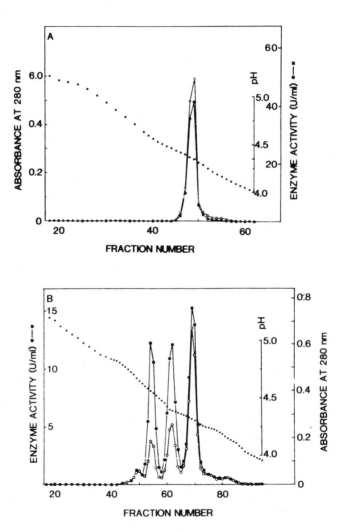

Figure 5-11. Panel A: Column chromatofocusing on PBE 94 of Anabaena sp. strain 7119 FNR purified in the presence of the protease inhibitor PMSF. A sample containing 3.5 mg of protein was applied to a PBE 94 column (1 × 25 cm), and the enzyme was eluted by using a pH gradient (●) generated with Polybuffer 74-HCl, pH 4.0 (7.5 μmol/pH unit/ml). Fractions of 2 ml were collected. Absorbance at 280 nm (○) and enzyme activity (■) were measured for each fraction. Panel B: Column chromatofocusing on PBE 94 of Anabaena sp. strain 7119 FNR purified in the absence of the protease inhibitor PMSF. A sample containing 13.5 mg of protein, purified from the same batch of cells used in A, was chromatographed under the conditions described in the legend of A.

5.7.3 Isolation of Human Very-Low-Density Lipoproteins

In this example, Weisweiler et al.[46] have utilized highly porous (> 500 A), mono-disperse polymer particles (10 μm) recently developed by Ugelstad[47] and derivatized by Pharmacia (Mono P) for chromatofocusing to improve their separations of human very-low-density lipoproteins (VLDL). Serum was prepared from hypertriglyceridemic patients by a combination of low-speed centrifugation and ultracentrifugation (160,000 g for 24 h) at 4°C. VLDL was removed from the top of the supernatant, dialyzed against 0.01 M ammonium bicarbonate buffer (pH 8.6) and lyophilized. Lipids were removed by extraction with diethyl ether and ethanol/diethyl ether. The VLDL apolipoproteins were solubilized in the chromatofocusing equilibration buffer which consisted of 0.025 M Bistris (pH 6.3) containing 6 M urea. Sample protein (up to 10 mg) was applied to a 4 ml prepacked Mono P column (5 mm i.d. × 20 cm) in column equilibration buffer and the descending pH gradient was developed with 30 ml of (10-fold diluted) Polybuffer 74 (pH 4.0) at a flow rate of 1 ml/min. Figure 5-12 shows the chromatofocusing elution profile of the VLDL apolipoproteins. Seven peaks were resolved and collected. Poly-buffers and urea were removed by dialysis and gel filtration (Sephadex G–50 fine). Overall recovery was reported to be 78%. Individual VLDL protein fractions were identified by double-immunodiffusion in 2% agar using monoclonal antibodies, before further analysis by isoelectric focusing and SDS-PAGE. The speed of VLDL apolipoprotein separation by chromatofocusing (30 min) on MonoBeads (Mono P) was considered a significant improvement over previously utilized, more time consuming procedures.

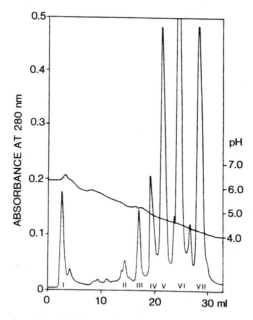

Figure 5-12. Fractionation of soluble VLDL apolipoproteins by chromatofocusing on 10 μm monodisperse beads. Apolipoproteins (approx. 8 mg protein) were applied to a prepacked column (pH 6.3) in 6 M urea (Mono P). The pH gradient was developed with 30 ml of 1/10 diluted polybuffer 74 (pH 4.0) containing 6 M urea. The absorbance at 280 nm and the pH of each fraction are as shown. Seven fractions were collected as indicated.

5.8 Selected References

1. L. A. AE. Sluyterman, J. Wijdeness, in *Proc. Int. Symp. on Electrofocusing and Isotachophoresis*, B. J. Radola, Graesslin, eds., Walter de Gruyter, Berlin, 1977, pp. 463–466.
2. L. A. AE. Sluyterman, O. Elgersma, *J. Chromatogr.*, *150*, 17–30 (1978).
3. L. A. AE. Sluyterman, J. Wijdenes, *J. Chromatogr.*, *150*, 31–44 (1978).
4. L. A. AE. Sluyterman, J. Wijdenes, *J. Chromatogr.*, *206*, 429–440 (1981).
5. L. A. AE. Sluyterman, J. Wijdenes, *J. Chromatogr.*, *206*, 441–447 (1981).
6. D. W. Darnall, I. M. Klotz, *Arch. Biochem. Biophys.*, *166*, 651–682 (1975).
7. P. G. Righetti, T. Caravaggio, *J. Chromatogr.*, *127*, 1–28 (1976).
8. P. K. Righetti, G. Tudor, K. Ek, *J. Chromatogr.*, *220*, 115–194 (1981).
9. W. J. Gelsema, C. L. DeLigny, *J. Chromatogr.*, *130*, 41–50 (1977).
10. W. J. Gelsema, C. L. DeLigny, N. G. Van Der Veen, *J. Chromatogr.*, *140*, 149–155 (1977).
11. N. Ui, *Biochim. Biophys. Acta*, *229*, 567–581 (1971).
12. L. S. Rodkey, *Protides Biol. Fluids*, *34*, 745–748 (1986).
13. Chromatofocusing with Polybuffer and PBE, Pharmacia Fine Chemicals, Uppsala, Sweden (1980).
14. D. D. Van Slyke, *J. Biol. Chem.*, *52*, 525–570 (1922).
15. H. Svensson, *Acta Chem. Scand.*, *15*, 325–341 (1961.
16. H. Svensson, *Acta Chem. Scand.*, *16*, 456–466 (1962).
17. O. Vesterberg, *Acta Chem. Scand.*, *23*, 2653–2666 (1969).
18. P. G. Righetti, in *Laboratory Techniques in Biochemistry and Molecular Biology*, T. S. Work, R. H. Burdon, eds., Elsevier, Amsterdam, 1983.
19. H. T. S. Britton, R. A. Robinson, *J. Chem. Soc.*, 458–473 (1931).
20. H. T. S. Britton, R. A. Robinson, *J. Chem. Soc.*, 1456–1462 (1931).
21. D. A. Ellis, *Nature*, (London), *191*, 1099–1100 (1961).
22. R. L. Prestidge, M. T. W. Hearn, *Sep. Purif. Methods*, *10*, 1–28 (1981).
23. T. W. Hutchens, C. M. Li, P. K. Besch, *J. Chromatogr.*, *359*, 157–168 (1986).
24. T. W. Hutchens, C. M. Li, P. K. Besch, *J. Chromatogr.*, *359*, 169–179 (1986).
25. T. W. Hutchens, *Prot. Biol. Fluids*, *34*, 749–752 (1986).
26. T. W. Hutchens, C. M. Li, P. K. Besch, *Prot. Biol. Fluids*, *34*, 765–768 (1986).
27. N. Y. Nguyen, A. Chrambach, *Anal. Biochem.*, *74*, 154–163 (1976).
28. C. B. Cuono, G. A. Chapo, *Electrophoresis*, *23*, 65–75 (1982).
29. C. B. Cuono, G. A. Chapo, A. Chrambach, L. M. Hjelmeland, *Electrophoresis*, *4*, 404–407 (1983).
30. M. T. W. Hearn, D. J. Lyttle, *J. Chromatogr.*, *218*, 483–495 (1981).
31. D. D. Perrin, B. Dempsey, *Buffers for pH and Metal Ion Control*, Chapman and Hall, London, 1974.
32. D. H. Leaback, H. K. Robinson, *Biochem. Biophys. Res. Commun.*, *67*, 248–254 (1975).
33. M. Page, M. Belles-Isles, *Can. J. Biochem.*, *56*, 853–856 (1978).
34. J. L. Young, B. A. Webb, D. G. Coutie, B. Reid, *Biochem. Soc. Trans.*, *6*, 1051–1054 (1978).
35. J. L. Young, B. A. Webb, *Anal. Biochem.*, *88*, 619–623 (1978).
36. J. L. Young, B. A. Webb, *Science Tools*, *25*, 54–56 (1978).
37. J. L. Young, B. A. Webb, *Prot. Biol. Fluids*, *27*, 739–742 (1979).
38. J. P. Edmond, M. Page, *J. Chromatogr.*, *200*, 57–63 (1980).
39. T. W. Hutchens, W. E. Gibbons, P. K. Besch, *J. Chromatogr.*, *297*, 283–299 (1984).
40. A. H. Couffalik, M. Feldman, M. Platica, L. Toth, D. Dreiling, V. P. Hollander, *Cancer Res.*, *43*, 5235–5242 (1983).
41. FPLC Ion Exchange and Chromatofocusing, Pharmacia Fine Chemicals (undated).
42. T. W. Huchens, R. D. Wiehle, N. A. Shahabi, J. L. Wittliff, *J. Chromatogr.*, *266*, 115–128 (1983).
43. T. W. Hutchens, H. E. Dunaway, P. K. Besch, *J. Chromtogr.*, *327*, 247–259 (1985).
44. T. W. Hutchens, C. M. Li, P. K. Besch, *Biochemistry*, *26*, 5608–5616 (1987).
45. A. Serrano, *Anal. Biochem.*, *154*, 441–448 (1986).
46. P. Weisweiler, C. Friedl, P. Schwandt, *Biochim. Biophys. Acta*, *875*, 46–51 (1986).
47. J. Ugelstad, L. Söderberg, A. Berge, J. Bergström, *Nature*, *303*, 95–96 (1983).

6 High Resolution Reversed Phase Chromatography

Milton T. W. Hearn
Center for Bioprocess Technology
Monash University
Clayton, Victoria
Australia

6.1 Introduction

The past decade has witnessed spectacular advances in chromatographic methods for the analysis and purification of polypeptides and proteins. To a large extent, these developments have been catalysed by the recognition that the hydrophobic effect can provide important selectivity options not available to separation methods based solely on differences in solute net charge or size.

In ion-exchange and several other liquid-solid chromatographies the interactions between the protein and adsorbent only involve strong polar forces. The selectivity is thus mediated only by differences in the solute's net charge and charge anisotropy. In contrast when weak van der Waals-type force interactions between proteins and non-polar (or polar) surfaces is used, different and unique aspects of protein structure contribute selectively to the separation.

Hydrophobic interactions are now commonly accepted to be of great biological significance in the determination of tertiary and quaternary structural hierarchy of proteins and also in the dynamics of protein motion in solution. Most proteins in their native, biological functional shape possess hydrophobic regions as part of their surface topography. Extended regions of the surface of a typical membrane protein involve these hydrophobic patches. There they play major roles in the orientation of the protein in the membrane phospholipid bilayer, in the transduction of biological signals across the membrane boundaries and in protein-protein aggregative phenomena. Even with water soluble globular proteins, a considerable proportion of the surface is hydrophobic in the sense that the atoms in these regions are not able to form in solution hydrogen bonds with the surrounding water molecules (see Chapter 7). Specific protein-protein recognition phenomena involved in the assembly of multi-subunit proteins, multi-enzyme complexes or intracellular organelles are all believed to be consequences of the hydrophobic effect. Further, solvent- and salt-induced protein precipitation is mediated to a large extent through this effect. Is it little wonder then that much effort has been devoted to minic these biological models in protein fractionation so that the hydrophobic effect is exploited in well designed chromatographic systems?

This chaper is devoted mainly to the separation of polypeptides and proteins by reversed phase high performance liquid chromatography (RR-HPLC). Recent advances both in the theory and practice of RP-HPLC as it applies to the fractionation of these biomolecules are critically assessed. Selected examples of proteins purified by RP-HPLC methods are also used to illustrate different chromatograhic stratagems. Polypeptide size has been deliberately limited by selecting examples with molecular weight greater than *ca* 3,000. The reader is directed elsewhere for recent reviews on the separation by HPLC techniques of amino acids and small peptides, including the important area of HPLC mapping of enzymatic digests of proteins[1,2,3].

Historically, the chromatographic separation of polypeptides and proteins via hydrophobic mechanisms has evolved out of studies with chemically modified soft polymeric gels. As early as 1950 the concept that polar components could be separated through the participation of weak, van der Waals-type interactions between the solute and a non-polar stationary phase was established by Boscott[4], Boldingh[5] and Howard and Martin[6]. They also coined the descriptive acronym for this separation mode—*reversed phase chromatography*, RPC—in contrast to normal or *polar phase chromatography* in vogue at that time. Here, as in paper chromatography, the solutes elute in increasing order of polarity. In contrast the distribution processes in RPC result in the elution of solutes in *decreasing* order of polarity. The solutes migrate in decreasing order of net charge, extent of ionisation, and hydrogen bonding capabilities.

At an early stage RPC polymeric matrices were adsorptively coated with non-polar stationary phase components such as long chain alkanes. These systems lacked long term stability and exhibited low resolution with most polypeptides and proteins. With the advent of chemically bonded n-alkylagaroses and similar materials[7,8,9] difficulties associated with rapid loss of the non-polar ligands were mostly circumvented. However, it was with the emergence of HPLC-media—meso- and macro-porous chemically bonded, hydrocarbonaceous supports of small particle diameters, typically in the range 5–10 μm, with narrow particle and pore size distributions—that RPC became a viable,

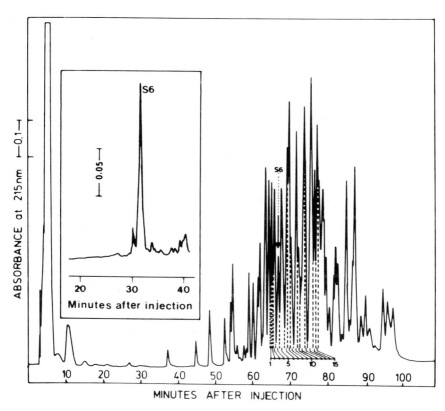

Figure 6-1. Resolution of proteins from whole rat liver ribosomes by reversed phase HPLC. A 6 M guanidinium hydrochloride extract of whole ribosomes (200 μg) was separated on a Bakerbond C_8 column using a 90 min linear gradient from 0.1 % TFA in water (A) to 0.1 % TFA-acetonitrile-water (50:50, v/v) (B) at a flow rate of 1 ml/min. A small amount of ^{32}P-labelled ribosomal protein S_6 was added to the extract. The elution position of protein S_6 (and ^{32}P-labelled S_6 under these conditions coincidental) is denoted by the arrow. Insert, fraction 4 was rechromatographed on a Bakerbond C_8 column using a linear gradient of 45 to 100 % B over 60 mins. After these two chromatographic procedures protein S_6 was homogeneous as revealed by N-terminal sequence analysis. Adapted from Ref. 10.

rapid, high resolution method for the analysis and purification of polypeptides and proteins (see for example Fig. 6-1).

Today silica-bound hydrocarbonaceous stationary phases are in wide use. This is due mainly to their excellent mechanical stability, important in the HPLC mode, and fairly robust chemical stability over a wide variety of elution conditions. Considerable versatility exists for changes in eluent composition thus allowing an extremely wide variety of substances to be resolved. Excellent selectivities can be achieved for polypeptides and proteins separated on chemically bonded n-alkylsilicas and other modern hydrocarbonaceous supports. With many 5- or 10-μm alkylsilicas now in use it is possible to achieve column efficiencies for polypeptides and proteins between 2,000—20,000 theoretical plates per metre, with short separation times of the order of minutes, and peak capacities approaching 250. This situation can be contrasted with conventional open column systems packed with soft gel matrices where efficiencies in the region of 100–200 theoretical plates per metre were obtained with separation times in the order of hours or even days.

6.2   Characteristics of Stationary and Mobile Phases; Sample Properties and Column Design

6.2.1   Properties of the Stationary Phase

The stationary phase in RPC typically consists of 5–10 μm diameter porous beads of silica, covered completely with covalently bound hydrophobic n-alkyl chains. The retention of proteins on the solid phase depends both on hydrophobic (solvophobic) and silanophilic effects. There have been discussions in the literature on the interplay between these effects[11–14]. Controversy still persists on the optimal stationary phase characteristics. An important reason is insufficient systematic data on a number of crucial stationary phase parameters. These include the uniformity of surface coverage with n-alkyl chains, ligand density, method used in the silica bonding treatment, and preparation history for the n-alkyl supports.

Generally, not all of the non-polar surface area is accessible to a polypeptide or protein, once it enters the pore. This is due to solute-ligand conformational constraints, surface tortuosity, ligand chain-length compression, solute solvation or other diffusion controlled factors.

A large variety of non-polar functional groups have over recent years been bonded to silica matrices, both as monolayers and as polymeric layers. The dimethylpropyl-dimethyloctyl, dimethyloctadecyl and the diphenyl phases are however normally used in polypeptide and protein separations. For compendia of available phases see Refs. 14–15. Some common solid phases are tabulated in Table 6-1.

Large pore silicas with specific surface areas as low as 50 m^2/g are very suitable support materials provided a narrow pore distribution exists. (See section 6.3.2 for an explanation). High coverage 5–10 μm particle diameter, 25 to 50 nm nominal pore diameter, n-alkylsilicas are currently the most popular. But, as is becoming increasingly apparent, these wide pore supports are not necessarily more suitable than smaller pore supports in every case of protein separation. It is now known[16] that the retention dependencies (but not relative retention) of polypeptides and protein on solvent composition are essentially independent of pore size at low sample loadings. With alkylsilicas of different porosity but identical alkylchain length and ligand coverage, the solute S-parameters are essentially constant although the extrapolated k_w values are different (the symbols are explained on page 189). The important difference between stationary phases of different porosities relates to the accessibility of surface area, a feature adequately predicted by theory.

It has been widely assumed in the literature that the unsatisfactory retention behaviour of polypeptides and proteins seen with some narrow pore (≤ 10 nm) n-alkylsilicas is due to their physical entrapment arising from restricted diffusion in the pore chambers. Ten-fold differences in surface area per gm of packing material are commonly found between small and large pore n-alkylsilicas. Hence eluents with significantly higher elutropic strength are needed to achieve the same retention, k-value, with small pore alkylsilicas (≤ 10 nm) compared to large pore alkylsilicas (≥ 30 nm).

Often these eluents are unacceptable as proteins solvents and do not permit elution development and acceptable recovery.

The alkylchain length has been found to have only a small influence on relative retention if the alkylchain densities are the same. However, it does affect recoveries. For some proteins with $M_r \geq 20,000$ recoveries can for example be improved when an octyl or butyl phase is substituted for an octadecyl phase.

The chemical characteristics of the parent silica, and the pretreatment history have significant effects on selectivity. The presence of trace amounts of adsorbed metal ions, the water content of the parent silica prior to bonding and the relative Brönsted acidity of the silica particles all affect the resolution in RP-HPLC separations of polypeptides and proteins. In order to minimise the free silanol group content a variety of strategies have been attempted. These include vigorous endcapping, sequential bonding and the use in combination of two monofunctional alkylsilanes of different chain length.

Theory predicts that improvement in chromatograhic performance will be obtained as the surface coverage becomes more uniform. This can be achieved, for example, through the use of chlorotrimethylsilane in combination with chloro-octyldimethyl-silane. Such an approach has been employed[17] with the ProRPC stationary phases. A number of other options are available, to decrease S-parameters, such as branched-chain phases[18,19,20] and hydrophilic phases sequentially bonded with non-polar groups[21–23].

6.2.2 Different Mobile Phases

All of the common water-soluble organic solvents have now found use in separation of proteins by reversed phase HPLC (see Table 6-1), although their efficacy in terms of elutropic strength and protein recovery differ considerably. Organic solvent modifiers not only affect the gross properties of the mobile phase (through changes in surface tension, dielectric constant, viscosity etc.) but also interact in specific ways with the proteins themselves. At concentrations below 0.1 % v/v, alcohols and other dipolar organic solvents can stabilise the structure of some proteins, e.g. chymotrypsinogen, whereas at higher concentrations these solvents usually induce either reversible or nonreversible structural deformation[24–26]. The denaturing effects of most organic solvents can be attributed to regional disruption of the hydrophobic interactions between nonpolar side-chains in the protein. Addition of an organic solvent modifier to an aqueous medium containing a protein in its native biological conformation will, in general, alter the hydration structure of this protein. Whether this change leads to denaturation depends on the specific equilibria associated with each protein. Where mobile phase-induced denaturation occurs, an increase in surface contact with the stationary phase is anticipated. The retention value k (the protein elution volume expressed in number of void volumes) becomes larger for proteins in the denatured (and carboxymethylated) form with low pH eluents[1,27].

The elutropic strength E^o of the eluent increases with its content of organic solvent (expressed in vol fraction) as confirmed by several experimental studies[27–29]. The relative retention k for a particular polypeptide decreases in the series of solvent modifiers methanol < ethanol < acetonitrile < 1-propanol or 2-propanol at the same volume percentage. The solvent composition range over which the log k value for a particular polypeptide or protein remains small (that is the protein elutes close to the volume expected for pure size exclusion and without evidence of selectivity reversals) is considerably greater with 1-propanol than for acetonitrile. This dependency, coupled

Table 6-1. Selected Examples of Polypeptides and Proteins Purified by Reversed Phase HPLC

Column	Hydrophobic chain	Mobile phase	Protein purified	Ref.
μBondapak C18	n-octadecyl	0.1% TFA, 12–70% CH_3CN	Acid Phosphatase	13
Nucleosil C18	n-octadecyl	0.1% TFA, EtOH-butanol-methoxyethanol	Aldolase	82
μBondapak C18	n-octadecyl	5% formic acid 40–80% EtOH	Bacteriorhodopsin	83
Nucleosil C18	n-octadecyl	50 mM KH_2PO_4, 10–50% 2-methoxyethanol	Bacteriorhodopsin	85
LiChrospher C8	n-octyl	400 mM Pyr-formate, 0–40% nPrOH	Bovine Serum Albumin	84
μBondapak alkylphenyl		1% TEAP, 10–50% CH_3CN	C-apolipoproteins	86
Aquapore RP300		0.1% H_3PO_4, 10 mM $NaClO_4$, 0–60% CH_3CN	Calmodulin	87
Ultrasphere C8	n-octyl	10 mM TFA, 0–45% nPrOH	Carbonic anhydrase	18
Ultrasphere C3	n-propyl	155 mM NaCl, pH 2.1, 0–75% CH_3CN	Carbonic anhydrase	18
μBondapak C18	n-octadecyl	0.1% TFA, 0–60% CH_3CN	Chorionic gonadotropin	88
LiChrospher C4	n-butyl	10 mM H_3PO_4, 0–45% nPrOH	Chymotrypsinogen	11
Bakerbond diphenyl	diphenyl	0.1% TFA, 0–50% CH_3CN	Collagen chains	89
U-ODS	n-octadecyl	0.2% HFBA, 0–50% CH_3CN	Epidermal Growth Factor	90
Nucleosil C18	n-octadecyl	50 mM KH_2PO_4, 10–50% 2-methoxyethanol	Ferritin	85
Pharmacia ProRPC	n-octyl	0.3% TFA, 39–50% CH_3CN	Globin chains	42

Column	Ligand	Mobile phase	Protein	Ref.
LiChrospher C4	n-butyl	100 mM NH_4HCO_3, 0–50% CH_3CN	Growth hormone	91
μBondapak C18	n-octadecyl	0.2% TFA, 0–50% CH_3CN	Histone proteins	92
LiChrospher C8	n-octyl	0.8 M Pyr-1 M formic acid, 0–60%	Human fibroblast interferon	93
Ultrasphere C3	n-propyl	0.1% TFA, 0–50% CH_3CN	Inhibin, follicular	50
LiChroprep RP8	n-octyl	0.9 M Pyr-acetate, 20–60% nPrOH	Interleukin-2	94
W-DP		50 mM KH_2PO_4 2-methoxyethanol-PrOH	Leukocyte interferon	95
N-ODS	n-octadecyl	33 mM NaOAc, 10–50% EtOH, pH 5.2	Macroglobulin, α_2-	96
μBondapak C18	n-octadecyl	12 mM HCl, 35–80% EtOH	Microglobulin, β_2-	97
unspecified C8	n-octyl	0.1% TFA, 0–60% iPrOH	Ovalbumin	98
LiChrospher C4	n-butyl	10 mM H_3PO_4, 0–45% nPrOH	Papain	56
LiChrospher C18	n-octadecyl	50 mM Tris-HCl, 0.1 mM $CaCl_2$, 0–70% iPrOH	Parvalbumin	99
Ultrasphere C3	n-propyl	155 mM NaCl-HCl, 0–50% nPrOH	Phosphorylase	19
LiChrospher C8	n-octyl	0.1% TFA, 0–50% iPrOH	Platelet derived Growth Factor	100
μBondapak C18	n-octadecyl	0.1% H_3PO_4, 10–75% CH_3CN	Proinsulin	47
Various		0.1% TFA, CH_3CN-PrOH	Rhodopsin	99
Ultrapore RPSC		0.1% TFA, 10–80% CH_3CN	Ribosomal 50S proteins	10
Synchropak RPP		0.1% TFA, 15–75% CH_3CN	Ribosomal proteins	10
ToyaSoda TMS250		0.2% TFA, 25–75% CH_3CN	Sendai viral proteins	82
LiChrospher C18	n-octadecyl	15 mM H_3PO_4, 0–60% CH_3CN	Thyroglobulin	16
Bakerbond C18	n-octadecyl	100 mM NH_4HCO_3, 0–50% CH_3CN	Thyrotropin and subunits	74

with the more rapid decrease in bulk surface tension at higher water content is advantageous for 1- and 2-propanol compared to acetonitrile or methanol. The propanols give shorter relative retention, generally higher mass recoveries and less sensitivity to alkylchain length effects with silica-based packing materials of nominally equivalent matrix characteristics.

Although relative retention for polypeptides and proteins decreases as the elutropic strength of the organic modifier increases it does not necessarily follow that overall peak capacity or even regional resolution will improve with more effective desorbing solvents. For example with gradients of the same slope, different peak capacities will occur when different organic solvents are employed as is evident with the acetonitrile- and 1-propanol-based elution profiles shown in Fig 6-2.

It is often advantageous to use ternary mixtures containing water and two organic solvents. When their compositions are chosen to maximise the log k versus φ slope this might lead to improved selectivity. With such systems hydrophobic proteins such as cytochrome c oxidase subunits[30] may be eluted with a mobile phase of higher water content than can be achieved with either organic solvent.

The addition to binary mobile phases of small amounts of polar and non-polar solvents such as 0.1–0.5 % v/v ethylene glycol, formamide, tetrahydrofuran, dioxane, butanol, n-pentanol or the non-ionic detergents is anticipated to influence the stationary phase surface. These composite phenomena will affect both the magnitude of the multisite binding and the accessibility of the hydrophobic regions on solute surface. Typical of this behaviour are the plots[31] of the dependence of the capacity factor of several polypeptides on the concentration of the detergent Brij 35 in the mobile phase. This and the corresponding gradient profiles are shown in Fig 6-3 and Fig. 6-4. When high adsorption isotherms are observed with binary mobile phases the use of ternary solvent modifiers may provide a remedy of these adverse sorption-desportion events. If no attention is given to these situation, erratic k values and peak shapes symptomatic of 'irreversible' binding will rise. In many cases though, with strongly retained solutes, protein solubility limits the use of a secondary solvent. Then further options must be sought other than simply increasing the column fraction of the primary or secondary organic solvent modifiers. Different pH conditions, or the use of polar or suface-active ionic modifiers should be contemplated immediately.

6.2.3   The Sample: Protein Solubility and Avoidance of Protein Denaturation

One feature of conventional n-alkylsilica stationary phases which has caused concern to many protein chemists has been the propensity of these hydrophobic supports to cause protein denaturation under some operational conditions. Although the more naïve investigator has tended to rush into hasty conclusions about the generality of this phenomenon, practice teaches that denaturation and/or impaired mass recovery are not a general effect with high coverage n-alkylsilicas and aquo-organic solvent eluents of low to intermediate pH and buffer capacity. Rather the opposite can be the case with denaturation as a consequence of inappropriately selected operation conditions. Sufficient experimental data are now available with large polypeptides and proteins encompassing the molecular weight range 5,000–65,000 to enable the following four scenarios to be identified. Each is time dependent and responsive to manipulation of thermodynamic and kinetic variables such as temperature or ligand surface heterogeneity.

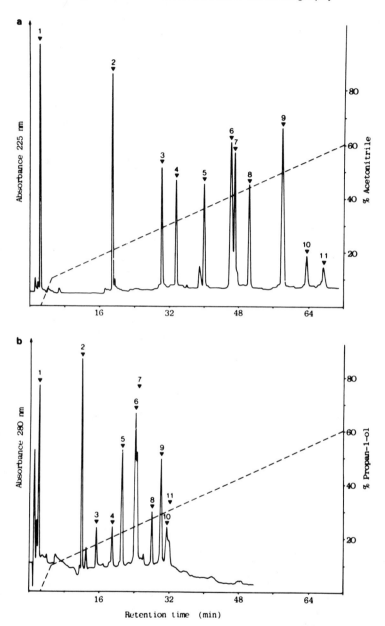

Figure 6-2. Comparative elution profiles using acetonitrile- or n-propanol-based gradients of equivalent rate of change of organic solvent modifier concentration. The protein key is 1, tryptophan; 2, ACTH (1–24); 3, ribonuclease; 4, human calcitonin; 5, lysozyme, 6, bovine serum albumin; 7, bovine β-lactoglobulin; 8, lactoglobulin A; 9, bovine prolactin; 10, rat prolactin; 11, rat growth hormone. Separations were carried out at a temperature of 45°C, a flow rate of 1 ml/min on a Ultrapore C_3 column (75 × 4.6 mm (i.d.)) and eluted with acetonitrile and a primary colvent of 0.155 M NaCl/0.01 M HCl (a) and 1-propanol and a primary solvent of 0.01 M TFA (b) using the gradient shapes indicated. Adapted from Ref. 19.

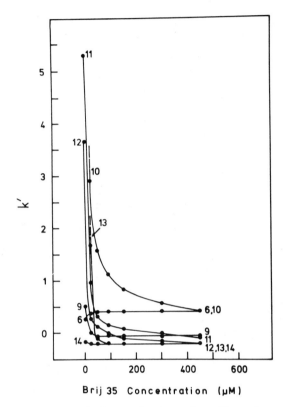

Figure 6-3. Dependence of the capacity factors of various polypeptides on the concentration of the nonionic detergent Brij 35 in the mobile phase. Chromatographic conditions: octadecylsilica, Flow rate, 1.0 ml/min; temperature, 18°C; mobile phase, acetonitrile-water (30:70)-50 mM KH_2PO_4, -15 mM H_3PO_4, pH 2.3 containing various concentrations of the detergent. Polypeptide key: 1, trityrosine; 2, angiotensin 1; 3, porcine glucagon; 4, bovine insulin B-chain; 5, bovine insulin; 6, cytochrome C; 7, hen lysozyme. Adapted from Ref. 53.

The first situation represents the optimal case history in which both the mobile phase composition and stationary phase do not cause observable changes in activity or recovery, e.g. serine proteases[1,27] inhibins[32]. Underlying this optimal behaviour may well be subtle conformational effects but these do not impinge on the chromatographic behaviour during the time required for mass transport through the chromatographic bed.

The second situation involves mobile phase mediated changes in conformation and is independent of the stationary phase per se, e.g. catalase[33] when the mole fraction of organic modifier is $> 15\%$. These solvent-induced denaturation effects with some proteins in acidic low ionic strength eluents reflect surface tension dependent changes in the solvated structure of the protein. They are thus not generally expected to respond to changes in incubation or dwell time at the stationary phase surface. Experimentally, characterisation of this effect is easy. From several small scale pilot experiments protein stability in mobile phases of different composition is studied.

The third situation involves ligand or matrix mediated changes in conformation and

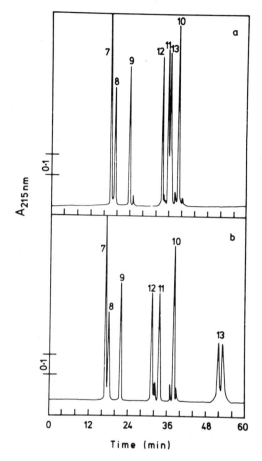

Figure 6-4. Gradient elution separation of several polypeptides on an octadecylsilica column in the absence (a) or presence (b) of 0.01 % Brij 35 (w/v). Chromatographic conditions: column 10 nm pore diameter octadecylsilica; flow rate, 1.0 ml/min; initial conditions, (A) acetonitrile-water (15:85 v/v)—50 mM NaH_2PO_4–15 mM H_3PO_4, final conditions, (B) acetonitrile-water (50:50 v/v)—50 mM NaH_2PO_4—15 mM H_3PO_4. The detergent was added at 0.01 % (w/v) to both the initial and final conditions in (b). Linear gradient from (A) to (B) over 60 mins. Polypeptide key: 1, angiotensin III; 2, angiotensin II; 3, angiotensin I; 4, bovine insulin; 5, bovine insulin B-chain; 6, cytochrome c; 7, porcine glucagon. Adapted from Ref. 53.

is independent of mobile phase composition e.g. pepsin[33], β-lactoglobulin[34]. This effect responds to column dwell phenomena and is thus depenent on the stationary phase surface area, porosity and ligand density, or on the elution time. Again remedies for this effect have been developed, viz. short columns, steep gradients and non-porous stationary phases.

Finally, the composite interplay of a particular mobile phase composition with a particular stationary phase may also give rise to protein denaturation whilst slight variation in the phase ratio or the eluent composition does not. The last situation represents the most troublesome circumstance for the novice who becomes disenchanted, although a perfectly suitable procedure might have been found if additional effort or imagination had been applied. The variation seen in recovery of a particular protein with stationary phases of the same base silica matrix, nominal chain length and surface coverage from different commercial sources can often be ascribed to this last effect.

Recently methods[1,32–36] to rapidly characterise these effects in reversed phase HPLC have been well established and a new approach to characterise conformational changes with polypeptides and proteins on-line developed. This latter technique—called chromatotopography[36]—provides a rapid method to follow mobile phase or stationary phase induced changes in protein 3-D structure provided intermediates have halflives >10 times the time required for the protein to traverse a theoretical plate. The theoretical basis for evaluating these interconversions is now firmly established[34–37] for both analytical and semi-preparative chromatographic separations.

An alternative approach to overcome mobile phase or stationary phase induced conformation changes in RP-HPLC whilst still utilising the hydrophobic effect involves the use of salt-mediated hydrophobic interaction stationary phases. These can be either of the ether-type[38] or the amide-type[39,40] ligand characteristics. Compared to the n-alkylsilicas, the peak capacities of these hydrophobic interaction phases are lower, typically by a factor of 5–10, and although bio- and mass recovery may be improved, the disadvantage of subsequent desalting must be overcome. This additional step frequently represents the greatest loss of material with micropreparative separations.

Post column sample manipulation represents a poorly investigated area of development. The regeneration of bioactivity by proper attention to the collection technique, eluent neutralisation and concentration can permit an otherwise inactivated protein to be reconstituted in active form, e.g. alkaline phosphates, horse radish peroxidase[33]. Again what is required is a thoughtful strategy which accommodates the vagaries of the protein of interest, rather than just a "hit or miss" attempt totally dictated by a cursory application of one or two of the currently popular recipes otherwise peak shapes symptomatic of 'irreversible' binding poor mass or bio-recovery and low resolution performance could extenuate.

6.2.4 Column Design

Advantage can be taken of the high affinity and attendant multisite binding of polypeptides and proteins for alkylsilicas in the design of column configuration. For example, a 10-fold increase in column length results in only a small improvement in resolution for many globular proteins and such gains in chromatographic performance are usually at the expense of protein recovery. As a consequence, columns with dimensions (10–15 × 0.6–0.8 cm) are gaining popularity both for analytical and semipreparative separations. In contrast to high performance ion-exchange, process scale purification of proteins by reversed phase procedures still remains underdeveloped. However, multi-gram purification of polypeptides using such procedure has been realised[41,2] for several years and these procedures will certainly find additional advocates.

The large adsorption isotherms manifested by polypeptides and proteins with n-alkylsilicas has another practical benefit. Trace enrichment from very dilute solutions, with (ultimately) relatively high loadings, can be readily achieved, i.e. loadings up to 50 mg of ribonuclease in volumes up to 500 ml did not significantly reduce retention or recovery on a conventional analytical column (μBondapak C_{18}, 25 × 0.4 cm). This feature permits the rapid processing of large volumes of a biological extract, particularly in situations when the chromatographic behaviour of the desired component is already known.

6.3   Basic Theory

6.3.1   General Considerations

In the majority of the published literature on RPC separations of polypeptides the selection of chromatographic parameters has generally been made on empirical criteria. Typical conditions for separation on a macroporous octyl- or octadecylsilica support is as follows: low pH organic solvent combination (usually encompassing the 0–50% (v/v) solvent), relatively low ionic strength (usually in the range 10–100 mM), a flow rate between 0.5–2.0 ml/min and ambient temperature. Figure 6-5 illustrates a separation representative of these conditions, namely the resolution of globin chains from a heterozygous haemoglobin carrying a β-chain nutation (36 Pro → Thr)[42,17].

Both the nature of the solute-stationary phase interaction and the wide structural diversity encountered with even a relatively simple protein mixture requires gradient elution. If the sample contains many substances of widely differing relative retentions isocratic elution has generally been found to be ineffective. Components with short retention times tend to be poorly resolved whilst strongly retained compounds often elute as asymmetric peaks with excessive band broadening. This chromatographic behaviour is a direct consequence of the co-operative multi-site interactions between the proteins and the heterogeneous stationary phase surface. Here gradient elution provides a simple expedient to reduce separation times, decrease peak volumes and improve peak symmetries.

The same variables which control retention, bandwidth and resolution in isocratic elution are also relevant in gradient elution. Thus, isocratic retention data can be used to calculate corresponding gradient retention data and vice versa. Gradients can yield equal bandwidths for all compounds separated and roughly comparable values of retention expressed as the effective median capacity factor, $\bar{k}$. Gradient steepness parameters can be easily varied to adjust bandspacing under gradient conditions as part of the general optimisation of resolution.

Available evidence supports the concept that the extent of retention for peptides and proteins on silica-based hydrocarbonaceous supports follows their topographic surface polarity. That is, with water-rich eluents the solutes elute in order of increasing effective hydrophobicity. Several predictive methods[1,2,43–45] have evolved for the estimation of peptide selectivity on alkylsilicas. With small polypeptides of known sequence, up to 5000 M_r, reasonable correlation is found between the predicted and experimentally-observed retention behaviour. With larger polypeptides and proteins these approaches are much less reliable. Here the data base does not adequately take into account amino acid positional effects and unique changes in the secondary or tertiary structure of a particular protein which arise under different elution conditions. Inherent in these treatments is the assumption that only a single, conformationally stable molecular species is involved in the distribution phenomena. In fact it is highly likely that time-dependent molecular orientation and re-orientation of all biopolymeric solutes occur at the stationary phase

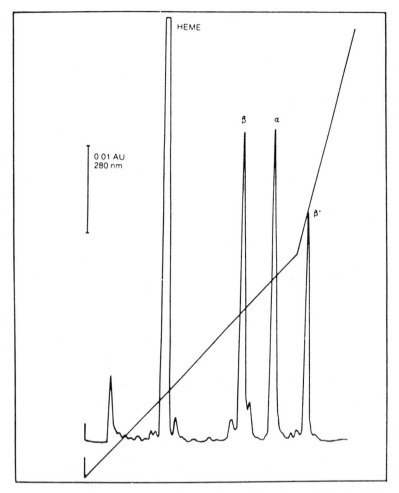

Figure 6-5. Separation of globin chains from a heterozygous haemoglobin carrying a $\beta 36$ (proline → threonine) chain mutation. Chromatographic conditions: column ProRPC HR 5/10, gradient, solvent A, 0.3 % TFA-water-acetonitrile (61:39); solvent B, 0.15 % TFA-water-acetonitrile (55:45); 60 min linear, flow rate, 0.2 ml/min. Detection, 280 nm. Adapted from Ref. 42. By permission of authors and publishers.

surface. As a consequence, the overall adsorption-desorption behaviour of these molecules on modern high performance chromatographic supports often reveals retention and kinetic phenomena not obvious with conventional soft gels.

6.3.2 Relative Retention

Solute retention is frequently expressed in terms of the capacity factor, k', (called k, see Chapter 2, p. 48). This is defined by the phase ratio of the chromatographic system, (equal to the ratio of the volumes of the stationary to the mobile phase V_s/V_m) and the adsorption (distribution) coefficient, K_D,

$$k = \phi \cdot K_D \qquad (6\text{-}1)$$

k can also be expressed in retention time, t_R (or volume, V_R) and the column void time, t_0 (or volume V_0)

$$k = \frac{t_R - t_0}{t_0} = \frac{V_R - V_0}{V_0} \tag{6-2a}$$

or from the column parameters

$$k = \frac{t_R \cdot v}{L} - 1 \tag{6-2b}$$

where v is the linear velocity of the eluent, and L the column length. Typical values for k are $1 \leq k \leq 20$, where $k = 1$ implies no interaction with the solid phase but complete permeation through the porous system.

Recent experimental evidence has indicated that retention mechanisms based solely on van der Waals interactions do not adequately describe all the different types of separation effects seen. Alternative models, derived from thermodynamic, mechanistic and non-mechanistic arguments, can be advanced. In all of these theoretical treatments it is recognised that the overall retention is the composite of several factors. These are size exclusion, interactive processes involving hydrophobic interactions, and coulombic and hydrogen-bonding interactions. Usually chromatographic conditions can be selected so that regular reversed phase retention behaviour dominates, i.e. contributions to retention from polar effects are not significant and k decreases monotonously as the volume fraction, φ, of an organic solvent modifier such as acetonitrile or n-propanol increases. Achieving the appropriate compromise between the primary distribution process and secondary equilibria mediated by coulombic effects or other forms of polar interaction is the key to high resolution and frequently to high recovery.

For a specific column, the capacity factor k for a specific solute can be expressed in terms of eluent composition, temperature and pH. In the absence of slow competing conformational equilibria, the retention of polypeptides and proteins on n-alkylsilicas has been found to follow the (linearised) form

$$k_i = k_{sec,i} + k_{w,i} \cdot e^{-S_i \varphi} + k_{o,i} \cdot e^{-D(1-\varphi)} \tag{6-3}$$

where the three terms are, in order, the size exclusion term, the solvophobic term and the silanophilic term. Here k_w and k_o are the capacity factors of the solute, i, in neat water and at the final organic solvent modifier concentration respectively. S and D are solute and condition dependent variables[29,46]. As the mole fraction of organic modifier increases k will decrease, reach a minimum at the φ-value where

$$\varphi = (\log k_{w,i} + \log k_{o,i} - D)/(S - D) \tag{6-4}$$

and then increase up to the point of solute precipitation. Figure 6-6 shows a schematic representation of the dependency of k on φ. As the magnitude of the S and D-variables increase then the elution window over which realistic k values can be achieved will progressively narrow. Experimental observations are in good agreement with this anticipated behaviour.

6.3.3   Isocratic Elution

Over the retention range consistent with reasonable reversed phase chromatographic practice, i.e. over the range $1 \leq k \leq 20$, the relationships between organic modifier volume fraction and retention can be approximated to

$$\log k_i = \log k_{i,w} - S\varphi \tag{6-5}$$

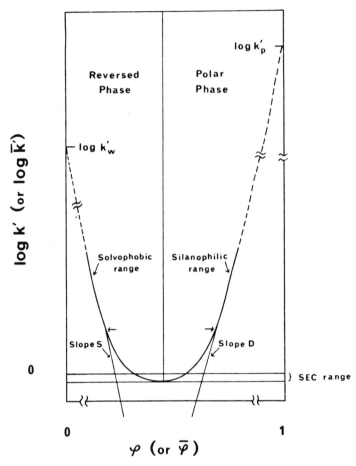

Figure 6-6. Schematic representation of the interplay between size exclusion, solvophobic and silanophilic phenomena on the retention dependency between the logarithmic capacity factor, $\log k'$, and the mole fraction, φ, of the organic solvent modifier. The extrapolated intercepts of $\log k'$ at $\varphi = 0$ and $\varphi = 1$ (dashed lines) are indicated as $\log k'_w$ and $\log k'_o$ respectively. The tangents to the curves at a specified k' value, for example, $\log k' = 2.5$, are given by S and D respectively. The $\log k'_{sec}$ range corresponds to the retention time of t_R (hydrodynamic) to t_R (total inclusion) for non-interactive solutes of very large and small molecular weight respectively. For convenience $t(\text{sec})$ of most polypeptides can be set at t_0 of the column when $\log k'_w$ is large.

where φ typically has the limits $0 \leq \varphi \leq 0.5$, i.e. between 0 and 50% organic solvent. Since the slope parameter S is solute and condition dependent, evaluation of S-value dependencies provides crucial information for both isocratic and gradient optimisation. Both non-mechanistic and thermodynamic treatments predict an inverse relationship of $\log k$ on φ over the range of mobile phase compositions consistent with reversed phase behaviour. Experimental verification of these relationships has been provided by several studies[16,29,47,48], very steep non-linear dependencies between $\log k$ and φ exist with polypeptides and proteins with S-values typically greater than 20. Small peptides in contrast exhibit S-values usually in the range 2–10.

Isocratic retention can thus be classified in various ways as summarised in Fig. 6-7. A typical plot of log k versus φ for a protein is shown by curve (a) where only a narrow range in solvent percentages (or solvent concentrations) exists between the φ-value at which the protein first begins to migrate in a column and the final φ-value at which k effectively becomes k_{sec}, i.e. not retarded at all.

6.3.4   Gradient Elution

Because of the interrelated molecular and chromatographic complexities, the separation of polypeptides and proteins by reversed phase HPLC is typically carried out by gradient elution. The gradient is described by the steepness parameter b, where

$$b = S\Delta\varphi t_o/t_G \tag{6-6a}$$

where $\Delta\varphi$ is the change in φ during the gradient of duration t_G. An alternative expression for steepness is

$$b = S \cdot V_m \cdot \frac{d\varphi}{dt} \bigg/ F \tag{6-6b}$$

which interrelates the flowrate F, the column volume V_m and the rate of change of φ with time.

For linear solvent strength gradients, that is with gradients which result in constant b-values for all compounds eluted, the effective k-value can be expressed, using the Snyder gradient model[5], by $\bar{k}$. This corresponds to the time taken for the solute to traverse halfway along the column. Under such elution conditions and when t_{sec} approached t_o, then $\bar{k}$ can be approximated by

$$\bar{k} = 1/1.15b = Ft_G/1.15\Delta\varphi SV_m \tag{6-7}$$

It can thus be seen that changes in b or $\bar{k}$ are equivalent to changes in the average φ-value ($=\bar{\varphi}$). Values of b for each component can be readily calculated for a particular chromatographic system and specified polypeptide or protein mixture from multiple gradient runs with different t_G, F or $\Delta\varphi$. From this computed b-data, values of S, k_w, $\bar{k}$ and $\bar{\varphi}$ can be calculated and plots of $\bar{k}$ versus $\bar{\varphi}$ (formally equivalent to k versus φ) generated. This in turn allows gradient retention time for any defined b-value to be predicted[51-55].

6.3.5   Peak Capacity of Reversed Phase System

Changes in b, $\bar{k}$ or $\bar{\varphi}$ will have profound effects on band spacing between eluted polypeptides and proteins and dramatically influence the average resolution of the system. The situation in which the resolution of polypeptides with small selectivity differences changes when the gradient slope is changed is probably familiar to many investigators (see Fig. 6-8 for one such example). With solutes of different S-values loss of resolution and even selectivity reversals can be anticipated from the above treatment when gradients of different rate of change of organic modifier ($d\varphi/dt$) or flow rate are employed. For convenience average resolution can be defined in terms of peak capacity PC (not to be confused with the capacity factor k) such that

$$PC = t_G/4\sigma_t = t_GF/4\sigma_v \tag{6-8}$$

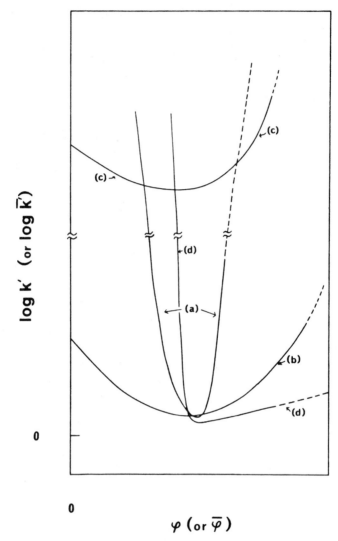

Figure 6-7. Schematic representation of the retention dependency of $\log k'$ on φ for solutes of different molecular weight and hydrophobicity. In case (a) both the S- and D- slopes are large and the corresponding $\log k'_w$ and $\log k'_o$ values are also large. This behaviour is typical of many polypeptides and proteins in water-organic solvent-low pH systems with some n-alkylsilicas. In case (b) the S- and D-slopes are small with corresponding small $\log k'_w$ and $\log k'_o$ values. This behaviour is typical of small peptides. In case (c), the S- and D-slopes are small but $\log k'_w$ and $\log k'_o$ values are large due to incorrect choice of chromatographic selectivity. In case (d), the S-slope is large but D-slope remains small. This behaviour has been observed with fibrous proteins and some proteins eluted with hydrogen donating solvents. Other permutations have been documented, i.e. small S-slope but large D-slope, etc. (see for example Refs. 21, 46, 47, 49, 50). Generally S increases with molecular weight and for smaller polypeptides takes the approximate form $S = 2.99 \, (\text{M.W.})^{0.21}$.

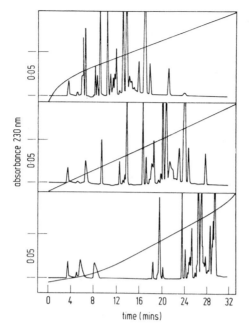

Figure 6-8. Change in selectivity and bandspacing induced in a chromatographic separation by variation of the gradient steepness parameter. Profiles show the separation of the tryptic peptides of haemoglobin A eluted an 0.1% H_3PO_4-water-acetonitrile gradient.

With well packed columns, containing 4–7 μm particles maximally bonded with C_8- or C_{18}-chains, peak capacities approaching 250 can be achieved with appropriate gradient steepness parameters. This equates with a peak width of *ca* 500 μl for a component eluting at 60 min at a mobile phase flow rate of 2ml/min.

The resolution equation in gradient elution takes the familiar form, namely

$$R_S = \tfrac{1}{4}(\alpha - 1)N^{1/2}(\bar{k}/1 + \bar{k}) \qquad (6\text{-}9a)$$

or

$$R_S = \left(\frac{k_1}{k_2} - 1\right)\left(\frac{\bar{k}}{1 + \bar{k}}\right)\sqrt{L}\,\bigg/\,\frac{1}{4\sqrt{H}} \qquad (6\text{-}9b)$$

where α is the selectivity factor (ratio of k values) for band-pairs when the bands traverse the column mid-point, and N is the theoretical plate number at the same point. Hence for a given chromatogram when selectivity is maintained constant, maximising peak capacity is synonymous with maximising average resolution. Both PC and R_S will thus show inverse dependencies on the b-term with maximum values of PC and R_S favoured by smaller values of b, that is for given solutes (S-values) and gradient values ($\Delta\varphi$) by larger t_G values (Fig. 6-9). With very long t_G values a compromise arises in the sense that peak height and sensitivity, both inversely proportional to bandwidth, will also decrease as the k of the band increases. Higher peak detection limits are thus favoured with steep gradients but obviously resolution will suffer.

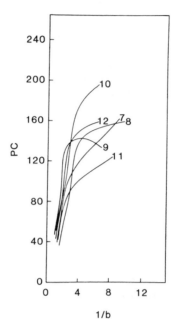

Figure 6-9. Plot of peak capacity , PC, versus the reciprocal of the gradient steepness parameter. As *b*-values increases PC decreases. However, the converse is only true at very small *b*-values when on anomalous kinetic effects occur with dramatically influence the bandwidth of the migrating peak. Data adapted from Ref. 55.

6.3.6   Kinetic Effects

Up to this point the discussion has assumed that protein retention in isocratic or gradient modes involves only a single, conformationally stable molecular species. The retention of proteins at n-alkylsilica surfaces is obviously far more complex than implied by this assumption. The reversible and non-reversible deformation of proteins in water/ organic solvent combinations at the hydrocarbonaceous interface must certainly be considered. The participation of other secondary equilibria processes mediated by the mobile phase or stationary phase characteristics is also likely. In many cases, these effects will be solute-specific and time-dependent. Inappropriate kinetics associated with these equilibria processes can lead to band broadening, at times bizarre in terms of peak skew, asymmetry or degeneracy. The participation of some secondary equilibrium processes such as ionisation or ion-pair formation can usually be exploited to enhance selectivity through appropriate choice of mobile phase conditions, i.e. by variation of pH or co-/ counter-ion concentration. Solute conformational changes or changes in subunit associa- tion/dissociation are generally undesirable secondary equilibrium processes in chromato- graphic separations, particularly when the kinetics of these phenomena are not rapid with regard to the time scale of the chromatographic separation process.

As an illustrative case, the first order reversible binding of a protein, P, with native globular structure to a hydrocarbonaceous stationary phase and the concomitant first order interconversion of the native form to an unfolded form $P*$ can be considered. Such a situation probably applies to the RP-HPLC of many proteins including ribonuclease[56],

α-chymotrypsinogen[52] and papain[57]. The interconversion cycle for such a system can be represented by[52,55]:

$$
\begin{array}{ccc}
P_m & \underset{k_{41}}{\overset{k_{14}}{\rightleftharpoons}} & P_m^* \\[2mm]
k_{21} \Big\| k_{12} & & k_{34} \Big\| k_{43} \\[2mm]
P_S & \underset{k_{32}}{\overset{k_{23}}{\rightleftharpoons}} & P_S^*
\end{array}
\qquad (6.10)
$$

where the subscripts, m and s, refer to the mobile and stationary phases respectively. According to this model the concentration of the protein, P, can change biphasically depending on the chromatographic conditions and whether denaturation occurs in the mobile phase and/or at the stationary phase surface. When the overall time constant for an interconversion cycle between P and P^* is comparable to the time of passage of P through the column bed then an apparently homogeneous protein can elute as composite, skewed peaks with different retention times for the native, partially unfolded and fully unfolded forms. The choice of column configuration, elution velocity and particularly the stationary phase characteristics (surface area, ligand density, pretreatment history) will influence these processes. The rate constant for interconversion of P and P^* at the stationary phase of the order of $8-20 \times 10^{-4}$/sec at $20°$ has been observed[52,57] for several proteins.

It can be recognised from the above discussion that column efficiency which reflects, inter alia, the adsorption/desorption kinetics will depend not only on the usual chromatographic parameters such as flow rate, column length, temperature etc., but also on the residence time the solute spends at the stationary phase surface. In a typical polypeptide or protein separation on alkylsilicas the plate height is expected to be dominated by stationary phase kinetics contributions[3,58-60].

6.4 Chromatographic Techniques

6.4.1 General Considerations

The aim of all high resolution chromatographic separations is to achieve maximum selectivity with minimum zone broadening. Further, solute elution should occur rapidly, in high yield and reproducibly.

Few laboratories have access to the necessary silicabonding technology or are directly involved in the preparation of specialty 'in-house' hydrocarbonaceous silica-based supports. Thus success with complex polypeptide or protein separations on commerical alkylsilicas depends to a very large extent on the ability to manipulate the mobile phase composition. In some cases, selections of different commerical support materials packed into columns of different configuration can be used on a "trial and error" basis with a single mobile phase composition. But when details of the stationary phase characteristics are poorly documented, such empirical approaches rarely shed much useful light on general trends to improve resolution and recovery.

There are obviously some physical and surface characteristics of n-alkylsilica stationary phases which may be considered more appropriate for polypeptide and protein separation than, for example, in the separation of biogenic amines. However, the same requirements for high resolution separations—small particle diameters and narrow particle diameters distribution, uniform porosity, stability of the bonded ligand, suitable ligand coverage, appropriate column packing practices—that apply to the small molecule world are also pertinent to high resolution separations of polypeptides and protein[61].

Similarly, appropriate selection of initial and final mobile phase compositions is important. Here variation of retention solely through changes in the water content of the eluent is but one option available. The chemical and physical properties of the organic modifier, the pH, the nature and concentration of ionic or non-ionic additives and buffers, the temperature and the flow rate can all be used to influence peak capacity, resolution and overall recoveries in most separations. Each of these options can be exercised individually or in concert.

The important goals of RPC can be summarised as methods which achieve (1) the optimal resolution R_S or peak capacity PC; (2) the smallest peak volume (equivalent to $4\sigma_v$) thus ensuring maximal detection sensitivity, (3) the shortest retention time (t_R or t_g) compatible with the determined resolution and (4) the highest recovery. The most effective strategy demands constant reappraisal of the limitations of all the separation and detection variables. This chromatographic diversity can generally be realised through rational manipulation of remarkably few separation variables. Clues for this selection of chromatographic conditions are provided by measurements on solubility and denaturation effects in eluents of different composition, from exploratory elution experiments carried out at the microanalytical scale, from evaluation of selectivity matrices and from the effects on protein function or structure at the stationary phase surface by changes in pH, ionisation or pairing ion reagents. Except in the most straightforward cases, rigorous adherence to only a single chromatographic protocol cannot be expected to achieve the above goals.

6.4.2   Choice of Organic Solvent Composition. Gradient Design

When applied to strictly aqueous eluents, most polypeptides are strongly adsorbed to modern microparticulate reversed phase packing materials. Because of the magnitude of the S- and k_w-values for these solutes, gradient elution provides the most powerful routine for working up a separation. Not only do gradient procedures permit rapid 'scouting' of mobile phase conditions for acceptable resolution, but they also permit ready assessment of solute 'ghosting'. The ghost is the protein carry over on a subsequent blank gradient, i.e. running the same gradient without sample[61].

Ideally, in exploratory runs, two different organic solvent modifiers should be used over the range 0–50% (v/v) with two different pH, ionic strength or ionic modifier conditions. This gives background information of the separation. Based on the results of these four gradient experiments, resolution optimisation is then attempted. As indicated in Section 6.2.1, resolution is dependent on the gradient steepness parameter, b, with peak capacity PC following an essentially inverse asymptotic dependence on b. With a defined alkylsilica support and mobile phase condition the average resolution of the system can be set (say at $R_S = 1.5$). An elution window corresponding to $k \leq 20$ for the desired component (s) is generated through adjustment of the initial and final mobile phase compositions. Band spacing can then be varied by changing the steepness of the gradient. Hence for very complex mixtures where only a few components (often only one) are required to be purified, changes in b can be exploited as aids to regional selectivity changes. Since the gradient steepness parameter b is related to the flow rate F as well as the rate of change of organic modifier $\Delta\varphi$, comparable resolution can be achieved at different flow rates or gradient rates provided $(d\varphi/dt)/F$ is held constant.

Commonly used gradient conditions such as 1% change in organic modifier per minute at a flow rate of 1–2 ml/min for a standard stainless steel analytical column (15–25 × 0.4 cm) should only be considered as initial points to establish the correct limits of these mobile phase variables. Once appropriate F, $d\varphi/dt$ and k values have been determined, significantly reduced separation times can be achieved using procedures outlined above. When the initial solvent strength is close to the elution concentration, when the dwell time is small, and when the optimal value of the gradient steepness parameter b is achieved then it has been the experience of this laboratory that optimal detection sensitivity, separation speed and purification factors are realised. Often in these circumstances, high recovery of mass and biological activity of globular proteins up to *ca* 60 Kdaltons are also achieved.

6.4.3   Choice of Ionic Strength and pH

Variation of eluent pH regulates protein retention to alkylsilicas through several processes. Changes in pH influence the extent of ionisation of the solute and affect the ionisation status of accessible silanol groups and other adsorbed ionisable groups on the stationary phase surface. Further, pH variations in the eluent will also affect protic equilibria of many ionisable components added to the mobile phase to enhance selectivity. Because of the chemical instability of n-alkylsilicas at high pHs, most reversed phase separations use mobile phases with pH ≤ 7.5 and most commonly in the range pH 2–3. With the advent of polymeric organic matrices of large pore size such as the TSK phenyl G5000 PW[62] the pH limitations imposed by silica-based supports have been to some extent circumvented. However, there remain a number of other reasons for the popularity of the low pH choice:

1. At low pH condition, suppression of carboxyl group ionisation occurs and the amino groups are essentially fully protonated. Thus competing equilibria are suppressed and the solute can behave as a single, averaged ionised species.
2. The isoelectric point of most polypeptides and proteins is above these low pH values. For reasons not yet fully clarified higher selectivity is obtained when running below the protein pI in reversed phase systems.
3. Many alkanoic and perfluoroalkanoic acids have proved effective mobile phase additives at low concentrations at these pH values. They increase the solubilisation of polypeptides and proteins and lead to advantageous ion-pairing/dynamic ion-exchange type retention phenomena. All of these organic acids can be removed readily by lyophilisation.
4. Low ionic strengths can be used at low pHs. This facilitates easier recovery, better peak shape and more reproducible retention[2,11,41].

Several codicils, however, must also be applied when these low pH conditions are contemplated. The possibility that a specific protein will be poorly soluble at or below its pI value must be taken into account when the pH value is chosen. Such behaviour will have dramatic effects on chromatographic performance, particularly evident in trace enrichment experiments or preparative separations. Similarly, pH-induced protein unfolding or subunit dissociation may occur at inappropriate pH values. In specific cases this effect can be utilised for the purification of the apoprotein moiety or individual subunits of a multimeric protein. Reconstitution by addition of the correct prosthetic group/co-factor or by subunit rehybridisation experiments may lead to a useful purification of the native protein.

Even with relatively modest increases in pH, significant band broadening of eluted peaks may occur. Because of the existence of multiple pH versus retention optima, optimisation of resolution by random pH adjustments usually presents difficulties. Iterative window diagram procedures have however shown considerable promise for localising optimal pH conditions with difficult separations. However, decreased resolution at higher pH values due to loss of column efficiency is often observed with polypeptides, particularly in the absence of suitable co- or counter-ions capable of stabilising solute structure in solution. High resolution separation of polypeptides and proteins in alkylsilicas solely by pH step or gradient adjustment under constant ionic strength conditions has not been employed extensively. Rather, the synergistic effects of suitable buffer additives and pH are usually combined (often inadvertently) through the use of so called ion-pair or dynamic liquid-liquid ion exchange procedures.

6.4.4 Addition of Counter Ions. Hetaeric Chromatography

The introduction[28,63-65] during the late 1970's of a variety of ionic modifiers, many of them novel to peptide and protein purification, resulted in a new era in column liquid chromatography. The emergence of high resolution reversed phase separations as a viable technique has undoubtably stimulated renewed interest in high performance ion-exchange and affinity support media. During the ensuring years, solvophobic theory and the principles of secondary solution equilibria, as they apply to so-called "ion-pair" phenomena, have been successfully applied so that it is now possible to rationalise the effect of many added co- and counter-ions adequately. Following the proposal of Horvath and co-workers[66], these complexing ions can be considered to be hetaerons and thus hetaeric chromatography denotes a technique in which the complexing ion is added to the mobile phase in order to affect selectivity of the chromatographic system by secondary equilibria.

Both non-polar and polar hetaerons can be used to manipulate selectivity. Currently more than 100 inorganic and organic anions and cations have been reported to be suitable as pairing ions in reversed phase separations. Table 6-2 lists a selection of these species suitable for polypeptide or protein fractionation in reversed phase system. Retention and selectivity can be varied independently with these additives, the extent of this variation being defined by the retention modulus or capacity modification factor. As a consequence both increases and decreases in retention with constant (or varying) selectivity can be achieved with suitable hetaeric systems. Although the list of pairing ions continues to grow (for a compendium see Ref. 28) two main classes have become favoured[67-81], that is the bulky, surface-active ions of molecular weight ca 200, such as the alkylsulphates and the quaternary alkylammonium salts, and the smaller, more polar ions with low absorption isotherms for the hydrocarbonaceous stationary phase, such as phosphate, perfluoroacetate, bicarbonate or tetrabutylammonium salts. Within each class subdivision can be achieved whether or not the ion is volatile or removable by simple extraction techniques. Controversy still surrounds the issue whether hetaeric

Table 6-2. Selection of Ionic Species which at pH < 7.0 Modify the Retention Characteristics of Amino Acids, Peptides and Proteins on Chemically Bonded Hydrocarbonaceous Stationary Phases

Cationic pairing ion	Ref.	Anionic pairing ion	Ref.
Group I & II inorganic cations	2, 41, 44, 62, 86	$H_2PO_4^-$	11, 27, 45, 62, 82, 86, 87
Pyridinium salts	44, 50, 84, 87, 92	Cl^-	18, 66, 67, 96, 99
		ClO_4^-	2, 63
		SO_4^{2-}	47
NH_4^+	27, 50, 63, 69	HCO_3^-	27, 91
$CH_3N^+H_3$	27	BO_3^{3-}	70
$C_3H_9N^+H_3$	27	Tartrate	27
		HCO_2^-	28, 84
$C_{12}H_{25}N^+H_3$	27, 63, 64, 65	$CH_3CO_2^-$	50
$HOCH_2CH_2N^+H_3$	27	$C_2H_5CO_2^-$	70
$C_6H_{13}N^+H_3$	27	$CF_3CO_2^-$	18, 27, 41, 68, 70, 71, 88, 89
$(HOCH_2CH_2)_3H^+H$	28, 74	$C_3F_7CO_2^-$	42, 75, 77
$(HOCH_2CH_2)_3N^+H$	28, 75	$C_3F_7CO_2^-$	75, 77, 90
$(C_2H_5)_3N^+H$	47, 48, 69, 76, 86	$C_6F_{13}CO_2^-$	28, 41
Morpholinium salts	54		
N-Methylpiperidinium	70	$C_4H_9SO_3^-$	28, 78
Piperazinium	70	$C_5H_{11}SO_3^-$	28, 41
TEMED	70	$C_6H_{13}SO_3^-$	79, 80
Triammonium propane	70	$C_7H_{15}SO_3^-$	70, 81
$(CH_3)_4N^+$	28, 63	$C_8H_{17}SO_3^-$	28, 70
$(C_2H_5)_4N^+$	41, 64	$C_{12}H_{25}SO_4^-$	79, 80
$(C_3H_7)_4N^+$	41, 64	Camphor-10 sulphonate	78, 79, 80
$(C_4H_9)_4N^+$	75, 94	p-Toluene sulphonate	28, 41

This selection of ionic species is generally compatible with UV detection below 235 nm or in the case of those reagents with an aromatic nucleus, with fluorometric or UV (above 235 nm) detection. The concentration of these compounds in the mobile phase is usually < 5 mM or ca 0.1%. Adapted from Ref. 28. See also Table 6-1 for selected applications.

chromatography with bonded reversed phases involves the retention of the complex of the solute-pairing ion formed via ion-pair interactions in the bulk mobile phase followed by adsorption of the complex to the solvated non-polar ligand or by dynamic, stoichiometric exchange processes involving transient interactions between the solutes and co- and counter-ion moieties adsorptively bound to the stationary phase. Certainly many of the hetaerons now popular are surface-active and exhibit stationary phase binding properties as described by Langmuir or Freudlich isotherms.

6.4.5 Choice of Temperature and Flow Rate. Column Pressure Drop

Temperature variation as a means of controlling average resolution of polypeptides and proteins on reversed phase columns have not been extensively investigated at this stage. Usually, ambient temperature is employed. Although hydrophobic interactions have been postulated to decrease with decreasing temperatures, often capacity factors for polypeptides and proteins show the opposite effect with n-alkylsilicas. Higher temperatures favour more rapid mass transfer of these solutes, more rapid unfolding between native and disorganised species and faster dynamics associated with ion- and solvent-interaction phenomena. All of these effects will result in improved efficiency, higher solute solubility (up to the critical temperature associated with gel-sol, helical-random coil, etc. phase changes) and regional selectivity changes. These dependencies with temperature variation may well fall into the class of so-called "entropy-compensated" effects which correlate with differences in molecular shape of the solute molecules, and their orientation and dynamics at the liquid-solid interface. The dependence of k on temperature, T, can be expressed as:

$$\ln k = -\frac{\Delta H^{\circ}_{assoc}}{RT} + \frac{\Delta S^{\circ}_{assoc}}{RT} + \log \phi \qquad (6\text{-}11)$$

where ΔH°_{assoc} and ΔS°_{assoc} are the standard enthalpy and entropy change for the transfer of the solute to the stationary phase.

From plots of $\ln k$ versus reciprocal temperature (van't Hoff plots), it is possible to derive H°_{assoc} values and hence the change ΔH°_{assoc} and in ΔS°_{assoc} values associated with polypeptide and protein unfolding at the stationary phase surface. One codicil which has arisen from studies with polypeptides has been the demonstration that although column efficiency improves with increasing temperature (ca 2-fold for a 20°C increase) the average resolution or peak capacity in the gradient mode with solvent-compensated separations is lower at temperatures above 37–40° than is the case with temperatures at 15–20°.

Conventional chromatographic wisdom tells us that column efficiency depends on the linear flow velocity of the mobile phase and this dependence can be visualised in the form of van Deempter or Knox plots. Improvement in efficiency is thus anticipated as the flow rate is decreased. Changes in column and eluent parameters which directly influence lateral diffusion of the solute at the mobile phase—stationary phase interface will have major effects on solute band broadening. This is explained by the fact that the reduced plate height (efficiency) is proportional to peak variance and inversely proportional to particle diameter and column length. Because resolution changes only as the square root of column length, it is thus much more advantageous to manipulate eluent composition and flow rate in tandem.

A further factor which must also be considered in the design of column configuration is the pressure drop over the column bed given by

$$\Delta p = \frac{v \cdot \eta \cdot L}{\rho^* \cdot dp^2} \qquad (6\text{-}12)$$

where η is the eluent viscosity, v the linear flow velocity of the eluent and ρ^* the permeability. With silica-based hydrocarbonaceous phases there is a linear relationship between Δp and v. The consequences of the dependencies inherent in Eq. 6-12 for scale-up should not be overlooked since they bear directly on the choice of column dimensions, particle diameter and particle porosity. Although low flow rates are advisable in view of the slower diffusion rates of polypeptides and proteins, in gradient elution it is advisable to adjust the gradient slope so that relatively small b-values are obtained. Proteins can show dramatic changes in h over very small k ranges, with peak assymetry factors typically larger for small pore n-alkylsilicas rather than for large pore n-alkylsilicas (see Fig. 6-10). This behaviour demands that adequate consideration is given to the total

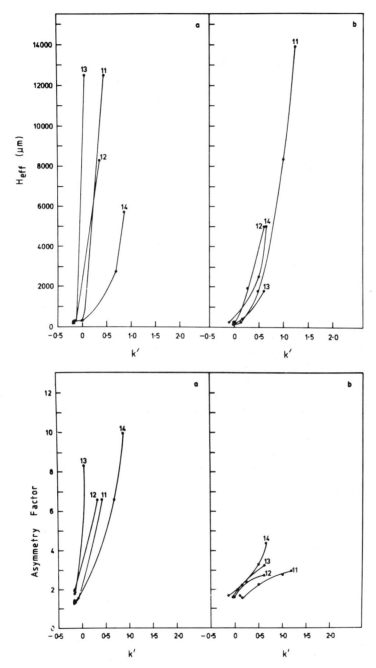

Figure 6-10. Effect of mobile phase strength on the plate height equivalent, H_{eff} and peak asymmetry, a_s, for several polypeptides separated on a 7.3 nm pore size (a) or a 30 nm pore size (b) octadecylsilica at 2.0 ml/min. The polypeptides are; 11, hen lysozyme; 12, sperm whale apomyoglobin; 13 porcine trypsin; 14, bovine serum albumin. The divergence of H_{eff} and a_s evident between the small and large pore octadecylsilica is particularly noteworthy and may indicate the fundamental reason why protein recoveries on large pore silicas are often significantly higher than those obtained with small pore n-alkylsilicas. Adapted from Ref. 58.

separation time as an integral part of the design of column configuration, as anticipated on the basis of the discussion in section 6.2. Paradoxically, resolution may be increased with shallow gradients of low flow rate but recovery often falls off disastrously.

6.4.6   Detection and Sample Recovery

The availability of suitable detection methods also has significant bearing on the choice of the organic solvent modifier and, in general, the mobile phase composition. Biological *in vitro* or *in vivo* response assays, radioligand assays and radio-immunoassays can all detect under suitable conditions a specific polypeptide or protein at a level several orders of magnitude more sensitively than can be obtained by spectrophotometric methods or by classical procedures of chemical analysis. All the popular primary organic solvent modifiers—methanol, ethanol, acetonitrile, propanols etc. can be obtained with UV transparency at 210–215 nm of 0.005 a.u. At this wavelength the isobestic absorption of the polypeptide backbone occurs with molar adsorptivities essentially independent of conformation, *ca*, 10^3 $M^{-1}dm^{-3}$ per residue. As a consequence, with UV detectors of small cell volume ($\leq 5\,\mu l$) and short path length (≤ 8 nm) the limits of detectability are *ca* 20 ng/ml at a signal to noise ratio larger than 5. For most polypeptides and proteins fluorescence detectors, being more selective, do allow in specific cases a significant increase in the level of detectability (*ca* 10 fold) since the detection is less susceptible to background perturbations due to solvent contaminants or other UV-absorbing artefacts. Where possible, endogeneous tryptophan fluorescence (215 nm excitation, 340 nm emission) can be monitored. Elsewhere, post-column derivatisation will be required but in these circumstances it must be remembered that even with high derivatisation efficiency, broader peaks (i.e. increased σ_v) and smaller peak heights (i.e. decreased $1/\sigma_v$) will always be obtained, thus decreasing detectability.

Careful pre- and post-chromatographic handling of samples is often required with labile proteins exposed to aqueous organic solvent mixtures at low pH values. A useful remedy in some cases, such as prostatic acid phosphatase[73] and protein hormones[55,74] is the direct collection of the chromatographic fractions in tubes containing buffers of more appropriate pH. Deamidation and residue cleavage of proteins at acid labile sites have also been documented following reversed phase separations.

6.5   Applications

The main application area for reversed phase chromatography of proteins is the purification of polypeptides for structural studies, represented by the two examples given in this section. Due to the extreme conditions often used in RPC, some proteins will not be eluted in their native, biologically active state. Proteins which will retain their activities and native structures most often are low molecular weight containing several disulphide bonds, e.g <60 kDa.

6.5.1   Ribosomal Proteins

Eucaryotic and procaryotic ribosomes are composed of one large and one small subunit containing a large number of different proteins. Eucaryotic ribosomes contain about 80 proteins, including several phosphoproteins, of which about 33 belong to the

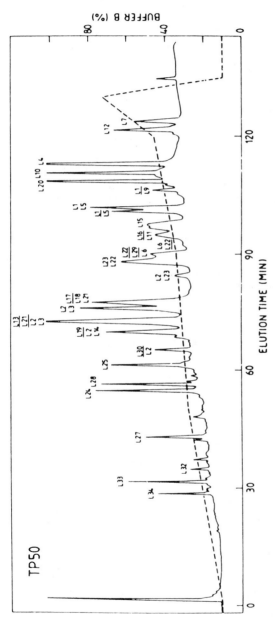

Figure 6-11. Purification of 50 S ribosomal proteins from E. coli by reversed phase HPLC on Ultrapore RPSC. Two mg TP50 were injected into Ultrapore RPSC (5 μm particle size; 300 Å pore size; column size, 75 × 4.6 mm; 35°C; flow rate 0.5 ml/min). The eluent was: buffer A, 0.1 % TFA in water, buffer B, 0.1 % TFA in acetonitrile. The gradient applied was: 10 % B to 30 % B for 50 min; 30 % B to 37 % B for 40 min; 37 % B to 50 % B for 30 min; 50 % B to 80 % B for 10 min; 80 % B to 10 % B for 5 min; and reconditioning for 30 min at initial conditions. Measurements were made at 220 nm, range 0.64. The proteins were identified by two-dimensional polyacrylamide gel electrophoresis and micro-sequencing as indicated in the figure. Adapted from Ref. 72. By permission of the authors and publishers.

small subunit and about 45 to the large subunit. The small and large subunits of the *Escherichia coli* ribosome contain 21 and 32 different proteins, respectively. The molecular weights range between 5 kd and 25 kd for the large subunit and between 8 kd and 27 kd (except S1 which is around 61 kd) for the small subunit.

Reversed phase chromatography has proved extremely useful for the separation and purification of ribosomal proteins for structural studies. From the *E. coli* ribosome, 15 S proteins and 23 L proteins were isolated in sequencer purity using this technique[101]. In Fig. 6-11 is shown one representative example of the separation of the 50 S ribosomal proteins from *E. coli*.

6.5.2   Haemoglobin Varieties

The major haemoglobin protein, HbA, has the subunit structure $\alpha_2\beta_2$. The α chain contains 141 amino acid residues and the β chain contains 146 residues. There are more than 300 known point mutations in the human haemoglobin β chain. About 100 of these give rise to symptoms with the individual, the most common being fatigue due to stronger oxygen binding than in normal haemoglobin. In 1984 a new point mutation was discovered in Sweden. In this case, β 36 Pro had been replaced by Thr. In Fig 6-5 is shown the separation of the haemoglobin chains of this individual by reversed phase chromatography on an octyl bonded silica gel. Under the conditions of this experiment, the chains and the heme groups are completely dissociated. In this case the individual was heterozygot with respect to the point mutation and explains why he had both normal and mutated chains in his haemoglobin.

Acknowledgement

The support of the National Health and Medical Research Council in Australia is gratefully acknowledged.

6.6   References

1. M. T. W. Hearn, M. I. Aguilar, in *Modern Physical Methods in Biochemistry*, A. Neurberger, L. L. M. van Deenan, eds., Part B, Elsevier Sci. Publ., 1988, pp. 107–142
2. M. T. W. Hearn, in *High Performance Liquid Chromatography: Advances and Perspectives*, Cs. Horvath, ed., Academic Press, New York, 1983, vol. 3, pp. 87–155
3. M. T. W. Hearn, in *Chemical Separations: Separation Science and Technology*, C. J. King, J. D. Navratil, eds., Litarvan Press, Co., 1986, pp. 77–98
4. R. J. Boscott, *Nature*, *159*, 342 (1947).
5. J. Boldingh, *Experimentia*, *4*, 270 (1948).
6. G. A. Howard, A. J. P. Martin, *Biochem. J.*, *46*, 532 (1950).
7. S. Shaltiel, *Meth Enzymol.*, *34*, 126 (1974).
8. Z. Er-el, Y. Zaidenzaig, S. Shaltiel, *Biochem. Biophys. Res. Commun.*, *49*, 383 (1972).
9. B. H. J. Hofstee, *Biochem. Biophys. Res. Commun.*, *50*, 751 (1973).
10. H. P. Nick, R. E. H. Wetherall, M. T. W. Hearn, F. J. Morgan, *Anal. Biochem.*, *148*, 93 (1985).
11. K. A. Cohen, K. Schellenberg, B. L. Kerger, B. Grego, M. T. W. Hearn, *Anal. Biochem.*, *140*, 223 (1984).
12. B. Grego, M. T. W. Hearn, *J. Chromatogr.*, *266*, 75 (1983).
13. L. A. Witting, D. J. Gisch, R. Ludwig, R. Eksteen, *J. Chromatogr.*, *296*, 97 (1984).
14. J. D. Pearson, N. T. Lin, F. E. Regnier, *Anal. Biochem.*, *124*, 217 (1982).
15. C. T. Wehr, in *CRC Handbook of HPLC*, CRC Press, Boca Raton, 1984, 31.
16. M. T. W. Hearn, B. Grego, *J. Chromatogr.*, *282*, 541 (1983).

17. G. Lindgren, B. Lundstrom, I. Kallman, K. A. Hansson, *J. Chromatogr.*, *296*, 83 (1984).
18. N. H. C. Cooke, B. G. Archer, M. J. O'Hare, E. C. Nice, M. Capp, *J. Chromatogr.*, *255*, 115 (1983).
19. M. J. O'Hare, M. W. Capp, E. C. Nice, N. H. C. Cooke, B. G. Archer, *Anal. Biochem.*, *126*, 17 (1982).
20. K. A. Cohen, J. Chazaud, G. Galley, *J. Chromatogr.*, *282*, 423 (1983).
21. Y. Kato, T. Kitamura, T. Hashimoto, *J. Chromatogr.*, *266*, 49 (1983).
22. J. L. Fausnaugh, E. Pfannkoch, S. Gupta, F. E. Regnier, *Anal. Biochem.*, *137*, 462 (1984).
23. M. Shin, N. Sakihama, R. Oshino, H. Sasaki, *Anal. Biochem.*, *138*, 259 (1984).
24. J. P. Harrington, T. T. Herskovits, *Biochemistry*, *14*, 4972 (1975).
25. T. Askura, K. Adachi, E. Schwartz, *J. Biol. Chem.*, *253*, 6423 (1978).
26. T. T. Herskovits, H. Jaillet, *Science*, *163*, 282 (1969).
27. M. T. W. Hearn, M. I. Aguilar, *J. Chromatogr.*, *359*, 31 (1986).
28. M. T. W. Hearn, in *Ion Pair Chromatography*, Marcel Dekker, New York, 1984, pp. 1–296.
29. X. Geng, F. E. Regnier, *J. Chromatogr.*, *296*, 1 (1984).
30. S. D. Powers, M. A. Lochrie, R. O. Poyton, *J. Chromatogr.*, *266*, 585 (1983).
31. M. T. W. Hearn, B. Grego, *J. Chromatogr.*, *296*, 309 (1984).
32. M. T. W. Hearn, *J. Chromatogr.*, *418*, 3 (1987).
33. R. Janzen, K. K. Unger, H. Giesche, J. N. Kinkel, M. T. W. Hearn, *J. Chromatogr.*, *397*, 81 (1987).
34. X. M. Lu, K. Benedek, B. L. Karger, *J. Chromatogr.*, *359*, 19 (1986).
35. M. T. W. Hearn, M. I. Aguilar, *J. Chromatogr.*, *397*, 47 (1987).
36. M. T. W. Hearn, M. I. Aguilar, T. Nguyen, M. Fridman, *J. Chromatogr.*, *435*, 271 (1988).
37. M. T. W. Hearn, M. I. Aguilar, *High Performance Liquid Chromatography of Peptides, Proteins and Polynucleotides*, M. T. W. Hearn, VCH Press, 1989, pp. 345–370
38. N. T. Miller, B. Feibush, B. L. Karger, *J. Chromatogr.*, *316*, 519 (1984).
39. H. Engelhardt, D. Mathes, *J. Chromatogr.*, *185*, 305 (1979).
40. A. J. Alpert, *J. Chromatogr.*, *359*, 85 (1986).
41. M. T. W. Hearn, *Adv. Chromatogr.*, *20*, 1 (1982).
42. S. O. Jeppsson, I. Kallman, G. Lindgren, L. G. Fagerstam, *J. Chromatogr.*, *297*, 31 (1984).
43. J. L. Meek, Z. Rossetti, *J. Chromatogr.*, *211*, 15 (1981).
44. S. J. Su, B. Grego, B. Niven, M. T. W. Hearn, *J. Liquid Chromatogr.*, *4*, 1745 (1981).
45. H. Stotzel, G. J. Hughes, *Biochem. J.*, *199*, 31 (1981).
46. M. T. W. Hearn, B. Grego, *J. Chromatogr.*, *203*, 349 (1981).
47. M. T. W. Hearn, B. Grego, *J. Chromatogr.*, *255*, 125 (1983).
48. M. T. W. Hearn, B. Grego, *J. Chromatogr.*, *266*, 75 (1983).
49. M. J. O'Hare, E. C. Nice, *J. Chromatogr.*, *171*, 209 (1979).
50. B. Grego, M. T. W. Hearn, *J. Chromatogr.*, *366*, 28 (1984).
51. L. R. Snyder, in *High Performance Liquid Chromatography*, Cs. Horvath, ed., Academic Press, New York, 1980, vol. 1, p. 208.
52. M. T. W. Hearn, A. N. Hodder, M. I. Aguilar, *J. Chromatogr.*, *327*, 47 (1985).
53. M. T. W. Hearn, M-I. Aguilar, *J. Chromatogr.*, *392*, 33 (1987).
54. M. A. Stadalius, H. S. Gold, L. R. Snyder, *J. Chromatogr.*, *296*, 31 (1984).
55. M-I. Aquilar, A. M. Hodder, M. T. W. Hearn, *J. Chromatogr.*, *352*, 115 (1986).
56. S. A. Cohen, S. Dong, K. Benedek, B. L. Karger, *J. Chromatogr.*, *317*, 227 (1984).
57. X. M. Lu, K. Benedek, B. L. Karger, *J. Chromatogr.*, *359*, 19 (1986).
58. M. T. W. Hearn, B. Grego, *J. Chromatogr.*, *296*, 61 (1984).
59. M. T. W. Hearn, A. N. Hodder, M-I. Aguilar, *J. Chromatogr.*, *327*, 115 (1985).
60. F. E. Regnier, *Science*, *238*, 319 (1987).
61. C. Jilge, R. Janzen, H. Giesche, K. K. Unger, J. N. Kinkel, M. T. W. Hearn, *J. Chromatogr.*, *397*, 71 (1987).
62. V. Kato, T. Kitamura, T. Hashimoto, *J. Chromatogr.*, *292*, 418 (1984).
63. W. S. Hancock, C. A. Bishop, R. L. Prestidge, D. R. K. Harding, M. T. W. Hearn, *Science*, *200*, 1168 (1978).
64. M. T. W. Hearn, W. S. Hancock, *Trends Biochem. Sci.*, *4*, 58 (1979).
65. M. T. W. Hearn, *Adv. Chromatogr.*, *18*, 59 (1980).
66. C. Horvath, W. Melander, I. Molnar, P. Molnar, *Anal. Chem.*, *49*, 2295 (1977).
67. A. F. Bristow, C. Wilson, N. Sutcliffe, *J. Chromatogr.*, *270*, 285 (1983).
68. E. C. Nice, M. Capp, M. J. O'Hare, *J. Chromatogr.*, *147*, 413 (1979).
69. J. Spiess, J. Rivier, C. Rivier, W. Vale, *Proc. Natl. Acad. Sci. USA.*, *78*, 6517 (1981).

70. B. Grego, M. T. W. Hearn, *J. Chromatogr.*, *336*, 25 (1984).
71. N. E. Tandy, R. A. Dilley, F. E. Regnier, *J. Chromatogr.*, *266*, 599 (1983).
72. R. M. Kamp, B. Wittman-Liebold, *Febs Letts*, *167*, 59 (1984).
73. M. P. Strickler, J. Kintzios, M. J. Genski, *J. Liquid Chromatogr.*, *5*, 1921 (1982).
74. P. G. Stanton, M. T. W. Hearn, *J. Biol. Chem.*, *262*, 1623 (1987).
75. H. P. J. Bennett, *J. Chromatogr.*, *266*, 249 (1983).
76. S. D. Power, M. A. Lochrie, R. O. Poyton, *J. Chromatogr.*, *266*, 585 (1983).
77. J. R. Walsh, H. D. Niall, *Endocrinology*, *107*, 1258 (1980).
78. S. Terabe, K. Konaka, K. Inouye, *J. Chromatogr.*, *172*, 163 (1979).
79. M. T. W. Hearn, B. Grego, W. S. Hancock, *J. Chromatogr.*, *185*, 429 (1979).
80. M. T. W. Hearn, B. Grego, *J. Chromatogr.*, *218*, 497 (1981).
81. B. Grego, F. Lambrou, M. T. W. Hearn, *J. Chromatogr.*, *266*, 89 (1983).
82. R. van der Zee, G. W. Welling, *J. Chromatogr.*, *244*, 134 (1982).
83. G. E. Gerber, R. J. Anderegg, W. C. Herlihy, C. P. Gray, K. Biemann, H. G. Khorana, *Proc. Natl. Acad. Sci. USA.*, *76*, 227 (1979).
84. R. V. Lewis, A. Fallon, S. Stein, K. D. Gibson, S. Udenfriend, *Anal. Biochem.*, *104*, 153 (1980).
85. W. Monch, W. Dehnen, *J. Chromatogr.*, *147*, 415 (1978).
86. W. S. Hancock, C. A. Bishop, A. M. Gotto, D. R. K. Harding, S. M. Lamplugh, J. T. Sparrow, *J. Lipid. Res.*, *16*, 250 (1981).
87. K. J. Wilson, M. W. Berchtold, P. Zumskin, S. Klause, G. J. Hughes, in *Methods in Protein Sequence Analysis*, M. Elizinga, ed., Hamana Press, Clifton, New Jersey, 1982, p. 260
88. G. J. Putterman, M. B. Spear, K. S. Meade-Cobun, M. Widra, C. V. Hixson, *J. Liquid Chromatogr.*, *5*, 715 (1982).
89. S. J. M. Skinner, B. Grego, M. T. W. Hearn, C. G. Liggins, *J. Chromatogr.*, *308*, 113 (1984).
90. A. W. Burgess, J. A. Knesel, L. G. Sparrow, W. A. Nicola, E. C. Nice, *Proc. Natl. Acad. Sci. USA.*, *79*, 5753 (1982).
91. B. Grego, G. S. Baldwin, J. A. Knessel, R. J. Simpson, F. J. Morgan, M. T. W. Hearn, *J. Chromatogr.*, *297*, 21 (1984).
92. L. R. Gurley, J. A. D'Anna, M. Blumenfeld, J. G. Valdez, R. J. Sebring, P. R. Dohahue, D. A. Prentice, W. D. Spall, *J. Chromatogr.*, *297*, 147 (1984).
93. S. Stein, C. Kenny, H. J. Freisen, J. Shively, U. Del Valle, S. Pestka, *Proc. Natl. Acad. Sci. USA.*, *77*, 5716 (1980).
94. R. A. Wolfe, J. Casey, P. C. Familletti, S. Stein, *J. Chromatogr.*, *296*, 277 (1984).
95. S. W. Herring, R. K. Enns, *J. Chromatogr.*, *266*, 249 (1983).
96. L. Sottrup-Jensen, T. M. Stepanik, C. M. Jones, P. B. Lonblad, T. Kristensen, D. M. Wierzbicki, *J. Biol. Chem.*, *259*, 8293 (1984).
97. V. L. Alvarez, C. A. Roitsch, O. Henriksen, *Anal. Biochem.*, *115*, 353 (1981).
98. W. Kopaciewicz, F. E. Regnier, *Anal. Biochem.*, *129*, 472 (1983).
99. M. W. Berchtold, K. J. Wilson, C. W. Heizmann, *Biochemistry*, *21*, 6552 (1982).
100. C. N. Chesterman, T. Walker, B. Grego, K. Chamberlain, M. T. W. Hearn, F. J. Morgan, *Biochem. Biophys. Res. Comm.*, *116*, 809 (1983).

7 Hydrophobic Interaction Chromatography

Kjell-Ove Eriksson
Department of Biochemistry
Uppsala University
Uppsala Biomedical Center
Box 576, S-751 23 Uppsala, Sweden.

7.1 Introduction

Hydrophobic molecules in an aqueous solvent will self-associate. This association is due to hydrophobic interaction. The hydrophobic interaction is of prime importance in biological systems. It is a major driving force behind the folding of globular proteins, the association of protein subunits, the binding of many small molecules to proteins as in enzyme catalysis, regulation and transport across surfaces. It is also responsible for the self-association of phospholipids and other lipids to form the biological membrane bilayer and the binding of integral membrane proteins.

In hydrophobic interaction chromatography (HIC) the hydrophobic interaction is utilized for the binding of proteins to adsorbents with hydrophobic ligands. Our present, rather detailed, knowledge of protein three-dimensional structure has revealed that the surfaces of globular proteins can have extensive hydrophobic patches in addition to the expected hydrophilic groups. It is these hydrophobic regions that bind to hydrophobic ligands, alkyl or aryl side chains on the gel matrix, in media favouring hydrophobic interaction, e.g., an aqueous solution with a high salt concentration. Elution (and separation), according to differences in the strength of interaction between the proteins and the amphiphilic gel, is in general brought about by decreasing the salt concentration of the eluent. In some cases a decrease of the solvent polarity is also needed.

HIC has been developed during the last two decades. The first gels of practical use for HIC were of a mixed hydrophobic-ionic character[1-3]. Neutral adsorbents (alkyl and aryl ethers) were later prepared by Porath et al.[4] and Hjertén et al.[5], the latter leading to the introduction of Octyl- and Phenyl-Sepharose®. It was also Hjertén who introduced the now generally accepted name of the technique: hydrophobic interaction chromatography[6]. The term hydrophobic chromatography should be avoided, since it is the interaction between the solute and the gel which is hydrophobic, and not the chromatographic procedure[7]. Lately, the HIC technique has been adapted to the HPLC mode using both the traditional gel material agarose[8], as well as organic polymers[9] and silica based matrices[10]. New methods for immobilization of hydrophobic ligands, e.g. attachment of alkyl sulphides to oxirane-activated agarose[11], have been developed.

Salting-out chromatography and other types of chromatography related to HIC will be discussed briefly, as will other types of chromatographic adsorbents developed more recently[12].

Adsorbents for reversed phase chromatography (RPC) and HIC both contain hydrophobic ligands. In the RPC adsorbents the ligand density is much higher than in those used for HIC. Although the separation on both types of adsorbents is based on hydrophobic interaction, the mechanism on the molecular level is different. Whereas an RPC adsorbent can be regarded as a continuous hydrophobic phase, the ligands on an HIC adsorbent are interacting individually with the solutes. As a result, globular proteins very often denature when applied on RPC columns, which are therefore used mainly for peptides and small proteins. RPC also requires more drastic conditions for elution, such as a gradient of organic solvents, as compared to HIC. HIC thus has a more general field of application. RPC is described in Chapter 6.

In order to give the reader a general understanding of the principle of hydrophobic interaction chromatography, this chapter discusses the technique from both a theoretical and a practical point of view. Finally, some applications of the HIC technique will be described.

7.2   The Hydrophobic Interaction

7.2.1   Theory

Hydrophobic interactions in aqueous solvents are driven primarily by interactions within the solvent and to a lesser extent by interactions between the non-polar solutes. For a detailed treatment of hydrophobic interaction see Refs. 13 and 14. Below follows a short theoretical discussion as a background to the HIC technique.

Water is a poor solvent for non-polar solutes. Dissolving a non-polar substance in water is thermodynamically unfavourable. Due to its hydrogen bonding capability water has a unique structure, causing a high surface tension. A non-polar solute forces the water to form a cavity in which the solute fits (Fig. 7-1A). In the process, many hydrogen bonds between the water molecules are broken, but new hydrogen bonds are formed among the water molecules surrounding the cavity, leading to a negative change in enthalpy, ΔH.

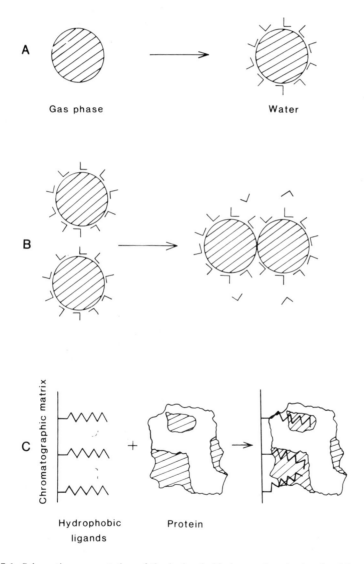

Figure 7-1. Schematic representation of the hydrophobic interaction. A, the dissolving of a non polar solute in water; B, the interaction of two non-polar solutes in water and C, a hypothetical representation of an interaction of ligands on an amphiphilic gel and hydrophobic surfaces on a protein. ⌐ represents water molecules.

This change is only to a small extent due to the weak van der Waals interaction between the solute and the surrounding water. The increase in the degree of order of solvent molecules near the solute explains the negative change in entropy, ΔS, that results.

The well known expression:

$$\Delta G = \Delta H - T\Delta S \qquad (7\text{-}1)$$

describes the change in free energy ΔG, of a process. A negative value of ΔG implies that the process is thermodynamically possible. The dissolution of a solute is thus favoured either by a negative change in enthalpy or by a positive value of the entropy change. For the cavity-forming process described above the net change in free energy is positive, which makes it unfavourable.

In hydrophobic interaction (Fig. 7-1B), the increase in entropy ($\Delta S > 0$) originating from water molecules leaving the more ordered structure around the non-associated solutes for the more unstructured bulk water is the main driving force. The net decrease in cavity surface area that occurs when two non-polar solutes associate explains why some water molecules go over to the bulk phase. The positive enthalpy term ΔH is smaller than the entropy term and does not influence the spontaneous association to a great extent[15].

The interaction between an amphiphilic gel and hydrophobic areas on a protein (Fig. 7-1C) can be explained in the same way as the interaction between non-polar solutes in water (Fig. 7-1B).

Both entropy and enthalpy change with temperature for hydrophobic interactions. A theoretical treatment of the temperature dependence of the hydrophobic interaction is thus complicated, but one can conclude that the strength of hydrophobic interactions should increase with an increase in temperature, at least in the temperature range that is of interest for HIC.

Two factors of great importance for HIC are the type and concentration of salt used and additives which change the polarity of the solvent. The latter, exemplified by ethylene glycol, decrease the interaction between the HIC gel and proteins by changing the overall structure of water slightly towards a structure resembling an organic solvent.

The influence of different salts on hydrophobic interaction follows the well known Hofmeister (lyotropic) series[16]. Salts that promote hydrophobic interaction most are to the left in the series. The anions to the right—ClO_4^-, I^-, SCN^-—are called chaotropic.

Anions:
$$SO_4^{2-} > Cl^- > Br^- > NO_3^- > ClO_4^- > I^- > SCN^-$$

Cations: $\qquad\qquad\qquad\qquad\qquad\qquad\qquad\qquad\qquad (7\text{-}2)$
$$Mg^{2+} > Li^+ > Na^+ > K^+ > NH_4^+$$

Melander and Horvath[17] have shown that the effectiveness of different salts in promoting hydrophobic interactions can be explained by their contribution to the surface tension of the solution. The formation of a cavity in water with a salt that gives a high surface tension needs a bigger input of energy than does the formation of a cavity with the same surface area in water containing a salt giving lower surface tension. The higher the salt concentration the stronger the interaction. It is the salts that increase the surface tension most that give the strongest hydrophobic interaction.

Other factors than surface tension can also effect the hydrophobic interaction. Protein hydration and specific interactions between the protein and the salt ions seem to be factors that influence the strength of interaction[18].

7.2.2 The Hydrophobicity of Amino Acids and Proteins

Although most hydrophobic amino acids are buried in the interior of globular proteins, and hydrophilic amino acids have a tendency to be exposed on the surface, some hydrophobic amino acids also appear on the surface. The hydrophobicity of protein surfaces is the sum of the hydrophobicities of the exposed amino acids and parts of the backbone. A discussion of protein surface hydrophobicity is thus based on the hydrophobicity of amino acids.

Two approaches have been used for the estimation of the hydrophobicity of amino acids. The first is based on direct measurements of the solubilities of individual amino acids in water and organic solvents[19,20]. A hydrophobicity scale for the different amino acids has thus been constructed on the basis of the free energy of transfer for the amino acids from ethanol or dioxane to water.

The second approach is based on empirical inspection of known protein structures[21–25]. Here, several hydrophobicity scales are based on, e.g., the environment of the different amino acids, the fraction of amino acids that is buried in the protein, a side chain interaction parameter or a fractional accessibility to the surrounding solvent of the different residues. For a list of different scales proposed, see J. L. Cornette *et al*[26].

Table 7-1 is a comparison of two scales of hydrophobicity, the first based on amino acid solubility[19,20] and the second based on the fraction of amino acids buried within

Table 7-1. Hydrophobicity scales for amino acids. Scale I is based on the solubility of the different amino acids, expressed as the free energy of transfer for the amino acids from ethanol or dioxane to water[19,20]. Scale II is based on the fraction of the number of the different amino acids buried within proteins[22] (average values obtained from 20 proteins with known structures).

Scale I kcal/mol		Scale II Fraction buried	
Trp	3.77	Phe	0.87
Ile	3.15	Trp	0.86
Phe	2.87	Cys	0.83
Pro	2.77	Ile	0.79
Tyr	2.67	Leu	0.77
Leu	2.17	Met	0.76
Val	1.87	Val	0.72
Met	1.67	His	0.70
Lys	1.64	Tyr	0.64
Cys	1.52	Ala	0.52
Ala	0.87	Ser	0.49
His	0.87	Arg	0.49
Arg	0.85	Asn	0.42
Glu	0.67	Gly	0.41
Asp	0.66	Thr	0.38
Gly	0.10	Glu	0.38
Asn	0.09	Asp	0.37
Ser	0.07	Pro	0.35
Thr	0.07	Gln	0.35
Gln	0.00	Lys	0.31

proteins[22] (average values obtained from 20 proteins with known structures). Some differences are striking. Proline is a rather hydrophobic amino acid, but its secondary structure-breaking properties make it appear in bends typically on the surface of proteins. Lysine is also classified as a rather hydrophobic residue in the scales based on solubility studies, although it is the most exposed of the amino acids. The four methylene groups of the lysine side chain disfavour the solubility of this amino acid in water, whereas the amino group with its hydrogen binding capability has a strong tendency to be exposed on the surface of proteins.

To get more accurate values for the hydrophobicities of amphiphilic amino acids, a recent report suggests a scale based on the hydrophobicity of each individual atom[27].

There is a linear relationship between the logarithm of the solubilities of hydrocarbons and the surface area they form in water[28]. The cavity area is the same as the accessible surface area of the hydrocarbons.

Protein surfaces are not smooth, but are rather rough and complex. Analyses of protein surfaces often use the method of Lee and Richards[29,30]. Figure 7-2 shows a part of a hypothetical protein surface.

It is the accessible surface (see legend to Fig. 7-2), with a water molecule as the probe, that is used when protein surfaces are discussed. The non-polar surface is defined as the area containing side chains with carbon and sulphur and main-chain carbon atoms (hydrogen atoms are not considered). The proportion of non-polar surface area does not differ much among the globular proteins examined. A report on the subject gives figures of 41 per cent for lysozyme, 48 per cent for myoglobin and 46 per cent for ribonuclease[29].

In a recent report[31] 46 monomeric globular proteins with known structures was studied, and their surfaces and interiors examined. Also in this report it was found that the proportion of hydrophobic surface area does not differ much among the proteins. With the definition of non-polar compounds used in this report, the non-polar fraction of the surfaces varied between 50 and 68 per cent.

7.2.3   Interaction Between Protein and the HIC Gel

No general and definite answer can be given to the question of how a protein surface should look to be able to interact with hydrophobic ligands on a HIC gel. It can, however, be concluded that more than one ligand on the adsorbent must be involved to get any adsorption[5], so-called multipoint attachment.

Studies of the kinetics of the binding of a protein, phosphorylase b, to butyl Sepharose have shown that the binding is a multi-step reaction[32]. The rate-limiting step is not the collision between the protein and the amphiphilic gel, but rather a slow conformational change or reorientation step of the protein on the HIC gel.

When α-chymotrypsinogen is activated by cleavage of four peptide bonds, Ile-16 and Val-17 are buried and Met-192, Gly-193 and Arg-145 exposed[33]. α-Chymotrypsin thereby becomes more hydrophilic than its zymogen. In HIC experiments, the enzyme elutes earlier than the zymogen[34], as expected from this change.

Lysozyme is more retarded than myoglobin on a HIC adsorbent during a chromatographic experiment, but considering the proportion of non-polar surface area (see above) the order should be the opposite. The molecular weights of these proteins are similar, so the number of ligands that can come into contact with them are expected to be similar.

The retention of seven different avian lysozymes has been studied on a HIC column[18], and the effect of amino acid substitutions was investigated. Chromatographic

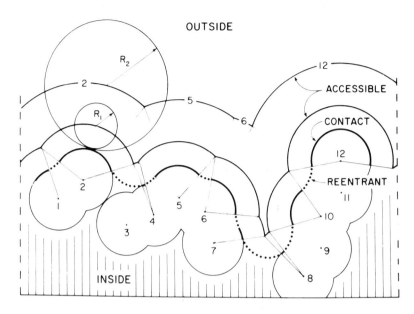

Figure 7-2. Possible molecular surface definitions. A section through part of a hypothetical protein is shown. Each atom (1 to 12) is represented by a sphere of its van der Waals envelope, and covalently bonded atoms are truncated. When a probe of radius R_1 is allowed to roll on the outside of the protein, having contact with the van der Waals surface of the different atoms, it will never contact atoms 3, 9 or 11. Choosing a probe with a larger radius R_2, even fewer atoms of the protein surface can make contact. These atoms are considered to be a part of the interior of the protein. There are several ways to measure the surface area. A straightforward procedure is to use the sheet defined by the locus of the centre of the probe. This sheet or surface is called the accessible surface. Those parts of the molecular van der Waals surface that can actually be in contact with the surface of the probe are called the contact surface. This provides a series of disconnected patches. The re-entrant surface, also a series of patches, is the area on the protein with which the interior-facing part of the probe comes into contact when it is in contact simultanously with more than one atom. The sum of the contact and re-entrant surfaces is called the molecular surface. Reproduced, with permission, from ref 30.

retention is only affected by substitutions on the lysozyme surface opposite the catalytic cleft and not by substitutions close to it. Substitutions of hydrophobic as well as hydrophilic amino acid residues have an effect on the retention[18]. These observations indicate that the strength of interaction between a HIC gel and a protein is not only determined by the proportion of non-polar surface area of a protein, but also by the distribution of non-polar areas on the surface of a protein. The HIC technique might thus be more specific than generally thought.

Hydrophobic interaction chromatography is, in general, a mild method, due certainly to the stabilizing influence of salts, and recoveries are often high[35]. This can be illustrated by an experiment with model proteins chromatographed on an octyl agarose column[8] (Fig. 7-3). In the first chromatography the elution was started immediately after the application and in the second the elution was delayed 16 hours. No difference can be seen between the two chromatograms, indicating that these proteins did not alter their

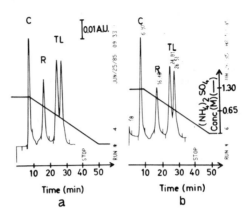

Figure 7-3. The influence of residence time of model proteins on a HPLC-HIC adsorbent (octyl agarose). *a*, elution started immediately after application and in *b*, elution delayed 16 hours. Reprinted with permission from S. Hjertén *et al.*[8]

structures upon binding to the gel, e.g., exposing their more hydrophobic interior, at least not irreversibly.

But more labile proteins, upon contact with an HIC adsorbent, can change their structures. This has been shown for α-lactalbumin[36] (after removing the Ca^{2+} ions that normally stabilize the structure). A longer residence time on the column increased the size of a broad peak, that eluted after the native protein and consisted of several more or less unfolded species of α-lactalbumin. A rerun of the second peak again gives two peaks, showing that the unfolding is reversible. The higher the temperature the larger the proportion of the α-lactalbumin that elutes with a partly unfolded structure.

The retention also increases with temperature. This is also the case for the more stable lysozyme[37], but to a lesser extent. One concludes that the temperature effect in HIC is due to two factors, (1) the increase in hydrophobic interaction with temperature and (2) a temperature dependent alteration of the structures of proteins, especially labile ones.

7.3 Gels for Hydrophobic Interaction Chromatography (HIC) and Gels Related to HIC

7.3.1 Gels for HIC

Many types of matrices are suitable for preparing gels for HIC, but the most extensively used is agarose. When the technique has been adapted for HPLC, silica and organic polymer resins have also been employed.

A number of commercially available HIC-gels are shown in Table 7-2 and some of them are shown in Fig. 7-4 (together with two types, D and E, that are not commercialized). All structures shown (except C) are octyl gels, but any chain length can be chosen if one prepares the gels oneself. The first (7-4A) was introduced by Shaltiel[2] and is based on cyanogen bromide activation[38] of an agarose gel, after which coupling with an alkylamin

Table 7-2. Commercially available gels and columns for HIC. The list may not be complete.
(A) Conventional packings.

Supplier	Product name	Functional group(s)	Remarks
Pharmacia LKB AB (Uppsala, Sweden)	Phenyl-Sepharose CL-4B	Phenyl	Agarose based (Sepharose CL-4B)
	Octyl-Sepharose CL-4B	Octyl	Agarose based (Sepharose CL-4B)
	Phenyl-Sepharose FF	Phenyl	Agarose based (Sepharose FF), large scale chromatography
Tosoh Corp. (Tokyo, Japan)	TSK-GEL Butyl-Toyopearl 650 (Fractogel TSK Butyl-650)	Butyl	Organic polymer resin
Miles-Yeda (Rehovot, Israel)	Alkyl-agarose	(Agarose-C_n, n = 2, 4, 6, 8, 10)	Agarose based
J. T. Baker (Phillipsburg, NJ, USA)	Bakerbond WP-HI-Propyl	Propyl	Silica based, bead size 40 μm, pore size 275 Å

(B) HPLC (FPLC) packings

Supplier	Product name	Functional group(s)	Remarks
Pharmacia LKB AB (Uppsala, Sweden)	Phenyl-Superose™	Phenyl	Agarose based (Superose 12), bead size 10 μm
	Alkyl-Superose™	Neopentyl	Agarose based (Superose 12), bead size 10 μm
Tosoh Corp. (Tokyo, Japan)	TSK-GEL Phenyl-5PW	Phenyl	Organic polymer resin, bead size 10 μm, pore size 1000 Å
SynChrom, Inc. (Linden, IN, USA)	SynChropak	Hydroxypropyl, Methyl, Benzyl, Butyl, Pentyl	Silica based, bead size 6.5 μm, pore size 300 Å
Supelco, Inc. (Bellafonte, PA, USA)	SUPELCO LC-HINT	Stabilized polar phase	Silica based, bead size 5 μm
Brownlee Labs (Santa Clara, CA, USA)	Aquapore HIC		Silica based, bead size 7 μm, pore size 300 Å
	Polybore Phenyl HIC	Phenyl	Resin based, bead size 10 μm, wide pore material
J. T. Baker (Phillipsburg, NJ, USA)	Bakerbond WP-HI-Propyl	Propyl	Silica based, bead sizes 5 and 15 μm, pore size 300 Å

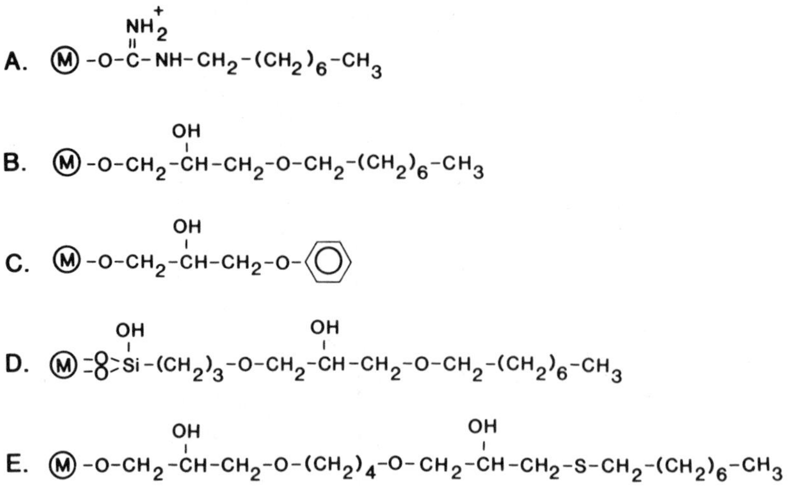

A. $\overset{+}{\underset{\text{M}}{\bigcirc}}\!-\!O\!-\!\overset{\overset{\displaystyle +}{\overset{\displaystyle NH_2}{\|}}}{C}\!-\!NH\!-\!CH_2\!-\!(CH_2)_6\!-\!CH_3$

B. $\overset{\text{M}}{\bigcirc}\!-\!O\!-\!CH_2\!-\!\overset{\overset{\displaystyle OH}{|}}{CH}\!-\!CH_2\!-\!O\!-\!CH_2\!-\!(CH_2)_6\!-\!CH_3$

C. $\overset{\text{M}}{\bigcirc}\!-\!O\!-\!CH_2\!-\!\overset{\overset{\displaystyle OH}{|}}{CH}\!-\!CH_2\!-\!O\!-\!\langle\!\bigcirc\!\rangle$

D. $\overset{\text{M}}{\bigcirc}\!\overset{-O}{\underset{-O}{>}}\!Si\!-\!(CH_2)_3\!-\!O\!-\!CH_2\!-\!\overset{\overset{\displaystyle OH}{|}}{CH}\!-\!CH_2\!-\!O\!-\!CH_2\!-\!(CH_2)_6\!-\!CH_3$

E. $\overset{\text{M}}{\bigcirc}\!-\!O\!-\!CH_2\!-\!\overset{\overset{\displaystyle OH}{|}}{CH}\!-\!CH_2\!-\!O\!-\!(CH_2)_4\!-\!O\!-\!CH_2\!-\!\overset{\overset{\displaystyle OH}{|}}{CH}\!-\!CH_2\!-\!S\!-\!CH_2\!-\!(CH_2)_6\!-\!CH_3$

Figure 7-4. Different hydrophobic ligands coupled to a gel matrix, M.

is performed. This coupling procedure will create ligands with a positive net charge. The structure in 7–4A is the isourea derivative proposed by Kohn and Wilchek[39].

In Fig. 7-4 are the B and C types based on the glycidyl ether (with an epoxide, oxirane, functional group) coupling procedure[5] (also used for the production of Octyl- and Phenyl-Sepharose). This coupling method is the most widely used, so it will be described in some detail (see also section 10.4.1.2). Some glycidyl ethers are commercially available, but they can also be prepared according to Ulbrich et al[40]. The reaction scheme for the coupling of the ligand to the gel (usually agarose) is as follows:

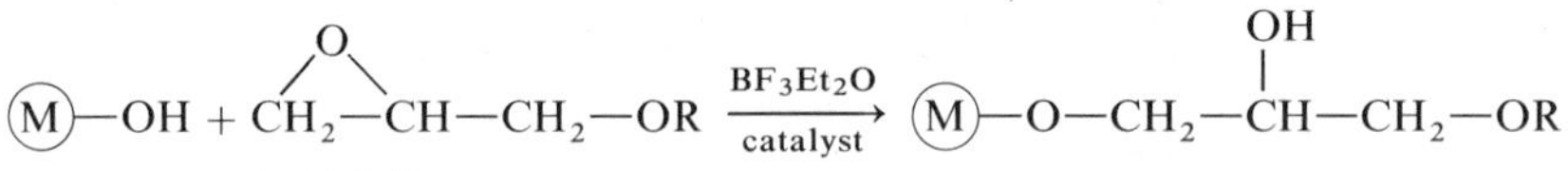

$$\overset{\text{M}}{\bigcirc}\!-\!OH + \overset{O}{\overset{/\backslash}{CH_2}}\!-\!CH\!-\!CH_2\!-\!OR \xrightarrow[\text{catalyst}]{BF_3Et_2O} \overset{\text{M}}{\bigcirc}\!-\!O\!-\!CH_2\!-\!\overset{\overset{\displaystyle OH}{|}}{CH}\!-\!CH_2\!-\!OR$$

$$\text{glycidyl ether} \hspace{6cm} (7\text{-}3)$$

R is an alkyl or aryl group. The gel should be transferred to an organic solvent, as dioxane. This is done in steps (100 ml portions to 100 ml sedimented gel), on a Büchner funnel:

(1) one washing with water-dioxane (4:1)
(2) one washing with water-dioxane (3:2)
(3) one washing with water-dioxane (2:3)
(4) one washing with water-dioxane (1:4)
(5) seven washings with dioxane

The gel is transferred to a reaction vessel equipped with a stirrer.

100 ml dioxane is added to 100 ml of sedimented gel, and 2 ml of a 48 percent solution of boron trifluoride etherate in diethyl ether is added. Stir for 5 minutes. 1 ml of glycidyl ether dissolved in 10 ml dioxane is then added dropwise from a separatory funnel. The reaction takes about 40 minutes. After the reaction the gel is transferred back to water by the above scheme, but in reverse order, and is finally washed with water. The amount of ligand to be coupled to the gel can be controlled by varying the amount of glycidyl ether added. The glycidyl ether coupling method produces gels that are charge

free, and thus should have no other interaction with proteins than hydrophobic interaction. However, the phenyl group shown in 7-4C, (and other aromatic groups) also have a potential for π-π interaction.

A recently introduced coupling method which leads to the structure shown in Fig. 7-4D was used for coupling hydrophobic ligands to agarose gels used as an HPLC packing[8,41]. The agarose is first activated with γ-glycidoxypropyltrimethoxy silane in water. The immobilization of the ligands is then performed in the alcohol that is to be coupled to the gel. The resulting gel is non-charged and also contains a spacer, making the ligands more available for the proteins.

The ligand structure shown in Fig. 7-4E was introduced by Maisano et al[11]. The agarose is first activated with a bis-epoxide, 1,4-butanediol diglycidyl ether, and then coupled with an alkyl mercaptan. This gel, also charge free, contains a spacer arm. The ligand density can be regulated easily by varying the amount of alkyl mercaptan.

7.3.2 Ligand Density and Capacity of HIC Gels

The density of ligands is important for the strength of the interaction between the gel and the proteins as well as for the capacity. For the commercially available Octyl- and Phenyl-Sepharose gels the ligand density is approximately 40 μmoles/ml gel bed, corresponding to a degree of substitution of approximately 0.2 moles hydrophobic substituent per mole galactose[42]. The capacity of Phenyl- and Octyl-Sepharose CL-4B in 0.01 M sodium phosphate buffer, pH 6.8, containing 1 M ammonium sulphate, is approximately 15 to 20 mg human serum albumin or 3 to 5 mg β-lactoglobulin per ml of gel[42]. Several procedures for determining ligand density have been employed[43,44] including NMR, gas chromatography and elementary carbon analyses (see also section 10.3.4). Ligand density determination for HIC adsorbents containing aromatic groups, e.g., phenyl, can also be made by derivative ultraviolet spectroscopy[45] and adsorbents of the type shown in Fig. 7-4E can be analysed by sulphur determination[11].

7.3.3 Other Types of Gels Related to HIC

At high salt concentrations, matrices used for gel filtration show adsorption of proteins[46–49]. The interaction, which is of a hydrophobic nature, can be used for the separation of proteins. Matrices like cellulose, dextran, agarose, organic polymers and modified silica can be utilized. The technique is called salting-out chromatography.

Spacers used in affinity chromatography are often of hydrophobic nature and can in some cases make a considerable contribution to the interaction between solutes and the adsorbents[50].

HIC gels with aromatic ligands show, besides the hydrophobic interaction, also the so-called π-π interaction. This effect is explicitly utilized in the charge-transfer chromatographic technique[51], using different kinds of ligands.

Another method related to HIC is the so-called thiophilic adsorption technique[52,53]. The ligand is divinylsulfone substituted with, e.g., mercaptoethanol. Although the technique is related to HIC, the adsorption characteristics of the thiophilic adsorbents are different from those of HIC ones.

The binding of proteins to dye-ligand substituted adsorbents[54,55] is due partly (although far from predominantly) to hydrophobic interaction. Also, binding to other ligands utilized in affinity chromatography can be partly of hydrophobic nature.

7.4 Chromatographic Technique

7.4.1 Introduction

Most proteins bind to HIC adsorbents, and the technique can thus be used in many purification procedures. HIC is based on a different separation principle from most other separation techniques and can thus, in combination with these, afford a high degree of purification. The mass recoveries of proteins are mostly excellent, as is the recovery of enzyme activity.

The high capacity of HIC adsobents makes them suitable for use at an early stage in a purification scheme[35]. The similarities to salt precipitation, also based on protein surface hydrophobicity, make a combination of these two methods less useful. But after an initial precipitation step to remove the most hydrophobic proteins, the usual second precipitation can be replaced by an HIC purification step, especially because HIC gives sharper separation than does salt precipitation.

Owing to the rather high salt concentrations of the eluate from the HIC column, a subsequent ion-exchange chromatography step is not possible withouth an intermediate desalting step. The reverse is easier to perform. A HIC step can, of course, be followed directly by gel filtration, whereby the salt and the other additives are automatically removed.

HIC adsorbents can also be used for batch procedures, making them suitable for purification of protein on a large scale[35]. Rigid gels allowing high flow rates can be used for large scale chromatography.

7.4.2 Choice of gels

Adsorbents used for HIC should preferably be charge free. At low ionic strengths protein can interact with the positively charged amino groups on HIC gels with the structure shown in Fig. 7-4A. In some cases this can lead to an irreversible binding of the protein to the gel[6,56].

The strength of interaction between a protein and hydrophobic ligands on a HIC adsorbent increases with the increase in length of the ligand[43] (alkyl type of ligands). Ligands containing between 4 and 10 carbon atoms are suitable for most separation problems. For proteins with poor solubility in buffers of high salt concentration, e.g., membrane proteins, HIC adsorbents with rather long ligands are recommended[7]. A phenyl group has about the same hydrophobicity as a pentyl group, although it can have a quite different selectivity compared to pentyl ligands owing to the possible π-π interaction. Aromatic groups on protein surfaces can interact specifically with the aromatic ligands.

HIC gels with a high ligand density interact more strongly with proteins than do gels with lower ligand densities[43]. With a low ligand density isocratic elution can be employed, as shown in Fig. 7-5. For isocratic elution the instrumental setup is simpler than that required for gradient elution.

7.4.3 HPLC-HIC

The adaption of HIC, as well as other classical protein purification techniques, to the high performance mode has been made possible by the production of beads of small and

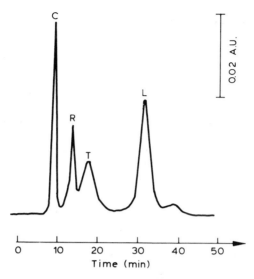

Figure 7-5. Isocratic separation of model proteins on a pentyl agarose column in the HPLC mode. Reprinted with permission from S. Hjertén *et al.*[8]

uniform size. The smaller the bead size the better and faster the separations can be done. On columns packed with HIC beads with a size of 1.5 μm, a separation of model proteins can be achived in less than 3 minutes[57]. Smaller bead sizes limit the number of matrices that can be utilized successfully, since smaller beads have a higher flow resistance that leads to higher pressures. Only rigid matrices can be used as high performance packings. High performance HIC packings have been based on silica[10], organic polymers[9] as well as the traditional gel material agarose[8,58]. The capability of HPLC-HIC is illustrated in Figs. 7-5 and 7-6, which show separations of model proteins and serum proteins on a pentyl agarose column.

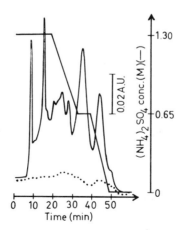

Figure 7-6. High performance HIC of human serum on a pentyl agarose gel. Elution with a negative salt gradient. Reprinted with permission from S. Hjertén *et al.*[8]

7.4.4 Conditions for Adsorption

Adsorption of proteins to a HIC gel is favoured by a high salt concentration, but owing to differences in the strength of interaction between the gel and different proteins the concentration of salt needed for adsorption can vary considerably. The concentration of salt used for adsorption should be below the concentration that precipitates different proteins in the sample to be chromatographed. For HIC gels with a ligand density similar to the Phenyl- and Octyl-Sepharose[R], the salt concentration is usually between 0.75 and 2 M (100 percent saturation = 4.05 M) with ammonium sulphate or 1 and 4 M with sodium cloride. For the most hydrophobic proteins the salt concentration can be lower. The salt should be dissolved in a buffer solution with a concentration of 0.01 to 0.05 M.

Different salts give rise to differences in the strength of interaction between proteins and the HIC gel. The strength of interaction as well as the capacity follows the series[16]:

$$Na_2SO_4 > NaCl > (NH_4)_2SO_4 > NH_4Cl > NaBr > NaSCN \qquad (7-4)$$

Na_2SO_4, NaCl and especially $(NH_4)_2SO_4$ are the most utilized salts in HIC. Besides the differences in the strength of interaction the different salts also show some additional selectivity[8].

The pH of the buffers used in HIC experiments has a decisive influence on the adsorption of proteins to the gel[8] (Fig. 7-7). Some proteins with high pI values bind strongly to HIC gels at elevated pH values, although no general trend in the interaction

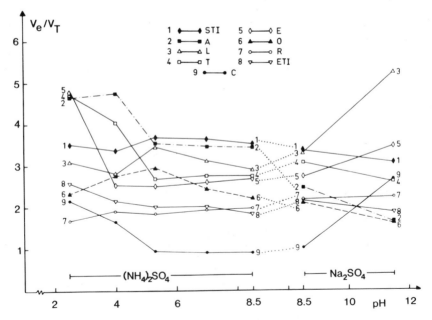

Figure 7-7. The pH dependence of the interaction between proteins and an octylagarose gel expressed as V_e/V_T (V_e is the elution volume of the different proteins and V_T is the elution volume of a non-retarded solute). Elution by a negative linear gradient of salt. The model proteins used were STI = soy trypsin inhibitor, A = human serum albumin, L = lysozyme, T = transferrin, E = enolase, O = ovalbumin, R = ribonuclease, ETI = egg trypsin inhibitor and C = cytochrome c. Adopted from Ref. 8.

strength and protein pI values has been observed. The change in retardation with pH is large for most proteins so it can be worthwhile to test different pH values for adsorption. The only limitation is the stability of the protein to be purified and the stability of the chromatographic matrix (e.g., silica is not stable at high pH).

As discussed above, the temperature dependence of HIC is not simple, although, generally, a decrease in temperature decreases the interaction. Labile proteins should be chromatographed at low temperatures. It is important that the column is equilibrated with the same buffer as the sample. If the sample is opalescent it should be centrifuged or filtered before application to the column. The sample volume can be large because the proteins generally become bound to the column before the elution starts.

7.4.5 Elution

Elution, whether done stepwise or with a gradient, can generally be achieved in three different ways:

1. Changing the salt concentration: The elution of solutes, in the order of increasing hydrophobicity, is accomplished by decreasing the salt concentration.
2. Changing the polarity of the solvent: A decrease in the interaction is achieved by adding solvents such as ethylene glycol or (iso)propanol. Up to 80% ethylene glycol may be used, whereas propanol should be used at lower concentrations. The addition of the polarity decreasing agents can be made after the salt has been removed from the column or concomitantly with the decrease of salt concentration.
3. Adding detergents. Detergents work as displacers of the proteins. They have been used mainly for purification of membrane proteins by HIC.

In theory, elution of proteins from HIC gels can be accomplished by changing the salt species to a chaotropic ion, e.g. SCN^-, but this should be avoided due to the protein structure breaking properties of these ions. A change in pH may also be used for elution, but it is (so far) impossible to predict how the strength of interaction between a protein and a HIC gel will be affected upon changing the pH.

7.4.6 Regeneration of HIC Gels

HIC adsorbents can be re-used several times. After each chromatographic run the adsorbents can be washed with 6 M urea or guanidine hydrocloride to remove the strongest adsorbed proteins. After washing, the gel can be equilibrated with starting buffer and used immediately for the next run. If detergents have been used a regeneration procedure involving washing with different alcohols has to be applied[42]. Gels can be stored at 4°C in the presence of 20% ethanol.

7.5 Applications

7.5.1 Purification of an Acid Phosphatase from Bovine Cortical Bone Matrix[59]; Elution by Decreasing Salt

In the purification of an acid phosphatase that displays phosphotyrosylprotein phosphatase activity several purification methods were employed, including HIC[59].

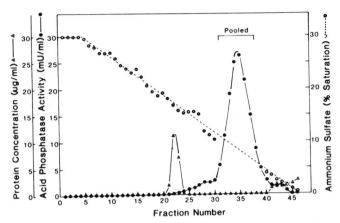

Figure 7-8. Phenyl-Sepharose chromatography of an acid phosphatase from bovine cortical bone matrix. Reprinted with permission from K.-H. W. Lau *et al.*[59]

Bovine long bones (*tibia*) were washed carefully to remove non-bone tissues and blood cells. Bone cubes were ground to powder which was then homogenized and extracted. The extract was subjected to CM-Sepharose ion-exchange chromatography, cellulose phosphate affinity column chromatography, Sephacryl S-200 gel filtration chromatography and HIC.

For the last step, HIC on a Phenyl-Sepharose column (Fig. 7-8) was used. Active fractions from the gel filtration column were pooled and adjusted to 30% saturation with ammonium sulphate by slowly adding solid ammonium sulphate at 4°C and stirring gently for at least 2 hours (buffer: 100 mM sodium acetate pH 6.5). The enzyme fraction was then applied to the Phenyl-Sepharose column equilibrated with sodium acetate buffer containing ammonium sulphate at 30% saturation. The chromatography was done at room temperature. The column was washed with this buffer and then eluted with a negative salt gradient, from 30% saturation to buffer without salt. The active fractions were concentrated and pooled. The phosphatase was stable for at least 4 months when stored at 4°C. The recovery of the enzyme in the HIC step was 92% and the purification was 13-fold. The enzyme was pure according to SDS-gel electrophoresis.

7.5.2   Purification of Human Pituitary Prolactin[60]; Elution by Decreasing Polarity

In the isolation of human pituitary prolactin, HIC on Phenyl-Sepharose has been utilized[60]. The entire preparation was performed at 5°C, and the starting material was frozen human pituitary glands. The glands were homogenized and extraction was done at elevated pH. The first chromatographic step was gel filtration on Sepharose CL-6B at pH 9.8. The fractions from the gel filtration column which contained prolactin were pooled and chromatographed on a Phenyl-Sepharose column (Fig. 7-9). To the column (3.2 × 25 cm) was applied a sample containing 200–400 mg of protein. The column was equilibrated with 0.2 M glycine/NaOH buffer (pH 9.8) and the elution was carried out at pH 9.8 by a stepwise decrease in the buffer concentration (to 0.02 M glycine/NaOH) and

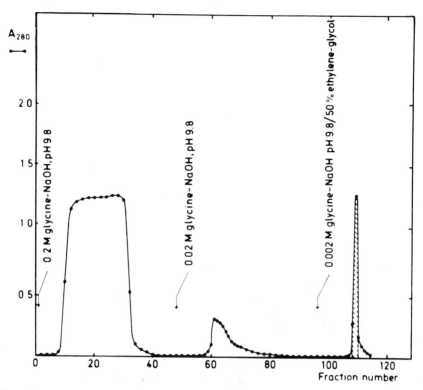

Figure 7-9. Chromatography on Phenyl-Sepharose of a prolactin preparation. The hatched area represents the prolactin-containing fractions. Reprinted with permission of P. Roos *et al.*[60]

finally, by inclusion of ethylene glycol (50%, v/v). Fractions of about 15 ml were collected and the flow rate was 45 ml/hour. The prolactin activity was eluted by the buffer containing ethylene glycol and recovery was 95%. The purification of the prolactin was completed by an additional gel filtration (Sephadex G-100 Superfine) and an ion-exchange step on DEAE-Sepharose CL-6B.

7.5.3   Purification of a Phospolipase C from *Trypanosoma brucei*[61]; Elution by Detergent

In the purification of phospholipase C from *T. brucei*, HIC was used at an early stage[61].

Trypanosomes were lysed and the membranes were solubilized and extracted with n-octyl glucoside containing buffers. A fraction containing the phospholipase activity was precipitated with an equal volume of saturated ammonium sulphate solution. After centrifugation, the supernatant (50 ml, containing 8.7 mg protein) was applied to a Phenyl-Sepharose column (1.2 × 14 cm) equilibrated with 50% saturated ammonium sulphate in 25 mM sodium succinate, pH 6.0. A linear gradient (100 ml) with decreasing salt concentration from 50% saturation to buffer without ammonium sulphate was applied, followed by further washing with 50 ml buffer. The phospolipase was eluted with

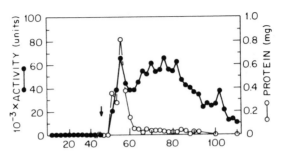

Figure 7-10. HIC of a phopholipase C from *Trypanosoma brucei*. The vertical arrow shows the point at which elution with 1 % CHAPS was initiated. Fraction numbers are shown on the abscissa; note the change in scale after fraction 44. Reprinted with permission from D. Hereld *et al.*[61]

buffer containing detergent, 1 % CHAPS (3-[(3-cholamidopropyl)-dimethylammonio]-1-propanesulfonate). The chromatogram, Fig. 7-10, shows that most of the protein eluted in a sharp peak after introducing the detergent. The phospolipase on the other hand, eluted in a broad peak. The trailing fractions of the activity peak, which contained about 70 % of the activity but relatively little protein, were pooled. The yield in the HIC step was 62 % and the purification was 22-fold. The HIC step was followed by an ion exchange step (CM-Sephadex C-25) and gel filtration chromatography (Sephacryl S-200).

7.5.4 Exchange of Detergents Bound to Membrane Proteins[62]

Unfortunately, with most intrinsic membrane proteins, no single detergent is usually well suited for all different steps in a purification scheme. Methods for detergent-exchange are therefore needed. One method is Phenyl-Sepharose mediated detergent-exchange chromatography. The alkyl detergents; lauryl maltoside, octyl glucoside and dodecyl sulphate were each successfully exchanged for Triton X-100, Triton N-101, or Nonidet P-40 present in a solution of either cytochrome *c* oxidase, a mixture of inner

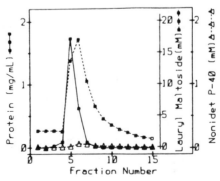

Figure 7-11. Exchange of lauryl maltoside for protein-bound Nonidet P-40. Reprinted with permission from N. C. Robinson *et al.*[62]

mitochondrial membrane proteins, or a mixture of erythrocyte membrane proteins. Below follows a description of an exchange of lauryl maltoside for cytochrome c oxidase-bound Nonidet P-40 (Fig. 7-11).

A 0.5×10 cm bed of Phenyl-Sepharose was presaturated with 13 ml of 10 mM lauryl maltoside in a pH 9.0 buffer at an ionic strength of 0.01 (Tris-HCl containing 0.1 mM EDTA), followed by 8 ml of 2 mM lauryl maltoside in the same buffer; a 0.6 ml protein sample was applied that contained 3 mg of protein/ml and 26 mM Nonidet P-40 in the above buffer to which enough lauryl maltoside was added to make the concentration 2 mM. Elution was effected by the above buffer containing 2 mM lauryl maltoside, and fractions were collected and analyzed. At pH 9 and an ionic strength of 0.01, only 5–15% of the Nonidet P-40 solubilized cytochrome c oxidase complex was bound to the column that had been saturated with lauryl maltoside. Less than 0.1% of the original amount of Nonidet P-40 remained in the complex after the detergent exchange.

7.6 References

1. R. J. Yon, *Biochem. J. 126*, 765 (1972).
2. Z. Er-el, Y. Zaidenzaig, S. Shaltiel, *Biochem. Biophys. Res. Commun. 49*, 383 (1972).
3. B. H. J. Hofstee, *Anal. Biochem. 52*, 430 (1973).
4. J. Porath, L. Sundberg, N. Fornstedt, I. Olsson, *Nature 245*, 465 (1973).
5. S. Hjertén, J. Rosengren, S. Påhlman, *J. Chromatogr. 101*, 281 (1974).
6. S. Hjertén, *J. Chromatogr. 87*, 325 (1973).
7. S. Hjertén, In *Methods of protein separation*; Catsimpoolas N. ed.; Plenum publishing: New York, 1976, vol. 2, Chapter 6.
8. S. Hjertén, K Yao, K.-O. Eriksson, B. Johansson, *J. Chromatogr. 359*, 99 (1986).
9. Y. Kato, T. Kitamura, T. Hashimoto, *J. Chromatogr. 360*, 260 (1986).
10. J. L. Fausnaugh, E. Pfannkoch, S. Gupta, F. E. Regnier, *Anal. Biochem. 137*, 464 (1984).
11. F. Maisano, M. Belew, J. Porath, *J. Chromatogr. 321*, 305 (1985).
12. J. Porath, *J. Chromatogr. 376*, 331 (1986).
13. C. Tanford, *The hydrophobic effect: formation of micelles and biological membranes*; Wiley: New York, (1973).
14. T. E. Creighton, *Proteins, structures and molecular properties*; W. H. Freeman: New York, (1984).
15. S. Lewin, *Displacement of water and its control of biochemical reactions*; Academic: New York, 1974, p. 71.
16. S. Påhlman, J. Rosengren, S. Hjertén, *J. Chromatogr. 131*, 99 (1977).
17. W. Melander, C. Horvath, *Archiv. Biochem. Biophys. 183*, 200 (1977).
18. J. L. Fausnaugh, F. E. Regnier, *J. Chromatogr. 359*, 131 (1986).
19. Y. Nozaki, C. Tanford, *J. Biol. Chem. 246*, 2211 (1971).
20. D. D. Jones, *J. Theor. Biol. 50*, 167 (1975).
21. P. Manavalan, P. K. Ponnuswamy, *Nature 275*, 673 (1978).
22. D. H. Wertz, H. A. Scheraga, *Macromolecules 11*, 9 (1978).
23. C. Chotia, *J. Mol. Biol. 105*, 1 (1976).
24. W. R. Krigbaum, A. Komoriya, *Biochim. Biophys. Acta 576*, 204 (1979).
25. G. D. Rose, A. R. Geselowitz, G. J. Lesser, R. H. Lee, M. H. Zehfus, *Science 229*, 834 (1985).
26. J. L. Cornette, K. B. Cease, H. Margalit, J. L. Spouge, J. A. Berzofsky, C. DeLisi, *J. Mol. Biol. 195*, 659 (1987).
27. D. Eisenberg, A. D. McLachlan, *Nature 319*, 199 (1986).
28. R. B. Hermann, *J. Phys. Chem. 76*, 2754 (1972).
29. B. Lee, F. M. Richards, *J. Mol. Biol. 55*, 379 (1971).
30. F. M. Richards, *Ann. Rev. Biophys. Bioeng. 6*, 151 (1977).
31. S. Miller, J. Janin, A. M. Lesk, C. Chotia, *J. Mol. Biol. 196*, 641 (1987).
32. H. P. Jennissen, *J. Colloid. Interface Sci. 111*, 570 (1986).
33. S. T. Freer, J. Kraut, J. D. Robertus, H. T. Wright, N. H. Xuong, *Biochemistry 9*, 1997 (1970).
34. J. L. Fausnaugh, L. A. Kennedy, F. E. Regnier, *J. Chromatogr. 317*, 141 (1984).

35. R. K. Scopes, *Protein purification, principles and practice*; Springer: New York, (1982).
36. S.-L. Wu, A. Figueroa, B. L. Karger, *J. Chromatogr. 371*, 3 (1986).
37. S.-L. Wu, K. Benedek, B. L. Karger, *J. Chromatogr. 359*, 3 (1986).
38. R. Axén, J. Porath, S. Ernback, *Nature 214*, 1302 (1967).
39. J. Kohn, M. Wilchek, *Applied Biochemistry and Biotechnology 9*, 285 (1984).
40. V. Ulbrich, J. Makes, M. Jurecek, *Collect. Czech, Chem. Commun. 29*, 1466 (1964).
41. S. Hjertén, K. Yao, Z.-h. Liu, D. Yang, B.-l. Wu, *J. Chromatogr. 354*, 203 (1986).
42. Pharmacia Fine Chemicals. Octyl-Sepharose CL-4B and Phenyl-Sepharose CL-4B.
43. J. Rosengren, S. Påhlman, M. Glad, S. Hjertén, *Biochim. Biophys. Acta 412*, 51 (1975).
44. B.-L. Johansson, I. Drevin, *J. Chromatogr. 346*, 255 (1985).
45. B.-L. Johansson, I. Drevin, *J. Chromatogr. 391*, 448 (1987).
46. J. Porath, *Nature 196*, 47 (1962).
47. L. G. Hoffmann, P. W. McGivern, *J. Chromatogr. 40*, 53 (1969).
48. N. Sakihama, H. Ohmori, N. Sugimoto, Y. Yamasaki, R. Oshino, M. Shin, *J. Biochem. 93*, 129 (1983).
49. K. Adachi, *Biochim. Biophys. Acta 912*, 139 (1987).
50. S. Shaltiel, In *Methods Enzymol.*; W. B. Jakoby and M. Wilchek, eds.; Academic: New York, 1974, Vol XXXIV, p. 126.
51. J. Porath, *J. Chromatogr. 159*, 13 (1978).
52. J. Porath, F. Maisano, M. Belew, *FEBS Lett, 185*, 306 (1985).
53. T. W. Hutchens, J. Porath, *Anal. Biochem. 159*, 217 (1986).
54. R. L. Easterday, I. M. Easterday, *Adv. Exp. Med. Biol. 42*, 123 (1974).
55. R. K. Scopes, *J. Chromatogr. 376*, 131 (1986).
56. L. Hammar, S. Påhlman, S. Hjertén, *Biochim. Biophys. Acta 403*, 554 (1975).
57. R. Janzen, K. K. Unger, H. Giesche, J. N. Kinkel, M. T. W. Hearn, *J. Chromatogr. 397*, 91 (1987).
58. B.-L. Johansson, Ö. Jansson *J. Chromatogr. 363*, 387 (1986).
59. K.-H. W. Lau, T. K. Freeman, D. J. Baylink, *J. Biol. Chem. 262*, 1389 (1987).
60. P. Roos, F. Nyberg, L. Wide, *Biochim. Biophys. Acta 588*, 368 (1979).
61. D. Hereld, J. L. Krakow, J. D. Bangs, G. W. Hart, P. T. Englund, *J. Biol. Chem. 261*, 13813 (1986).
62. N. C. Robinson, D. Wiginton, L. Talbert, *Biochemistry 23*, 6121 (1984).

8 Immobilized Metal Ion Affinity Chromatography

Lennart Kågedal
Research & Development
Pharmacia LKB Biotechnology AB
S-751 82 Uppsala, Sweden

8.1 Introduction

In 1975, Porath *et al.* published a paper entitled "Metal chelate affinity chromatography, a new approach to protein fractionation"[1]. The principle of the suggested method was based on differences in the affinity of proteins for metal ions bound in a 1:1 complex of iminodiacetic acid (IDA) immobilized on a chromatographic support. The principle was not new—it had been suggested already in 1961 by Hellferich who named it "ligand exchange chromatography"[2] (LEC) and the technique has been widely used since then[3]. However, Porath's paper showed for the first time that immobilized metal ions could be used with advantage to fractionate and purify proteins. As will be demonstrated in this review the technique has become an important tool for the isolation of many proteins.

227

In the past, the term "metal chelate affinity chromatography" (MCAC) has been the accepted term for LEC when used in biochemistry. Other names have also been used, e.g. "metal chelate interaction chromatography" (MCIC). Following a proposal by Porath *et al.*[4,5], I will use the term "immobilized metal ion affinity chromatography" (IMAC) in this review.

A comprehensive review of LEC was published in 1977 by Davankov and Semechkin[3]. The theory they presented is to a large extent relevant also for IMAC and we refer to them for a more complete coverage of the literature up to 1977. Some general aspects of the technique as applied to proteins have been reviewed by Porath and Belew[4], Lönnerdal and Keen[6], Sulkowski[7], and Porath and Olin[8].

Most proteins can form complexes with metal ions. Many of these are multi-dentate complexes (chelates) and allow for the purification of the proteins by IMAC. The strength of the complexes formed varies from protein to protein which, in many cases, gives rise to the high specificity of IMAC.

The chromatographic sorbent used in IMAC (see scheme in Fig. 8-1) consists of a suitable chromatographic support to which a metal chelating substance (B) has been attached by a leash or linkage group (A). The structure of the complex formed when metal ions are added must be such that some coordination sites are left free for the binding of solvent or solute molecules (ligands). Alternatively, the complex should be able to rearrange itself to allow incoming ligands to participate in the formation of chelates or complexes with the metal ion. Solvent or buffer molecules will occupy "free" coordination sites of the metal in the absence of ligands with higher affinity for the metal ion.

Muzzarelli *et al.* have suggested the following definition of LEC[9]: "Ligand exchange chromatography is based on the principle that a molecule or ion, which is part of a complex fixed on a support, can be released because a different molecule or ion enters to form a more stable complex, or because the complex collapses when the medium is

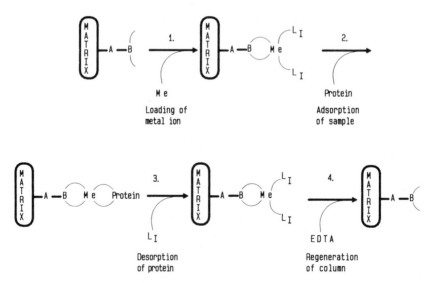

Figure 8-1. Principle of immobilized metal ion affinity chromatography. A = linkage (spacer) group, B = chelating group, Me = metal ion, L_I = solvent or buffer molecule.

altered". The definition may be somewhat limited compared to that proposed by Davankov and Semechkin (Ref. 3, p. 314) but it gives a sufficiently clear basis for the understanding of IMAC as reviewed in this article.

Some of the special features of IMAC of proteins can be summarized as follows:

— Exposure of certain amino acid residues (histidine, cysteine, tryptophan) on the "surface" of the proteins is required for the adsorption of proteins.
— The steric arrangement of the protein chain plays an important role, which means that molecules with closely similar properties with respect to charge, molcular size, amino acid composition, but with differences in their secondary and tertiary structure, can be separated.
— Simple ionic adsorption and other complicating factors can be suppressed or modified by buffers of high ionic strength.
— Binding is influenced by pH. Low pH causes elution of adsorbed substances. Exceptions to this are known[10].
— Several elution techniques are available (pH gradient, competitive ligands, organic solvents, chelating agents).
— IMAC is a general technique for purifying proteins. Metalloproteins do not bind specifically at their metal coordination sites but rather through amino acid residues exposed at the protein surface.

IMAC is an established technique for the purification of proteins on a laboratory scale and is beginning to find industrial applications. Its acceptance may have been rather slow initially but its usefulness and versatility has been unequivocally demonstrated sufficient to convince biochemists that it is a valuable addendum to the arsenal of preparative methods.

The apparent complexity of the method because of the number of factors influencing the IMAC process may have added to the reluctance of biochemists to adopt the method. Although there still are several question marks left concerning the exact mechanisms and chemical structures involved there is sufficient information available for a systematic optimization of preparative experiments. For example, we know how to modify adsorption and desorption by correct choice of chelating group, metal ion, pH, and buffer constituents.

The specificity of IMAC obviously depends on the exploitation of the combined effects of primary structure (occurrence and position of a limited number of metal binding amino acid residues) and secondary and tertiary structure (exposure of certain amino acids on the protein surface). It may well be that the proximity of the metal binding amino acids are also important.

8.2   The Metal Chelate Gels

8.2.1   Choice of Chromatographic Support

Basically, the requirements of the support in affinity chromatography of biological molecules apply also to IMAC (Ref. 11, p. 20). Ideally the support should

— be easy to derivatize,
— exhibit no unspecific adsorption,

— have good physical, mechanical and chemical stability,
— be of high porosity to provide easy ligand accessibility,
— permit high flow rates,
— be stable to eluants including, for example, denaturing additives,
— allow regeneration of column without deterioration of the gel bed,
— provide a stable gel bed with no shrinking/swelling during the chromatographic process.

Beaded agarose is the support predominantly used for IMAC of proteins and published protocols in most cases describe the use of Sepharose® 4B or Sepharose 6B to which iminodiacetic acid (IDA) had been coupled by the original bisoxirane method described by Porath et al[1].

Silica-based IDA gels have been constructed for use in preparative work[12]. Gels for use in HPLC and FPLC™ have been prepared using hydrophilized resins[13,14], cross-linked agarose[15,16] (10 μm), and silica[17] (5 μm) using IDA as the chelating ligand.

The properties of the resin-based gel used with Zn^{2+} parallel those of the agarose type of gels, whereas the silica based material shows complications due to electrostatic and hydrophobic interactions.

8.2.2   Chelating Ligand and Strength of Chelates; Capacity for Metal Ions

IMAC in general can be carried out using various types of ion exchangers[3]. In IMAC of proteins, however, only chelating groups have been used to fix the metal ion to the support, the reason being that metal ions bind more strongly to chelating groups. The binding energy of transition metal cations, as calculated by Schmuckler, is 2–3 kcal/mol with ordinary cation exchangers and 15–25 kcal/mole with chelating groups[18]. Table 8-1 lists the formation constants for 1:1 chelates of some chelating compounds. Compounds I and II have a structure similar to that of the chelating group in IDA Sepharose and compound III is a close analogue of the so-called TED group, introduced by Porath and Olin[8]. Studies of the formation constants of the IDA groups of Dowex A-1 indicated that formation constants of the same order of magnitude are obtained for analogous compounds free in solution or bound to a chromatographic support[3]. The relative stability of complexes formed with an IDA derivative of cellulose and divalent metal ions are in the order Cu(II) > Ni(II) > Zn(II) ≥ Co(II) ≫ Ca(II), Mg(II), as shown by Horváth and Nagydiósi,[19] which agrees well with data in Table 8-1. They also concluded that Ca(II) and Mg(II) ions do not form chelates with the IDA group since their presence did not affect the shape of the potentiometric acid-base titration curve.

The fixing of the IDA-ligand to the support has in most cases been effected by Porath's original method of coupling to agarose into which reactive epoxy groups have been introduced by reaction with 1,4-bis-(2,3-epoxypropoxy)-butane[1]. The method has the dual advantage of providing a chemically stable ether linkage to the ligand and a twelve-atom spacer. A three-atom spacer is obtained if activation of the agarose matrix is carried out with epichlorohydrin[8]. In Chelating Sepharose Fast Flow, Table 8-2, a seven atom spacer is used.

Results obtained by Hansson et al.[15] show that the metal ion capacity of the IDA-gel influences the retention of proteins (Fig. 8-2). IDA derivatives of Superose® 12 carrying varying amounts of the chelating moiety were used. The resolution of the sample proteins was also affected indicating that the extent of the influence on retention is dependent on the properties of the proteins.

Table 8-1. Formation constants[a] ($\log K$) for 1:1 complexes of chelating compounds[b] and metal ions. (I) N-Methyliminodiacetic acid, (II) N-(hydroxymethyl)iminodiacetic acid, (III) N-(hydroxyethyl)ethylenediaminetriacetic acid.

Chelating compound		Metal ion						
No.	Formula	Ca^{2+}	Fe^{2+}	Fe^{3+}	Co^{2+}	Ni^{2+}	Cu^{2+}	Zn^{2+}
I	CH₃—N with CH₂COOH, CH₂COOH	3.8	6.3		7.6	8.7	11.1	7.6
II	CH₂—N with OH, CH₂COOH, CH₂COOH	4.8	6.8		8.1	9.4	11.7	8.5
III	N—CH₂CH₂—N with CH₂COOH, CH₂COOH, CH₂CH₂OH, CH₂COOH	8.2	12.2	19.5	14.5	17.1	17.5	14.6

[a] From critical stability constants vol. 1-4, E. Martell & R. M. Smith, Plenum Press, New York & London 1974–1977.
[b] In the chelate the ligand is present in its ionized form.

Table 8-2. Commercially available chromatographic supports for IMAC

Designation	Manufacturer	Matrix
Chelating Sepharose™ 6B[a]	Pharmacia LKB AB Uppsala, Sweden	Agarose
Chelating Sepharose Fast Flow[a]	Pharmacia LKB AB Uppsala, Sweden	Crosslinked agarose
Chelating Superose™ [a]	Pharmacia LKB AB Uppsala, Sweden	Crosslinked agarose
Immobilized Iminodiacetic Acid I	Pierce, Rockford, USA	Agarose
Immobilized Iminodiacetic Acid II	Pierce, Rockford, USA	Sephadex®
Immobilized Iminodiacetic Acid	Pierce, Rockford, USA	Fractogel TSK HW-65F
Immobilized Tris(carboxymethyl)ethylenediamine	Pierce, Rockford, USA	Agarose
Immobilized Tris(carboxymethyl)-ethylenediamine	Pierce, Rockford, USA	Fractogel TSK HW-65F
Iminodiacetic acid-agarose	Sigma Chemical Comp. St. Louis, USA	Agarose
Iminodiacetic acid-epoxy activated Sepharose 6B	Sigma Chemical Comp. St. Louis, USA	Agarose
Zink chelate affinity adsorbent	Boehringer Mannheim, West Germany	Agarose
TSKgel Chelate-5PW[b]	Toyo Soda, Japan	Polymer resin

[a] Capacity ca 30 μmole Zn^{2+}/ml gel
[b] Capacity ca 20 μmole Zn^{2+}/ml gel

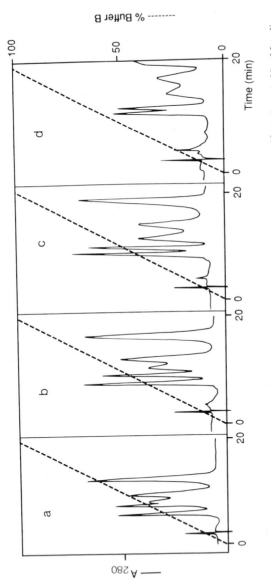

Figure 8-2. Cu²⁺-IMAC of proteins on IDA Superose[R] 12 gels with different metal ion capacity[15]. Buffers: A, 20 mM sodium phosphate, 1 M NaCl, pH 7.2; B, 20 mM sodium phosphate, 1 M NH₄Cl, pH 7.2. Column: 5 × 50 mm (HR 5/5 Pharmacia AB). Sample: 50 μl of a solution consisting of cytochrome C (peak 1), lysozyme (peak 2), β-lactoglobulin (peak 3), lens lectin (peak 4), and α-lactalbumin (peak 5) dissolved in half-diluted buffer A. Flow rate: 0.6 ml/min. Conditioning of columns before sample application: Elution with water, 5 ml; 0.25 M Cu₂SO₄, 5 ml; water, 5 ml; buffer B, 20 ml. Elution of proteins: From 100% A to 100% B in 20 min. Cu²⁺-capacity of packed gels (μmole/ml): Panel a, 18; panel b, 26; panel c, 32; panel d, 43.

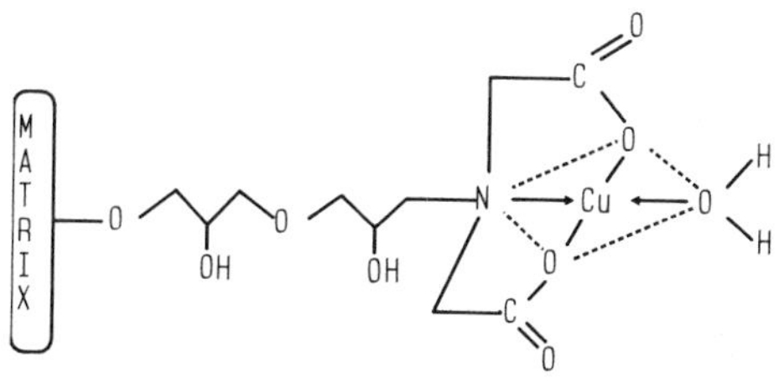

Figure 8-3. Postulated planar Cu^{2+} chelate with iminodiacetic acid as chelating group.

Commercially available column packings for IMAC of proteins are listed in Table 8-2.

Figure 8-3 shows the postulated planar structure of the chelate formed between Cu^{2+} and Chelating Sepharose Fast Flow including the linkage group. One coordination site is occupied by a water molecule (or possibly by some buffer constituent) that will be substituted by a sample molecule during the course of the chromatographic experiment.

8.3   Factors Influencing Adsorption and Desorption

8.3.1   Chelate Structure and Metal Ion

The exact structure of the stationary complex depends upon the metal ion used and on the composition of the eluent buffer. According to Davankov et al. Zn^{2+} and Cu^{2+}, Fig. 8-3, will leave one and Ni^{2+} three sites for solvent or buffer molecules when the three-dentate IDA is used[3]. Others seem to assume higher coordination numbers than four for the Zn^{2+} and Cu^{2+} ions. Porath and Belew have used N,N,N-(tricarboxymethyl)ethylene diamine coupled to Sepharose (TED Sepharose)[8]. Since the TED group is a five-dentate chelating group, it can occupy five of the six coordination sites of the Ni^{2+} ion. Consequently, Ni^{2+}-TED Sepharose is a weaker adsorbent than the corresponding IDA-column[8,20,21]; this is demonstrated by, for example, the adsorption on Ni^{2+}-IDA Sepharose of proteins which pass through a Ni^{2+}-TED Sepharose column. Recently, Hochuli et al.[59] have claimed that the Ni^{2+} chelate of a nitrilotriacetic acid derivative has a high specificity for proteins containing adjacent histidine residues.

It is reasonable to assume that the sorption of ligands from the sample will cause rearrangement of the stationary complex (Ref. 3, p. 326), which in turn may alter the stability of some of the stationary complexes. Such a supposition is supported by the observation that when a protein sample is applied to a column of chelating gel incompletely charged with copper ions, i.e. the lower part of the column is still white, the blue zone of the copper complex will move down the column to some extent[22].

In 50 recorded published protocols for the purification of proteins by IMAC, Zn^{2+} was used in 27 cases, Cu^{2+} in 15, Ni^{2+} in 3, Co^{2+} and Fe^{3+} in 2, and Ca^{2+} once. Cu^{2+}, Ni^{2+}, Zn^{2+}, and Co^{2+} form a series of metal ions with decreasing binding strength as shown by Porath and others[4,7,23]. Other ions such a Cd^{2+} may be used in a similar way[11].

Ca^{2+} and Fe^{3+} seem to represent special cases. Ca^{2+} has been used for the purification of Dolichos biflorus seed lectin[24]. The lectin has very high affinity for Ca^{2+} and was eluted from the column using EDTA. The special character of the interaction of

Fe^{3+}-IDA gels with proteins is reflected in the requirement for special elution modes, see section 8.3.3.

8.3.2 Protein Structure

The amino acid residues responsible for the adsorption of proteins in IMAC have been discussed in a number of papers[8,23,25-32]; yet there are several questions left open. According to Porath et al. some studies with amino acids indicated that histidine and cysteine would be most likely to cause strong interactions with metals in IMAC[1]. Tryptophan, which contains the indole structure, was also thought to contribute to the binding. Porath has later demonstrated that Fe^{3+} bound to IDA Sepharose has a rather specific affinity for tyrosine-containing proteins[27].

The finding by Sulkowski[32] that there is an abrupt change in the adsorption of bovine serum albumin to a Ni^{2+}-IDA gel around pH 6.5 supports the notion that histidine residues are involved in the sorption process in this case.

Desorption of ribonuclease with an imidazole gradient was found to be influenced by the pH of the eluent, a higher pH requiring a lower imidazole concentration.

Using the α-chymotrypsinogen A and α-chymotryspin pair and glycosylated and non-glycosylated ribonuclease, Sulkowski[32] has shown that the microenvironment of the histidine residues influences the retention of proteins significantly.

Sulkowsky et al. have used IMAC as a tool for the study of the surface topography of interferons[23]. The same group has also used some model proteins to establish the influence of various potential complex-forming amino acid residues on the behaviour of proteins in IMAC. Their results showed that for angiotensin I, angiotensin II and pancreatic ribonuclease A, the strength of binding to different metals decreases in the following order: $Cu^{2+} > Ni^{2+} > Co^{2+} > Zn^{2+}$. The difference between the strength of binding to Co^{2+} and Zn^{2+} was rather small. In the absence of tryptophan and cysteine on the protein surface, an increasing number of histidine residues causes stronger retention. Pancreatic tryspin inhibitor, which lacks histidine, tryptophan and cysteine on the protein surface, was adsorbed only onto Cu^{2+} at pH 7.4, "presumably due to coordination via amino groups".

In a recent study by Hammacher et al.[16] IMAC was used in the structural characterization of platelet-derived growth factor (PDGF) purified from human platelets.

PDGF is the general name for any one of several dimers which differ in composition of the two homologous polypeptide chains denoted A and B. The chains are linked by disulphide bridges into homodimers of either the PDGF-AA or -BB type or a heterodimer of the PDGF-AB type. Depending on the cell type used as a source of PDGF, the purified product is either a single isoform (PDGF-AA, PDGF-AB or PDGF-BB) or a mixture of different isoforms. The A and B chains differ in the content of histidine, B having one and A three residues. When applied to IMAC (Cu^{2+}-IDA), PDGF-BB was not adsorbed whereas PDGF-AA was retained on the column. The unknown sample, PDGF purified from human platelets, was partly retained. The retained material eluted before the A homodimer and was concluded to be a heterodimer (PDGF-AB). PDGF purified from human platelets thus consists of PDGF-AB and PDGF-BB, see section 8.6.6.

Using carboxypeptidase A as a model protein Muszyńska et al.[20] have studied the possibilities of employing IMAC for the characterization of metalloproteins and their preparation in metal-free forms.

In general it seems that whether a protein is a metalloprotein or not has no bearing on the binding or non-binding of the protein to the chromatographic support in IMAC. For example, Lönnerdal *et al.* showed that Fe-lactoferrin, Cu-lactoferrin and apo-lactoferrin exhibited identical sorption and desorption properties; the Fe-lactoprotein was recovered from a Cu^{2+}-chelate gel in its initial iron-saturated form[6].

Nucleoside diphosphatase bound to Zn^{2+} and Cu^{2+} chelate gels exhibited almost full catalytic activity[29], thus indicating that the binding site is separated from the catalytically active site.

At present there is no well-founded rationale for predicting the conditions under which a protein will be adsorbed onto a metal chelate column. Obviously it is related to the extent of exposure on the protein of residues such as histidine, cysteine, tyrosine, tryptophan and lysine or possibly non-amino acid residues with an affinity for metals. It is, for example, not understood why albumin, which contains many histidine and lysine groups, displays a rather weak binding in IMAC. Albumin does bind under certain conditions, however, and it seems that, for most proteins, conditions can be found which afford binding. The possibility of using IMAC as a "negative" purification step should not be neglected, i.e. to remove contaminating proteins by adsorption to the metal chelate column while the desired protein elutes in the starting buffer.

When the amino acid compositions of the proteins to be fractionated are known, the following classification adapted after Sulkowski[7,32] with respect to histidine and tryptophan residues may be used to predict results:

Presence of histidine or tryptophan on "surface" of protein	Metal ions providing adsorption
No His/Trp	—
One His	Cu^{2+}
More than one His	Cu^{2+}, (Stronger adsorption) Ni^{2+}
Clusters of His	Cu^{2+}, Ni^{2+}, Zn^{2+}, Co^{2+}
Several Trp, no His	Cu^{2+}

No fool-proof rules can be set as yet. Even in the absence of the three amino acids considered to be most important for adsorption, namely histidine, tryptophan and cysteine, adsorption or protein has been observed[23].

The fact that histidine residues are of great importance in IMAC has been exploited by Hochuli *et al.* using a genetic approach[59]. They fused DNA elements coding for adjacent histidine residues to the carboxy or the amino terminus of a gene of a recombinant protein and showed that the hybrid proteins could be very efficiently extracted and purified from the disintegrated cell paste after expression in *E. coli*. A similar approach has been presented by Smith *et al.*[60].

Belew *et al.*[33] have made quantitative studies of the interaction between proteins and Cu^{2+}-IDA on Chelating Sepharose Fast Flow using both frontal analysis and equilibrium binding analysis. Closely similar results were obtained with both methods. The dissociation constant (K_d) at pH 7.0 was $33-37 \mu M$ for lysozyme and $3.5-6.8 \mu M$ for ovalbumin. The measured binding capacities (L_t) were 6.8 and 1.7 μmole/ml gel for the same proteins. See also Hutchens *et al.*[58].

Table 8-3. Behaviour of Cohn IV fraction proteins from human serum in IMAC

Metal ion	Buffers in sample and starting buffer	Albumin	Transferrin	α_2-HS glyco-protein	α_1-Lipoprotein
			Behaviour of proteins		
Zn^{2+}	50 mM Tris-HCl pH 8.0 150 mM NaCl	no retention	retained	retained	retained
Cu^{2+}	100 mM Na acetate pH 7.7 500 mM NaCl	complete retention	retained	retained	retained
Cu^{2+}	20 mM phosphate pH 7.7 500 mM NaCl	complete retention			
Cu^{2+}	50 mM Tris-HCl pH 8.0 150 mM NaCl	partial retention			
Cu^{2+}	50 mM Tris-HCl pH 8.0 150 mM NH_4Cl	major part unretained[a]	retained	retained (partial?)	retained

[a] Albumin oligomers and minor part of albumin monomer were retained.

8.3.3 pH, Types of Buffers and Ionic Strength

There has been no comprehensive study of how buffers of different types affect sorption of proteins to Me^{n+}-IDA gels. Tables 8-3 and 8-4 give some indications how conditions can be varied to give a suitable strength of adsorption.

Fe^{3+}-IDA gels present a special case as demonstrated by Andersson and Porath[10]. They showed that such a gel could be used to isolate phosphoproteins by an ascending (sic!) pH gradient or by inclusion of phosphate in the elution buffer. Sulkowski[34] has extended the study to other types of proteins and also shown that selective desorption can be effected with an increasing concentration of sodium chloride.

It is customary to include sodium chloride, 0.1-1.0 M, in buffers used in IMAC. One effect of this is to suppress ionic interactions between sample and the gel. Sulkowski[32] has, however, shown that at least in some cases the ionic strength of the eluent affects the retention of proteins on Cu^{2+}-IDA columns. Cytochrome c from horse heart, pI 10.6, eluted earlier in a pH gradient when ionic strength was high whereas the opposite was observed with calmodulin, pI 4.1. The disparate behaviour of these proteins was accounted for by assuming that the net charge of Cu^{2+}-IDA is negative.

Table 8-4. Qualitative influence of various factors on the binding of proteins in IMAC

System parameter	Weaker binding			Stronger binding	
Stationary phase:					
Metal ion	Ca^{2+}	Co^{2+}	Zn^{2+}	Ni^{2+}	Cu^{2+}
Chelating ligand					
Type	TED				IDA
Amount bound	Low				High
Mobile phase:					
pH	5		7		8
Buffer ions	(EDTA) citrate		ammonia tris ethanolamine		acetate phosphate

High salt concentrations have also been used to suppress the protein-protein interactions during IMAC[35-37]. Such associations may make other chromatographic methods such as ion exchange difficult if not impossible, and the use of IMAC as an early step allows ion exchange or gel filtration to be used later in the purification scheme.

Wunderwalt et al.[38] used α_2-macroglobulin bound to Cu^{2+}-IDA Sepharose for the removal of endoproteinase from fluids in a "sandwich technique". They discovered that much less bound α_2-macroglobulin was removed from a column eluted with several column volumes of buffer of high ionic strength when sodium chloride was added to generate the ionic strength than when sodium phosphate was used.

8.3.4 Detergents and Other Additives

Detergents can be used with advantage when necessary in IMAC. Collagenase, a difficult enzyme to purify, was isolated with Zn^{2+} IMAC as one step of a purification scheme. A detergent, Brij 35, was used in all buffers[39]. Tween 80 was included in buffers used in a procedure for the isolation of plasminogen activator from human melanoma cells[40]. The use of detergent gave better yields than were obtained in an earlier isolation of the activator from human uterine tissue[41]. In both cases, Zn^{2+} IMAC was used as one of the steps, elution being effected by an imidazole gradient.

In the purification of human factor XII a trypsin inhibitor and a surface inhibitor were used in all buffers[42].

Human placenta mitochondrial membrane proteins could be fractionated using 0.4% octaethylene glycol dodecylether in the eluents[43].

A 1:1 mixture of ethanol-water was used to remove coloured contaminants before eluting the adsorbed protein in the purification of albumin from a Cohn IV extract[44].

8.4 Chromatographic Conditions

8.4.1 Planning of Experiments

For a protein with unknown properties regarding its binding to metal chelate columns it is advisable to use Cu^{2+} and a neutral phosphate or acetate buffer containing 0.15-0.5 M NaCl for initial experiments. Ideally, total binding of the desired protein should occur with direct elution of contaminating proteins or the reverse. If inappropriate binding is obtained a number of variations as outlined in Table 8-4 can be tried.

A few experiments with variation of metal ion, pH and buffer composition should be sufficient to establish suitable starting conditions. To vary the degree of substitution is more laborious. However, the amount of bound metal can be reduced by leaching the metal-primed column with a citrate buffer, at least when Cu^{2+} is used[45].

The following section deals in some detail with practical aspects of IMAC for the optimization and systematization of experiments.

8.4.2 Loading of Metal Ion and Regerenation of Column

Metal ions such as Cu^{2+}, Zn^{2+}, Co^{2+}, Ni^{2+} and Ca^{2+} can be loaded in neutral or weakly acid solutions to avoid hydroxy precipitates with, for example, Cu^{2+}. For Cu^{2+}, the saturation of the column can be followed simply by observing the blue colour of the column. When Zn^{2+} is used formation of a precipitate on addition of Na_2CO_3 to the eluent indicates saturation of the column.

Table 8-5. Purification of proteins by IMAC. Selected examples.

Protein	Metal ion(s)	Starting buffer	Elution buffer	Remarks	Ref.
Plasma amyloid P component	Zn^{2+}	20 mM Tris-HCl 150 mM NaCl 1 mM benzamidine pH 8.0	Linear gradient of histidine, 0–30 mM, in starting buffer	Dialyzed sample from Ba^{2+} citrate and $(NH_4)_2SO_4$ precipitation	49
Thiol proteinase inhibitor	Cu^{2+}	25 mM Tris-HCl 100 mM NaCl pH 8.2	Enzyme in flow through fraction (80%)	"Negative" chromatography. Flow through fraction used in next step	50
Hu IFN-γ	Ni^{2+}	20 mM Tris-HCl 150 mM NaCl pH 8.5	100 mM Na acetate 1.0 M NaCl pH 4.5	22° gave better resolution and recovery. Typical recovery >90%. Very dilute sample could be used	51
Granule proteins	Cu^{2+}	40 mM Tris 5 mM phosphate 500 mM NaCl pH 8.2	Phosphate-acetic acid gradient pH 7.7–2.8 containing 500 mM NaCl	High ionic strength eliminates aggregation	36
Nucleoside di-phosphatase	Cu^{2+} Zn^{2+}	20 mM Tris-HCl pH 7.5	10 mM L-histidine 10 mM maleate pH 6.6 or with EDTA	Enzyme is active when bound to chelating gel	29

Dolichos bi-flourus seed lectin	Ca^{2+}	50 mM Tris-acetate 500 mM NaCl pH 8.2	10 mM EDTA in starting buffer	No desorption with 100 mM glycine-HCl, pH 2.2. Subunit with carbohydrate binding activity binds to gel	24
Collagenase (pig synovial)	Zn^{2+}	25 mM Na borate 150 mM NaCl pH 8.0	50 mM Na acetate 800 mM NaCl pH 4.7	Cu^{2+} and Ni^{2+} failed	39
Ovalbumin	Fe^{3+}	100 mM acetate pH 5.0	Gradient from A to B A: 50 mM Mes-NaOH pH 5.7 B: 50 mM Pipes-HCl pH 7.2		10
Pepsin	Fe^{3+}	100 mM Na acetate pH 5.0	100 mM Na acetate 20 mM K phosphate pH 5.0		10
Pituitary growth factor	Cu^{2+}	10 mM phosphate 500 mM NaCl glycol 20% 1 mM imidazole pH 7.4	Gradient with starting buffer containing 1–10 mM imidazole		52

Some bleeding of metal ions from the column is likely to occur during most IMAC experiments. In many cases this does not matter, but the leakage can be suppressed if necessary by washing the primed column with the final eluent before equilibration with the starting buffer (the buffer in which the sample is applied). Alternatively the column can be charged with metal ion to only 70–90% of its maximum capacity; this will cause migrating metal ions to be captured by chelating groups at the bottom of the column, equivalent to having an unloaded small column in series with the separation column, and to extracting metal ions from the eluate with a small amount of chelating gel after the chromatographic run.

As a general rule it may be advisable to strip columns of metal ions between runs using chelating agents, such as EDTA, and to recharge with metal ion before the next experiment. Stripping and recharging is essential when ions of weak complexing strength, such as Zn^{2+} and Co^{2+}, are used in descending pH gradient protocols.

With Cu^{2+}-IDA columns, many chromatographic cycles can, at least under certain conditions, be performed without regeneration of the column. The experiment depicted in Fig. 8-2c was repeated 20 times without regeneration of the column. The last chromatogram was indistinguishable from the first one[15].

8.4.3 Sample Application

One obvious advantage of IMAC is that it can be used to concentrate a protein from very dilute solutions. It may, however, be necessary to equilibrate the sample carefully with the starting buffer by, for example, dialysis to remove components, e.g. amines, that may interfere in the adsorption step. Depending on the properties of the samples one may wish to include salts, urea, detergents, glycol etc. in the sample and elution buffers. Such additives seem to have no or only minor influence on the adsorption of the protein but may serve to improve performance and yield. It is advisable to use 0.15–1.0 M NaCl or other salts in all buffers to get consistent results.

8.4.4 Elution Modes

After washing away unbound material, bound substances are recovered by changing conditions to favour desorption. A gradient or step-wise reduction in pH to 3 or 4 is often suitable. Alternatively, competitive elution with a gradient of increasing concentration of, for example, ammonium chloride, glycine, histamine, histidine or imidazol may be used. It has been pointed out[7] that when, for example, imidazole is used for elution, the columns should be saturated with imidazole prior to adsorption of the sample protein, and that imidazole should be included in the starting buffer. This is not essential for good results, however, as shown in the purification of plasminogen activator[41].

A third method of elution is to include a chelating agent such as EDTA in the eluent. In this case, all adsorbed proteins will be eluted indiscriminately along with the metal ion. In all cases, high ionic strength should be maintained.

8.5 Areas of Use of IMAC

The predominant use of IMAC to date has been in the isolation and purification of proteins and peptides as discussed in this review. Future use of IMAC will include HPLC of proteins[16,43,46].

The chromatography of nucleotides, dinucleotides and related compounds on metal chelates has also been shown to be possible[47,48]. Pyrimidines show little interaction with the metal ions but purines are resolved on Cu^{2+} chelates. The technique is potentially very useful for the large scale purification of these compounds.

The technique as discussed above has also been used in the study of the structure of proteins, whereby the exposure of certain amino acid residues on the surface of the proteins can be explored, and for the immobilization of proteins to be used as active solid phase enzymes[29] or affinity adsorbents[38].

Table 8-5 lists some data from published purifications. Care has been taken to select examples exploiting various modifications of IMAC rather than showing only very typical procedures. The list, which is not exhaustive and does not include applications presented in Section 8.6, demonstrates that a wide range of adsorption and desorption buffers can be used.

8.6   Applications

8.6.1   Purification of Copper, Zinc Superoxide Dismutase (Cu, Zn SOD) from Human Erythrocytes

A two-step purification protocol for the purification of Cu, Zn SOD has been designed by Weselake et al.[53] The first step is ion exchange chromatography of a filtered lysate from human erythrocytes on a column of DEAE Sepharose CL-6B from which the Cu, Zn SOD containing fraction is eluted in 10 mM potassium phosphate, pH 6.4, containing 100 mM NaCl.

An IDA Sepharose 6B gel was prepared according to Porath[1]. It contained 25–30 μmole of IDA/ml of sedimented gel as determined by titration of immobilized carboxyl groups of the immobilized IDA. A 4 × 5 cm (diameter) column was packed and charged to saturation with 50 mM copper sulphate. A second column, 2 × 5 cm, packed with uncharged IDA-gel was connected after the charged column to serve as a scavenger of any copper ions leaking from the first column.

The columns were equilibrated with 10 mM potassium phosphate, pH 6.4, containing 100 mM NaCl and charged at a rate of 150 ml/h with the Cu, Zn SOD solution obtained from the ion exchange step. The columns were then washed sequentially at a flow rate of 150 ml/h with

— 10 mM potassium phosphate, 1.0 M NaCl, pH 6.4, 3500 ml
— 10 mM potassium phosphate, pH 6.4, 400 ml
— 10 mM sodium acetate, pH 5.0, 300 ml
— 20 mM sodium citrate, pH 5.0 (elution of Cu, Zn SOD)

Fractions containing Cu, Zn SOD obtained with the citrate buffer were collected as shown in Fig. 8-4, concentrated to 30 ml by pressure ultrafiltration, dialyzed against distilled water, freeze-dried and stored dessicated at −20°C. The volume of the collected fraction was 365 ml.

The specific activity of the purified Cu, Zn SOD was 3800 U/mg. The overall purification factor was 2000 and the yield 58 %. The purification in the IMAC step was 60 fold and the yield 63 %.

Attempts to use higher concentrations of citrate buffer and/or lower pH resulted in sharper elution profiles but with concomitant release of impurities.

The purified Cu, Zn SOD was analyzed by SDS-gel electrophoresis, gel filtration, and isoelectric focussing and found to be essentially pure.

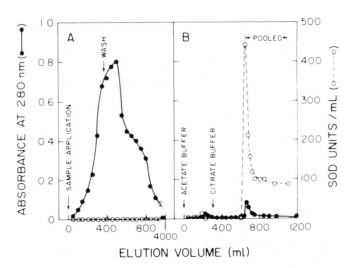

Figure 8-4. Cu^{2+}-IMAC of Cu, Zn SOD. (A) Application of SOD solution (365 ml), washing with 3500 ml 10 mM potassium phosphate buffer, pH 6.4, containing 1 M NaCl, 400 ml of same buffer without NaCl and (B) with ca 300 ml of 10 mM sodium acetate. Enzyme elution with 20 mM citrate buffer, pH 5.0. Flow rate 150 ml/h. Reproduced with permission[53].

The authors concluded that

— the release of SOD from the copper ion column was related to the chelating properties of the citrate ion,
— Cu^{2+}-IMAC appears to be a simple and rapid procedure for purifying human erythrocyte Cu, Zn SOD, which avoids solvent- and heat-treatment steps and that
— the procedure did not appear to deactivate SOD by removing copper from the active site as judged from the retained enzymatic activity.

8.6.2 Purification of Human Plasma α_2-Macroglobulin and α_1-Antitrypsin

A fine example of plasma protein purification by IMAC was given by Kurecki *et al.* in their preparation of α_2-macroglobulin (α_2M) and α_1-antitrypsin (α_1AT), Table 8-6[54]. They used dialyzed samples obtained by fractional ammonium sulphate precipitation. Large amounts of α_2M bound tightly to a zinc column at pH 6 allowing removal of contaminating proteins. At pH 5, α_2M eluted in a sharp concentrated peak, Fig. 8-5. Their results show that a 2.5 × 14 cm zinc chelate column can accommodate at least 1.0 g of α_2M. An electrophoretically homogeneous inhibitor was obtained in good yield by this mild two-step procedure.

α_1AT was isolated from the same batch of pooled human plasma, Fig. 8-6. Since the major contaminating plasma protein, albumin, was not retained by the zinc chelate at the pH of the starting buffer, pH 8, this step was very efficient giving a 20-fold purification by elution at pH 6.5. Two ion exchange steps were required in addition to IMAC to obtain homogeneous α_1AT. According to the authors, this method increases the yield of α_1AT 2.5-fold as compared to other methods.

Table 8-6. Purification of α_1AT and α_2M

Purification step	Total protein (mg)	Total activity	Specific activity	Recovery (%)	Purification (fold)
α_1-Antitrypsin					
Plasma (1000 ml)		665000	16	100	1
Ammonium sulfate (50–80%)		612000	28	92	2
Zn^{2+}-IDA gel		396000	550	60	34
DE-52, pH 6.5		276000	996	42	62
DE-52, pH 7.5	354	231000	1305	35	82
α_2-Macroglobulin					
Plasma (1000 ml)		86121	2.1	100	1
Ammonium sulfate (40–55%)		23603	6.4	27	3
Zn^{2+}-IDA gel	342	18605	61	22	29

[a] From Ref. 24 with permission.

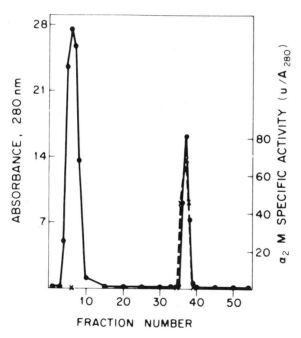

Figure 8-5. Purification of α_2-macroglobulin (α_2M) by Zn^{2+}-IMAC. A 140 ml sample of α_2M in 50 mM sodium phosphate, 20 mM NaCl, pH 6.0, prepared by $(NH_4)_2SO_4$ precipitation from 1 l of human plasma was charged on a 2.5 × 14 cm Zn^{2+} chelate column equilibrated with 20 mM sodium phosphate, 150 mM NaCl, pH 6.0 at a flow rate of 100 ml/h. Elution was then begun at 50 ml/h with the same buffer and 20 ml fractions were collected. At fraction 30 the buffer was changed to 20 mM sodium cacodylate, 150 mM NaCl, pH 5.0, and 11 ml fractions were collected, ●, A_{280}; ×, α_2M specific activity. The α_2M material in fractions 36–38 was homogenous in polyacrylamide gel electrophoresis. Reproduced with permission[54].

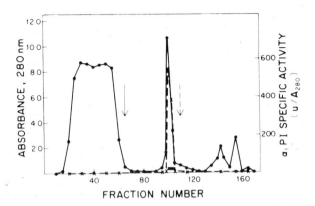

FRACTION NUMBER

Figure 8-6. Purification of α_1-antitrypsin (α_1AT) by Zn^{2+}-IMAC. A 700 ml sample of α_1AT in 50 mM sodium phosphate, 150 mM NaCl, pH 8.0, prepared from one 1 of human plasma by $(NH_4)_2SO_4$ precipitation was charged on a 2.5 × 90 cm Zn^{2+} chelate column equilibrated in the sample buffer at a flow rate of 100 ml/h. Elution was begun with the same buffer, and 22 ml fractions were collected. At fraction 65 the buffer was changed to 50 mM sodium phosphate, 150 mM NaCl, pH 6.5 and 12 ml fractions were collected. The column was stripped with 50 mM EDTA, 500 mM NaCl, pH 7.0, at fraction 110. ●, A_{280}; ×, α_1AT specific activity. Bar marks pooled fractions. Reproduced with permission[54].

8.6.3   Purification of Albumin from a Cohn IV Fraction of Human Plasma

While purifying albumin from the Cohn IV fraction, Hansson et al. carried out a study of the adsorption-desorption characteristics of albumin and contaminating proteins using zinc and copper chelates[44]. The results, summarized in Table 8-3, showed that, for a given metal ion, the choice of buffer and salt in the elution medium governs the behaviour of the proteins.

Buffer constituents such as acetate and phosphate ions were found to favour the adsorption of the proteins while the use of tris or ammonium salts caused direct elution of the protein.

These findings were used to design the following purification procedure.

Sample:
An extract, 65 ml, of a Cohn IV fraction containing 0.64 g of albumin and 0.76 g of other proteins, mainly transferrin.

Buffers:

A   0.1 M sodium acteate, 500 mM NaCl, pH 7.7
B   Ethanol-water 1:9
C   Ethanol-water 1:1
D   50 mM Tris-HCl, 150 mM NH_4Cl, pH 8.0
E   50 mM EDTA, 500 mM NaCl, pH 7.0

Coloured impurities were eluted with aqueous ethanol before elution of the albumin with a tris buffer containing ammonium chloride.

Highly purified albumin was recovered with a 65% yield in one chromatographic step from the very crude and strongly discoloured Cohn IV extract, Fig. 8-7.

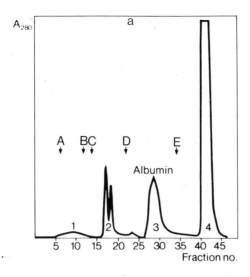

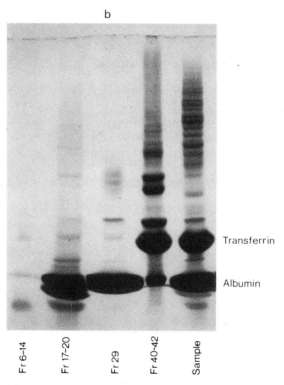

Figure 8-7. (a) Isolation of albumin from Cohn fraction IV extracts by Cu^{2+}-IMAC. The copper loaded column (K 16/20 , Pharmacia LKB AB, 65 ml gel bed) was charged with sample and eluted with eluents A-E as indicated in the chromatogram. See text regarding composition of sample and buffers. Fraction size was 11 ml and flow rate 1.1 ml/min. (b) Polyacrylamide gel electrophoresis of fractions in Fig. 8-7a. Electrophoresis in a 4/30 gradient gel was carried out according to the manufacturers instructions (Pharmacia AB) after concentration of fractions in Minicon™ concentration cells (Amicon, Lexington, MA, USA). Reproduced with permission[44].

Albumin in peak 3, Fig. 8-7a, was 99.9% pure according to cellulose acetate electrophoresis.

8.6.4   Human γ-Interferon

Several types of interferons have been successfully purified by IMAC. The method is cheap and very efficient in isolating pure human fibroblast interferon (Hu IFN-β). Papers by Edy et a.[55] and Heine et al.[56,57] illustrate the methods. Using controlled porous glass in a batch procedure as a first step, they obtained interferon with an activity of 3.0×10^5 units/mg. From 44 mg of this they isolated ca. 7 μg of homogeneous Hu IFN-γ in one IMAC step using a Zn^{2+}-chelate column. The specific activity was 1.7×10^9 units/mg and the yield in the IMAC step was 91.4%. Overall recovery was 52.6%[57].

Of methodological interest is the use of the end buffer to wash the Zn^{2+} chelate column before equilibration with the starting buffer to remove excess Zn^{2+}.

The purification was performed at 4°C using a 1.5×16 cm column (K9/15, Pharmacia LKB) packed with IDA Sepharose 6B prepared by epoxy-coupling as described by Porath[1]. Before each run the column was regenerated. The flow rate was 15–20 ml/h. During elution steps, 1 ml fractions were collected and protein content, pH, and interferon activity determined.

Buffers:

A   20 mM phosphate, 1 M NaCl, 50 mM EDTA, pH 7.4
B   20 mM phosphate, 1 M NaCl, pH 7.4
C   100 mM sodium acetate, 1 M NaCl, pH 4.0
D   100 mM sodium acetate, 1 M NaCl, 1 mM $ZnCl_2$, pH 4.0
E   100 mM sodium acetate, 1 M NaCl, pH 5.9
D   100 mM sodium acetate, 1 M NaCl, pH 4.2
F   100 mM sodium acetate, 1 M NaCl, pH 4.0

Regeneration of column:

1 Removal of Zn^{2+} with A, 5 bed volumes.
2 Removal of EDTA with B, 5 bed volumes.
3 Equilibration with buffer C.
4 Introduction of Zn^{2+} with buffer D. Saturation tested by the formation of a precipitate when a drop of the eluate was mixed with a Na_2CO_3 solution.
5 Removal of excess Zn^{2+} with buffer C, 5 bed volumes.

Sample preparation:

6 Prepurification by adsorption to porous glass and dialysis against B. Final volume 100 ml.

Chromatography:

7 Application of dialysed sample.
8 Wash with B, 1.5 bed volumes.
9 Wash with E, 5 bed volumes.
10 Elution with D, 1 bed volume.
11 Elution with F, 2 bed volumes.

Table 8-7. Purification of human fibroblast interferon; Zn^{2+}-IMAC step

Fractions	Volume (ml)	Total protein (mg)	Total activity (units)	Specific activity (units/mg)
Sample	100	44	13×10^6	3.0×10^5
Void + pH 7.4 wash	130	42	$<0.2 \times 10^3$	—
pH 5.9 wash	150	1.6	0.8×10^6	0.5×10^6
pH 5.6–4.2 eluate	6	0.0007	12×10^6	1.7×10^9

The interferon peak fraction eluted at pH 5.2. Fractions with pH 5.6–4.2 contained interferon and were pooled.

Table 8-7 summarizes the results from one typical run (simplified from Ref. 57).

8.6.5 Lactoferrin from Human Milk

The use of high ionic strength buffers to eliminate any non-specific interactions with the gel is common in IMAC. High salt concentrations were used to prevent the association of lactoferrin from human milk[35] and granulocytes[36] with other proteins during IMAC, Fig. 8-8.

8.6.6 Separation of Iso Forms of Human Platelet-Derived Growth Factor

Some of the background for the use of IMAC in the study of platelet-derived growth factor (PDGF) has been given in section 8.3.2. The study by Hammacher et al.[16] showed that PDGF purified from human platelets consists of PDGF-AB and PDGF-BB.

An IDA-derivative of Superose® 12 with a capacity for Zn^{2+} around 30 μmole/ml gel, packed in a 0.5×5 cm column was used. The packed column was a non-commerical sample from Pharmacia LKB AB. The column was connected to an LKB dual pump HPLC instrument.

The chromatographic experiments were carried out at ambient temperature with a flow rate of 0.5 ml/min. Adsorbed material was eluted by the simultaneous application of an ascendant imidazole gradient and a falling sodium chloride gradient. The latter was introduced because it was found to increase resolution.

Buffers:

A 100 mM sodium acetate, 1 M NaCl, pH 3.8
B 100 mM sodium acetate, 1 M NaCl, 12.8 mg/ml $CuSO_4$, pH 3.8
C 20 mM phosphate, 50 mM imidazole, 0.5 M NaCl, pH 7.4
D 20 mM phosphate, 1 mM imidazole, 1 M NaCl, pH 7.4

Samples were prepared as described in the paper by Hammacher et al.[16] and references given therein:

1 PDGF was purified from human platelet lysate.
2 The A chain homodimer (PDGF-AA) was purified from medium conditioned by a human glioma cell line.

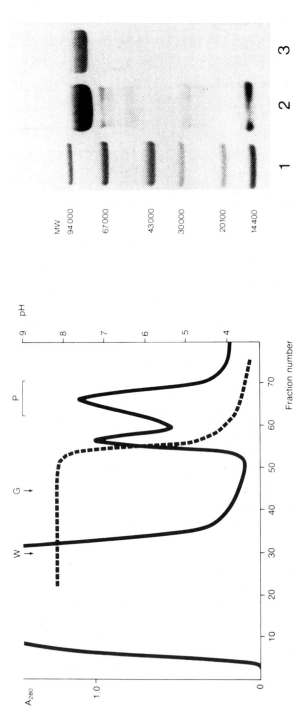

Figure 8-8. Isolation of lactoferrin by Cu^{2+}-IMAC. Sample: 70 ml defatted, casein-free human milk equilibrated with the starting buffer. Column: C 10/20. Bed height: 16.3 cm, charged with Cu^{2+} for 10.3 cm. Eluents: W = wash with starting buffer 50 mM Tris-acetate pH 8.2. 500 mM NaCl, G = development to final buffer 50 mM Tris-acetate, 500 mM NaCl, pH 2.8. Flow rate: 25 ml/h. Fraction size: 2.5 ml. Insert shows gradient gel electrophoresis in SDS. Gel: PAA 4/30. Lane 1. LMW calibration kit. Lane 2. Defatted casein-free milk. Lane 3. Pooled material (P). (Work from Pharmacia AB).

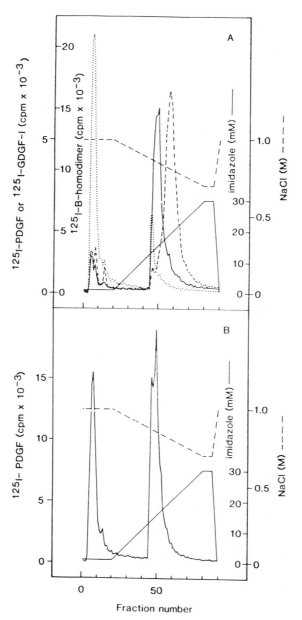

Figure 8-9. Analysis by IMAC of PDGF dimers. See text for details. (A) . . . . PDGF-BB, PDGF-AA. (B) _____ PDGF purified from human platelets.

3 The B chain homodimer (PDGF-BB) was purified from medium conditioned by a CHO cell line transfected with a B chain cDNA construct.

Chromatography:

1 Equilibration with buffer A.
2 The column was loaded to *ca.* 80% with Cu^{2+} using buffer B.
3 Washing of the column with buffer A.
4 Equilibration with buffer C.
5 Equilibration with buffer D.
6 Application of sample (250 μl) in buffer D.
7 Elution with buffer D for 10 minutes.
8 Elution with a linear gradient from 100% buffer D to 60% buffer C in 30 minutes (increasing imidazole concentration 1 mM/min, decreasing NaCl concentration 10 mM/min).
9 Equilibration with 60% buffer E for 3 minutes.
10 Linear gradient from 60% buffer C to 100% buffer D in 2 minutes followed by equilibration with buffer D for chromatography of next sample (step no 6).

8.5 References

1. J. Porath, J. Carlsson, I. Olsson, *et al. Nature 258*, 598–599 (1975).
2. F. Hellferich, *Nature, 189*, 1001 (1961).
3. V. A. Davankov, A. V. Semechkin, *J. Chromatogr., 141*, 313–353 (1977).
4. J. Porath, M. Belew, I. M. Chaiken, *et al. Acad. Press.* 173–190 (1983).
5. J. Porath, B. Olin, B. Granstrand, *Biochem. Biophys. 225*, 543–547 (1983).
6. B. Lönnerdal, C. L. Keen, *J. Appl. Biochem., 4*, 203–208 (1982).
7. E. Sulkowski, *Trends Biotechn. 3*, 1–7 (1985).
8. J. Porath, B. Olin, *Biochemistry, 22*, 1621 (1983).
9. R. A. A. Muzzarelli, A. F. Martelli, O. Tubertini, *Analyst, 94*, 616–624 (1969) .
10. L. Andersson, J. Porath, *Anal. Biochem. 154*, 250–254 (1986).
11. W. H. Scouten, "Affinity Chromatography, Bioselective adsorption on inert matrices", *Chemical Analysis, 59*, John Wiley & Sons, 1981.
12. L. Fanou-Ayi, M. Vijayalakshmi, *N.Y. Acad. Sci. 413*, 300–306 (1983).
13. Y. Kato, K. Nakamura, T. Hashimoto, *J. Chromatogr. 354*, 511–517 (1986).
14. M. Gimpel, K. Unger, *Chromatography 17*, 200–204 (1983).
15. K-A. Hansson, G. Moen, L. Kågedal, unpublished results.
16. A. Hammacher, U. Hellman, A. Johnsson, A. Östman, K. Gunnarsson, B. Westermark, Å. Wasteson, C.-H. Heldin, *J. Biol. Chem. 263*, 16493–16498 (1988).
17. Z. E. Rassi, C. Horváth, *J. Chromatogr. 359*, 241–253 (1986).
18. G. Schmuckler, *Talanta, 12*, 281–290 (1965).
19. Z. Horváth, G. Nagydiósi, *J. Inorg. Nucl. Chem., 37*, 767–769 (1975).
20. G. Muszyńska, Y. J. Zhao, J. Porath, *J. Inorg. Biochem. 26*, 127–135 (1986).
21. L. Andersson, *J. Chromatogr. 315*, 167–174 (1984).
22. L. Kågedal, unpublished data.
23. E. Sulkowski, K. Vastola, D. Oleszek, W. von Muenchhausen in "*Affinity Chromatography and related techniques*", Veldhoven, The Netherlands, June 22–26 1981. T. C. J. Gribnau, J. Visser, R. F. J. Nivard, eds. Elsevier Scientific Publishing Company 1982, pp. 313–322.
24. C. A. K. Borrebaeck, B. Lönnerdal, M. E. Etzler *FEBS Lett., 130*, 194–196 (1981).
25. M. Conlon, R. F. Murphy, *Biochem. Soc. Trans., 4*, 860–861 (1976).
26. M. Yoshimoto, M. Laskowski, Sr. *Preparative Biochemistry, 12*, 235–254 (1981).
27. J. Porath, *Pure and Appl. Chem., 51*, 1549–1559 (1979).
28. M. F. Scully, V. V. Kakkar, *Biochem. Soc. Trans. 94*, 335–336 (1981).
29. I. Ohkubo, T. Kondo, N. Taniguchi, *Biochim. Biophys. Acta, 616*, 89–93 (1980).

30. E. Bollin, E. Sulkowski, *Arch. Virol.*, *58*, 149–152 (1978).
31. A. Gozdzicka-Józefiak, J. Augustyniak, *J. Chromatogr.*, *131*, 91–97 (1977).
32. E. Sulkowski, in "Protein Purification: Micro to Macro", UCLA Symposia on molecular and cellular biology new series, *Vol. 68*. R. Burgess, ed., Alan R. Liss, Inc. 1987, pp. 149–162.
33. M. Belew, T.-T. Yip, L. Andersson, J. Porath. *J. Chromatogr. 403*, 197–206 (1987).
34. E. Sulkowski, *Macromol. Chem.*, *Macromol. Symp*, *17*, 335–348 (1988).
35. B. Lönnerdal, J. Carlsson, J. Porath, *FEBS Lett.*, *75*, 89–92 (1977).
36. A. R. Torres, E. A. Peterson, W. H. Evans *et al. Biochim. Biophys. Acta, 576*, 385–392 (1979).
37 H. Kikuxhi, M. Watanabe, *Anal. Biochem.*, *115*, 109–129 (1981).
38. P. Wunderwald, W. J. Schrenk, H. Port, G-B. Kresze, *J. Appl. Biochem. 5*, 31–42 (1983).
39. T. E. Cawston, J. A. Tyler, *Biochem. J. 183*, 647–656 (1979).
40. D. C. Rijken, D. Collen, *J. Biol. Chem. 256*, 7035–7041 (1981).
41. D. C. Rijken, G. Wijngaards, M. Zaal-De Jong, *et al. Biochim. Biophys. Acta, 580*, 140–153 (1979).
42. R. A. Pixley, R. W. Coman, *Affinity Chromatography Thrombosis Res. 41*, 89–98 (1986).
43. Y. Kato, T. Kitamura, K. Nakamura, A. Mitsui, Y. Yamasaki, T. Hashimoto, *J. Chromatogr. 391*, 395–407 (1987).
44. H. Hansson, L. Kågedal, *J. Chromatogr. 215*, 333–339 (1981).
45. K. C. Chadha, P. M. Grob, A. J. Mikulski, et al. *J. Gen Virol.*, *43*, 701–706 (1979).
46. M. Belew, T. T. Yip, L. Andersson, R. Ehrnström, *Anal. Biochem. 164*, 457–465 (1987).
47. P. Hubert, J. Porath, *J. Chromatogr. 198*, 247–255 (1980).
48. P. Hubert, J. Porath, *J. Chromatogr. 206*, 164–168 (1981).
49. I. Ohkubo, W. Sahashi, C. Nanikawa, K. Tsukada, T. Takeuchi, M. Sasaki, *Clin. Chim. Acta. 157*, 95–102 (1986).
50. H. C. Ryley, *Biochem. Biophys. Res. Comm. 89*, 871–878 (1979).
51. D. D. Coppenhaver, *Meth. Enzymol. 119*, 199–204 (1986).
52. J. M. Rowe, S. F. Henry, H. G. Friesen, *Biochemistry*, *25*, 6421–6425 (1986).
53. R. J. Weselake, S. L. Chesney, A. Petkau, A. D. Friesen, *Anal. Biochem. 155*, 193–197 (1986).
54. T. Kurecki, L. F. Kress, M. Laskowski, *Anal. Biochem.*, *99*, 415–420 (1979).
55. V. G. Edy, A. Billiau, P. De Somer, *J. Biol. Chem.*, *252*, 5934–5935 (1977).
56. J. W. Heine, M. De Ley, J. Van Damme, A. Billau, P. De Somer, *Ann. N.Y. Acad. Sci.*, *350*, 364–373 (1980).
57. J. W. Heine, J. van Damme, M. De Ley, A. Billau, P. De Somer, *J. Gen. Virol.*, *54*, 47–56 (1981).
58. T. W. Hutchens, T.-T. Yip, J. Porath, *Anal. Biochem. 170*, 168–182 (1988).
59. E. Hochuli, W. Bannwart, H. Döbeli, R. Gentz, D. Stüber, *Biotechnology* Nov. 1988, p. 1321–1325.
60. M. C. Smith, T. C. Furman, T. D. Ingolia, C. Pidgeon, *J. Biol. Chem.*, *263*, 7211–7215 (1988).

9 Covalent Chromatography

Lars Rydén
Department of Biochemistry,
Uppsala University
Uppsala Biomedical Center
S-751 23 Uppsala, Sweden

Jan Carlsson
Research & Development
Pharmacia Diagnostics AB
S-751 82 Uppsala, Sweden

9.1   Introduction

Chromatographic techniques for the isolation of proteins and other biological macromolecules, such as ion exchange and hydrophobic interaction chromatography, are based on non-covalent interactions between the molecules to be separated and the adsorbent. Covalent chromatography[1-3], on the other hand, as indicated by the name, involves the formation (and breaking) of covalent bonds between the solute and the stationary phase (Fig. 9-1). As is usually the case in the chromatography of biologically active proteins, specificity and mild conditions are required. These requirements can, as we will see, be fulfilled.

Potential sites for chemical reactions in proteins are primarily the amino acid side chains with their various functional groups. In particular the amino- and carboxyl functions have been used for the immobilization of proteins to insoluble polymers. In these cases the chemistry has been designed to give stable bonds and it is in practice difficult to release the immobilized protein without destroying it. The only functional group for which conditions are available for the formation of a stable covalent bond which can also be split under mild conditions is the thiol group. Covalent chromatography, which was discussed as a potential technique already a long time ago, has thus, with a few exceptions[3] been applied to the isolation of thiol-containing substances.

Patchornik[4,5] and co-workers have, however, designed methods for reversible covalent attachment of methionyl and tryptophanyl side chains to activated polyacrylamide matrices. The chemistry involved requires long incubation times and low pH (dilute acetic acid) and the methods have not received general attention. Probably they would be useful for the isolation of peptides containing methionine or tryptophan in special cases.

In this chapter we will deal exclusively with covalent chromatography based on thiol-disulphide exchange. Thiols frequently occur in proteins and often participate directly in their functions as enzymes, hormones, receptors etc.[6,7]. Thiol groups in proteins vary in reactivity and can also be introduced chemically. Both circumstances make covalent chromatography a widely applicable technique.

To exploit fully the possibilities and avoid some pitfalls of covalent chromatography a thorough knowledge of the chemical properties of the thiol groups is necessary. We shall therefore give a short description of the main types of reactions in which thiols and some of their derivatives take part, with an eye to their relevance for the practical application of the technique. In particular, their reactions with the so-called reactive disulphides utilized in covalent chromatography are treated in some detail.

We will also briefly describe some of the properties of thiol groups in proteins from various biological sources. Special emphasis is placed on how the reactivities of thiol groups change with the conditions used.

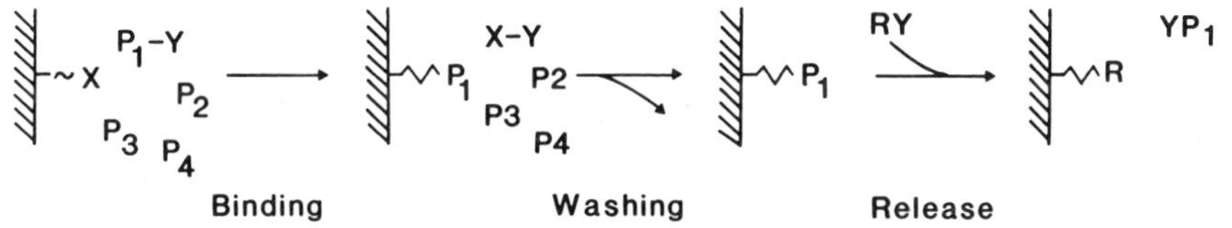

Figure 9-1. General scheme for covalent chromatography. A protein containing group Y reacts with a solid phase containing a reactive group X. After washing out non-bound proteins the bound protein is released by a low-molecular-weight compound RY.

We hope that these introductory sections will make the main part of the chapter—
the properties of the gel derivatives used in covalent chromatography and the principles
and practice of the chromatography itself—easy to follow. The chapter is concluded with
descriptions of a few practical applications.

9.2   Chemical Properties of Thiol Groups

Thiol groups are normally the most reactive groups found in proteins and they can
participate in a large number of reactions[6-8]. This is due to the high nucleophilicity of the
corresponding thiolate ions, which exist at reasonable concentrations at neutral to
weakly alkaline pH-values. Thiols easily oxidize to give disulphides, which are comparatively unreactive and important for the stabilization of protein tertiary and quaternary
structure.

9.2.1   Ionization, Oxidation, Metal Ligation and Alkylation

Thiols can dissociate into a proton and a thiolate anion:

$$RSH \longrightarrow RS^- + H^+ \tag{9-1}$$

Their pK_a values generally lie in the range 8–10.5. Thiolate-dependent reactions thus
proceed readily at pH 8 and above, and at moderate speed in the pH range 6–8. Some
respresentative pH-values are given in Table 9-1. The described pH dependence applies
for low molecular weight alifatic thiols and many protein-bound thiols. The latter,
however, occasionally show a completely different rate-pH dependence due to microenvironmental effects.

The thiol group oxidizes to form disulphide (RSSR) which under more extreme
conditions (higher pH, presence of oxidative agents, etc.) can be further oxidized to e.g.
sulphenic ($RSOH$), sulphinic (RSO_2H) and sulphonic (RSO_3H) acids.

$$2\ RSH \longrightarrow RSSR \ \text{(disulphide)} + 2\ H^+ + 2\ e^- \tag{9-2a}$$

$$RSSR + 2\ OH^- \longrightarrow 2\ RSOH \ \text{(sulphenic acid)} + 2\ e^- \tag{9-2b}$$

$$RSOH + OH^- \longrightarrow RSOOH \ \text{(sulphinic acid)} + H^+ + 2\ e^- \tag{9-2c}$$

Table 9-1. Dissociation constants for
some low-molecular-weight thiol compounds. Only the dissociation constant of
the thiol proton is given.

Compound	pK_{SH}
Glutathione	9.2
β-mercaptoethylamine	8.3
β-mercaptoethanol	9.5
β-mercaptopropionate	10.3
β-mercapto N,N'-diethylamine	7.8

The rate of oxidation increases with increasing concentration of thiolate ion, that is, with increasing pH. At low pH and in the absence of strong oxidation agents such as peroxides, on the other hand (pH < 5) the thiol group is stable and can tolerate several days of incubation even in the presence of air and at room temperature. Heavy metal ions catalyze the oxidation reactions even in trace amounts. The presence of a chelating agent, such as EDTA, is thus advisable in situations where oxidation is a potential problem.

Thiols are avid ligands for heavy metal ions. In some cases, as in the cupric ion complex, the thiolate oxidizes rapidly in the presence of oxygen. Other complexes, such as the zinc and even more so the mercury complex, are stable. The latter was used in some of the first applications of covalent chromatography[9], where the gel contained an immobilized organic mercury compound:

$$Polymer - HgCl + R_1SH \longrightarrow Polymer - HgSR_1 + HCl \qquad (9\text{-}3a)$$

$$Polymer - HgSR_1 + RSH \quad (\text{excess}) \longrightarrow Polymer - HgSR + RSSR_1 \quad (9\text{-}3b)$$

Alkylation that leads to a stable thioether can be used to block a thiol group permanently. In the most common procedure iodoacetate or its amide is used and a carboxymethyl derivative is formed[10].

$$ICH_2COO^- + RSH \longrightarrow RSCH_2COO^- + HI \qquad (9\text{-}4)$$

This reaction, like other nucleophilic displacements, depends on the thiolate ion and the rate therefore increases with increasing pH.

9.2.2 Thiol-Disulphide Exchange

Thiol-disulphide exchange is a special form of alkylation (S-alkylation). This reaction is easily reversible and is used in covalent chromatography. It will therefore be discussed in some detail.

The reaction is a two-step nucleophilic displacement in which a mixed disulphide is formed as an intermediate:

$$R_1SSR_1 + RSH \longrightarrow R_1SH + R_1SSR \qquad (9\text{-}5a)$$

$$R_1SSR + RSH \longrightarrow R_1SH + RSSR \qquad (9\text{-}5b)$$

The sum of these two reactions is:

$$R_1SSR_1 + 2 RSH \longrightarrow 2 R_1SH + RSSR \qquad (9\text{-}5)$$

The reaction can also be described as a redox process, since the oxidation state of the sulphur atoms changes in the direction of greater electron deficiency in the disulphide.

As in the alkylation reactions, the rate increases with increasing pH. At physiological conditions (pH 7 and 25°) rate constants up to 600 M^{-1} min^{-1} are typical for reactions involving alifatic thiols and disulphides. The corresponding equilibrium constants are usually near unity. In general, therefore, a large excess of thiol must be used to reduce a disulphide and, vice versa, a large excess of disulphide is required for the oxidation of a thiol.

Thiol-disulphide exchange is often utilized for the reduction of both intra- and intermolecular disulphides in proteins. When low molecular weight thiols are used, it is important to note that the necessary excess at a given pH increases with the pK of the thiol. A commonly used thiol is β-mercaptoethanol ($pK_a = 9.5$):

$$RSSR + 2\ HOCH_2CH_2SH \quad (excess) \quad \longrightarrow \quad 2\ RSH + HO(CH_2)_2SS(CH_2)_2OH \quad (9\text{-}6)$$

When one of the two isomeric thiol-diols, dithioerythritol (DTE) or dithiothreitol (DTT), is used for the reduction the situation is quite different. The mixed disulphides formed from these compounds and an alifatic disulphide are unstable and undergo subsequent internal thiol-disulphide exchange leading to the formation of an internal disulphide in form of a stable six-membered ring, which drives the reaction toward completion. Thus DTE or DTT will reduce an alifatic disulphide in equimolar amounts.

$$(9\text{-}7a)$$

$$(9\text{-}7b)$$

It should also be mentioned that thiol-disulphide exchange reactions play an important role in biochemical processes, such as the biosynthesis of proteins, the aggregation and polymerization of some proteins, possibly the regulation of the activities of some intracellular enzymes, effector-receptor interactions and membrane transport.

9.2.3   Reactions with Reactive Disulphides

A particularly interesting thiol-disulphide exchange is that between an alifatic thiol and a so-called reactive disulphide (Fig. 9-2). These are compounds in which the corresponding thiol forms are stabilized either by resonance or by thiol-thione tautomerism. This results in decreased nucleophilicity and a correspondingly low reactivity toward disulphides. Reactive disulphides can be either homogenous or mixed. In the first case, the two halves of the molecule are identical. In the mixed types, one of the halves (R-S-S-) is derived from an alifatic thiol (RSH, where R denotes the alifatic group). An

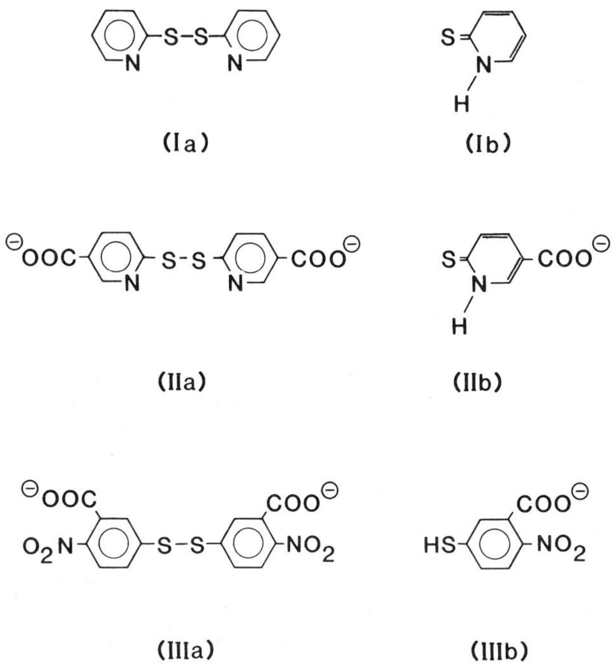

Figure 9-2. Reactive homogenous disulphides with their corresponding reduced forms.
(Ia) 2,2'-dipyridyldisulphide, (Ib) 2-thiopyridone, (λ_{max} 343 nm, $\varepsilon = 8.08 \times 10^3$ M^{-1} cm^{-1})
(IIa) (5-carboxy-2-pyridyl) disulphide, (IIb) 5 carboxy-2-thiopyridone (λ_{max} 344 nm, $\varepsilon = 10^4$
M^{-1} cm^{-1}) (IIIa) 5,5'-dithiobis-(2-nitrobenzoate), (IIIb) 2-nitro 5-mercapto benzoate. ($\lambda_{max} =$
412 nm, $\varepsilon = 14.14 \times 10^3$ M^{-1} cm^{-1})[3].

example is the reaction between an alifatic thiol and the reactive disulphide 2,2'-dipyridyl-disulphide (2-PDS):

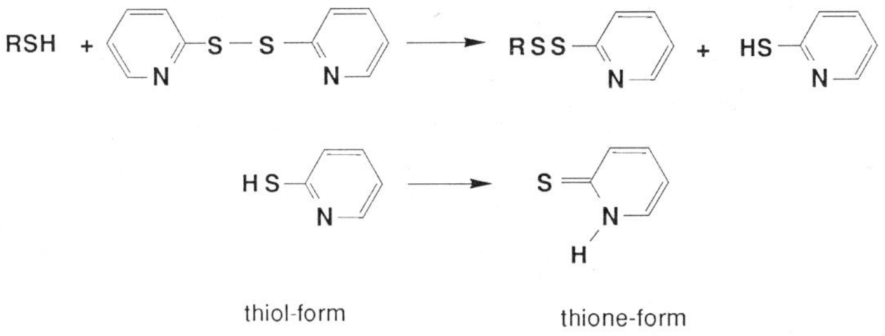

In contrast to the thiol-disulphide exchange reaction involving an alifatic disulphide, where the equilibrium lies toward the "middle", these reactions are driven essentially to completion by formation of the thione form (K_{eq} around 3×10^4). An alifatic thiol can thus be quantitatively transformed into a mixed reactive disulphide concomitantly with formation of equimolar amounts of thione.

If a mixed reactive disulphide is used, the disulphide formed will be an ordinary alifatic one:

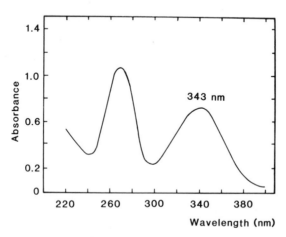

$$RSH \ + \ R_1S-S-\langle N \rangle \longrightarrow RSSR_1 \ + \ HS-\langle N \rangle \tag{9-9a}$$

$$HS-\langle N \rangle \longrightarrow S=\langle N \atop H \rangle \tag{9-9b}$$

thiol-form   thione-form

Both reactions 9-8 and 9-9, which involve 2-thiopyridyl-compounds, will proceed at pH-values where alifatic thiol-disulphide exchanges are slow or non-existent. In fact, thiol-disulphide exchange with 2-pyridyldisulphides can be carried out at pH-values in the range 1–9. The low nucleophilicity of the alifatic thiol at acidic pH is compensated for by an increased electrophilicity of the 2-pyridyldisulphide as a result of the protonation of its ring nitrogen (pK_a around 3). It should be noted that in a mixed 2-pyridyl-disulphide the sulphur atom on the alifatic side is the more electrophilic and will thus react with the incoming thiol sulphur to form the new disulphide, whereas the 2-thiopyridyl structure will leave as a thione. These facts explain the advantageous properties of the 2-thiopyridyl structure as a leaving group and why the binding step in covalent chromatography can proceed at low pH.

The thiol-disulphide exchange described with a mixed pyridyldisulphide produces an alifatic disulphide that is stable towards simple alifatic monothiols in excess at low pH. However, if DTT is used in the exchange, the intermediate form will rearrange to form the corresponding cyclic disulphide (Eq. 9-7b) even at low pH. The combined properties of the thiopyridyl compounds and DTT thus allow the formation of a new thiol over a wide pH range.

The properties of reactive disulphides have been exploited in a number of applications. The homogenous types have been used for the titration of thiol groups in proteins.

Figure 9-3. Spectrum of the reduced form (thione) of 2,2′-dipyridyl-disulphide. The absorption at 343 nm fades at higher pH due to the loss of a proton (pK_a 9.5).

The released thiones can be quantified easily by spectrophotometry, since their absorption maxima lie above the region where ordinary proteins absorb, and the molar absorptivity is above 8000 (Fig. 9-2 and 9-3). The most important application of the mixed type of reactive disulphide is in covalent chromatography. This will be discussed in greater detail below.

9.3 Thiol-Containing Proteins

9.3.1 Redox State in Biological Tissues

In tissues, the redox state within the cells is entirely different from that in the extracellular matrix and fluid. Inside cells a number of reactions are critically dependent on free thiol groups. Co-enzyme A and lipoic acid are examples of essential thiol compounds. The intracellular environment must thus be kept in a reduced state. This is achieved by the cysteine-containing tripeptide glutathione (GSH) (Fig. 9-4), which accounts for about 90% of all thiols in the cell and is present in total concentrations in the millimolar range. Glutathione serves as a scavanger of free radicals, oxidizing compounds and other molecules which react with thiol groups.

The high concentration of reduced glutathione in the cell is maintained by new production through glutathione synthetase and by reduction of the disulphide form (oxidized glutathione). The reduction is catalyzed by the NADPH-requiring enzyme, glutathione reductase. In an erythrocyte the ratio GSH/GSSG is about 500.

On the other hand, the extracellular environment is rich in oxygen, which is incompatible with the presence of exposed thiol groups.

9.3.2 Intra- and Extracellular Proteins

An overview of intracellular proteins[6-7], which have been studied with regard to their content of thiol groups, shows that most of them have exposed thiol groups, whereas very few, if any, contain disulphide bridges. In many instances the thiols participate in the catalytic reactions and alkylation of these groups leads to complete inactivation of the enzyme. This is in particular often the case of oxido-reductases and transferases. Sometimes the thiol groups serve as ligands to metal ions, as in alcohol dehydrogenase.

For extracellular proteins the reverse situation applies[6-7]. Thiol groups are exceptions and occur only when they are required for a special purpose, whereas disulphide bridges are common. Some examples are the extracellular hydrolases, such as the serine proteases trypsin, chymotrypsin and several of the blood coagulation factors. All have several disulphide bridges and no thiol group. On the other hand, some intracellular proteases, the cathepsins, depend on thiols for their activity.

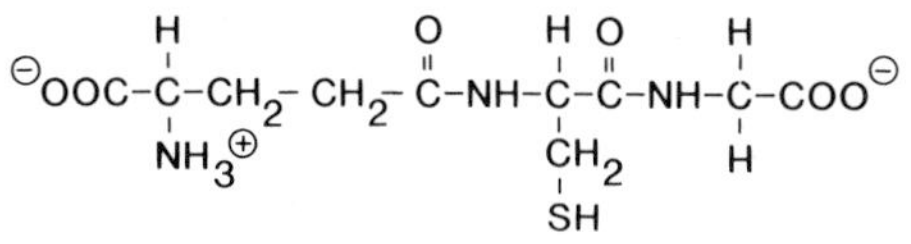

Figure 9-4. The structure of glutathione (γ-glutamyl-cysteinyl-glycine).

Small extracellular proteins, such as venom neurotoxins and the protease inhibitors, are especially rich in disulphide bridges, which are required for stabilization of the structure[11]. Similarly, many peptide hormones contain disulphides but none has a thiol group.

Exceptions to the rule are the plant proteases ficin, bromelain and papain, which are extracellular thiol-dependent proteases. In these the thiol group occurs in an active site pocket where it is somewhat protected from oxidation. The mammalian plasma protein albumin sometimes carries a free thiol group (mercaptalbumin). This thiol group can trap thiol- or disulphide containing compounds in serum under the formation of mixed disulphides.

9.3.3   Demasking Protein Thiol Groups

The reactivity of a protein thiol depends on the conditions. Many proteins contain thiols that are buried in a hydrophobic environment and denaturants are needed for reaction with a hydrophilic reagent. In native human haemoglobin two thiols per molecule react, whereas an additional four are unreactive[12]. The known three-dimensional structure of the protein indicates that these four are partly buried. In aldolase one thiol per subunit reacts readily, but an additional three groups react if a small amount of detergent is added[13]. Still higher concentrations of detergent are required to abolish the enzymatic activity. In the native form of the copper-enzyme ceruloplasmin, present in mammalian serum, no thiol group is accessible, but a slight modification—apparently the "nicking" of a single peptide bond—exposes one thiol[14]. This occurs without loss of activity.

In many cases complete unfolding of the protein is necessary to allow all thiol groups to react. For example, in ceruloplasmin an additional three thiol groups are reactive in 8M urea or 6M guanidine containing EDTA to trap released copper[14]. This is also an example of a protein in which the thiols are liganded to a metal ion.

9.3.4   Reduction of Protein Disulphides

Thiols can be created in proteins by reducing disulphide bonds. This is usually done under conditions where all disulphides present will be reduced (use of denaturants and large excess of reducing reagent). But sometimes one can find conditions under which one or a few disulphide links can be reduced selectively. This is particularly true for interchain disulphide. In the lectin ricin, consisting of two peptide chains, and the immunoglobulins, containing two identical halves each with two peptide chains, the interchain disulphide links can be reduced selectively, without destroying the tertiary structure of the proteins, by a reducing agent such as DTT[15,16].

In some cases an internal disulphide can be reduced without destroying the gross conformation or even the activity of the protein. One of the disulphides in bovine pancreatic trypsin inhibitor was reduced specifically by an equimolar amount of DTT in the absence of urea[17]. Most often one finds, however, that reduction of intrachain bridges proceeds co-operatively in zipper fashion.

9.3.5   Introduction of Thiol Groups

One can introduce thiol groups into proteins thereby making them available to covalent chromatography. Several reagents are available for these thiolation reactions.

The most commonly used is N-acetyl homocysteine thiolactone, which reacts with protein amino groups[18]. Reagents have also been designed for the introduction of protected thiol groups. One of them is N-succinimidyl 1,2'-pyridyl dithiopropionate (SPDP).

$$\text{(SPDP)}$$

$$\text{(9-10)}$$

The 2-pyridyl disulphide introduced by this reagent is a mixed reactive disulphide and can be reduced under mild conditions, under which protein-disulphides are not split, by the use of an equimolar amount of DTT (an alifatic monothiol is not suitable, as this leads to the formation of a mixed disulphide). The 2-thiopyridone liberated can be determined spectrophotometrically. The pyridyldisulphide-substituted protein can also be immobilized on a thiol gel.

9.3.6   Derivatization of Thiol Groups

Thiol groups can be converted either to a blocked form, e.g. by alkylation, or to a reactive form by treatment with a reactive homogenous disulphide. We shall describe the latter of these possibilities, since it has been used to activate thiol proteins for covalent chromatography on a thiol matrix (see below).

The reaction that one wants to achieve is illustrated in Eqs. 9-8. It is important to use an excess of 2,2'-dipyridyl-disulphide. Otherwise the newly activated thiol groups on the protein will be able to react with neighbouring thiols to form disulphide links, either internally or intermolecularly to give dimers or polymers of the protein.

The reaction can be performed either in ordinary buffers or in the presence of strong denaturants, in which case all thiols will become activated. The pH and other conditions are as described above. The reaction with a thiolated gel will be discussed on p. 268.

9.4   Gels for Covalent Chromatography

9.4.1   Principles

The most important application of the mixed reactive disulphides is as reactive groups in covalent chromatography. If a solid phase with a mixed reactive disulphide is incubated with a thiol-containing molecule the latter becomes attached to the solid phase as a result of the thiol-disulphide exchange reaction. The disulphides shown in Fig. 9-2 have been used with success for this purpose. Several others are likely to work as well.

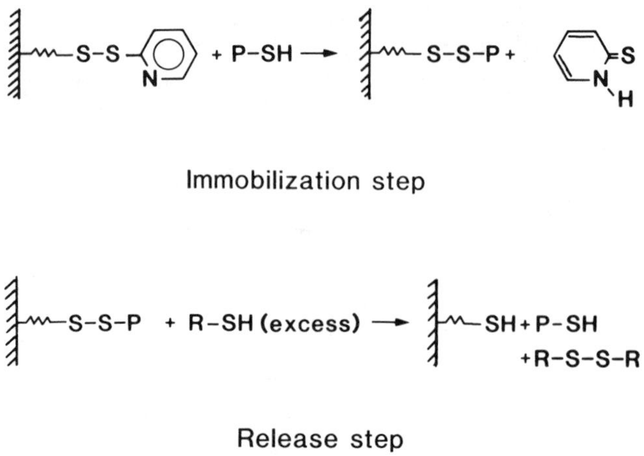

Immobilization step

Release step

Figure 9-5. Principle of 2-pyridyldisulphide based covalent chromatography.

However, in most of the work reported so far, 2-pyridyl disulphide substituted solid phases have been used. The principle of this type of chromatography is shown in Fig. 9-5.

Any material that can be properly substituted is potentially usable. In fact, cross-linked dextran, agarose, cellulose, polyacrylamide, porous glass and various silica derivatives have all been used as matrices in covalent chromatography. The same requirements as in other types of chromatography still apply: good packing and flow properties, low non-specific adsorption, etc. As in most affinity methods, the beaded agarose gels are often preferred. They can be derivatized without changing their excellent properties for chromatography of proteins. They can also be transferred into non-polar solvents without shrinking, which is a valuable property in some instances.

9.4.2 Introduction of Reactive Disulphides into Gels

This is usually done by first introducing thiol groups into agarose gels and subsequently converting the gel-bound thiols into 2-pyridyldisulphide groups. In an inverted form of covalent chromatography the thiols gels are used without further modification (see below). An early method for the introduction of thiols into agarose gels was the coupling[1,21] of glutathione to a cyanogen bromide activated agarose gel:

$$\left|\!\!\!-\, O\!-\!C\!\equiv\!N \;+\; H_2N\!-\; \text{glutathione}\!-\!SH \longrightarrow\right.$$

activated matrix

$$\left|\!\!\!-\, O\!-\!\overset{\overset{\displaystyle NH_2^{+}}{\|}}{C}\!-\!NH\!-\; \text{glutathione}\!-\!SH \right. \tag{9-11}$$

The presence of two carboxylate functions on the glutathione tripeptide (Fig. 9-4) will make the "arm" that links the thiol to the matrix negatively charged. This derivative can be activated subsequently by reaction with 2,2'-dipyridyldisulphide.

A second way to obtain a thiol-substituted agarose starts with the introduction of oxirane structures into the gel by the use of epichlorohydrin or bis-epoxides[22]. Oxirane structures (the three membered rings) are reacted with thiosulphate to form a Bunte salt derivative ($RSSO_3$), which is subsequently reduced to a thiol by an excess of a low molecular weight thiol. The derivative obtained can be activated in the same way as described above.

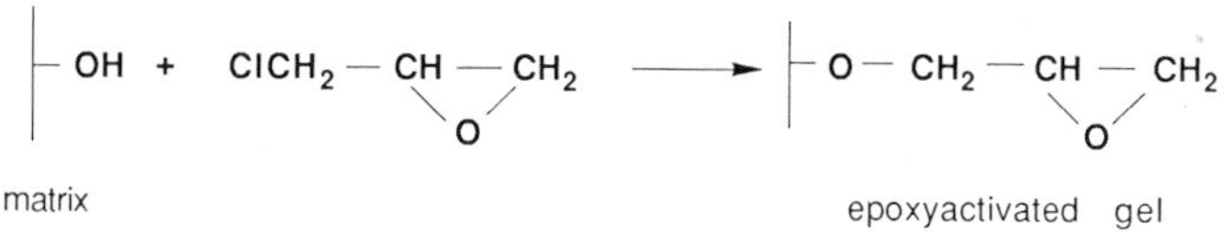

$$\vdash OH + ClCH_2 - CH - CH_2 \longrightarrow \vdash O - CH_2 - CH - CH_2 \qquad (9\text{-}12)$$

matrix   epoxyactivated gel

$$\vdash O - CH_2 - CH - CH_2 + Na_2S_2O_3 \longrightarrow \vdash O - CH_2 - CH - CH_2 - SSO_3 \quad (9\text{-}13)$$
$$\qquad\qquad\qquad\qquad\qquad\qquad\qquad\qquad\qquad OH$$

$$\vdash O - CH_2 - CH - CH_2 - SSO_3 + RSH \longrightarrow \vdash O - CH_2 - CH - CH_2 - SH$$
$$\qquad\quad OH \qquad\qquad\qquad\qquad\qquad\qquad\qquad\qquad OH$$

thiopropyl gel

$$(9\text{-}14)$$

$$\vdash O - CH_2 - CH - CH_2 - SH + \text{(pyridyl-S-S-pyridyl)} \longrightarrow$$
$$\qquad\quad OH$$

$$\vdash O - CH_2 - CH - CH_2 - S - S - \text{(pyridyl)} + S = \text{(pyridone)}$$
$$\qquad\quad OH$$

activated thiopropyl gel

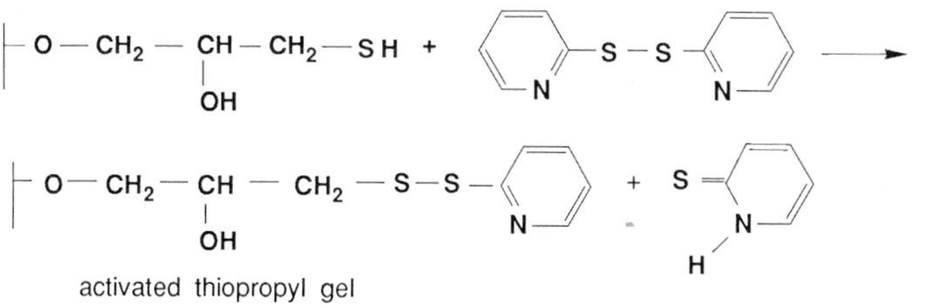

$$(9\text{-}15)$$

In contrast to glutathione, the hydroxypropyl spacer contains no charged structures.

Both gels are stable and can be stored as suspensions at cold room temperature ($+4°C$) and pH 6–7 for several months. The use of sodium azide as a bacteriostatic should be avoided. The azide is a good nucleophile and reacts with the pyridyl disulphide group to form an immobilized labile sulphenylazide and thiopyridone, thereby consuming the active structures. In the presence of proper additives, such as dextran and galactose, the agarose derivatives can be lyophilized.

The methods described can also be used to introduce thiol groups into a number of other hydroxyl group-containing matrices. Other, less used methods for the immobilization of thiol compounds are also available. These include the thiolation of matrices containing amino groups by any of the methods available for thiolation of proteins (see above).

9.4.3 Different Gels. Degree of Substitution and Capacity

Some of the gels at present available on the market are listed in Table 9-2. The degree of substitution varies considerably. The range is 0.1–20 micromoles per ml packed bed, although it generally is in the order of 1–5 micromoles/ml gel.

Table 9-2. Thiol gels available at some larger companies. The table is not comprehensive.

Gel	Structure	Method of preparation	Available from	Brand name
Thiol gels				
Cysteamine-agarose	$\mid$O—C(=$\overset{+}{N}H_2$)—NH—CH$_2$—CH$_2$—SH	CNBr-coupling of cysteamine	Sigma Chemicals	
Cysteine-agarose	$\mid$O—C(=$\overset{+}{N}H_2$)—NH—CH—CH$_2$—SH, COOH	CNBr-coupling of L-cysteine	Sigma Chemicals	
N-acetylhomo-cysteine agarose	$\mid$spacer—CH$<$(CH$_2$—CH$_2$—SH)(NHCOCH$_3$)		BioRad	Affigel 401
Glutathione-agarose	$\mid$glutathione—SH	Bisoxirane-coupling with and without spacer	Sigma Chemicals	
Activated gels				
Thiopropyl agarose	$\mid$O—CH$_2$—CH—CH$_2$—SS—pyridine, OH	Epichlorohydrine followed by Na$_2$S$_2$O$_3$ (Eq. 9-12–9-15)	Pharmacia	Activated Thiopropyl Sepharose 6 B
Glutathione agarose	$\mid$glutathione—SS—pyridine	CNBr-coupling of glutathione (Eq. 9-11 and activation)	Pharmacia	Activated Thiol Sepharose 4 B

Table 9-3. Capacities of two different activated thiol gels. The degree of substitution refers to the content of 2-pyridyl-disulphide structures (2-PySS) reactive towards low-molecular-weight thiol compounds. The practical capacity refers to the amount of protein that can be bound to the gels[30].

Gel type	Degree of substitution (μmoles/ml gel)	Practical capacity
Activated glutathione agarose-gel. (agarose-glutathione-2-pyridyldisulphide)	1	2–3 mg HSA/ml gel (0.05 μmoles/ml gel)
Activated thiopropyl-agarose (agarose-hydroxy-propyl-2-pyridyl-disulphide)	20	14 mg Ceruloplasmin/ml gel (0.1 μmoles/ml gel)

The degree of substitution achieved are very different for the two methods of synthesis described above (Table 9-3). Only a low degree of substitution is obtained by immobilization of glutathione, whereas a very highly substituted gel can be obtained by the second method. The difference can be close to one hundred fold. This, however, does not imply that the capacities of the gels will differ by the same factor. The *degree of substitution* is obtained by reacting the gel with a low molecular weight thiol compound and measuring the released thiopyridone by spectrophotometry. The *practical capacity* obtained in a chromatography experiment is in general much lower than the degree of substitution, especially for a highly substituted gel. This is due to the fact that the binding of high molecular weight substances such as proteins is limited by the space available on the polymer rather than by the concentration of active groups. The amount of gel needed for a specific application should ideally be worked out in a pilot experiment.

9.5 Chromatographic Techniques

9.5.1 Preparatory Experiments

We recommend that the thiol content of the sample for covalent chromatography is analyzed in advance. This is to make sure that the capacity of the gel is not exceeded. Very often biological samples contain low molecular weight thiols such as glutathione. These should be removed by dialysis or gel filtration before covalent chromatography.

Thiol titration is conveniently done by spectrophotometric determination of the 2-thiopyridone released when a small amount of the sample (1–5 mg in 1–3 ml) reacts with 2,2'-dipyridyldisulphide (0.1–0.2 mM). The conditions can be chosen to suit the sample in question. Buffers between pH 3 to 8 (formate, acetate, phosphate and Tris) in the concentration range 0.05–0.4 M, with or without strong denaturants such as 8 M urea or 6 M guanidine HCl, can be used. Under standard conditions (pH 7.5) a reaction time of a few minutes is usually enough for complete reaction. The addition of 0.05 M EDTA is recommended to trap transition metal ions.

A similar sample can also be reacted with the activated thiol gel. In the case of thiopropyl agarose, 0.3 g of a swollen and equilibrated gel dried by aspiration on a filter is incubated batchwise with the same amount of protein as above in 2–3 ml in a small tube that is closed and rotated end over end for the prescribed time (about 1 h). The tube is then centrifuged and a spectrum is run on the supernatant. The amount of protein thiols bound by the gel can be calculated from the absorbance at 343 nm after the background level has been substracted. It is important to run a spectrum and not just to read at a single wavelength because the background is sometimes considerable.

A small amount of the thiopyridone liberated upon incubation of the activated gel is released by mechanisms other than the binding of protein[23]. This "leakage" is of the order 0.02% of the active structures per hour in a buffer without denaturants, but is higher when high concentrations of urea or guanidine are present. The release corresponds to an absorbance at 343 nm of 0.004 per hour at pH 4.

The results of the preliminary experiments are used to determine proper conditions for the chromatography and the practical binding capacity of the gel. In many cases this is not more than about 1% of the active structures present in a highly substituted thiopropyl agarose gel (Table 9-3).

9.5.2   Binding of Sample Proteins

The coupling of the sample to the gel can be performed either batchwise, i.e. by suspending the gel in the sample solution, or columnwise, i.e. by letting the sample pass through a column packed with the gel equilibrated with the chosen buffer. The packing and dimensions of the column are not critical. It is, however, often better to use a long, thin column than a short, wide one and to adjust the flow rate such that the sample is in contact with the gel for the time chosen (at least one hour in the standard procedure). The absorbance at 343 nm of the effluent can be used to estimate the amount of thiol groups that have reacted with the gel. If the reading at 280 nm is used to estimate the amount of non-bound protein, the contribution from released thiopyridone (Fig. 9-3) has to be subtracted. The absorbance of 2-thiopyridone at 280 nm and 343 nm is roughly equal.

9.5.3   Washing

Unbound and non-specifically adsorbed proteins should be washed off the column. The choice of washing buffers depends on the stability and intended use of the sample that is covalently bound to the column. Normally, a high ionic strength buffer is recommended to neutralize charge interactions (buffer containing 0.1–0.3 M NaCl). In the simplest case the same buffer is used for application, washing and elution. If necessary, washing with detergents like Triton and Tween can also be included. The washing should be monitored by measuring the absorbance of the effluent. 1–2 column volumes is usually sufficient. The washing operation is completed by equilibrating the column with the buffer to be used for the reductive elution.

If the coupling is done batchwise, the washing can be performed either on a glass filter or in a column, the latter procedure being most convenient for a small amount of gel.

9.5.4   Reductive Elution

Reductive elution is normally done at pH 8. The low-molecular weight thiol which will reduce the bound sample as well as residual thiopyridyl structures can be either 10–25 mM DTT or 25–50 mM 2-mercaptoethanol. If cysteine is used one must remember that its oxidized form, cystine, is less soluble and might precipitate in the column after some time. As above, the elution can be followed by measuring the absorbance of the effluent at 343 nm.

Since the thiopyridyl groups are reduced much more easily it is possible to avoid the contamination of eluted proteins by thiopyridone from non-used structures on the gel by doing two separate reductive elutions. In the first the residual thiopyridyl structures are removed by an equimolar amount of reducing agent. After appropriate washing, the bound protein is then released by an excess of thiol. The first step can be performed at either alkaline or acidic pH.

In some cases, a series of solutions of thiols of different reducing power has been used to achieve specific release of bound proteins (see application 9.6.3).

9.5.5 Recovery of Thiol Proteins

Often the released protein is eluted together with a considerable amount of thiopyridone, (i.e., only one reductive elution step is employed) excess reducing compound, and its disulphide form. These low molecular weight compounds should preferably be separated from the proteins before further handling, such as derivatization of thiol groups, activity measurement, etc. This is done most easily by a gel filtration step. If it is possible to use a low pH buffer the risk of oxidation of thiols is minimized. If the solutes are low molecular weight peptides rather than proteins, desalting is still possible by the use of a slightly hydrophobic gel, such as Sephadex LH-20, at an acidic pH[23]. The peptides will then elute in the void volume, whereas thiopyridone, mercaptoethanol disulphide and salts are retarded (Fig. 9-6).

The eluted material can also be recovered by lyophilization if appropriate volatile buffers are used. This is convenient if the eluted sample is recovered in a large volume (which is often the case). When lyophilization is to be performed we recommend that the unused thiopyridyl structures be removed in a special reductive elution step as described above. If 2-mercaptoethanol and a volatile buffer salt such as ammonium acetate are used, all low molecular weight compounds will evaporate in the lyophilization step.

9.5.6 Reactivation of Thiol Gels

After chromatography the gel is in its thiol form and has to be reactivated before reuse. This is best done in a batchwise fashion. The gel is then washed on a glass filter funnel.

The gel is first incubated for 45 minutes with a 5 mM solution of DTT in 0.1 M sodium phosphate, pH 7.5. Then all alifatic disulphides that might have been formed in the elution are reduced. Washing with buffer will remove excess DTT. The gel is then incubated for 45 minutes with a saturated (1.5 mM) solution of 2,2'-dipyridyldisulphide in 0.1 M phosphate buffer, pH 8.0. With high capacity gels it is necessary to use a 20 mM solution of 2,2'-dipyridyldisulphide to obtain complete reactivation. In this case, the reaction and the subsequent washing is done in buffer containing 20–30% ethanol in order to ensure that the reagent is dissolved. Excess reagent is finally removed by extensive washing. The reactivated gel can be stored in the same way as the fresh gel.

The presence of DTT in the eluates is conveniently checked by thiol titration with reagents such as 2,2'-dipyridyldisulphide or 5,5'-dithiobis-(2-nitrobenzoate). Similarly, the 2,2'-dipyridyldisulphide can be assayed by addition of a small amount of thiol compound followed by absorbance measurement at 343 nm.

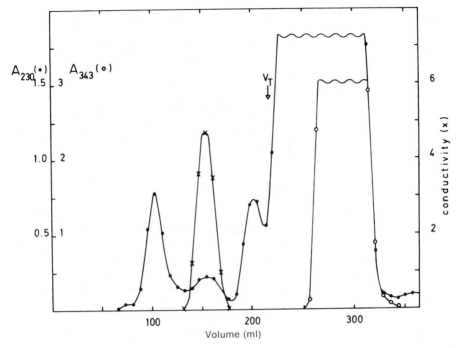

Figure 9-6. Chromatography of 20 ml of eluate from the reductive elution of an 11 ml bed of thiopropyl-agarose containing coupled thiol peptides on a column (3.2 × 27 cm) of Sepharose LH-20. Fractions of 7.3 ml were collected and analyzed for absorption at 230 nm (●), 343 nm (○) and conductivity (×). The fractions in the first peak (elution volume 100 ml), which contained the thiol peptides were pooled and lyophilized. The Tris buffer eluted at about 150 ml and the components with elution volumes of about 200 ml, 240 ml and 290 ml are believed to be mercaptoethanol, mercaptoethanol disulphide and thiopyridone, respectively. Reproduced from Ref. 23 with permission.

9.5.7 Chromatography of Activated Proteins or Peptides

Covalent chromatography can also be done in the reverse fashion to that described above. Then the sample is treated with a homogenous reactive disulphide and subsequently run on a gel containing immobilized thiol groups. In this approach it is important to remove excess reagent e.g., by a gel filtration, after the activation step.

The coupling should be done with a low capacity gel to minimize unwanted thiol-disulphide exchange reactions on the gel, which would lead to immediate release of the newly linked protein. Such side reactions can be diminished by performing the coupling at a slightly acidic pH where alifatic thiol-disulphide exchange is minimized. The ensuing steps in the procedure are the same as in the conventional approach.

This reversed covalent chromatography is particularly useful for the isolation of peptides obtained by proteolytic digestion of large proteins (see below).

9.6 Applications

Covalent chromatography has found its greatest use in the separation of thiol-containing molecules from nonthiols at an early stage in the fractionation of complex protein mixtures. Under certain conditions a higher degree of specificity can be obtained such that different thiol-containing molecules can even be separated from each other. Moreover, the technique can be used to concentrate thiols from solutions in which they are present at very low concentrations.

These applications and others are illustrated by the following examples.

9.6.1   Isolation of Urease from Jack Bean[24]

Urease is an enzyme which catalyzes the hydrolysis of urea into ammonia and carbon dioxide. It is a thiol-rich protein consisting of six identical subunits non-covalently associated to an aggregate of molecular weight 500,000. Many of the thiol groups are non-essential and can be modified without loss of the urease activity.

The starting material was Jack bean meal which is commerically available. The meal, 60 g, was mixed with 300 ml 0.05 M Tris-HCl buffer containing 36 % ethanol, 0.1 M KCl and 1 mM EDTA, pH 7.2. The mixture was stirred for 5 minutes at $+28°C$ and filtered. The filtrate was centrifuged (500 g, 20 minutes). The supernatant (about 210 ml) was diluted to 300 ml with 0.05 M Tris-HCl buffer, pH 7.2, and the pH was adjusted to 7.2 by 0.5 M NaOH.

A column with a total volume of 6.3 ml (1×8 cm) was prepared from activated thiopropyl-agarose, a high-capacity gel of the type described above in Table 9-3. The column was equilibrated with 0.05 M Tris-HCl buffer, pH 7.2, containing 0.1 M KCl and 1 mM EDTA. 250 ml of the Jack bean meal extract was passed through the column at a flow rate of 20 ml/h. Most of the UV-absorbing material passed through the column unretained. The first 150 ml of eluate contained no urease activity. The activity gradually increased to that of the applied sample within the next 100 ml of eluate. The column was washed with the Tris-HCl buffer until the absorbance at 280 nm of the eluate was less than 0.04. The urease activity was released by eluting the column with 20 mM dithiothreitol (20 ml) dissolved in 0.05 M Tris-HCl buffer, pH 8, containing 0.1 M KCl and 1 mM EDTA.

The specific urease activity (units/mg dry material) of the eluted material (after removal of low molecular weight substances on Sephadex G-25) was 167 times that of the starting material. This figure increased to 280 after an additional gel filtration on Sepharose 6B, which removed some high molecular weight material of low specific activity.

The capacity of the activated thiopropyl agarose to bind urease active material was 5.1 mg/ml gel. The purified urease preparation was stable for several weeks when stored at $+4°C$. Before the reductive elution step the column could hydrolyze urea very efficiently when a solution of the substrate was passed through it[25]. The column was thus an effective urease reactor.

9.6.2   Purification of Papain[21]

Papain is a protease that occurs in the latex of the tropical fruit, *Carica papaya*. It is a single-chain protein with a molecular weight of about 23,500. Papain has a single thiol group which is essential for its activity. The procedure described below worked both with a commerical crystallized papain preparation and an ammonium sulphate precipitate of dissolved dried papaya latex.

In the preparation based on commercial enzyme, 200 mg of papain was dissolved in 0.1 M Tris-HCl, pH 8, containing 5 mM dithiothreitol (DTT). The reduction activates the enzyme by converting blocked active-site cysteines to the thiol form. Excess DTT was then removed on Sephadex G-25 equilibrated with 0.1 M Tris-HCl at pH 8 or 0.1 M sodium acetate at pH 4, both containing 0.3 M NaCl and 1 mM EDTA. The sample for covalent chromatography was this protein (usually containing 0.4–0.6 mole of thiol per mole of protein) in 100 ml of the Tris or acetate buffer. The preparation based on crude papain was obtained by dissolving 100 g of dried papaya latex containing 0.1–0.2 mole of thiol per mole protein in about 200 ml of either of the above mentioned buffers.

Either sample was applied on a column (1.8 × 30 cm) of activated thiol agarose (a low capacity gel, see Table 9-3). The column was eluted with the application buffer until the absorbance of the eluate at both 280 nm and 343 nm was less than 0.03. The gel was then equilibrated with 0.1 M Tris-HCl, pH 8, containing 0.3 M NaCl and 1 mM EDTA. The papain was eluted from the column with 50 mM L-cysteine in the same buffer. Fractions were read at 280 nm and 343 nm and were tested for esterolytic activity towards N-benzoyl-L-arginine ethyl ester (BAEE). The activity peak was pooled and the protein was precipitated by addition of 30 g of $(NH_4)_2SO_4$ per 100 ml of solution. The precipitate was re-dissolved in a minimum volume of pH 8 buffer and the protein was separated from low molecular weight material by chromatography on Sephadex G-25 in 0.1 M KCl containing 1 mM EDTA. To prevent the formation of papain-L-cysteine mixed disulphide during gel filtration, DTT was added immediately before application to the Sephadex G-25 column to a final concentration of 5 mM. Using these conditions about 100 mg of pure papain with a thiol content of 1 mole per mole of protein was obtained.

The chromatography described above at pH 8 is an example of the separation of proteins on the basis of the presence of an exposed thiol. The specificity is even more pronounced at pH 4 when, owing to microenvironmental effects, the thiol group of native papain reacts much faster with 2-pyridyldisulphide groups than does the thiol of denatured papain or low molecular weight compounds. When a mixture of fully active papain (0.1 mM) and L-cystein (up to approx. 5 mM) was subjected to covalent chromatography at pH 4, all of the papain reacted with the mixed disulphide gel and essentially all of the L-cysteine passed through the column.

9.6.3 Sequential Elution Covalent Chromatography[26]

Hillson succeeded in separating the two enzymes—protein disulphide isomerase (PDI) and protein disulphide oxidoreductase, also called glutathione-insulin transhydro-genase (GIT)—involved in the in vivo formation of protein disulphides by reductive sequential elution of a thiopropyl-agarose column on which they had been immobilized. The starting material was a sample of partially purified protein disulphide isomerase from beef liver, which apart from PDI also contained several thiol-oxidoreductase activities. Samples were prepared by two different techniques, one involving ion exchange chromatography (partically purified preparation) and the other ammonium sulphate precipitation (crude preparation).

The protein samples (25–125 mg at 10 mg/ml) were pre-treated with 0.1 mM dithiothreitol in 50 mM Tris-HCl, pH 7.5, 25 mM KCl, 5 mM $MgCl_2$/ 1.25 mM EDTA/ 0.1 M NaCl (TKM/EDTA/NaCl buffer) at 30°C for 30 minutes. This gentle reduction unmasks any buried thiol groups which may be present as mixed disulphide. After reduction the sample was centrifuged and separated from DTT on a Sephadex G-25 column (2 × 25 cm) using the above-mentioned buffer. The reduced protein was then applied directly on a column of activated thiopropyl-agarose (30–45 g wet weight of high capacity gel). The elution was then interrupted for 30–60 minutes at 30°C to allow binding to occur.

Alternatively, the sample was incubated batchwise with the gel for 16 hours with gentle shaking, after which the gel was poured into a column. The batch incubation gave a higher level of coupling and is therefore the preferred sample loading method.

After application the column was cooled to 4°C and washed with TKM/EDTA/NaCl buffer to remove unbound and non-specifically adsorbed protein. Bound proteins were then displaced from the gel by successive elution with different low-molecular weight thiol compounds used in order of increasing reducing power: 20 mM L-cysteine, 50 mM glutathione, and 20 mM DTT, each in TKM/EDTA/NaCl buffer, pH 7.5. In each step one void volume of the reducing buffer was run into the column, which was then incubated at 30°C for 30 minutes to allow reaction to occur. Elution was then continued at 4°C at flow rates of 2–10 ml/h, followed by a wash with TKM/EDTA/NaCl buffer to rinse the column. Fractions of 5 ml were collected and monitored for protein at 280 nm and for displaced 2-thiopyridone at 343 nm.

In each step, fractions containing protein were pooled and solid DTT was added to a final concentration of 5 mM. The pooled fractions were incubated at 25°C for 30 minutes to reduce any mixed disulphide formed between protein and eluent and the solution was then dialysed extensively against TKM buffer, pH 7.5, at 4°C. The procedure resulted in preparations of four protein fractions; namely, unbound protein washed through the column and material displaced by L-cysteine, glutathione and DTT, respectively.

The break-through peak contained both PDI and GIT, probably due to overloading or incomplete reaction. PDI activity was found only in the cysteine fraction, with no associated GIT-activity. The glutathione fraction showed no detectable activities except for a small amount of GIT in one run and finally the DTT fraction contained GIT-activity but no detectable PDI-activity.

The degree of purification and the yields differed significantly depending on whether the starting material was a partially purified (ion exchange) or a crude protein mixture (ammonium sulphate-precipitate). In the latter case the percentage yields of enzyme activities in the bound fractions were 70–98% of PDI-activity in the cystein fraction and 89–100% of GIT-activity in the DTT fraction. Therefore sequential-elution covalent chromatography is a powerful tool for the rapid isolation and separation of protein-disulphide-isomerase and protein-disulphide-oxidoreductase which had not been achieved previously.

9.6.4 Isolation of Band 3 Proteins from Human Erythrocyte Plasma Membrane[27]

Band 3 proteins are the predominant polypeptide components of the human erythrocyte membrane. They appear to be a family of glycoproteins of 90 000 to 100 000 molecular weight, which are involved in anion-, cation- and possibly glucose-transport. The proteins contain free thiol groups.

The starting material was membrane ghosts treated with dimethyl-maleic anhydride (DMMA) (to remove two proteins associated with band 3) and extracted with 0.5% Triton X-100. The extract was centrifuged and dialyzed against 0.1 M Tris-HCl, pH 7.2, containing 0.3 M NaCl, 1 mM EDTA and 0.5% Triton X-100. The dialyzed extract (60 ml prepared from packed ghosts equivalent to about 100 mg of membrane protein) was applied to a column (19 ml bed volume) containing agarose-glutathione-2-pyridyl-disulphide gel (low capacity gel) equilibrated with the same buffer. The flow rate was 5–5.5 ml/h. Unbound and unspecifically adsorbed material was washed through the

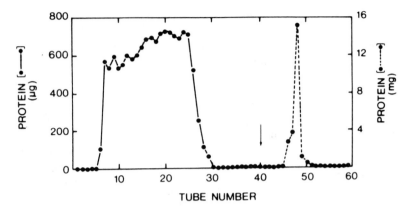

Figure 9-7. Covalent chromatography on activated thiol agarose. The Triton X-100 extract (60 ml) of membrane ghosts previously extracted with dimethylmaleic anhydride was applied to a column (0.9 × 28 cm) of activated thiol agarose at a flow rate of 5.5 ml/h and eluted with column buffer. (See application 9.6.4). At the arrow the specifically bound proteins (band 3) are eluted by 50 mM L-cysteine. Reproduced from Ref. 27 with permission.

column with 60 ml of the Tris-buffer. Reductive elution of covalently bound material was then performed with 75 ml of the same buffer containing 50 mM L-cysteine (Fig. 9-7).

Gel electrophoresis showed that the latter eluate contained 95% or more of pure band 3 protein. The recovery was around 90% in several experiments, which far exceeds previous results. The amount of protein obtained was 20–25 mg. Covalent chromatography can thus be used to isolate thiol-containing membrane proteins, where the presence of a detergent is necessary to dissolve the membrane structure.

9.6.5 Purification of Thiol Peptides[28,29]

Covalent chromatography affords facile group specific isolation of thiol peptides from protein digests. Svensson and co-workers[28] isolated peptides from papain digests of reduced ribonuclease and of mercaptalbumin. Their approach was to react the proteins with 2,2'-dipyridyldisulphide before digestion and then to apply the activated peptide mixture to a column of reduced thiopropyl agarose. The coupling to the column proceeded in 0.2 M ammonium acetate, pH 8.0. The reductive elution was done with the same buffer containing 50 mM 2-mercaptoethanol. The eluted peptide mixture was lyophilized, since only volatile buffer substances and mercaptoethanol had been used. The single thiol peptide from albumin was obtained directly in pure form.

In the procedure of Rydén and Norder[23,29] the protein, human ceruloplasmin, was immobilized to the column via its thiol groups before the protease was added. This allowed the use of large amounts of protease and permitted the use of two proteases (pepsin and trypsin) sequentially to obtain subfragments of the original peptides. In these experiments, the reaction with the column was done at pH 4.0 in sodium acetate buffer and elution was done with Tris HCl, pH 8.0 containing 50 mM mercaptoethanol. The thiol peptides eluted were purified further by gel filtration and HPLC. The recoveries in these experiments were around 70% for the coupling step.

Covalent chromatography is clearly the method of choice for rapid isolation of thiol peptides from large proteins, which normally give very complex peptide mixtures upon digestion. Serum albumin contains 589 residues and ceruloplasmin 1046.

9.6.6 Characterization of Subunit Proteins

Another application of covalent chromatography is the characterization of thiol-containing proteins composed of non-covalently linked subunits. Then the protein is bound through one of its subunits (single point attachment) to a low capacity gel (<0.1 micromole/ml gel). The amount of gel-bound protein is determined as well as the gel-bound enzyme activity if the protein is an enzyme. The subunits not bound to the gel are then removed by use of a denaturing agent such as urea. Finally, the amount of protein (and possibly activity) remaining on the gel is measured.

The figures found for untreated and treated gel-protein conjugate are compared. By this approach it was possible to show that urease from Jack bean is a hexameric enzyme and that each subunit is active in isolated immobilized form and contributes 1/6 of the activity of native urease[32].

References

1. K. Brocklehurst, J. Carlsson, M. J. K. Kierstan, E. M. Crook, *Methods in Enzymology*, Vol. 34, pp. 531–544 Academic Press, New York. L. A. AE. Sluyterman, J. Wijdenes, *ibd.* pp. 544–547. A. Ruiz-Carrillo, *ibd.* pp. 547–552. J. C. Nicolas, *ibd.* pp. 552–554. H. F. Voss, Y. Ashani, I. B. Wilson, *ibd.* pp. 581–591. (1974).
2. V. I. Lozinskii, S. V. Ragozhin, *Russian Chemical Reviews* 49(5), 460 (1980).
3. K. Brocklehurst, J. Carlsson, M. J. K. Kierstan, in *Topics of Enzyme and Fermentation Biotechnology* vol. 10. A. Wiseman, ed. Ellis Horwood Limited: Chichester, 1985; p 146–188.
4. M. Rubinstein, Y. Schechter, A. Patchornik, *Biochem. Biophys. Res. Commun.* 70(4), 1257 (1976).
5. Y. Schechter, M. Rubinstein, A. Patchornik, *Biochem.* 16(7), 1424 (1977).
6. M. Friedman , *The Chemistry and Biochemistry of the Sulfhydryl group in amino acids, peptides and proteins.* Academic Press, London, 1972.
7. P. C. Jocelyn, *Biochemistry of the SH-group.* Academic Press, London, 1972.
8. Y. M. Torchinsky, *Sulfur in proteins.* Pergamon Press, 1981, pp. 6–7
9. L. Eldjarn, E. Jellum, *Acta Chem. Scand.* 17, 2610 (1963).
10. C. H. W. Hirs, in *Methods in Enzymol.* Vol. 11; Academic Press, New York, 1967; p. 199.
11. M. O. Dayhoff, *Atlas of protein sequences and structure.* Vol. 5 National Biochemical Research Foundation, Washington D.C., 1978.
12. R. Cecil, M. A. W. Thomas, *Nature, 206,* 1317 (1965).
13. J. M. Nicolau, M. Bacila, *Arch. Biochem. Biophys. 129,* 357 (1969).
14. L. Rydén, D. Eaker, *FEBS-Lett 53,* 279 (1975).
15. S. Olsnes, A. Pihl, *Biochem. 12,* 3121 (1973).
16. J. B. Fleishman, R. M. Pain, R. R. Porter, *Arch. Biochem. Biophys. suppl. 1,* 174 (1961).
17. W. K. Liv, J. Meienhofer, *Biochem. Biophys. Res. Commun. 31,* 467 (1968).
18. F. H. White, Jr. in *Methods in Enzymology.*; Academic Press; New York, 1972, b, 25 B, 541.
19. M. Wilchek, T. Miron, *J. Chromatogr. 357,* 315 (1986).
20. J. Carlsson, H. Drevin, R. Axén, *Biochem. J. 173,* 723 (1978).
21. K. Brocklehurst, J. Carlsson, M. P. J. Kierstan, E. M. Crook, *Biochem. J. 133,* 573 (1973).
22. R. Axén, H. Drevin, J. Carlsson, *Acta Chem. Scand.* B29, 471 (1975).
23. L. Rydén, H. Norder, *J. Chromatogr. 215,* 341 (1981).
24. J. Carlsson, I. Olsson, R. Axén, H. Drevin, *Acta Chem. Scand.* B30, 180 (1976).
25. J. Carlsson, R. Axén, K. Brocklehurst, E. M. Crook, *Eur. J. Biochem.* 44, 189 (1974).

26. D. A. Hillson, *J. Biochem. Biophys. Methods.* *4*, 101 (1981).
27. A. Kahlenberg, C. Walker, *Anal. Biochem.* *74*, 337 (1976).
28. A. Svensson, J. Carlsson, D. Eaker, *FEBS-Lett.*, *73*, 171 (1977).
29. L. Rydén, D. Eaker, *Eur. J. Biochem.* *132*, 241 (1983).
30. In Pharmacia Fine Chemicals Booklet. Affinity Chromatography Principles and Methods. pp. 36–37.
31. Y. M. Torchinsky, *Sulfur in Proteins.* Pergamon Press. pp. 122–125 (1981).
32. J. Carlsson, R. Axén, T. Sato, Unpublished results.

10 Affinity Chromatography

Jan Carlsson
Research & Development
Pharmacia Diagnostics AB
S-751 82 Uppsala
Sweden

Jan-Christer Janson
Research Department
Pharmacia LKB Biotechnology AB
S-751 82 Uppsala
Sweden

Marianne Sparrman
Research Department
Pharmacia LKB Biotechnology AB
S-751 82 Uppsala
Sweden

10.1   Introduction

All biological processes depend on specific interactions between molecules. These interactions might occur between a protein and low molecular weight substances e.g., between substrates or regulatory compounds and enzymes; between bioinformative molecules—hormones, transmittors etc.—and receptors, and so on. But even more often biospecific interactions occur between two or several biopolymers, in particular proteins. Examples can be found from all areas of structural and physiological biochemistry such as in multimolecular assemblies, effector-receptor interactions, DNA-protein interactions and antigen-antibody binding. Affinity chromatography (see Refs. 1–3 for general and earlier references) owes its name to the exploitation of these various biological affinities for adsorption to a solid phase. One of the members of the pair in the interaction, the *ligand*, is *immobilized* on the solid phase, while the other, the *counterligand* (most often a protein), is *adsorbed* from the extract that is passing the chromatographic column. Examples of such affinity systems are listed in Table 10-1.

Table 10-1. Examples of biological interactions used in affinity chromatography.

Ligand	Counterligand
Antibody	antigen, virus, cell
Enzyme	substrate analogue, inhibitor, co-factor
Lectin	polysaccharide, glycoprotein, cell surface receptor, membrane protein, cell
Nucleic acid	nucleic acid-binding protein (enzyme or histone)
Hormone, vitamin	receptor, carrier protein
Sugar	lectin, enzyme or other sugar binding protein

In many cases affinity chromatography is a very powerful method. This is particularly true when the protein of interest is a minor component of a complex mixture. The extraction of the vitamin B_{12} transport protein transcobalamin II from blood serum is given as an impressive example of the purification of 10 mg of active protein from 40 kg of plasma in a simple two-step procedure, using a column with immobilized cobalamin as ligand[4].

The term affinity chromatography has been given quite different connotations by different authors. Sometimes it is very broad including all kinds of adsorption chromatographies based on non-traditional ligands, in the extreme all chromatographies except ion exchange. Often it is meant to include immobilized metal ion affinity chromatography (IMAC), covalent chromatography, hydrophobic interaction chromatography, etc. In other cases it refers only to ligands based on biologically *functional pairs*, such as enzyme-inhibitor complexes. Here we will use the term not only to include functional pairs but also the so called *biomimetic ligands*, in particular dyes, whose binding apparently often occurs to active sites of functional enzymes, although the dye molecules themselves of course do not exist in the functional context of the cell.

Since affinity chromatography proper relies on the functional properties, active and inactive forms can often be separated. This is, however, not unique to affinity methods. Covalent chromatography (Chap. 9) can do the same thing when the activity depends on a functional thiol group in the protein. By affinity elution ion exchange (Chap. 4) is also able to separate according to functional properties. These are, however, exceptions to what is a rule for the affinity methods.

Affinity chromatography has proved to be of great value also in the fractionation of nucleic acids, where complementary base sequences can be used as ligands, and in the separation of cells, where cell surface receptors are the basis of the affinity. Its main use has, however, been in the context of protein purification.

A field that has been so successful that it is often treated separately is affinity based on antigen-antibody interactions, called *immunosorption*. Sometimes this is the only available route to the purification of a protein and is especially attractive when there is a suitable monoclonal antibody at hand. This technique will be dealt with in some detail in this chapter.

Very often the use of affinity chromatography requires that the investigator him- or herself synthesises the adsorbent. The methods for doing this, to be described below, are today well worked out and easily adopted also for others than synthetic organic chemists.

To further simplify the task, activated gel matrices ready for the reaction with a ligand are commerically available. The immobilization of a ligand can in the best cases be a very simple affair. Immobilizations are in addition just as easy for proteins as for small molecules.

A property that needs special consideration is the *association strength* between ligand and counterligand. If it is too weak there will be no adsorption, if it is too strong it will be difficult to elute the protein adsorbed. It is always important to find conditions, such as pH, salt concentration or inclusion of, e.g., detergent or other substances, that promote the dissociation of the complex without destroying the active protein at the same time. It is often here that the major difficulties with affinity methods are encountered.

As in the example given above, ligands can be extremely selective, but they may also be only group specific. In the latter type we can include glycoprotein-lectin interactions, several dye-enzyme interactions and interactions with immobilized cofactors. However, these interactions have also proved to be extremely helpful in solving many separation problems. Good examples are ligands which are group selective against immunoglobins, e.g. staphylococcal protein A or streptococcal protein G.

In this chapter we shall first give an overview of the various interactions that have been exploited in affinity chromatography. Later the details of how to prepare an affinity adsorbent will be dealt with, followed by the practicalities of the chromatography. Finally, applications are described both for the immobilization reactions and the chromatographies.

10.2   Affinity Interactions

A good affinity ligand should possess the following characteristics:

— The ligand must be able to form reversible complexes with the protein to be isolated or separated.
— The specificity must be appropriate for the planned application.
— The complex constant should be high enough for the formation of stable complexes or to give sufficient retardation in the chromatographic procedure.
— It should be easy to dissociate the complex by a simple change in the medium, and without irreversibly affecting the protein to be isolated or the ligand.
— It should have chemical properties that allow easy immobilization to a matrix.

For any particular protein intended to be purified by affinity chromatography there is often a choice of several different ligands. Besides the obvious choice of using a monoclonal antibody of appropriate affinity, which is generally applicable to all immunogenic solutes, one may look for components of naturally occurring biospecific pairs such as enzyme-substrate (analogues), enzyme-cofactor (analogues) and enzyme inhibitor complexes. For glycoproteins, there is the possibility of using immobilized lectins, and the latter are often isolated by adsorption to immobilized carbohydrates. Considerable interest has been focused on immobilized biomimetic dyes which show a wide variety of specificities applicable to several groups of enzymes, plasma proteins and other proteins such as interferons.

In order to adsorb a protein counterligand to an affinity gel the binding constant K_A for the interaction needs for most practical purposes to exceed or equal 10^5–10^6 M (corresponding to a dissociation constant K_D of 1–10 μm; the K_D is equal to the inverse of K_A). However, interactions in the order of millimolar to micromolar ($K_A = 10^3$–10^6) will also, with a reasonable ligand density, cause retardation of the interacting protein. In these cases isocratic elution chromatography with small sample volumes can be very useful. Note that heterogenous immunoassays such as ELISA, require higher association constants, and thus other detection methods are needed in these cases. This is why one tends to loose monoclonal antibodies which have lower affinities, but which—precisely because of that–are very useful for purification purposes. The high affinity interactions often require drastic and sometimes denaturing conditions for elution, i.e., decrease of binding constant. Interactions with binding constants exceeding 10^{10}–10^{11} M are sometimes impossible to use since the conditions required to dissociate the complex are often the same as those that unfold the protein.

Generally, ligands may be classified as either mono-specific or group-specific, each of which in turn may be divided into low-molecular weight or macro-molecular.

10.2.1   Mono-specific Low Molecular Weight Ligands

In this group we find ligands such as steroid hormones, vitamines and certain enzyme inhibitors. The term monospecific refers to the fact that these ligands bind to a single or a very small number of proteins in any particular cell extract or body fluid. Thus, lysine binds only plasminogen from blood plasma samples[5] and vitamin B_{12} will bind only its transport proteins: intrinsic factor from pure gastric juices, and transcobalamin II from plasma[4].

In spite of the high specificity, non-specific adsorption might occur. This can be due to interaction with the ligand or with residues from the immobilization reaction or the spacer arm. One way to cope with this problem is to make a second adsorbent lacking only the ligand itself and to allow the desorbed material from the ligand-containing adsorbent to pass this under identical conditions. Another way is to use a specific displacer, e.g. the ligand itself in soluble form (section 10.5.4.3, p. 311), followed by a more harsh, general displacement agent for the regeneration of the adsorbent.

Generally, monospecific ligands bind more strongly and require harsher eluents than group specific ligands which can usually be eluted under mild conditions. Examples of extremely strong binding are the steroids and steroid receptors, which have association constants in the range 10^8–10^{10} M. Here it is often impossible to find elution conditions that allow the protein to be recovered in native form. A possibility is to use steroid analogues that have lower binding constants as ligands. Another example of a monospecific low molecular weight ligand with very high binding constant to its counterligand is biotin which binds avidin[6] with a K_A of 10^{15}.

10.2.2   Group-specific Low Molecular Weight Ligands

This is the largest group of ligands containing a wide variety of enzyme cofactors and their analogues. To this group are also assigned the biomimetic dyes, boronic acid derivatives and a number of amino acids and vitamins. A representative list of group specific ligands and their target proteins is given in Tables 10-2 and 10-3. The target proteins are most often enzymes and the most thoroughly studied are the NAD^+ and $NADP^+$ dependent dehydrogenases and the kinases.

Table 10-2. Examples of group specific low molecular weight ligands and their target proteins.

Ligand	Target proteins	Ref.
5'-AMP	NAD$^+$-dependent dehydrogenases	1
	ATP-dependent kinases	
2',5'-ADP	NADP$^+$-dependent dehydrogenases	1
ATP	ATP-dependent kinases	1
NAD$^+$	NAD$^+$-dependent dehydrogenases	1
NADP$^+$	NADP$^+$-dependent dehydrogenases	1
Benzamidine	Serine proteases	8
Phenylboronic acid	Glycoproteins	9
Cibacron Blue F3G-A	See Table 10-3	
Procion Red HE-3B	See Table 10-3	

Table 10-3. Examples of proteins with affinity for two broadly specific dye ligands, Cibacron Blue F3G-A and Procion Red HE-3B (from Ref. 9).

Cibacron Blue F3G-A	Procion Red HE-3B
Kinases and phosphatases	*Dehydrogenases*
Adenylate cyclase	Aldehyde reductase (NADP$^+$)
Adenylate kinase	Dihydrofolate reductase (NADP$^+$)
Amino acyl tRNA synthetase	Glucose-6-phosphate dehydrogenase (NADP$^+$)
cAMP-dependent protein kinase	Glutamate dehydrogenase (NADP$^+$)
Creatine kinase	Glutathione reductase (NAD$^+$/NADP$^+$)
DNA polymerase	3-hydroxybutyrate dehydrogenase (NAD$^+$)
Fructose diphosphatase	Isocitrate dehydrogenase (NAD$^+$)
cGMP-dependent protein kinase	Lactate dehydrogenase (NAD$^+$)
Nucleoside kinase	Malate dehydrogenase (NAD$^+$)
Phosphofructokinase	6-phosphogluconate dehydrogenase (NADP$^+$)
Phosphoglycerate kinase	
Phosphorylase A	*Other proteins*
Protein kinase	Carboxypeptidase G
Restriction endonucleases	Dopamine beta-monooxygenase
Succinyl-CoA transferase	Inhibin
	Interferon
	3-methylcrotonyl-CoA carboxylase
Dehydrogenases	Plasminogen
Alcohol dehydrogenase (NAD$^+$)	Propionyl-CoA carboxylase
Glutathione reductase	
Hydroxysteroid dehydrogenase	
Isocitrate dehydrogenase (NAD$^+$)	
Lactate dehydrogenase (NAD$^+$)	
Malate dehydrogenase (NAD$^+$)	
Phosphogluconate dehydrogenase	
Other proteins	
Albumin	
Blood coagulation factors II, IX	
Interferon	

There are a large number of affinity chromatography adsorbents based on group specific ligands coupled to a variety of carrier matrices commerically available from several sources. Two of the most widely used adsorbents are N^6-(6-amino-hexyl)5'-AMP coupled to beaded 4% agarose and the biomimetic textile dye Cibacron Blue F3G-A coupled to crosslinked beaded 6% agarose (e.g. Blue Sepharose™ CL-6B). The 5'-AMP gel shows affinity for a variety of NAD^+-dependent dehydrogenases with a binding capacity of approximately 10 mg enzyme per ml gel. In spite of its relatively broad specificity, very high purification factors may be obtained using specific elution protocols with either soluble cofactors or by ternary complex formation using a combination of cofactor and substrate[10]. Alternatively, when the ligand-enzyme association constants are sufficiently far apart, gradient elution with soluble cofactor may result in adequate separation as has been shown for LDH isoenzymes.

The blue dye ligand is an analogue of adenylylcontaining cofactors. Consequently, the adsorbent can be used to purify a very wide range of enzymes requiring such cofactors[11], including both NAD^+- and $NADP^+$-dependent enzymes, although it shows some selectivity for NAD^+-dependent enzymes. In this respect it resembles the 5'AMP ligand. However, the blue ligand binds a wider range of proteins and has been used for the isolation of several quite disparate proteins as shown in Table 10-3.

A red dye, Procion Red HE-3B, coupled to Sepharose CL-6B has been used for the purification of a variety of $NADP^+$-dependent enzymes and a number of other unrelated proteins such as interferon, inhibin, plasminogen, dopamine-β-monooxygenase etc., suggesting that binding may depend not only on specific steric factors but also on ionic and hydrophobic interactions. A large variety of other dyes have been exploited as affinity ligands by Scopes[12]. It is also possible to use chemical modification of the textile dye structures to improve their specificities and affinities. Lowe et al.[13] thus improved the interaction between horse liver alcohol dehydrogenase and Cibacron Blue F3G-A.

Another ligand type which appears to fit in the category low molecular weight and group specific is the one described by Porath as the T-gel (T = thiophilic)[14]. It contains sulfone and thio-ether groups and shows high selectivity for immunoglobulins.

10.2.3 Mono-specific Macromolecular Ligands

Specific protein-protein interactions are common and essential in biology. Examples are subunit interactions in quarternary structures, interactions in multienzyme complexes, and hormone-receptor protein interaction. Few of these have, however, been exploited in affinity chromatography so far. Exceptions are e.g. the binding of fibronectin to gelatin[15], of antithrombin to thrombin and heparin[16], a polysaccharide, and of the transferrin receptor to transferrin. The reader is referred to the relevant references for further information.

A group of specific protein-protein interactions that have large general significance is the antibody-antigen binding which is described in more detail below. A general description of immunological concepts is given in Chapter 14.

10.2.4 Immunosorbents

The high specificity of antibodies makes them extremely useful ligands for affinity chromatography, especially where the substance to be purified has no immediately apparent complementary binding substance other than its antibody. Both antigens and antibodies can be used as affinity ligands and the immobilized protein is known as an *immunoadsorbent* or *immunosorbent*. Immunoadsorbents can be used to purify soluble proteins and peptides, solubilized membrane proteins, viruses and even whole cells.

The traditional immunoadsorbents based on polyclonal antibody preparations have today largely been replaced by adsorbents based on monoclonal antibodies. Modern hybridoma technology allows highly specific antibodies to be obtained against a predefined antigen present at concentrations much less than 1% of total immunogen (Section 14.2). By using suitable screening methods one can obtain rare hybridomas producing antibody of virtually any desired specificity and affinity and which can be immobilized for use in purifying the antigen.

The advantages of using monoclonal antibodies for immunosorbents are several. For minor protein components single-step purification factors of several thousand fold are possible. High to moderate affinities ($K_D = 10^{-6}$–10^{-9} M) preserve the antigens (yields > 90%) and increase the life span of the adsorbent often to several hundred cycles. A pH of 2–3 will normally displace the antigen and the capacity for binding is normally at least 10-fold higher than that of an adsorbent based on polyclonal antibodies. The binding capacity would probably be still higher if all monoclonals were immobilized via their F_c-moieties. Uniform binding of the antigens allows sharp desorption peaks and consequently high concentration of antigen in the eluate.

In principle monoclonal antibodies also allow a constant supply of highly uniform antibody, which gives rise to high reproducibility from batch to batch of immunoadsorbent. The degree of substitution (surface density) is an important optimization factor.

One serious disadvantage with monoclonal antibodies is their high cost. This makes the use of a relatively small column with repetitive operation almost mandatory. The next disadvantage, which is shared with all adsorbents based on immobilized proteins, is the high risk of fouling and irreversible chemical denaturation, notably proteolytic degradation. General fouling by non-specific adsorption of various biopolymers and lipids is best prevented by a preliminary purification step. Thus, in order also to prevent proteolytic attack, one should never apply crude extracts directly to columns packed with adsorbents based on monoclonal antibodies.

Monoclonal antibodies have been covalently attached to several different matrices, using a variety of coupling methods. The use of 4% agarose and CNBr-method is normally to be recommended. The use of monoclonal antibodies in affinity chromatography has been thoroughly discussed by Goding[17].

Chase and co-workers[18] have shown that there is a linear relationship between the binding capacity of an immunoadsorbent and the amount of immobilized antibody. The ability of a particular immunoadsorbent to bind low concentrations of antigen depends on the dissociation constant, K_D, of the immobilized antibody (Table 10-4). No effect on the antibody loading could be registered for the dissociation constant, but the kinetic properties of the adsorbent were substantially improved at low loadings. This means that the immunoadsorption experiments can be run at higher flow-rates. However, the reduced capacity has to be compensated for by using larger bed volumes, which leads to larger volumes of wash and eluent buffers, dilution of desorbed antigen, greater risk for non-specific adsorption to the matrix and higher matrix costs. This is why immunoadsorbents with low antibody loading are primarily suggested for immunosubtraction procedures, i.e. for the removal of low concentrations of known impurities from protein products.

Table 10-4. Influence of dissociation constant on the adsorption of low concentration of antigen to immobilized antibody. The table shows the concentration of antigen ($\mu g/ml$) which results in a 50% utilization of the adsorption capacity of an immunoadsorbent when in equilibrium with the sample solution. (From Ref. 18).

K_D of immobilized antibody (M)	M_r of antigen						
	300	1,000	3,000	10,000	30,000	100,000	10^6
10^{-6}	0.3	1	3	10	30	100	1,000
10^{-7}	0.03	0.1	0.3	1	3	10	100
10^{-8}	0.003	0.01	0.03	0.1	0.3	1	10
10^{-9}	0.0003	0.001	0.003	0.01	0.03	0.1	1
10^{-10}	0.00003	0.0001	0.0003	0.001	0.003	0.01	0.1

Finally, it is appropriate to remember the most important general limitation of immunoadsorption as a tool for the isolation and purification of proteins. The immobilized antibody will only recognize and bind to the corresponding antigenic determinant of the actual protein. It will not discriminate between protein molecules which have been modified or partially degraded in other parts. This means that in the majority of cases other separation techniques have to be applied after the immunoadsorption step to remove molecules with possible immunogenic neodeterminant structures irrespective of whether these molecules are biologically active or not.

10.2.5   Group-Specific Macromolecular Ligands. Lectins and IgG-binding Proteins

In this group we find several ligands which have found widespread popularity, e.g. lectins such as Con A and lentil for the isolation of glycoproteins, staphylococcal Protein A and streptococcal Protein G for the purification of IgG, and calmodulin for the isolation of a wide variety of calcium-dependent enzymes. An important member of this group is also the sulphated polysaccharide heparin which is frequently used for the purification of several coagulation proteins and other plasma proteins in addition to a variety of enzymes and other unrelated proteins such as steroid receptors and virus surface antigens. In Table 10-5 are listed examples of proteins which have been shown to have affinity to immobilized heparin.

The ability of immobilized lectins to interact specifically with sugars makes them excellent tools for purifying both soluble and membrane-derived glycoproteins and polysaccharides such as enzymes, hormones, blood plasma proteins, antigens, antibodies and blood group substances. Table 10-6 lists the most commonly used lectins together with their specificities. Immobilized concanavalin A (Con A) has been the most widely used because of its specificity for the commonly occurring α-D-mannose and β-D-glucose and since binding of soluble glycoproteins to the gel is easily reversed by the addition of low molecular weight sugars or sugar derivatives. Secretory glycoproteins which contain much N-acetylglucosamine are usually purified on immobilized wheat germ lectin. Immobilized lentil lectin has the same specificity as Con A but with a lower binding constant. This makes it more suitable for the purification of membrane glycoproteins which often possess very strong binding affinity to Con A.

Table 10-5. Proteins with affinity to immobilized heparin. (Adapted from Ref. 93).

Coagulation proteins	Enzymes which act on nucleic acids
Antithrombin III	Restriction endonucleases
Factor VII	RNA polymerase
Factor IX	RNA polymerase I
Factor XI	RNA polymerase III
Factor XII, XIIa	DNA polymerase
Thrombin	DNA ligase
	Polynucleotide kinase
Other plasma proteins	
Properdin	*Protein synthesis factors*
Complement Cl	Initiation factors
Complement factor B	Elongation factor (EF-1)
BetaIH	Ribosomes
Complement C2	
Complement C3	*Receptors*
Complement C4	Steroid receptors
C3b inactivator	Oestrogen receptor
Inter-α-trypsin inhibitor	Androgen receptors
Gc globulin	
Protein HC	*Other proteins*
Fibronectin	Platelet factor 4
Beta$_2$-glyco-protein 1	Beta-thromboglobulin
C-reactive protein	SV 40 tumour antigen
Lipoprotein lipase	Hepatitis B surface antigen
Hepatic triglyceride lipase	Trehalose phosphate synthetase
Lipases	Hyaluronidase
Lipoprotein lipase	Collagenase inhibitor
Hepatic triglyceride lipase	Neurophysin
Lipoproteins	
VLDL, LDL	
VLDL apoprotein	
HOLP	

Table 10-6. Lectins and their sugar specificities.

Source of lectin	English name	Specificity
Dolichos biflorus	(Anti A lectin; horse gram)	α-N-acetyl-D-galactosamine
Bandeirea simplicifolia	(Sunn Hemp)	α-D-galactose
Ricinus communis	(Castor bean)	D-galactose, N-acetyl-D-galactosamine
Canavalia ensiformis	(Con A)	α-D-glucose, α-D-mannose
Helix pomatia	(Snail)	N-acetyl-D-galactosamine
Lens culinaris	(Lentil)	α-D-mannose, α-D-glucose
Pisum sativum	(Pea)	α-D-mannose, α-D-glucose
Arachis hypogea	(Peanut)	Galactose, β1-3 N-acetyl-D-galactosamine
Phaseolus vulgaris	(Phytohaemagglutinin)	N-acetyl-D-galactosamine
Glycine max	(Soybean)	N-acetyl-D-galactosamine
Triticum vulgaris	(Wheat germ)	Tri-N-acetyl-D-glucosamine

10.3   Preparation and Evaluation of Affinity Adsorbents

A general description of gels used in chromatography is given in section 2.2. A tabulation of commercially available matrices intended for ligand immobilization is provided in Table 10-7.

10.3.1   Choice of Matrix

As in all adsorption chromatography an adsorbent with a large surface area is desirable to maximize the capacity of the affinity adsorbent. The hydrophilic gels with a high surface-to-volume ratio (Chapter 2) are very suitable as matrices. For affinity chromatography applications the ideal gel material should meet the following characteristics:

— *Macroporous* to accommodate the free interaction of large molecular weight proteins with ligands which could themselves be proteins or other macromolecules.
— *Hydrophilic and neutral* to prevent the proteins from interacting non-specifically with the gel matrix itself.
— *Contain functional groups* to allow derivatization by a wide variety of chemical reactions.
— *Chemically stable* to withstand harsh conditions during derivatization, regeneration and maintenance.
— *Physically stable* to withstand hydrodynamic stress in packed beds and, when applicable, sterilization by autoclaving.
— *Readily available at low cost* to facilitate industrial applications.

These characteristics point to gels based on polymers highly substituted with alcohol hydroxyls and thus to polysaccharides. Among the latter, the spontaneously gel forming galactan *agarose* indeed possesses most of the characteristics of an ideal matrix for affinity chromatography. The major weakness of native agarose is its chemical and physical instability which, however, has been largely compensated for by chemical crosslinking of the physically cross-linked so-called junction zones in the agarose gel structure (see section 2.2.1). Ever since its introduction by Cuatrecasas, Wilchek and Anfinsen 20 years ago[19], 4% agarose has been the most popular matrix for affinity chromatography. A contributing reason for this popularity, besides the advantageous matrix properties as such, is that there were early simple and convenient coupling methods developed for agarose (see section 10.4.1) and even commerically available preactivated matrices (Table 10-7).

Less frequently used gel matrices for affinity chromatography are cellulose, crosslinked dextran, polyacrylamide, and silica. To this group also belong the potentially interesting and commercially available matrices made of mixtures of polyacrylamide and agarose (Ultrogel) and polymerized trishydroxymethyl-acrylamide (Trisacryl; Reactifs, IBF, France). *Cellulose* has found its niche as a carrier for ligands in the affinity chromatography of oligonucleotides and nucleic acids. This area has been throughly treated in the book by Schott[20]. Since cellulose is much cheaper than agarose, but used with the same immobilization methods, it is an alternative in large scale industrial applications of affinity methods. Gels based on *cross-linked dextran* and *polyacrylamide* both suffer from the serious disadvantage of having too small pore diameters which

Table 10-7. Commercially available activated matrices.

Ligand to be coupled	Functional group	Type of gel	Name of product	Name of manufacturer
Proteins	—NH$_2$	Beaded agarose with cyanate ester groups	*CNBr-activated Sepharose 4B	Pharmacia LKB Biotechnology
Proteins (peptides)	—NH$_2$	Beaded agarose with reactive ester on spacer	Activated CH—Sepharose 4B	Pharmacia LKB Biotechnology
Carbohydrates	—OH	Beaded agarose with epoxy (oxirane) groups on short spacer (low capacity)	Epoxi-activated Sepharose 6B	Pharmacia LKB Biotechnology
Thiol compounds (e.g. proteins)	—SH			Pharmacia LKB Biotechnology
Amines (peptides, proteins)	—NH$_2$			
Thiol compounds (thiol proteins and low mol. weight thiols)	—SH	Beaded agarose with reactive disulphide groups (short spacer) (high capacity)	Thiopropyl Sepharose 6B	Pharmacia LKB Biotechnology
Amines and thiols (including proteins)	—NH$_2$ —SH	Beaded agarose with reactive sulfonic ester groups	Tresylactivated Sepharose 4B	Pharmacia LKB Biotechnology
Aldehydes (low and high mol. weight)	—CHO	Beaded agarose with adipic acid hydrazide groups	Agarose-Adipic acid hydrazide	Pharmacia LKB Biotechnology
Amines (esp low MW type)	—NH$_2$	1,1'-carbonyldimidazole activatted 6% crosslinked beaded agarose, part. diam. 45-165 μm	Reacti-Gel (GX)	Pierce
Amines (esp low MW type)	—NH$_2$	1,1'-carbonyldimidazole activated Fractogel TSK, part. diam. 32-65 μm, frac. range: 50,000-5,000,000 MW	Reacti-Gel (HW-65F)	Pierce
Amines (esp low MW type)	—NH$_2$	1,1'-carbonyldimidazole activated beaded crosslinked dextran. Dry part. diam. 20-80 μm, frac. range: 1,000-5,000 MW	Reacti-Gel (25DF)	Pierce
Amines (esp low MW type)	—NH$_2$	1,1'-carbonyldimidazole activated Trisacryl	Reacti-Gel (GF-2000)	Pierce
Amines (esp proteins)	—NH$_2$	Glutaraldehyde-activated Ultrogel (2% polyacrylamide, 2% agarose). Part. diam. 60-140 μm Exd. limit: 3 × 10^6 Daltons	Act-Ultrogel ACA 22	IBF

Target compounds	Group	Description	Product	Manufacturer
Amines (esp proteins)	—NH₂	Derivatized crosslinked agarose gel containing N-hydroxysuccinimide ester groups	AffiGel 10 Gel	BioRad
Amines	—NH₂		AffiGel 15 Gel	BioRad
Thiol compounds (in principle also carbohydrates but matrix will hydrolyze under the harsh conditions necessary)	—SH	Possibly silica-based oxirane derivative groups part. diam. 32 63 µm	AF-Epoxy 650	Fractogel TSK
Amines (esp. low MW)	—NH₂	Possibly silica-based derivative with imidazoyl-carbonate groups, part. size: 32–63 µm	AF-CDI 650	Fractogel TSK
Amino acids, ketoacids, carboxylic acids	—COOH	Beaded agarose (4%) with (6 carbon) aminospacer	AH-Sepharose 4B	Pharmacia-LKB Biotechnology
Amino acids, ketoacids, carboxylic acids	—COOH	Beaded agarose (4%) with amino terminal OW 6-carbon spacer attached to matrix via highly stable ether bond thus more stable for leakage than AH-variety)	EAH-Sepharose 4B	Pharmacia-LKB Biotechnology
Amino acids, peptides	—NH₂	Beaded agarose (6%) with (6 carbon) carboxylspacer	CH-Sepharose 4B	Pharmacia-LKB Biotechnology
Amino acids, peptides	—NH₂	Beaded agarose (6%) with carboxyl terminal or 6 carbon spacer attached to matrix via highly stable ether bond (thus more stable to leakage than CH-type)	ECH-Sepharose 4B	Pharmacia-LKB Biotechnology
Amino acids, ketoacids, carboxylic acids	—COOH	Amino terminal agarose gel with 6-atom, hydrophilic spacer	Affi-Gel 102 Gel	BioRad
Amino acids, ketoacids, carboxylic acids	—COOH	Aminoethylpolyacrylamide matrix	Aminoethyl Bio-Gel P₂ and P-100 gels	BioRad
Amines, amino acids, peptides	—NH₂	Carboxy terminal agarose beads without spacer arm	CM = Bio Gel A Carboxymethyl Agarose	BioRad
Carboxyl cont. ligands (esp low MW)	—COOH	Possibly silica substituted with amino groups, part. size: 32–63 µm	AF-Amino 650	Fractogel TSK
Amino group cont. ligands (esp. low MW)	—NH₂	Possibly silica substituted with carboxylic groups, part. size: 45–90 µm	CM-650 (M)	Fractogel TSK

become still smaller after derivatization with affinity ligands. Also fewer methods are available for immobilization on polyacrylamide.

In traditional low pressure affinity chromatography systems, beads with a diameter of approximately 100 micron are usually standard. However, in recent years beads with diameters in the range 5 to 15 micron are used in so-called high performance liquid affinity chromatography (HPLAC). In this variety of affinity chromatography higher pressure drops are often required which also means a demand for higher gel rigidity. This is why the first HPLAC-applications were based on modified and derivatized porous silica[21]. The major reason for using smaller particles is to increase the chromatographic efficiency by decreasing the diffusion path lengths and increasing the interphase area between the stationary and mobile liquids. The most serious drawback with silica based stationary phases is their solubility at pH above 7.5 which prevents their regeneration and maintenance under alkaline conditions[22]. A recently introduced alternative to silica for HPLC-applications, which also should be useful in HPLAC, is a small diameter agarose bead, with less porosity than conventional[23]. *Synthetic organic polymers*, highly substituted with alcohol hydroxyls and with adequate porosity and rigidity, should also present interesting matrices. Some of these are now commerically available (Table 10-7).

Some notable differences between the different matrices exist with respect to their chemical properties. The majority of immobilization methods depend on the presence of hydroxyl groups on the matrix and have been adapted for use in aqueous solvents. Agarose, however, retains its macroporosity also in organic solvents. Thus activation procedures that require an organic milieu can be performed as well as coupling of ligands not soluble in water. As most organic chemistry is based on work in apolar solvents, this means that a wealth of ligand immobilization methods is potentially available for beaded agarose.

10.3.2   Properties of Ligand

For the preparation of the affinity adsorbent the ligand should:

— be compatible with the solvents used during the coupling procedure.
— possess at least one functional group by which it can be immobilized to the matrix. Commonly used groups are; $-NH_2$ (amino), $-COOH$ (carboxyl), $-CHO$ (aldehyde), $-SH$ (thiol), $-OH$ (hydroxyl).
— possess a functional group for coupling which is non-essential for its binding properties, i.e. the binding properties of the ligand should not be adversely affected as a result of its immobilization.

Ligands of high molecular weight type, e.g. proteins, with a large number of suitable functional groups, can normally be immobilized without adversely influencing structure or function. In low molecular weight ligands the coupling of course results in a relatively large change in the molecule. If the affinity interaction decreases, a chemical modification of the ligand, to provide it with an appropriate functional group for immobilization, might be necessary.

The functional group used should permit the formation of a stable covalent bond so that the ligand is not released from the matrix. This is particularly important for small ligands where "single point attachment" is often the case. For proteins "multi-point attachment" between ligand and matrix is rather common. In such affinity adsorbents the stability of each individual bond is less critical.

It is, of course, also essential that the ligand remains intact during the immobilization procedure and that it is sufficiently stable to allow the planned affinity chromatography to be carried out. This might be a problem when proteins are coupled at high pH. It is essential that the ligand reagent is as pure as possible and in particular does not contain substances with functional groups that can react competitively in the immobilization. Proteins should be subjected to gel filtration in order to remove low molecular weight substances such as ammonium sulphate.

10.3.3 Choice of Spacer Arm

Occasionally, an affinity adsorbent might show poor function due to low steric availability of the ligand. This rarely happens with high molecular weight ligands, but may occur with low molecular weight ligands. The use of a "spacer arm" in many cases solves this problem. Commonly used spacer arms are aliphatic, linear hydrocarbon chains with two functional groups located at each end of the chain. One of the groups (often a primary amine, $-NH_2$) is attached to the matrix, while the group at the other end is selected on the basis of the ligand to be bound. The latter group which also is called the terminal group is usually a carboxyl ($-COOH$) or amino group ($-NH_2$).

The most common spacers are 6-aminohexanoic acid ($H_2N-(CH_2)_6-COOH$), hexamethylene diamine ($H_2N-(CH_2)_6-NH_2$) and 1,7-diamino-4-aza-heptane (3,3-diamino-dipropylamine)[24-26]. The spacer arm is introduced into the matrix by the same immobilization methods, which are described below for ligands.

Longer spacer arms can be introduced by first immobilizing a spacer arm with a terminal primary amine and then increasing the length of the arm by reaction with succinic anhydride[25]. Another possibility is to immobilize a spacer arm with a terminal carboxyl-group and then increase its length by reaction with 1,7-diamino-4-aza-heptane by aid of a condensation reagent (see below).

A drawback with the hydrophobic arms, especially the longer ones, is that they can give rise to unwanted non-specific interactions. Polypeptides, particularly glycine oligomers, are examples of hydrophilic spacers. These, however, might bind proteins by non-specific ionic interactions[26].

Sometimes it is stated that a spacer should be used for ligands with molecular weight of less than 5000. But due to the risk of introducing non-specific binding sites, either by side reactions in the gel during coupling or by the arm itself, or both, we always recommend to first try to prepare an affinity adsorbent by direct coupling of the ligand to the matrix. It should also be remembered that in several of the ligand immobilization methods described below, like the bisepoxirane and glutaraldehyde method, the ligand will automatically be provided with a spacer as a result of its immobilization[27,28].

10.3.4 Evaluation of the Prepared Affinity Adsorbent

Before an attempt is made to use the prepared adsorbent in affinity chromatography one should always make sure that the ligand immobilization has succeeded and if possible determine the ligand density (degree of substitution as micromoles or mg of ligands per ml of affinity gel).

The analysis can be carried out in several ways. A simple method is "*indirect evaluation*", i.e. the amount of immobilized ligand is calculated as the difference between the amount of ligand originally added to the matrix and the amount of ligand recovered in the liquid phase and pooled washings after finished coupling. If the ligand absorbs light of a suitable wave length ($\lambda_{max} > 250$ nm) with an acceptable molar absorptivity ($\varepsilon > 5000$ cm^{-1} M^{-1}) this analysis can simply be carried out with a photometer. This method, however, often gives erronous values and should only be used for a rough estimation.

A very useful method is *elementary analysis*. This technique can be used if the ligand contains elements such as nitrogen, sulphur, halogen or phosphorus provided that these elements are not present in the matrix or become introduced into the matrix as a result of the activation and coupling procedures. Peptide and protein ligands can, after hydrolysis, be quantitatively determined by amino acid analysis. Another possibility is to label the ligand with a suitable gamma-emitting radioisotope before coupling. In the case of small ligands having carboxyl or amino groups, acid base titration is sometimes an easy method to determine the ligand density.

Finally, activity determination for enzyme ligands can be used. Then one has to take into account that immobilized enzymes often have changed kinetic properties, due to steric and diffusional restrictions in the interaction with substrate; i.e. comparison with free enzyme might not be valid.

An analysis of the *adsorption characteristics* of the affinity gel should also be performed.

The capacity of an affinity adsorbent is defined as mg or μmoles of counterligand that can be adsorbed per ml sedimented gel. In a 90 micron average diameter beaded 4% agarose gel, the total surface area is approximately 5 m^2 per ml bed volume, of which only about 8 cm^2 refer to the outer particle surface. (The external matrix surface area of a 4% agarose bead is only 2% of the surface area of a corresponding solid sphere). The maximum theoretical binding capacity of a 60 000 dalton molecular weight protein should thus be approximately 80 mg/ml. For several reasons this value cannot be achieved.

It is appropriate to distinguish between static and dynamic binding capacitities, respectively. The *static capacity* is measured in batch experiments which allow ample time for equilibrium to establish. The static capacity depends on the density of immobilized ligand and its availability for interaction with a particular protein. Some of the immobilized ligands might be inaccessible to a particular protein as a result of steric exclusion due to their location within the gel matrix. This is particularly true when small ligands are used for the binding of high molcular weight proteins. Thus, the functional binding capacity is often much lower than the nominal binding capacity as calculated from the measured ligand density.

The *dynamic capacity* of the affinity adsorbent is the binding capacity under operating conditions, i.e. in the packed affinity chromatography column during the sample application and washing procedures. The factors which influence the dynamic binding capacity are discussed in Chapter 2.

For affinity systems based on low molecular weight ligands which bind high molecular weight proteins, the matrix bound affinity complexes sometimes sterically shield neighbouring ligands from interacting with unbound protein molecules. In such cases adequate binding capacity can be achieved at substantially lower ligand substitution. In fact, a high degree of substitution should be avoided since it may cause undesired non-specific adsorption. When the ligand and corresponding binding protein are of approximately the same size or, more unusually, when the ligand is much larger than the protein to be isolated, the problem of gel porosity primarily concerns the immobilized ligand. An alternative way to achieve a large surface area besides using beads of high porosity is to use smaller beads as in HPLAC.

10.3.5 Storage of Affinity Adsorbents

The conditions for storage of the prepared affinity adsorbents of course depend on the stability of the matrix, the ligand and the covalent bond by which the ligand is attached to the matrix.

Polysaccharide matrices such as beaded agarose hydrolyze with a significant rate at acidic pH < 4 and oxidize with formation of matrix bound carboxyls at high pH > 9. Protein ligands might change their conformation and lose their activity as a result of exposure to extreme pH, high temperatures and denaturing agents (organic solvents, urea etc.). The commonly employed CNBr-method leads to the formation of an isourea linkage[30] which is split at rather high rate through hydrolysis and aminolysis at alkaline pH (>8).

The immobilization procedure might decrease or increase the stability of the system. Several procedures, notably CNBr and epichlorohydrin coupling[31], lead to the introduction of covalent crosslinkages into the matrix and thus render it more stable. Certain ligands may also be stabilized as a result of their immobilization. Thus aggregation and autodigestion which occur in solution with, for example, proteases might be prevented.

Normally the affinity adsorbent can be stored as a suspension in an appropriate buffer at physiological pH at +4°C for long periods of time. It is however, advisable to add an anti-microbial substance such as sodium azide to prevent bacterial growth. For very labile adsorbents sometimes lyophilization can be used to increase the storage time. To make sure that the beads will reswell properly on reconstitution with water or buffer it is necessary to prevent them from irreversibly collapsing as a result of the lyophilization. Dextran or polyethyleneglycol, PEG, is then added before lyophilization and, after reconstitution of the beads, washed away on a glass filter.

10.4 Immobilization Techniques

In general the immobilization procedure consists of three steps:

1. Activation of the matrix to make it reactive towards the functional group of the ligand.
2. Coupling of the ligand.
3. Deactivation or blocking of residual active groups by large excess of a suitable low molecular weight substance such as ethanolamine.

The activation normally consists of the introduction of an electrophilic group into the matrix. This group later reacts with nucleophilic groups such as —NH$_2$ (amino), —SH (thiol) and —OH (hydroxyl) in the ligand. Alternatively one can of course use a matrix with nucleophilic groups to immobilize a ligand containing an electrophilic group, although such an approach is less common (see section 10.4.1.6, p. 303). The activated structure is sometimes stable enough for the activated matrix to be isolated and stored until coupling of ligand is performed. In other cases the coupling procedure has to be performed immediately after activation.

The ligand is either coupled directly to the activated matrix, or the matrix is first provided with a spacer arm to which the ligand is subsequently attached. Coupling to the spacer arm is often performed in a one-step procedure by use of a condensation reagent which forms amide bonds between carboxyl and amino functions present in ligand and spacer arm, but spacer arms containing terminal carboxyl group can also be activated in a separate step.

When affinity adsorbents are prepared one normally tries to establish a bond as stable as possible between the matrix and the ligand to prevent leakage of the ligand. In certain cases, however, it might be useful to have the ligand attached through a bond which is stable but can be cleaved when so desired. An example of such a bond is the aliphatic disulphide which can be both formed and split under mild conditions by thiol disulphide exchange reactions. Procedures for carrying out such reversible covalent immobilizations of ligands are described in Chapter 9.

An overview of the various immobilization methods to be presented is found in Table 10-8. The more useful of these will be described here according to matrix—agarose, polyacrylamide and silica—and properties of the ligands. The methods used for agarose matrices can equally be utilized for other polysaccharide matrices, with the exception of methods requiring organic solvents.

In Section 10-6 several applications of the techniques are given in some detail.

10.4.1 Methods for Agarose and Other Polysaccharide Matrices

10.4.1.1 The CNBr and CDAP Cyanylating Procedures; Protein Ligands with —NH₂ Groups

These methods are suitable for —NH_2 containing ligands, especially polypeptides and proteins.

The original CNBr-technique as developed by Axén et al.[32,33] is a classical two-step method with activation and coupling. A water-suspension of a polysaccharide (e.g. beaded agarose) is reacted with CNBr at high pH (11–12). This leads to the introduction of cyanate-ester and imidocarbonate groups into the matrix.

$$\text{Matrix} + \text{CNBr} \longrightarrow \text{Cyanate ester (very reactive)} \quad \text{or} \quad \text{Cyclic imidocarbonate} \qquad (10\text{-}1)$$

The relative amounts of the two groups depend on the type of polysaccharide used. E.g. the relative amount of cyanate-esters is higher for agarose than for cross-linked dextran in which imidocarbonate is pre-dominant. If the right conditions are used the activated matrix can be stored for a long time without significant loss of reactive groups either as a suspension or lyophilized. The cyanate esters are hydrolyzed at alkaline pH, while the imidocarbonates are converted to carbonates at acidic pH. Activated agarose is commerically available (Table 10-7).

Table 10-8. Overview of Immobilization Procedures.

Reagent for activation and coupling	Matrix	Functional group in matrix	Activation conditions	Activated structure	Ligand	Functional group in ligand	Coupling conditions	Ligand matrix bond	Refs.
CNBr (titration)	Polyoles (esp. polysacch.)	—OH	aq. pH 11-12	Eq. 10-1	amines (esp. proteins)	—NH$_2$	aq. pH 7-8.5	isourea	33, 35
CNBr (buffer)	Polyoles (esp. polysacch.)	—OH	aq./buffer	Eq. 10-1	amines (esp. proteins)	—NH$_2$	aq. pH 7-8.5	isourea	34
CDAP	Polyoles (esp. polysacch.)		org/aq.	Eq. 10-4	amines (esp. proteins)	—NH$_2$	aq. pH 7-8.5	isourea	35
DSC (N,N'-disuccinimidylcarbonate)	(esp. agarose)	—OH	organic	Eq. 10-12	amines (esp. proteins)	—NH$_2$	pH 6-8	carbamate	38
CDI (Carbonyldiimidazole)	(esp. agarose)	—OH	organic		amines (esp. low MW)	—NH$_2$	pH 8-10	carbamate	39
Tosylchloride	(esp. agarose)	—OH	organic	Eq. 10-11	amines, thiols	—NH$_2$, —SH	pH 9-10	sec. amine thioether	37
Tresylchloride	(esp. agarose)	—OH	organic		amines, e.g. proteins, thiols	—NH$_2$, —SH	pH 8-9	sec. amine thioether	37
Bisoxiranes	polyols	—OH	aq. pH 13-14	Eq. 10-7	carbohydrates, amines, thiols	—OH, —NH$_2$, —SH	pH 11.5-13 pH 8-11	ether, thioether sec. amine	27
Epichlorohydrine	polyols	—OH	aq. pH 13-14	Eq. 10-9	carbohydrates, amines, thiols	—OH, —NH$_2$, —SH	pH 11.5-13 pH 8-11	ether, thioether sec. amine	36
DVS (Divinylsulphone)	polyols	—OH	aq. pH 13-14	Eq. 10-10	carbohydrates, amines, thiols	—OH, —NH$_2$, —SH	pH 10.5-12 pH 8-11	ether, thioether sec. amine	31
Carbodiimides	polyols	—COOH, —NH$_2$	aq.	Eq. 10-13	amines carboxylates	—NH$_2$, —COOH	pH 5	amides	9, 41-44
Esteractivated carboxyl	polyols	—COOH	organic	Eq. 10-14	amines (esp. proteins)	—NH$_2$	pH 5-9	amides	45
Reaction with matrix thiol	polyols	—SH	aq.		unsaturated compounds	—CH=CH—, =CO —CNH	pH 8-10	thioether	46
Thiol-disulphide exchange	polyols	—SH	aq.	Eq. 10-15	thiols e.g. proteins	—SH	pH 2-9	disulphide	Chapt. 9
Glutaraldehyde	polyamide	—CONH$_2$	aq.	Eq. 10-16	amines (esp. proteins)	—NH$_2$	pH 7	prob. sec. amine	52
Hydrazine (acylazide)	polyamide	—CONH$_2$	NaNO$_2$ in HCl	Fig. 10-2	amines (esp. low Mw)	—NH$_2$	pH 7-9	amide	53
Hydrazine	polyamide	—CONH$_2$	NaNO$_2$ in HCl	Fig. 10-13	aldehydes, ketones	—CHO, =CO	pH 7-9	amide	54, 55
Oxirane via silanization	silica, glass	—Si—OH	—		amines, thiols	—NH$_2$, —SH	pH 8	sec amide	57
Isocyanide	various	—COOH, —NH$_2$, —COH, =CO, —NC	aq. pH 6.5	Eq. 10-15	variety of comp.	—COOH, —NH$_2$, —COH, =CO	pH 6.5	amide	49, 51

293

Amino-containing ligands are covalently linked to the activated matrix in aqueous medium at close to physiological pH, 7–8.

$$\text{matrix}-O-C\equiv N \;+\; H_2N-\text{Ligand} \longrightarrow \text{matrix}-O-\underset{\underset{OH}{}}{C}-NH-\text{Ligand},\;\; \overset{+}{N}H_2$$

Isourea derivative   (10-2a)

The bonds formed between the ligand and the matrix are mainly of isourea type. When the ligand reacts with the imidocarbonates the products are N-substituted imidocarbonates as well as isourea derivatives. N-substituted carbamates also occur when the ligand reacts with cyclic carbonate (10-2c) formed by hydrolysis of the N-substituted imidocarbonates.

$$\text{matrix}\big\langle{}^{O}_{O}\big\rangle C=NH \;+\; N_2H-\text{Ligand} \longrightarrow \text{matrix}\big\langle{}^{O}_{O}\big\rangle C=N-\text{Ligand}$$

Substituted
Imidocarbonate   (10-2b)

$$\text{matrix}\big\langle{}^{O}_{O}\big\rangle C=O \;+\; H_2N-\text{Ligand} \longrightarrow \text{matrix}\big\langle{}^{OH}_{O-\underset{O}{\overset{\|}{C}}-NH-\text{Ligand}}$$

N - Substituted
carbamate   (10-2c)

The simplicity of the method, the fact that it works so well in combination with beaded agarose, a matrix with excellent chromatographic properties, and that it is mild enough for binding sensitive ligands such as proteins like antibodies and enzymes, has made the CNBr-technique by far the most used technique for the preparation of affinity adsorbents.

In the activation step the pH is kept constant either by addition of strong sodium hydroxide or by use of a buffer[33,34]. The high pH is needed to deprotonate the polysaccharide hydroxyl groups ($pK = 12$) to corresponding alkoxide ions which are sufficiently nucleophilic to react with the CNBr. The basic reaction medium causes the hydrolysis of CNBr to inert cyanate ions OCN^- thus consuming more than 90% of the initially added amount of CNBr (Fig. 10-1). The formed cyanate esters to a great extent also hydrolyze to inert carbamate groups and react with matrix hydroxyls to less active imidocarbonates[35]. This of course decreases the capacity of the activated matrix to bind ligand and introduces possible sites for non-specific interactions in the affinity-adsorbent.

In some cases the imidocarbonates (as well as the carbonates formed as a result of hydrolysis at acidic pH of the imidocarbonates) also act as covalent crosslinkages and stabilize the matrix mechanically and chemically without significantly changing the porosity. This is particularly useful when non-crosslinked agarose is used. In spite of the

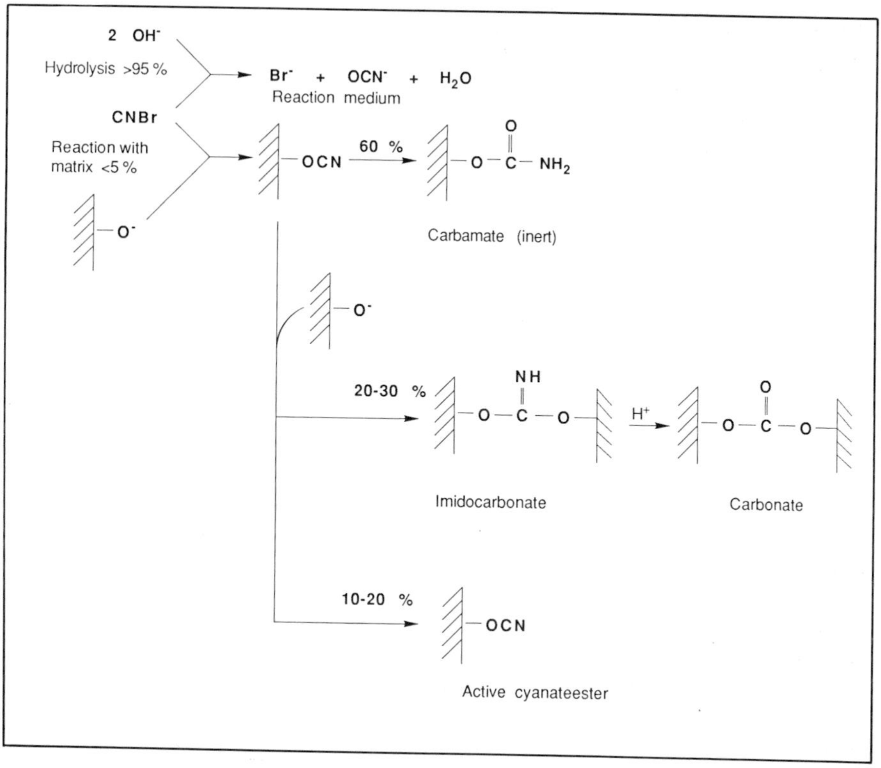

Figure 10-1. Reactions occurring when a polysaccharide matrix is activated by the classical CNBr-method at pH 11–12[35].

side reactions the original CNBr-method and varieties of it have been widely used in many successful applications of affinity chromatography, showing that by performing the activation and coupling in an accurate way, reproducible results can be obtained. The yield in the reaction is poor. Less than 2 % of the CNBr forms useful reactive groups. This does not present an economic problem, at least not in the small scale, since CNBr is a very cheap chemical. But more serious are the health hazards arising from dealing with large quantities of CNBr. Because of its toxicity and high vapor pressure all work with CNBr should be carried out in properly ventilated fume hoods.

Kohn and Wilchek[35] have devised a method to increase the electrophilicity of CNBr by forming a so-called cyanotransfer complex by CNBr and certain bases such as TEA (triethylamine) and DAP (dimethylamino-pyridine). The cyano transfer complex formed with TEA is not stable, but the one formed with DAP can be isolated as 1-cyano-4-(dimethylamino)-pyridinium bromide.

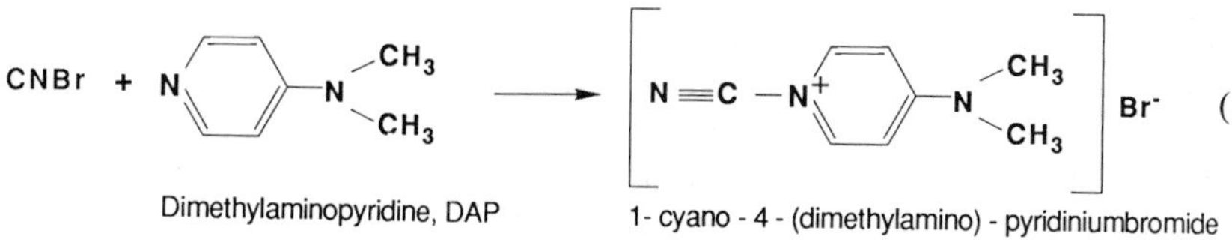

These cyanotransfer complexes are far more electrophilic than CNBr and thus able to cyanylate the matrix hydroxyl groups to cyanate esters at much lower pH than used in the conventional procedure. The hydrolysis of CNBr is thus avoided as is the transformation of cyanate esters to other products. As a result the yield of cyanate esters improves dramatically to 20–80% depending on the conditions.

$$(10\text{-}4)$$

However, the activation reaction requires that a mixed organic solvent-water systems is used, e.g. acetone—water = 6:4; in a pure water system irreproducible and low yields result. Low temperature, 0°C, typically also gives better results than room temperature.

The cyanotransfer reaction is best suited for agarose beads which can be transferred to mixed solvents without shrinking. Note that the matrices will not be reinforced by covalent crosslinkages to the same extent as with the original CNBr-method discussed above. Several cyanotransfer complexes have been described. One of the more useful reagents is 1-cyano-(4-dimenthylamino) pyridinium tetrafluoro borate (CDAP)[35], which is also available commercially.

$$(10\text{-}5)$$

1 - cyano - 4 - (dimethylamino) -
pyridiniumtetrafluoroborate

CDAP is a quite stable and non-hygroscopic salt that can be safely handled on the laboratory bench in open vessels without health hazards. It can be stored as solid at room temperature for long periods of time or dissolved in acetonitrile at −20°C for weeks. It hydrolyses rather slowly in 0.1 M HCl but at pH 7 complete hydrolysis occurs in a few hours.

In the coupling step, regardless of the activation method used, ligand amino-groups react with the matrix bound cyanate esters to form isourea bonds, Eq. (10-2a). In the case of crosslinked dextran and cellulose, which are most conveniently activated by the conventional CNBr-procedure, the majority of the activated groups are imidocarbonates which in the coupling step can be converted to both N-substituted isourea structures and N-substituted imidocarbonates, as mentioned above.

Isourea-derivatives have pK:s of about 9.5 and therefore are positively charged at neutral pH. The adsorbent thus becomes a weak anion exchanger. This does not usually present a problem. More serious is the fact that the isourea bond is reversible and can be cleaved e.g. by hydrolysis at weakly alkaline pH and by aminolysis with low molecular weight amines[30]. It can in fact be demonstrated that singlepoint attached ligands are released at a significant rate.

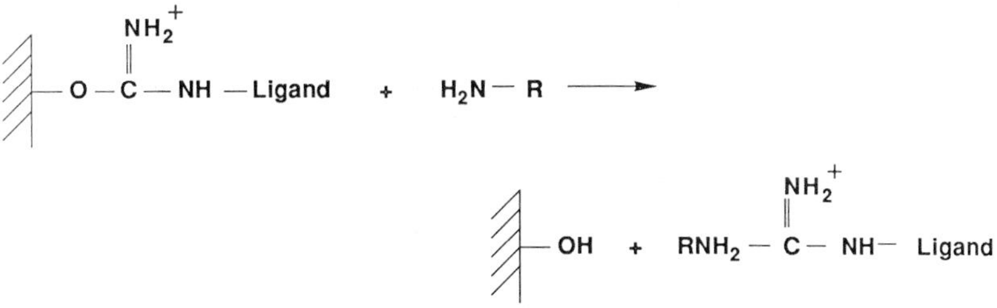

$$\text{matrix}-O-\overset{\overset{+}{NH_2}}{\underset{\|}{C}}-NH-\text{Ligand} \quad + \quad H_2N-R \longrightarrow$$

$$\text{matrix}-OH \quad + \quad RNH_2-\overset{\overset{+}{NH_2}}{\underset{\|}{C}}-NH-\text{Ligand} \quad (10\text{-}6)$$

Thus, the CNBr-technique is not the ideal immobilization method for such ligands. For ligands bound through multiple points like polypeptides the rate of release is not greater than for other commonly used immobilization methods and is more dependent on the stability of the matrix used.

10.4.1.2 Bisepoxirane, Epichlorohydrin and Divinylsulphone Methods; Low Molecular Weight Ligands with —OH, —NH₂ and SH groups

Activations based on bisepoxiranes permit immobilization of ligands containing hydroxyl, amine and thiol groups[27]. Especially useful is the possibility to couple sugar ligands (e.g. mono- and oligo-saccharides). An interesting characteristic is that the ligands will be provided automatically with a hydrophilic spacer arm. The reactions are described in Eq. (10-7).

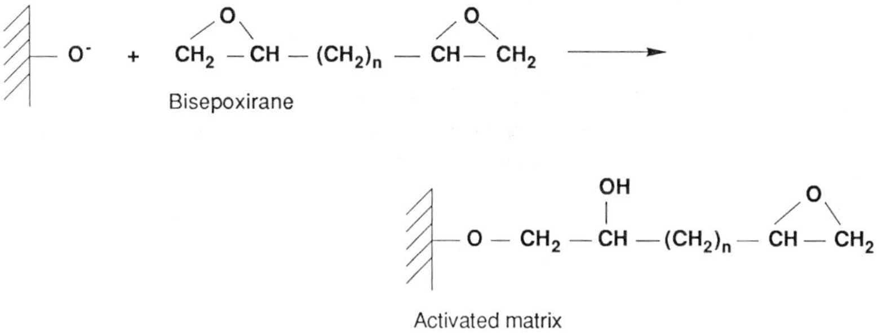

Bisepoxirane

$$(10\text{-}7a)$$

Activated matrix

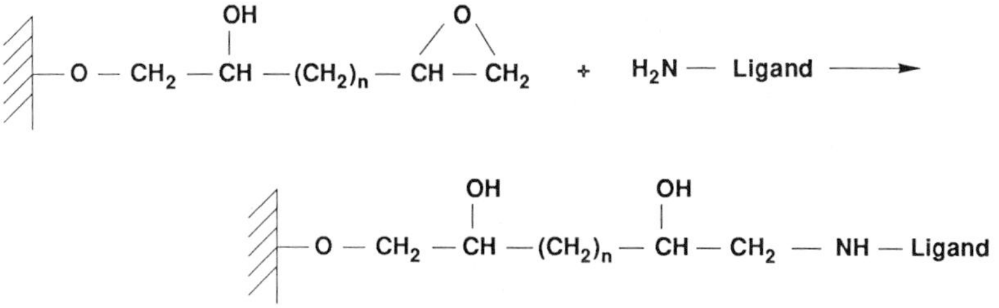

$$(10\text{-}7b)$$

This activation also introduces covalent crosslinkages in the matrix. Although this might lead to a certain decrease in porosity it is, at least in the case of agarose, advantageous since it increases the stability (e.g. thermostability) and rigidity of the gel.

The oxirane group is rather stable at pH below 8. The activated matrix can be stored, therefore, as a suspension for prolonged periods of time until used. This, however, also means that rather high pH have to be used to couple hydroxyl (pH 11–12) and amino ligands (preferably pH > 9). The method is therefore not suitable for unstable ligands (e.g. many proteins). It should, however, be possible to couple thiol-containing proteins at lower pH:s as the thiol (thiolate ion) is a better nucleophile than the other two functional groups. The poor reactivity of the oxirane groups also makes it difficult to eliminate residual activity (remaining groups) after coupling in the blocking step under conditions tolerable for the coupled ligand. Reactions with the commonly used deactivation reagent ethanolamine has to be performed at rather high pH to be efficient. Thiol reagents such as mercaptoethanol will work at lower pH, but can only be used if the ligand does not contain easily-reduced disulphide bonds. Residual oxirane groups can easily be determined by reaction with sodium thiosulphate[27]. The sodium hydroxide formed is simply titrated with acid.

$$\text{gel}-O-\!\!\sim\!\!\sim\!-CH\overset{O}{-}CH_2 + Na_2S_2O_3 \longrightarrow$$

$$\text{gel}-O-\!\!\sim\!\!-\overset{\overset{\displaystyle OH}{|}}{CH}-CH_2-S-SO_3^- + OH^- +2\,Na^+ \quad (10\text{-}8)$$

Typically a degree of substitution of 50 μmoles of active groups per ml of gel can be obtained on beaded 6% agarose by varying the excess of bisepoxirane used in the activation[27].

The most commonly used bisepoxiranes are 1,4-butanediol-bis-(epoxypropyl-ether), ethylene-glycol-bis-(epoxypropyl-ether), and 1,2:3,4-diepoxybutane.

The activation and coupling procedures with epichlorohydrin on polysaccharide gels are very similar to those used for bisepoxiranes[36].

$$\text{gel}-O^- + Cl-CH_2-CH\overset{O}{-}CH_2 \longrightarrow \text{gel}-O-CH_2-CH\overset{O}{-}CH_2 \quad (10\text{-}9a)$$

Epichlorohydrine   Activated matrix

$$\text{gel}-O-CH_2-CH\overset{O}{-}CH_2 + H_2N-\text{Ligand} \longrightarrow$$

$$\text{gel}-O-CH_2-\overset{\overset{\displaystyle OH}{|}}{CH}-CH_2-NH-\text{Ligand} \quad (10\text{-}9b)$$

Activation with epichlorohydrin leads to the introduction of gel bound oxirane groups and of crosslinkages in the matrix. The properties of the activated gel are thus very similar to those obtained with the bisepoxiranes, except for the fact that the spacer introduced is shorter.

As with the bisepoxiranes and epichlorohydrin, divinylsulphone (DVS) can also be used for the immobilization of amino-, hydroxyl- and thiol-containing ligands to hydroxyl containing matrices[31].

$$\text{—O}^{-} + \quad H_2C = CH - \underset{\underset{O}{\|}}{\overset{\overset{O}{\|}}{S}} - CH = CH_2 \quad \longrightarrow$$

Divinylsulfone , DVS

$$\text{—O} - CH_2 - CH_2 - \underset{\underset{O}{\|}}{\overset{\overset{O}{\|}}{S}} - CH = CH_2 \quad (10\text{-}10a)$$

$$\text{—O} - CH_2 - CH_2 - \underset{\underset{O}{\|}}{\overset{\overset{O}{\|}}{S}} - CH = CH_2 \quad + \quad H_2N - \text{Ligand} \quad \longrightarrow$$

$$\text{—O} - CH_2 - CH_2 - \underset{\underset{O}{\|}}{\overset{\overset{O}{\|}}{S}} - CH_2 - CH_2 - NH - \text{Ligand} \quad (10\text{-}10b)$$

Gels activated with DVS are more reactive than those activated with oxiranes. Ligand coupling can therefore be performed at 1 to 2 units lower pH. This is especially useful when coupling hydroxyl ligands such as sugars. However, the adsorbents formed are less stable at alkaline pH than those prepared by the oxirane method. Moreover, DVS is a highly toxic and expensive chemical.

10.4.1.3 Organic Sulphonyl Chlorides, Tosyl and Tresylchloride Methods; Ligands with —NH₂ and —SH Groups

These methods are most suitable for the immobilization of amino and thiol containing ligands to beaded agarose[37]. The reactions are described in Eq. (10-11):

$$\text{—OH} + Cl - \underset{\underset{O}{\|}}{\overset{\overset{O}{\|}}{S}} - \bigcirc - CH_3 \quad \longrightarrow \quad \text{—O} - \underset{\underset{O}{\|}}{\overset{\overset{O}{\|}}{S}} - \bigcirc - CH_3 \quad + \quad HCl$$

Tosylchloride   Activated matrix   (10-11a)

$$\text{—O} - \underset{\underset{O}{\|}}{\overset{\overset{O}{\|}}{S}} - \bigcirc - CH_3 + H_2N - \text{Ligand} \quad \longrightarrow$$

$$\text{—NH} - \text{Ligand} + {}^-O_3S - \bigcirc - CH_3 \quad (10\text{-}11b)$$

Organic sulphonyl chlorides react with hydroxyl groups forming good leaving groups, sulphonates, that allow binding of nucleophiles directly to the hydroxyl carbon. This principle can be utilized for the coupling of low molecular weight ligands, as well as high molecular weight ligands such as proteins, to hydroxyl group-carrying matrices. The method works with both amino- and thiol-containing ligands which become immobilized to the matrix through stable $-CH_2-NH-$ and $-CH_2-S-$ linkages.

The activation has to be performed in non-aqueous solvents, preferably acetone. Thus only matrices which swell in the solvents can be activated. Beaded agarose is most often used, but cellulose and silica derivatives containing hydroxyl-groups have also been activated successfully and used for immobilization of a variety of ligands. The activated matrices can be stored for several weeks as suspensions in 1 mM HCl without losing their coupling capacity.

The coupling can be performed both in an aqueous solvent and in an organic solvent such as DMF. The latter is used when the ligand is not soluble in water.

The reactivity of the sulphonate ester formed is strongly influenced by the R-group. Tosylates ($R = CH_3C_6H_5$) and especially tresylates ($R = CF_3CH_2$) seem to be the most suitable ones for the immobilization of ligands[37]. Tresylates in fact allow efficient immobilization even at neutral pH and at $+4°C$. Tosylated matrices however, are less reactive and require coupling at pH 9–10.5 and are thus used with ligands which can tolerate such conditions. Apart from being cheaper than tresylchloride, tosylchloride also has the advantage of releasing a chromphore upon reaction. This means that the coupling reaction can be followed photometrically.

10.4.1.4   Methods using N,N'-Disuccinimidyl-Carbonate (DSC) or Carbonyldiimidazol (CDI); Ligands with —NH₂ Groups

A two-step method, using N,N'-disuccinimidyl-carbonate, DSC, as activating agent, for the immobilization of amino-containing ligands to beaded agarose has been described by Wilchek and Miron[38].

N,N' - disuccinimidyl carbonate (DSC)

(10-12a)

Activated matrix

+ H₂N— Ligand ⟶

N - Substituted carbamate   N - hydroxy succinimide   (10-12b)

Hydroxy succinimide carbonate groups are first introduced into the gel by reaction of its hydroxyl groups (especially primary ones) with DSC in organic milieu with a base catalyst such as TEA. The carbonate subsequently reacts with amines under formation of carbamates. The coupling reaction runs both in aqueous systems under mild conditions (pH 6–8) and in organic solvent (if a base catalyst like TEA is used). It can be used to immobilize amino-containing ligands of both high molecular weight (such as sensitive proteins) and low molecular weight type. Unlike the isourea bond formed in the CNBr-methods, the carbamate bond is very stable and unchanged under conditions usually employed for affinity chromatography.

In aqueous systems the hydroxysuccinimide carbonate groups rapidly decompose by hydrolysis with regeneration of the gel hydroxyl groups and release of N-hydroxy-succinimide. The hydrolysis is faster at higher pH but proceeds at appreciable rate even at neutral and weakly acidic pH. The activated gel should therefore be protected from water before it is mixed with the ligand solution. The N-hydroxysuccinimide has a λ_{max} at 261 nm with a molar extinction coefficient of 10,000 M^{-1} cm^{-1}.

The degree of activation of the gel can thus be determined photometrically after complete hydrolysis of the hydroxysuccinimide carbonate groups at high pH, or after aminolysis with hydroxylamine. The hydroxy-succinimide groups also react to some extent with neighbouring hydroxyl groups on the gel under formation of gel bound carbonate groups. This seems to occur particularly in conditions favouring a high degree of substitution (large excess of DSC). The carbonate groups formed might, in spite of their low reactivity, cause problems, later on when the gel derivative is used as affinity adsorbent, by inadvertently immobilizing especially reactive substances in the sample. The carbonate groups in the gel can be eliminated by prolonged exposure to alkaline pH (about 11), conditions which might be detrimental to the ligand.

The carbonyl diimidazol (CDI) reagent can also be used to activate hydroxyl-containing matrices[39]. The imidazoyl carbonate groups thus introduced into the matrix react with amino-containing ligands with formation of carbamates. The activation has to be performed in organic solvent but the coupling can be run in aqueous systems. The imidazoyl carbonate groups are not as reactive as the groups introduced in the similar DSC-method. Thus the ligand coupling has to be performed at higher pH. The degree of activation (moles of imidazoyl-carbonate groups per ml gel) can be determined by keeping the activated gel at pH 3 for 4 hrs. This treatment leads to hydrolysis of the reactive groups and release of imidazol which can be determined by titration between pH 9 and 4[40].

10.4.1.5 Condensation Methods Based on Carbodiimides; Ligands with —COOH or —NH$_2$ Groups

For a long time the carbodiimides have been used for synthesis of peptides. They were also among the first reagents to be employed in the synthesis of affinity chromatography adsorbents and are still among the most widely used. Using these reagents, stable amide bonds can be formed between a ligand which contains an amino-group and a carboxyl-containing matrix (or vice versa) in a one-step procedure.

The reaction is performed by mixing the ligand with the matrix together with the reagent at slightly acidic pH (about 5) (pH adjustment with HCl and buffer).

$$
\begin{array}{ccc}
\text{—C—OH} & + & \text{R—N=C=N—R'} \\
\end{array}
\xrightarrow{\ H^+\ }
\text{—C—O—C—NH—R'}
$$

Carboxylgroup   Carbodimide   Isourea ester
containing gel   (10-13a)

—C—O—C—NH—R'   +   H$_2$N—Ligand   ⟶

—C—NH—Ligand   +   R—NH—C—NH—R'   (10-13b)

N,N'-dialkylurea

The first step in the condensation is the addition of the carboxylate to either C=N bond of the diimide, yielding the highly reactive and unstable O-acylurea (isourea ester). This, in turn, reacts chiefly with the amine to produce an amide and N,N'-dialkylurea. The major side reaction, the intramolecular rearrangement of the O-acylurea and formation of a stable N-acylurea, can be minimized if a large excess of amine-containing ligand is used. If that cannot be achieved some of the resulting N-acylurea will be bound to the matrix[26].

The reaction time is usually several hours and it is necessary to adjust the pH in the suspension during the first hour. Carbodiimides are relatively unstable compounds and must be handled with care because of their toxicity. The method is usually used to couple ligands to spacer arms with carboxyl or amino groups as terminal groups, but can of course also be used to immobilize ligands to any matrix containing amino or carboxyl groups.

Carbodiimides which have been used in the preparation of derivatives for affinity chromatography include 1-cyclohexyl-3-(2-morpholinoethyl)-carbodiimide-p-toluene-sulphonate (CMCL) and 1-ethyl-3-(3-dimethyl-aminopropyl)-carbodiimide hydrochloride (EDCL)[41-44].

An alternative way to activate carboxyl-containing matrices is to convert the carboxyl-groups to reactive esters by reaction with N-hydroxysuccinimide in anhydrous medium in the presence of a carbodimide e.g. dicyclohexylcarbodiimide (DCC)[45]. The matrix bound N-hydroxysuccinimide ester reacts easily at pH 5-8 with amines with the formation of an amide and release of N-hydroxy-succinimide.

N - hydroxysuccinimide   Reactive ester   (10-14a)

N - hydroxysuccinimide

The ester is labile and undergoes hydrolysis in aqueous solutions, especially at pH above 6. The activated gel should therefore be stored as a suspension in an anhydrous medium, e.g. dioxane. For shorter periods (minutes to a few hours) it can be stored in water at pH 5-6, provided that there are no amines present. As the technique is more laborious than those based on a direct use of a condensation reagent it is only recommended for ligands containing both primary amines and carboxyl-functions, e.g. proteins, where the use of a condensation reagent in the coupling would lead to unwanted inter- and intramolecular crosslinkages.

10.4.1.6 Methods for Thiol-containing Matrices; Ligands with Electrophilic Groups such as —CH=CH— and >C=O

So far two main principles for the immobilization of ligands to matrices have been discussed: (a) Introduction of electrophilic groups into the matrix which subsequently react with nucleophilic groups in the ligand, and (b) use of condensation agents to establish amide bonds between amino groups in the ligand and carboxyl groups in the matrix (or the other way around).

Another possibility is to immobilize a ligand by means of an electrophilic group (either already present or introduced prior to the immobilization) which reacts with a nucleophilic group in the matrix. Although not widely used, this approach has several merits. It is most easily applied to thiol-containing matrices. Thus, reaction of a thiol-containing matrix with alkyl or aryl halides and ligands containing C=O, and under certain conditions C=C bonds, will lead to stable thioether derivatives through nucleophilic displacement (in the first case) and addition (in the later case). Unsaturated compounds such as testosterone and oestradiol have been attached to beaded thiol agarose by means of gamma radiation[46]. Heavy metal ion containing ligands can also be bound to a matrix via thiol groups by mercaptide formation[47]. Activated halides with the halogen alpha to a carbonyl group (as in iodoacetic acid) react smoothly at weakly alkaline pH with thiols. These reactions usually take place in aqueous or polar organic solvents under rather mild conditions, but for unactivated halides higher pH have to be used. Oestradiol and testosterone, mentioned above, require higher pH values.

Most matrices suitable for affinity chromatography can be provided with aliphatic thiol groups by simple organic chemistry. Several methods for the thiolation of polysaccharides, especially beaded agarose, are described in detail in Chapter 9 on covalent chromatography. Porous glass and silica are also easily substituted with thiol groups by their silanization with γ-mercaptopropyl-trimethoxysilane[48].

10.4.1.7   The Isocyanide Method; Ligands with —NH$_2$, —COOH, =CO, —CHO and —NC Groups

Most of the ligand immobilization methods described above require that a reactive amino group is present in the ligand although in some cases they also work with hydroxyl or thiol compounds. Much more flexible is the so-called isocyanide (isonitrile) method, which in spite of the fact that it was described a long time ago has not attracted much interest[49]. The method is based on the four component condensation of amine, carboxyl, isonitrile and carboxyl compounds originally discovered and examined by Ugi et al.[50].

The principle is outlined in Eq. (10-15).

$$R'-\overset{\overset{\displaystyle O}{\|}}{C}-OH \qquad H_2N-R^2$$
Carboxyl   Amine

$$+$$

$$R^4-N\equiv C \qquad H-\overset{\overset{\displaystyle O}{\|}}{C}-R^3$$
Isonitrile   Aldehyde

$$\longrightarrow \qquad R^1-\overset{\overset{\displaystyle O}{\|}}{C}-\overset{\overset{\displaystyle R^2}{|}}{N}--\overset{\overset{\displaystyle R^3}{|}}{C}H-CONH-R^4$$

$$(10\text{-}15)$$

An ammonium ion structure is formed from an aldehyde or ketone and an amine. With the isocyanide this structure forms a highly reactive intermediate which is very susceptible to addition reactions with nucleophiles, such as carboxylic and hydroxylic ions. A stable amide is finally produced by intramolecular rearrangement. Although the reaction appears complicated, the technique is in practise very simple to use for immobilization purposes.

The ligand to be attached may contain either of the amino, carboxyl, aldehyde, ketone or isocyanide functions. The matrix may contain any of the others. The two remaining functional groups are added to the reaction mixture as low molecular weight substances. (E.g. aliphatic amines, carboxylic acids, aldehydes, ketones and isonitriles).

The reaction occurs in aqueous medium at pH 5-6. Matrices which have been used include cross-linked dextran and beaded agarose containing carboxyl groups, polyacrylamide, and agarose substituted with amino groups or carbonyl groups, and a large number of insoluble polymers containing isonitrile groups[49,51]. Immobilized substances are proteins, peptides, amino acids, biotin and steroids (which have been used as carbonyl compounds).

The coupling of a protein can be directed toward its amino or carboxyl groups by using an excess of low molecular weight amine or carboxyl compound, respectively.

10.4.2 Methods for Polyacrylamide Matrices

The glutaraldehyde and hydrazine methods are suitable for matrices having amide groups, such as polyacrylamide, and for the immobilization of amino-containing ligands[28,52]. These methods do not work on polysaccharide matrices. The mechanism of the activation and coupling is not completely understood but is supposed to follow the scheme outlined in Eq. (10-16).

Glutaraldehyde

Activated matrix   (10-16a)

Activated matrix + H_2N — Ligand ⟶

Immobilized ligand   (10-16b)

The yields obtained in this immobilization depends on the ligand. Advantages with the technique are that a spacer arm is introduced as a result of activation, that the coupling can be carried out with good yields at pH around neutrality, and that the bonds formed are stable enough to prevent leakage of ligands.

Hydrazine is a particularly useful reagent in combination with amide-containing matrices. When polyacrylamide is heated with hydrazine, hydrazide acrylamide (or polyacrylhydrazide) is formed. The hydrazide groups can be converted to reactive acylazides by treatment of the matrix with sodium nitrite in hydrochloric acid[53]. Amino-containing ligands can then be coupled to the activated matrix by formation of stable amide bonds (Fig. 10-2). It has been claimed that yields when coupling high molecular weight ligands are usually low, possibly because of a combination of instability of the acylazide group and steric hindrance. For low molcular weight ligands the technique seems to work well.

The hydrazide matrix can also be used for the immobilization of ligands containing aldehyde and ketone groups[55]. This reaction occurs at low pH (about 5) with formation of a hydrazone, which is then stabilized by reduction with alkaline sodium borohydride (pH about 9) (Fig. 10-2). Although ligand immobilization by means of hydrazines is most simply performed on amide-containing matrices, it can also be used in combination with other matrices such as agarose and porous glass, provided they are modified in a suitable way[55,56].

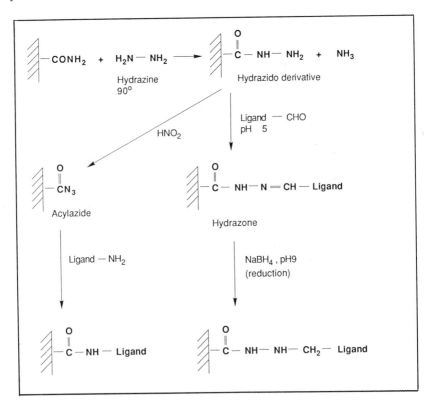

Figure 10-2. Immobilization of an amino group containing ligand to a polyacrylamide matrix by the hydrazine method.

10.4.3   Methods for Silica Gels and Porous Glass Matrices

Although silica and porous glass beads show good dimensional stability and thus can cope with the often rather high pressures used in HPLAC they do not fulfil several of the other requirements of an ideal affinity support (section 10.3.1). Their hydrophobic character and content of negatively charged silanol groups can be changed by chemical modification. The modification should, in addition to making the matrix surface more hydrophilic and masking the acidic silanol groups, provide the matrix with functional groups suitable for coupling of ligands either directly or after activation.

A commonly used procedure (Fig. 10-3) is silanization with reagents such as γ-aminopropylsilane and γ-glycidoxypropylsilane[21].

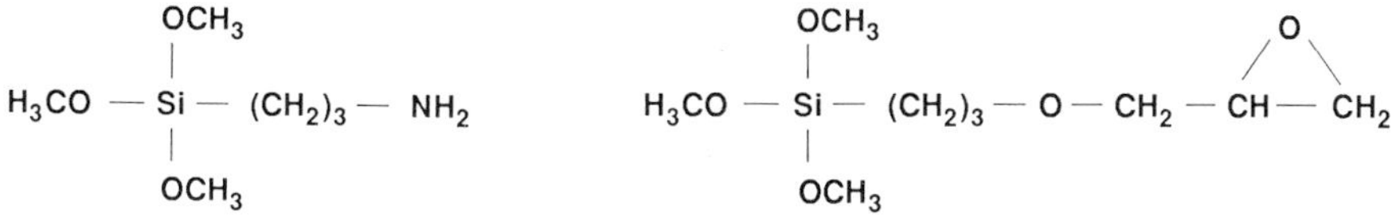

$$\gamma\text{ - aminopropylsilane} \qquad\qquad \gamma\text{ - glycidoxypropylsilane} \qquad\qquad (10\text{-}17)$$

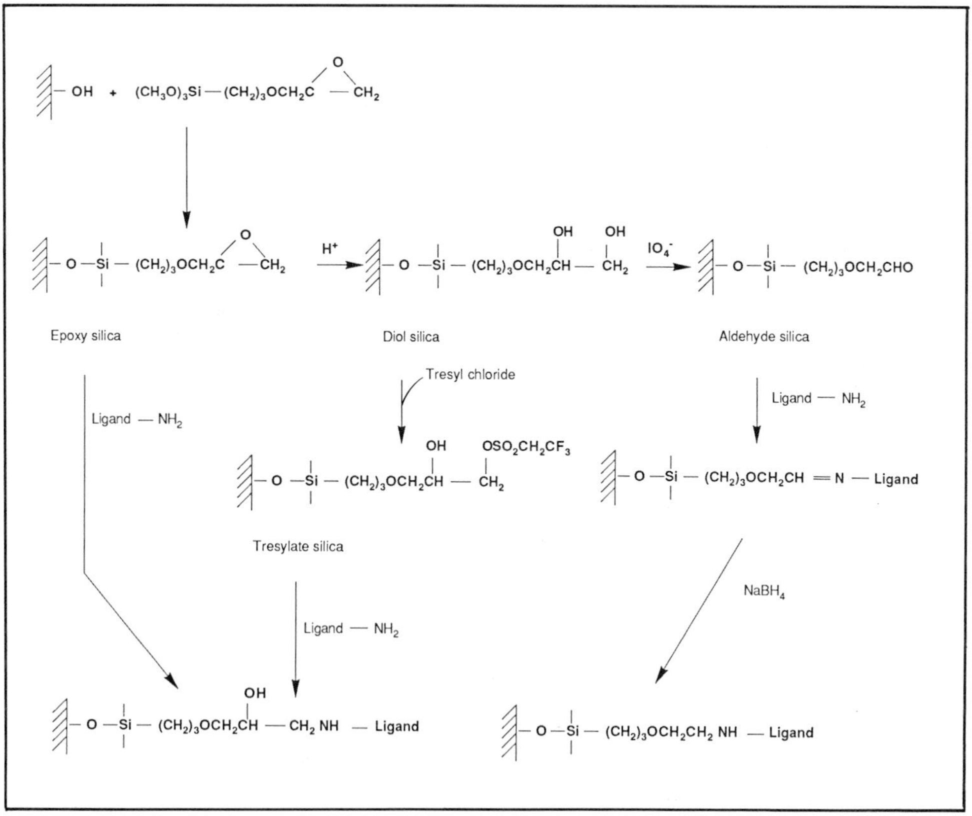

Figure 10-3. Immobilization of an amino group containing ligand to a silica matrix using activation with γ-glycidoxy-propylsilane.

Reaction with the first of these reagents gives an amino group functionalized matrix to which many affinity ligands are easily attached either directly by use of the condensation reagents discussed above or after further derivatization. The matrix may be converted to a carboxyl group containing matrix by reaction of the amines with a suitable anhydride. Unfortunately, these reactions turn the silica or glass into an ion exchanger.

Silanization with the γ-glycidoxylpropylsilane leads to the introduction of oxirane groups into the silica or porous glass[21]. As discussed above this group reacts with nucleophiles with formation of stable bonds and allows the coupling of ligands containing amino, hydroxyl or thiol groups to the activated matrix. Due to steric shielding only a small fraction of the oxirane groups are used up in the ligand immobilization. The excess groups may be hydrolyzed to diols by treatment with acid. These diols give the silica surface a hydrophilic character and decrease its tendency for unwanted protein binding.

10.5 Chromatographic Techniques

Having immobilized the ligand or selected the ready-made affinity gel, the next step is the chromatography. A typical separation by affinity chromatography consists of four stages: adsorption, washing, elution and column regeneration. The general practical aspects for affinity chromatography are described in this section. For detailed practical recommendations on individual adsorbents, reference should also be made to the manufacturer's instructions.

10.5.1 Sample Preparation. Prefractionation

Many affinity separations constitute one-step purification procedures. However, it may be advantageous to include a preliminary step using precipitation or ion exchange chromatography (Chapter 1). This removes some of the major contaminants, reduces the amount of material that must subsequently be processed and improves the resolution and concentrating effects of the affinity step. This preliminary separation step is recommended when large amounts of sample are to be processed or when the substance of interest is in the presence of very large amounts of precipitating contaminants such as lipoproteins, clotting factors etc. When a protein ligand is used in the affinity step there is also a risk of causing proteolysis by proteolytic enzymes in the extract. This is diminished by prefractionation.

For efficient adsorption the conditions (pH, ionic strength etc.) of the sample and the column must allow efficient binding, and preferably be optimal for binding. This can be achieved by dialysis of the sample, desalting and exchange of buffer by gel filtration, or by addition of chemicals or adjustment of pH.

When the starting material consists of tissue, cell culture, fermentation product or plant material, other steps, such as solubilization, homogenization, extraction, filtering and/or centrifugation are included in the fractionation scheme, as described in Chapter 1.

10.5.2 Column Operation

The column size is usually not a critical parameter. Rather the column size is governed by the capacity of the adsorbent and the amount of the substance to be purified.

Capacities of affinity gels are generally high, and short, wide columns are often used to obtain rapid separations on beds of usually 1–10 ml gel.

Typical capacities are 25 mg immunoglobulin/ml gel on Protein A-Sepharose CL-4B, 10 mg lactate dehydrogenase/ml gel on AMP-Sepharose, and more than 1 mg fibronectin/ml gel on Gelatin Sepharose 4B.

If the binding affinity of the ligand is markedly low ($K_D > 10^{-4}$ M) the protein to be purified will not bind but be retarded on the column (isocratic elution) and the separation of the desired protein from the unbound contaminants will become dependent upon the column length. Here longer columns are recommended. It is also advisable to use a small sample volumes, e.g. 5 % of the total bed volume. Ligand density should be maximized and flow rates reduced in isocratic elutions.

Column packing should follow the usual precautions for chromatographic techniques. Use the recommended flow rates and pack evenly to ensure straight bands.

To achieve binding-equilibrium several approaches apply. Sometimes batchwise procedures are preferred with gentle stirring during both adsorption and desorption or only during the adsorption step. When columns are used they can be left with reduced or stopped flow to make prolonged contact possible to achieve adequate adsorption and desorption respectively.

We can only give general recommendations with respect to flow rates. The investigator has to compromise between the acceptable level of resolution, the time required for the separation and the utilization of available adsorption capacity.

However, for sharp elution peaks and maximum recovery with minimal dilution of the purified protein, the lowest flow rate acceptable from a practical point of view should be used. This is especially important for adsorbents which bind several proteins and when competitive elution is used. Here the flow-rate used is governed by the rate of dissociation from the ligand.

For standard low pressure affinity chromatography flow rates 50 cm/h and higher have been used successfully. In HPLAC (See below) the improved mass transfer properties, giving higher dynamic capacities, allow the separation to be performed at much higher linear flow rates with sharper elution peaks giving smaller elution volumes. Typical flow rates used in HPLAC are between 50 and 125 cm/h. In some cases flow rates in excess of 200 cm/h have been used successfully also in industrial scale applications.

10.5.3   Adsorption of the Sample. Washing

For efficient adsorption the column and sample must be equilibrated with a buffer reflecting the conditions optimal for binding. The volume in which the sample is applied is not critical provided that the ligand binds specifically and effectively and that the total capacity of the column is not exceeded.

After sample application the column must be washed with several volumes of the starting buffer to remove all unbound material. The chromatography is usually monitored by UV-absorbance and the washing step is finished when the original base line is reached. It is recommended to wash with high ionic strength buffers when non-specific ionic binding is suspected. Non-specific hydrophobic binding is less easily handled since the specific binding often depends on hydrophobic interactions also.

10.5.4   Desorption

The principle of desorption is to change the binding equilibrium for the adsorbed substance from the stationary to the mobile phase. This can be achieved specifically or non-specifically.

Ligand-protein interaction is often based on a combination of electrostatic, hydrophobic and hydrogen bonds. Agents which weaken such interactions might be expected to function as effective non-specific eluants. Careful consideration of the relative importance of these three types of interaction as well as the degree of stability of the bound protein will help in the choice of a suitable eluant. A compromise may have to be made between the harshness of the eluant (or the combination of eluants) required for effective elution and the risk of denaturing the proteins to be purified. Elution can be the most difficult stage of affinity chromatography, especially if the dissociation constant of the ligand is very low.

Broadening of desorption peaks is often a problem. This can be due to slow diffusion mass transport, slow equilibration kinetics or a wide range of binding affinities in the system. The diffusion is usually not limiting in the HPLAC gels but might be so in the larger standard gel beads. Slow dissociation from the ligand can be a problem regardless of the gel. Peak shapes are often improved by reversing the direction of flow during desorption, since the distance in the column covered by the desorbed protein is minimized. This is called reversed elution.

Gradient elution often gives excellent results in affinity chromatography. A heterogeneity in the sample with regard to binding to the column is then utilized. For example isoenzymes can often be separated by gradient elution[57], and similarly glycoproteins can be separated according to affinity on immobilized lectins by a sugar gradient[58].

10.5.4.1 Change in pH or Ionic Strength

The most frequently used method for eluting strongly bound substances non-specifically is by decreasing the pH of the buffer[59,60]; sometimes an increase in pH can also be effective[61,62]. The chemical stability of the matrix, the ligand and the adsorbed substance determine how low in pH you can go. This is usually around pH 2–4. It is important to neutralize the fractions as soon as possible after elution. This is most easily done by pipetting a small volume of buffer in each tube of the fraction collector in advance.

An increase in the ionic strength of the buffer elutes proteins bound by predominantly electrostatic interactions[63,64]. Such interactions typically dominate binding to dye columns. Usually 1 M NaCl is sufficient but occasionally, 2 or 3 M salt is required[65]. Often continuous or stepwise salt gradients can resolve different proteins adsorbed to a particular dye column[66].

10.5.4.2 Change in Polarity

When the binding is very strong and dominated by hydrophobic interactions rather drastic methods of elution have to be used, such as reducing the polarity, or including a chaotropic salt or denaturing agent in the buffer. This type of elution is typical for immunosorbents based on immobilized polyclonal antibodies[67–69].

Commonly used chaotropic salts are KSCN, KCNO and KI in the concentration range 1 to 3 M. Urea and guanidine HCl in moderate concentrations (4–6 M) are sometimes to be preferred. The polarity can often be decreased enough to promote elution by including 20 to 40% ethylene-glycol in the buffer, although sometimes much higher values are required[70]. It is effective as well as being mild and less likely to denature proteins.

When eluting the most hydrophobic proteins, such as membrane proteins, decreasing the hydrophobic interactions with detergents is often to be preferred. Useful detergents are Lubrol, NP 40 or octylglycosides. Triton X-100 is disadvantageous because of its UV-adsorption. The concentrations to be used are just below the CMC (Table 1-5).

A low concentration of a detergent is sometimes included during the entire purification process (see Chapter 1 and Ref. 71). Thus, non-specific hydrophobic adsorption as well as aggregation can be suppressed—particularly in antigen-antibody purification or when handling membrane proteins. The detergents can also be left in the buffer when eluting with e.g. a moderate change in ionic strength or pH. An alternative is to increase the detergent concentration to achieve elution.

One example of an affinity system with very strong interactions is rat biotin-binding protein—biotin-AH-Sepharose[72]. This protein binds very tightly to the adsorbent and does not elute with saturating concentration of biotin (0.004 M) or with other desorption methods such as elevated temperature (40°C), 8 M urea, 8 M KI or low pH 3.6. However, the protein could be eluted by 3 M guanidine HCl or 2 M urea plus 0.004 M biotin in the equilibration buffer.

10.5.4.3 Specific Elution

In specific elution bound proteins are desorbed from the ligand by the competitive binding of the eluting agent either to the ligand or to the protein. Specific eluents are most frequently used with group specific adsorbents since the selectivity is greatly increased in the elution step. Glycoproteins can be desorbed from lectin columns by elution with competing carbohydrates[58,73]. Desorption of a lectin from an immobilized carbohydrate with a competing carbohydrate is described in application 10.6.2.1 and Ref. 74. Specific elution has also proved to be effective when isolating receptors[76].

Elution generally occurs at rather low concentrations of eluant (5–100 mM). A concentration of a single eluant or pulses of several different eluants can be used as well as gradients. Specific elution is often performed at neutral pH and is thus a mild desorption method causing little or no denaturation. If the eluting agent is bound to the protein it can be dissociated by desalting on a gel filtration column or, when the binding is stronger, by dialysis.

10.5.5 Regeneration of Adsorbent

All affinity adsorbents should be re-usable. The exact extent of re-use will depend on the nature of the sample and the stability of the ligand and matrix with respect to the elution and cleaning conditions used.

The most important aspect in regeneration is to remove any material still bound to the adsorbent. In most cases it is sufficient to re-equilibrate with several column volumes of starting buffer. If necessary, one can wash with buffers of alternately high and low pH and even include detergents and denaturing agents depending on the stability of the ligand. For ready-made adsorbents the manufacturer's instruction should be followed.

10.5.6 High Performance Liquid Affinity Chromatography, HPLAC

High performance liquid affinity chromatography (HPLAC) (for a review, see Ref. 21) is rapidly becoming a popular method both for the purification of proteins by affinity interaction and for analytical applications. HPLAC was introduced by Mosbach and co-workers[57] in 1978 by demonstrating separations on derivatized 10 μ macroporous silica gels. The reason for improved performance in adsorption chromatography using smaller beads is treated in Chapter 2. The limiting factor for many affinity systems seems not to be the adsorption kinetics but rather mass transport restrictions due to the slow diffusion in the pores of the gel particles[21,23].

Most often HPLAC is run using prepacked columns with different immobilized ligands, available from several manufacturers. Examples of such columns are Selecti-Spher-10™ Boronate, SelectiSpher-10™ Concanavalin A, SelectiSpher-10™ Protein A and SelectiSpher-10 Protein G (Pierce Chemical Company), Protein A Superose and Protein G Superose® (Pharmacia LKB Biotechnology AB), HiPAC™ Protein G (ChromatoChem), HPLAC columns with several ligands like biotin, Concanavalin A, Heparin etc. (available from Showa Denko K.K.).

Also commerically available are prepacked preactivated columns like SelectiSpher-10™ Activated Tresyl (Pierce Chemical Company) and Ultraffinity™ EP-columns (Beckman Instruments). Here the customer does the ligand immobilization *in situ*. Finally, it is possible to use an HPLC gel, both silica and cross-linked agarose, for activation and immobilization as described in sections 10.3 and 10.4. In this case one should be aware of the technical difficulties involved in obtaining well packed HPLC columns (see Section 2.5).

HPLAC has so far mainly been used for micropreparative and analytical purposes. Thus, cardiac myosin and actin were purified in one step using salicylate immobilized to a preactivated Ultraffinity EP-column[77]. The desorption was achieved by specific competitive elution and by increased ionic strength. Human IgG oligosaccharides were resolved by serial affinity chromatography on lectin-columns[78]. This case also nicely demonstrated the resolution of peaks obtained with small differences in k' during isocratic elution. Extremely fast (20 s or less) analytical separations have been obtained on affinity packings using minicolumns (<2 cm length)[79].

10.6   Applications

10.6.1   Immobilization of Ligands

Below are presented the experimental details of a number of commonly used procedures for the immobilization of different ligands to various matrices. Although most of the methods have been developed for beaded agarose as a matrix, they can often be used in combination with other polyol matrices provided that they retain their structure in the solvents used during the activation and coupling steps.

10.6.1.1   Some General Advice

Before and after activation and after coupling, the matrix is usually washed and equilibrated with different buffer solutions. These washing steps are most easily performed on a glass filter funnel fitted on a Büchner-flask (connected to a vacuum suction device). The matrix is suspended in a suitable volume of the desired medium which is subsequently filtered off or the reaction mixture is poured into the filter and the liquid with soluble components is sucked off. The matrix is resuspended in a new aliquot of medium and the liquid is again removed by filtration. This procedure is repeated the number of times required to remove unwanted soluble components or to equilibrate the matrix with a new medium. Allow ample time in each step for diffusion of soluble reagents and reaction products in and out of the beads. Often the filtration can be performed just by gravity flow or at a very small negative pressure. After each filtration a glass rod can be used to resuspend the matrix in a new portion of buffer.

Never let the matrix dry out completely on the filter during the filtration process (leave a few mm of liquid on top of the matrix surface) as this might lead to unwanted aggregation of the beads. This is especially critical when agarose is transferred from an aqueous to an organic medium such as acetone.

In order to obtain fast kinetics and efficient reactions the activation, coupling and deactivation should be carried out by suspending the non-reacted, activated or coupled matrix in a minimum volume of liquid containing the reagent, ligand and deactivation compound, respectively, to get a slurry that is dense but still can be agitated. As most beads, in particular agarose beads, are mechanically fragile agitation with stirrers, fleas and other harsh conditions should be avoided to minimize disintegration. For small volumes (up to 20 ml of suspension) end-over-end rotation can usually be employed by using sealed off tubes as reaction vessels (provided that there is no gas evolved as a result of the reaction). A shaking board can also be used.

When the coupling reactions are performed in a buffered medium it is essential to select buffers which do not react with the activated matrix in competition with the ligand to be immobilized. Thus, amino-containing buffers should be avoided when the ligand is coupled through an amino-function, carboxylic acids should not be used as buffers when the ligand is to be coupled via a carboxyl-group etc. When condensation agents such as carbodiimides are used, neither amines nor carboxyl containing buffers can be used. In most cases the reagents used for activation are highly toxic and often volatile.

It is therefore recommended that all work be performed in a ventilated fume hood until the washed and drained activated gel is incubated with the ligand. After coupling of the ligand, residual activated groups should be deactivated by reaction with an excess of a suitable low molecular weight compound (ethanolamin, mercaptoethanol etc.) under the same conditions as used for coupling the ligand.

The coupled matrix should then be washed very thoroughly to remove all non-covalently bound material. If possible, both alkaline and acidic buffers with varying ionic strength should be used. A tentative washing protocol could be:

1. 0.1 M sodium phosphate buffer, pH 8,
2. the same buffer with 0.5 M NaCl,
3. 0.1 M sodium acetate buffer, pH 5,
4. the same buffer with 0.5 M NaCl,
5. the buffer in which the adsorbent is to be stored or used in the affinity chromatography.

The procedure should, of course, be modified according to the special properties of the ligands. Thus, if an organic solvent has been used, the washing should start with this solvent. If necessary, an appropriate detergent can also be included in the washing buffers.

Until used, the prepared affinity adsorbents are best stored as suspensions at $+4°C$, the medium depending on stability of ligand, matrix and covalent bond employed. Preferably the suspension should also contain some anti-bacterial substance such as 0.02% sodium azide or merthiolate.

10.6.1.2 Immobilization of Amino-containing Ligands to Beaded Agarose by Means of the CNBr-Method

The method described is essentially according to Kohn and Wilchek[35], which is a modification of the original CNBr-buffer procedure originally presented by Porath et al.[34].

Activation step. Agarose gel (e.g. Sepharose 4B) is washed with distilled water. 10 g drained gel is then mixed with 10 ml distilled water and 20 ml of a 2 M sodium (or potassium) carbonate solution. The obtained suspension is cooled to 0°C by means of an ice-bath. An approximately 10 M solution of CNBr in acetonitrile, DMF or N-methyl pyrrolidone is prepared by dissolving 1 g of CNBr in 1 ml of organic solvent. This solution is added, all at once, to the gel suspension under vigorous agitation. After exactly 2 min the reaction mixture is transferred onto a glass filter funnel and washed with ice-cold water until all CNBr is removed. This washing procedure should be as quick as possible not to lose reactive groups by hydrolysis. The activated agarose beads should be used immediately for coupling. This procedure gives a coupling capacity for low molecular weight amines of about 10 μmoles/g drained agarose gel, which is about half the amount which can be obtained by the CNBr-titration procedure[33]. For proteins the differences are less pronounced. The method has the advantage of being reliable and easy to perform compared with the titration method which also requires an automatic titration equipment.

Coupling step. The ligand to be coupled is dissolved in 0.1 M $NaHCO_3$, pH 8.3, containing 0.5 M NaCl. This is the most commonly used coupling buffer. Other buffers with lower ionic strength in the range 8-10 (such as borate or phosphate buffers) can also be used. In this pH interval the amino-groups on the ligand are predominantly in the un-protonated and reactive form. The high salt concentration in the buffer (0.5 M NaCl) serves to minimize ionic protein-protein adsorption.

The amount of ligand added to the reaction mixture depends on what ligand density is desired. As a rule higher ligand concentration in the reaction medium leads to higher ligand content in the adsorbent. However, as discussed above, it might not always be advantageous to obtain a very high ligand content since this might have adverse effects on the affinity chromatography. For an efficient adsorbent 1-10 μmoles low molecular weight ligand per ml gel and for protein 5-10 mg protein/ ml gel is recommended. These figures can be obtained by adding two to three times excess of protein to the reaction mixture. A lower ligand concentration may in fact be more effective in e.g., immunoadsorbents, since it facilitates desorption.

The washed and drained activated gel is suspended in the ligand solution. If necessary, more buffer is added to make a slurry which can be efficiently mixed. The reaction mixture is then agitated for 2 hrs at room temperature ($+22- +25°C$) or overnight at $+4°C$. A number of residual active groups may remain on the gel after coupling (this is particularly the case after immobilizing high molecular weight ligands such as proteins). These groups can usually be removed by hydrolysis by leaving the gel for 2 hrs with Tris-HCl buffer, pH 8, or by addition of an excess of a small primary amine, e.g. ethanolamine, glycine or glutamic acid. The obtained gel product is finally washed and stored as described above.

10.6.1.3 Immobilization of Amino-containing Ligands to Beaded Agarose after Activation with the CDAP-Method[35]. (c.f. page 295)

Agarose gel (e.g. Sepharose 4B) is washed with water, then acetone:water (3:7) and finally with acetone:water (6:4). The gel is drained for a moment by mild suction (see above). 10 g of drained gel is transferred into a 50 ml glass beaker and 10 ml of acetone:water (6:4) is added. This should give a dense but easily stirrable slurry. The reaction mixture is cooled at 0°C. The desired volume of CDAP (CDAP is commercially available from e.g. Sigma Chemical Company, P.O.B. 14508, St Louis, MO 63178, U.S.A.), stock solution (see Table 10-9) is first added under vigorous stirring of the gel suspension and after 30 sec the corresponding volume of TEA solution is added dropwise over a period of 1–2 min.

The entire reaction mixture is then rapidly transferred into 200 ml ice-cold 0.05 M HCl. (This is to hydrolyze and remove from the gel the pyridinium isourea derivative which is formed as a byproduct in the activation. The active group, the cyanate ester, is stable toward dilute mineral acid and will thus not be effected). The gel is allowed to sediment for 15 min, is then washed on a glass filter funnel with ice cold water and is used for coupling immediately. Coupling of ligand as well as deactivation is performed as described above for the CNBr-buffer method.

With the CDAP-technique much higher degrees of activation can be obtained and thus more ligand, especially if low molecular weight, can be coupled to the agarose beads.

10.6.1.4 Immobilization of Amino-containing Ligands to Agarose with the N,N'-Disuccinimidyl Carbonate (DSC) Method

Essentially as described by Wilchek and Miron[38].

Beaded agarose (10 g of wet gel) is dehydrated (washed) and mixed slowly under agitation with 1.5 mmol of succinimidyl carbonate DSC (can be obtained from Sigma Chemical Co., St. Louis, MO, USA). To this suspension is added a 1.5-2-fold molar excess (with respect to DSC) of basecatalyst (either 0.38 ml TEA in 10 ml pyridine or 325 mg DAP, Dimethylamino pyridine (can be obtained commercially from e.g. Sigma) in 10 ml aceton slowly under agitation).

After agitation the suspension for 30–60 min at room temperature, the gel is washed successively with solutions of acetone, 5% acetic acid in acetone, methanol and 1 mM HCl (+4 C). If the gel is to be used within a few hours it can be stored in 1 mM HCl. As a suspension in acetone at +4°C the activated gel is stable for several weeks. The proper washing of the gel can be checked by diluting an aliquot of the methanol washings with 0.25 M NH_4OH. Remaining reagent will by this treatment turn into N-hydroxysuccinimide which can be detected photometrically at 260 nm as described above.

With the above given conditions an activated gel containing 20–40 μmol of hydroxysuccinimide carbonate groups/g gel is obtained. The degree of substitution depends among other things on the excess DSC used in the reaction.

Table 10-9. Amounts of CDAP and TEA employed for the activation of 10 g drained agarose (Sepharose 4B)[35].

Degree of activation	Approx. coupling capacity, μmole ligand/g gel)	0.1 g CDAP/ml dry acetonitrile (ml)	0.2 M aqueous solution of TEA (ml)
Weak	5	0.25[1]	0.2
Moderate	15	0.75[2]	0.6
Strong	30	1.50	1.2

Coupling of amino-containing ligands and proteins is performed at pH 6–9 by mixing the activated gel, after filtering of the 1 mM HCl or acetone, with the ligand dissolved in either 0.1–0.2 M phosphate buffer, pH 7.5, or fresh solutions of 0.1–0.2 M NaHCO$_3$, pH 8.3, for 4–16 hrs at $+4°C$. A ligand density of 37 μmoles/ml gel was obtained when aminocaproic acid was added to the activated gel suspension to give a final concentration of 0.5 M and 2.5 mg of protein A/ml gel was bound as a result of adding 3.0 mg of protein/ml gel. A higher degree of substitution is obtained by adding more protein to the activated gel. Deactivation is not necessary since the N-hydroxysuccinimide carbonate groups are rapidly hydrolyzed at the conditions of coupling.

10.6.1.5 Immobilization of Ligands to Beaded Agarose by the Tresylchloride Method

Essentially as described by Nilsson and Mosbach[37].

10 ml gel is washed with 3–4 volumes of water and then sequentially with acetone:water mixtures, 30:70, 60:40, 80:20 and finally with pure acetone; 10–20 ml per washing and 3–5 washings for each mixture.

The drained gel is suspended in 5 ml of dry acetone. Pyridine is added to twice the volume of the tresylchloride. Under agitation add dropwise 0.05–0.2 ml of tresylchloride (total addition time; 1 min). The reaction mixture is then agitated for another 10–15 min. Wash twice with 10 volumes of acetone and then wash sequentially with mixtures of acetone and 1 mM HCl (in water) 70:30, 50:50, 20:80 and finally with 3 volumes of 1 mM HCl.

The coupling is performed by suspending the gel in an equal volume of suitable buffer of pH 7.5–9.5, e.g. 0.1–0.2 M carbonate or phosphate, containing the ligand. The reaction mixtures is then agitated overnight at $+4°C$ or $+25°C$ depending on the stability of ligand. The gel is washed with the coupling buffer and is then resuspended in an equal volume of 0.1 M buffered ethanolamine, pH 7.5–8.5 and agitated for an additional 3–5 hrs at $+4°C$ or at room temperature. The obtained gel derivative is washed and stored as described above.

10.6.1.6 Immobilization of Amino- and Carboxyl-containing Ligands to Beaded Agarose Derivatives with Water-Soluble Carbodiimides

Essentially as described in Ref. 9, (see also Ref. 41–44).

An agarose derivative with carboxyl or amino groups substituted directly on the polysaccharide backbone or via a spacer is purchased or prepared. The gel should contain carboxyl groups if the ligand is an amine and amino groups if the ligand is to be coupled through a carboxyl-group. 10 ml of the selected drained agarose derivative is washed with 3–4 volumes of 0.5 M NaCl and then with distilled water. The gel is then transferred to a reaction vessel containing 5 ml of 0.04–0.1 M ligand dissolved in water or organic solvent (dioxane, ethyleneglycol, ethanol, methanol or acetone). The recommended concentration range will result in a molar excess, relative to groups on the gel, of low molecular weight ligands for which the method is recommended. Protein ligands are preferably immobilized by the technique based on activated matrix carboxyl groups, described below. The coupling is usually performed in unbuffered medium. In all cases, avoid using buffers containing amine, carboxyl or phosphate groups.

Adjust the pH to 4.7 with 0.1 M HCl or 0.1 M NaOH. Add 5.2 ml of 0.1 M EDCL (1-ethyl-3(3-dimethylamino-propyl) carbodimide hydrochloride) or CMCL (1-cyclohexyl-3-(2-morpholinoethyl)-carbodiimide-metho-p-toluene sulphonate) in water or mixed solvent. Up to 50% of organic solvents can be used when the ligand is poorly water-soluble. Maintain the pH at 4.5–5 for 1 hour (pH usually decreases under this time) by addition of dilute NaOH. The reaction mixture is agitated overnight at room temperature.

A blocking reaction is not usually necessary when an excess of ligand is used, but can be carried out with ethanolamine or glucose amine in the case of carboxyl-containing agarose and with acetic acid when the matrix is an amino agarose derivative.

The gel is washed and stored as described above.

10.6.1.7 Immobilization of Amino-containing Ligands to Activated Carboxyl Agarose

A modification of the procedure published by Cuatrecasas and Parikh[45].

10 ml of carboxyl containing gel is washed with deionized water and successively with 10×50 ml dioxane to remove all traces of water from the gel. The gel is then suspended in 15 ml of dioxane. 240 mg N-hydroxy-succinimide is added and agitation is performed until this compound is completely dissolved. 400 mg of cyclohexyl carbodiimide is added and agitation is continued for another 2 hrs. The gel is then washed with 4–10 volumes of dioxane and 4 times one volume of pure methanol (to eliminate the poorly soluble N,N'-dialkylurea derivative produced during activation). After washing the gel another 3 times with dioxane (or isopropanol) it can be stored as a suspension in a well-sealed vessel in the dark at $+4°C$ until used. This procedure usually gives a degree of activation of about 12 μmoles of ester groups/ml gel. The degree of substitution can be determined photometrically after release of N-hydroxy succinimide as described in section 10.4.1.4.

Before coupling of the ligand, excess solvent is removed. The gel is then suspended either directly or after being washed with 1 mM HCl in 100 ml of the chosen buffer at pH 5–9, e.g. 0.1 M NaHCO$_3$, pH 8.3 containing the ligand solution (if protein 2–20 mg protein per ml gel). The use of a lower pH has the advantage that hydrolysis of the active ester is minimized. Preferential reaction of α-amino groups, as opposed to ε-amino groups, can be obtained by coupling at low pH, due to the lower pK_a of α-amino groups. The reaction mixture is agitated for 10 min to 6 hrs at room temperature at $+4°C$ (coupling reaction is normally very rapid).

Excess activated groups usually hydrolyze at pH > 7, but blocking can also be performed by reacting the gel with e.g. 0.1 M ethanolamine buffered to pH 7.5–8.5 for 1 hr at $+4°C$.

The gel is washed and stored as described above.

10.6.1.8 Immobilization of Ligands to Beaded Agarose with Bisepoxirane and Epichlorohydrin

Essentially as described by Sundberg & Porath[27] and Porath & Fornstedt[36].

10 ml gel is washed with 3–4 volumes of deionized water and suspended in 5–10 ml of deionized water. One ml of the reagent (Bisepoxirane, e.g. 1.4-butane-diol-diglycidyl-ether, or epichlorohydrin) and 3 ml 2 M NaOH containing 20 mg of sodium borohydride are added under agitation. The activation is carried out during agitation at room

temperaure for 2 hrs. The activated gel is then carefully washed with deionized water until the reagents are completely removed.

The activated gel is suspended in an equal volume of a buffer of pH 9–13 (carbonate, borate or phosphate buffers can be used; higher pH for carbohydrate ligands) in which 0.5–1 mmoles/ml of small ligand and 5–10 mg/ml gel of macromolecular ligand is dissolved. If necessary up to 50% organic solvent (e.g. dioxane, DMF) may be used to dissolve the ligand. The mixture is agitated for 15–48 hrs at 20–45°C.

Oxirane groups not utilized for coupling of ligand are usually hydrolyzed at high pH. When lower pH (<10) has been used it may be necessary to block remaining oxirane groups with e.g. ethanolamine or mercaptoethanol as described above. The degree of activation obtained on 4% agarose beads with a bisepoxirane such as 1.4-bis-butane-diol-diglycidyl-ether is 10–20 μmoles of oxirane groups/ ml gel and with epichlorohydrin 30–40 μmoles/ml gel. In most cases a rather small number of these groups can be utilized for ligand immobilization.

The gel is then washed and stored as described above.

10.6.1.9 Immobilization of Ligands to Polyacrylamide with the Glutaraldehyde Method[28]

10 ml of gel is washed with 3 or 4 volumes of distilled water and then with 3–4 volumes of 0.5 M potassium phosphate buffer, pH 7.6 (other buffers with pH between 6.9 and 8.5 can also be used). The drained gel is transferred to a flask containing 100 ml of 25% aqueous glutaraldehyde, a treatment which also sterilizes the gel. Adjust the pH to 7.4. The suspension is agitated for 18 hrs at $+37°C$. The gel is then washed with 15–20 volumes of distilled water or 0.5 M phosphate buffer pH 7.7, and is then transferred to a flask containing 10 ml of 0.5 M potassium phosphate buffer, pH 7.6 in which the ligand, e.g. a protein, is dissolved. Buffers with pH from 6.9 to 8.5 containing 5–10 mg of protein per ml of gel should be used. After removal of free glutaraldehyde the activated polyacrylamide beads can also be stored at pH 7.7 and $+4°C$ for several days.

To remove remaining reactive aldehyde groups on the gel it is treated with 1 volume of 0.1 M buffered ethanolamine at pH 7.5 to 8.5 for 3 hrs at $+4°C$ (amino acids can also be used). Alternatively, the free aldehyde groups can be blocked by treatment with 1 volume of 0.1 M borate buffer, pH 8.5–9, containing 500 mg $NaBH_4$ for 15–20 min. (This treatment should not be used when the immobilized ligand contains disulphide bonds). The absolute amount of protein coupled to the polyacrylamide beads depends on the nature of the protein ligand and the excess of protein used in the reaction. 0.4–2 mg/ ml gel is typically obtained when 1–2 mg protein is added per ml of activated gel, that is, a yield of 20–100%. The gel is then washed and stored as described above.

10.6.1.10 Immobilization of Ligands to Silica via its Derivatization to γ-Glycidoxypropylsilica (Epoxy-Silica)[21]

10 g of Silica, Li Chrospher Si 1000, is briefly washed with 20% HNO_3, H_2O, 0.5 M NaCl, H_2O, acetone and ether and put into a 500 ml three-neck flask where it is dried for 4 hrs at 150°C under vacuum. The reaction flask is then cooled and sodium dried toluene (150 ml) is sucked into the flask. 2.5 ml of γ-glycidoxypropyl trimethoxy silane (Dow Corning Z6040) and 0.05 ml trimethylamine are added and the reaction mixture is agitated by an overhead stirrer and refluxed for 16 hrs, while a slow stream of dry nitrogen gas will ensure anhydrous conditions. The formed epoxy-silica is then washed on a glass filter with toluene, acetone and ether and dried under vacuum. The procedure gives an epoxy (oxirane) group content of about 50 μmoles/g as determined with the

method described above for bisoxirane and epichlorohydrine activated agarose derivatives.

Amino and thiol containing ligands can be coupled directly to the obtained silica derivative according to the procedure described above[21]. pH higher than 8 should not be used. This means that the coupling, at least for amino ligands, will be rather slow and if possible should be speeded up by using high concentrations of ligand and, if possible, elevated temperature.

The coupling is finished by converting excess epoxy-groups to more hydrophilic entities. The preferred way is acid hydrolysis, provided that the ligand is stable under these conditions, which will result in a diol structure.

Heating of γ-glycidoxypropyl-silica to 50°C at pH 2 for 3 hrs is enough for complete hydrolysis of the epoxide groups. Excess oxirane groups can also be converted into more hydrophilic structures by treating the silica derivative with 1 M mercapteothanol at pH 8.5 in room temperature for 2 hrs. This method is more suitable after immobilization of proteins, provided these do not contain disulphides.

Other ligand coupling methods described above such as the DSC and tresylchloride methods can also be used if the epoxy-silica is first converted to diol-silica by acid hydrolysis. This is performed by mild agitation of 10 g of epoxy-silica in 1000 ml of 0.01 M HCl at 90°C for 1 hr. Diol-silica can also be obtained commericaly from, e.g., E Merck AG. FRG (Li Chrospher DIOL).

10.6.2   Affinity Chromatography

10.6.2.1   Group-specific Adsorption and Biospecific Elution. Purification of Kinases and Dehydrogenases from Crude Yeast on Blue Sepharose CL-6B[75]

Affinity gels with group specificity have the potential advantage of being useful for the isolation of many compounds belonging to a particular group. This technique was nicely demonstrated by Easterday et al.[75], who purified enzymes from crude yeast extract on the dye column Blue Sepharose CL-6B and eluted them by competitive elution with enzyme-specific co-factors at different pH.

Dried baker's yeast was extracted in 1 M Na_2HPO_4 for 3 hrs at $+37°C$ and then centrifuged at $13700 \times g$ for 1 hr. After filtration and supernatant was precipitated with $(NH_4)_2SO_4$ (75% saturation) at 40°C. After centrifugation 220 mg of the precipitate was dissolved in 10 ml of starting buffer (0.02 M Tris/HCl, pH 6.4, containing 5 mM $MgCl_2$, 0.4 mM EDTA and 2 μM 2-mercaptoethanol). Blue Sepharose CL-6B was packed in a column with the dimensions 1.6×5 cm (bed volume 10 ml) and equilibrated with starting buffer before application of the sample. A peak of inactive material was eluted with starting buffer (Fig. 10-4). 5 mM NAD^+ and 20 mM $NADP^+$ dissolved in starting buffer were used to elute alcoholdehydrogenase (ADH) and glucose-6-phosphate dehydrogenase (Glu-6-PO_4-DH), respectively. Hexokinase (HK) was eluted when the pH of the eluent was raised to 8.6 and glyceraldehyde-3-phosphate dehydrogenase (Gly-3-PO_4-DH) was eluted with 10 mM NAD^+ at the same pH.

10.6.2.2   Group-specific Adsorption, Specific Competitive Elution and Elution with Decreased pH. Purification of Lectin from Falcata Japonica on a N-acetyl-D-galactoseamine-Sepharose 6B[74]

Ground seeds (20 g) were suspended in 200 ml of 0.01 M PBS (pH 7.4), stirred overnight at 4°C and then centrifuged at $2300 \times g$ for 15 min. The precipitate was

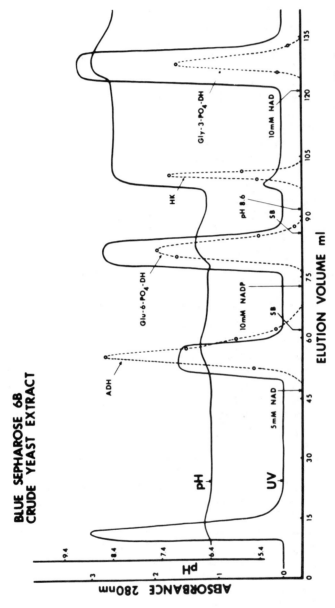

Figure 10-4. Purification of kinases and dehydrogenases from crude yeast extract by affinity chromatography on Blue Sep CL–6B. Reprinted with permission from Easterday et al.[75].

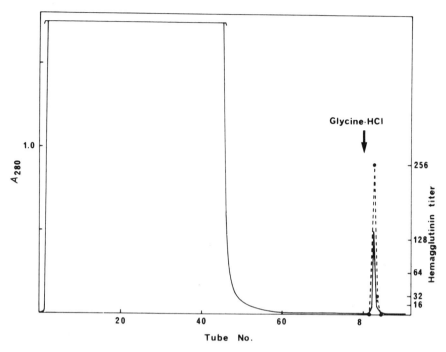

Figure 10-5. Specific elution of GalNAc-binding lectin from GalNAc-Sepharose 6B. Reprinted with permission from Nakajima *et al.*[74].

extracted with a half volume of starting PBS. Pooled supernatants were ultracentrifuged to remove the lipid materials. The crude extract (200 ml) was applied to a column of N-acetyl-D-galactoseamine (GalNAc) coupled to epoxi-activated Sepharose 6B (1.6 × 5.0 cm, 10 ml bed volume). The column was washed with 0.001 M PBS (pH 7.0) until the absorbance of the elute at 280 nm was less than 0.05. The bound lectin was desorbed by a specific competitive elution adding 25 mM GalNAc to the starting buffer (Fig. 10-5). Similar result was obtained by non-specific elution by decreasing the pH to 2.2 by glycine-HCl buffer. A 1000-fold purification was achieved in one step.

10.6.2.3   Group-specific Adsorbent. Binding of Immunoglobulins in Mouse Serum to Protein A Sepharose. Elution of IgG Subclasses by Step-wise Decrease in pH[81]

Protein A is a group-specific ligand with affinity for many different immunoglobulins[84]. The subclasses can be eluted separately according to binding strength by a step-wise decrease of pH.

All subclasses of mouse IgG can be purified on Protein A Sepharose[84,85]. Even IgG$_1$, which is known to have low affinity for Protein A[86–88], binds efficiently under certain buffer conditions[81]. The two parameters normally changed to effect binding of proteins to affinity adsorbents are pH and ionic strength. Biewenga *et al.*[89] studied the influence of these two parameters during binding of human myeloma IgA and human polyclonal IgG. They found that binding of IgA decreased with decreasing pH. At salt concentrations up to 2 M NaCl IgA binding decreased, while IgG binding was contant.

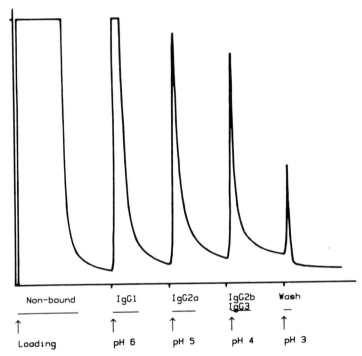

Figure 10-6. Elution of different IgG subclasses in mouse serum from a column of protein A Sepharose CL-4B by stepwise decreasing pH. Reproduced with permission from Fredriksson *et al.*[81].

Fredriksson *et al*,[81] studied the effect of ionic strength on the binding of mouse monoclonal antibodies of different IgG subclasses to Protein A Sepharose at high pH. They found that a high ionic strength (3 M NaCl) is necessary for efficient binding of IgG. This binding buffer is used (Fig. 10-6) to adsorb polyclonal immunoglobulins from mouse serum to the Protein A gel.

1 ml of Protein A Sepharose CL-4B was packed in a column with the bed dimensions 1 × 1.2 cm and run at a flow rate of 0.8 ml/min. The column was equilibrated with binding buffer (1.5 M glycin, 3 M NaCl adjusted to pH 8.9 with 5 M NaOH). 5 ml mouse serum was diluted with 5 ml of binding buffer and applied to the column. The column was then eluted using a series of buffers (0.1 M citric acid adjusted to pH 6.0, 5.0, 4.0 and 3.0). These pH values can also be used to elute the different IgG subclasses when purifying monoclonal mouse antibodies and thus extremely acid conditions are not necessary.

10.6.2.4 Specific Adsorption and Specific Elution at Neutral pH. Removal of Non-Specifically Bound Material by Increased Ionic-strength. Purification of Catechol O-Methyltransferase[80] on S-adenosyl-L-homocystein-Sepharose

Catechol O-methyltransferase (COMT) is a very labile protein. Previous methods for purification of this enzyme were often laborious and time-consuming. Veser *et al.*[80] described a rapid and specific purification method which combines ion exchange

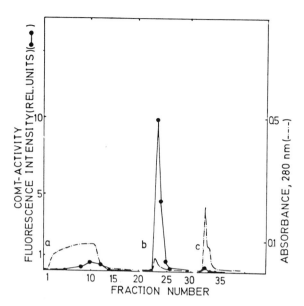

Figure 10-7. Affinity chromatography of catechol O-methyl transferase on a column of S-adenosyl-L-homocystein immobilized to AH-Sepharose 4B. Reprinted with permission from Veser *et al.*[80].

chromatography and affinity chromatography. A low molecular weight ligand, S-adenosyl-L-homo-cystein (AdoHcy) immobilized to a spacer arm-containing preactivated gel, AH-Sepharose 4B, was used as affinity adsorbent.

COMT was prepared from rat liver and partially purified by ion exchange chromatography[80]. AdoHcy-linked AH-Sepharose was packed into a column (1 × 10 cm, bed volume 8 ml) and equilibrated with 4 mM phosphate buffer, pH 6.4, containing 0.02 % sodium azide, 0.2 mM magnesium chloride, 1 mM mercaptoethanol. 0.3 mM dithio-threitol and 2 % glycerol. The COMT I-containing fractions from the ion exchange chromatography were pooled and dialyzed against the equilibration buffer mentioned above and applied to the affinity-column at a flow rate of 0.75 ml/min. The enzyme was eluted with equilibration buffer containing 0.1 mM S-adenosyl-L-homocysteine (AdoMet) (Fig. 10-7). The enzyme acticity could as well be eluted by a small increase in pH from 6.4 to 7.4. The column was regenerated and non-specific bound material eluted with 0.1 M NaCl in equilibration buffer. The recovery of the enzyme after the affinity chromatography step was 95 %.

10.6.2.5 Specific Adsorption. Non-specific Elution by Increased pH. Purification of the Angiotensin-converting Enzyme from Human Heart by Affinity Chromatography on an Immobilized Inhibitor[82]

The angiotensin-converting enzyme from human heart was isolated by a procedure including two chromatographic steps. After extraction the enzyme was partially purified by a batch adsorption to DEAE-cellulose followed by affinity chromatography on N-(1(S)-carboxy-5-amino-pentyl)-Gly-Gly-Sepharose linked to Sepharose 6B via a spacer arm. The affinity adsorbent (0.9 × 8.5 cm, bed volume 5 ml) was equilibrated with

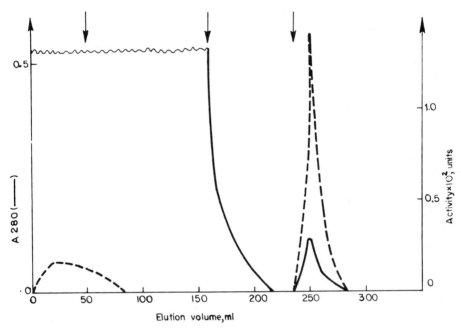

Figure 10-8. Affinity chromatography of human heart angiotensin converting enzyme on immobilized N(1-(S)-carboxy-5-aminopentyl)Gly-Gly Sepharose 6B. Elution by increase of pH to 8.9. Reproduced with permission from Sahkarov *et al.*[82].

20 mM Mes pH 6.0, containing 0.5 M NaCl, 0.1 mM zinc acetate and 0.1 % Nonidet P-40 at a flow rate of 8 ml/h. The enzyme was eluted by a buffer change to 50 mM sodium borate, pH 8.9 (Fig. 10–8).

10.6.2.6 Analytical HPLAC (High Performance Liquid Affinity Chromatography)[83]

This application demonstrates the use of HPLAC in an analytical mode for simultaneous detection of human serum albumin (HSA) and IgG in a single sample of serum by a dual-column system (Fig. 10-9). Reference 83 also presents a general scheme for design and optimization of such a multianalyte affinity system. The ligands chosen, Protein A and anti-HSA antibodies, were immobilized to activated diol-bonded Li-Chrospher Si-4000 and Si-500. (For a detailed description of the activation, see Refs. 21 and 90, and for the immobilization, see Ref. 83.) Quantitation of serum samples was performed on two minicolumns (6.35 mm length × 4.1 mm i.d.) connected in series and a 10 μl injection loop. The application buffer was 0.05 M phosphate, 0.05 M citrate buffer, pH 7. The elution buffer was 0.05 M phosphate, 0.05 M citrate, pH 3. The serum sample was diluted 1:5 with application buffer prior to injection. The anti-HSA column was placed before the Protein A-column to avoid non-specific adsorption of albumin to the Protein A-adsorbent (the details of this optimization experiment are given in Ref. 84.) Therefore, after the non-retained peak had been eluted from both columns the Protein A-column was switched off-line and the anti-HSA column was eluted with pH 3 buffer. After the HSA had been eluted, the protein A column was switched on line to elute the IgG. Standards with HSA and IgG were analysed. This method gave results in good agreement with commerically available methods, while requiring only 2 μl of serum and 6 min per cycle.

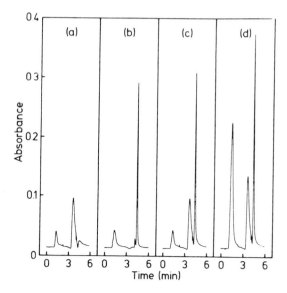

Figure 10-9. Chromatograms obtained after injections of (a) HSA, (b) IgG, (c) HSA plus IgG and (d) normal serum into the dual column system. The event sequence used was 0.00 min, switch from pH 3 to 7 buffer; 0.50 min, sample injection; 2.25 min, protein A column switched off-line, switch to pH 3 buffer; 4.00 min, protein A column switched on-line. Reproduced with permission from Hage and Walters[83].

10.6.2.7 Immunosorption. Use of the Avidin-Biotin System for Correct Orientation of the Immobilized Antibody. Non-specific Elution by Decreased pH and Chaotropic ions[91]

This application describes how the avidin-biotin system can be used to prepare immunosorbents on glass beads with directed immobilization of the antibodies. The methods described by Babashak and co-worker apply both the conventional (IAC) and high performance immuno affinity chromatography (HPIAC). Monoclonal antibodies (Mabs) were biotinylated with biotin hydrazine. This reagent couples the biotin to the carbohydrate moieties of the antibodies. The carbohydrate part of most antibodies is present in the F_c region and attachment of biotin thus ensures correct orientation of the antibody. For details of the biotinylation procedure, see Ref. 91. For silanilization and derivatization of the glass beads see Refs. 91 and 92. The streptavidin form of avidin was immobilized to the activated glass beads by incubation at pH 9 for 18 hrs at 4°C. The biotinylated Mabs were then attached to the streptavidin coated beads by incubation for 1 hr at 4°C in PBS on an overhead mixer. The beads were washed (PBS, pH 7) and slurry packed into 10 cm × 4.6 mm I.D. stainless steel columns at 250 p.s.i. The immunosorbent was used to isolate the B 27 human leucocyte antigen component from detergent solubilized human leucocyte membranes. The B 27 antigen is retained on the column, while most of the membrane components are washed through it (Fig. 10-10). The bound antigen is eluted by decreased pH or by chaotropic ions. The binding between steptavidin-biotin is so strong that no elution of biotinylated Mabs occurs in these conditions. More than 20 batches of streptavidin coated glass beads were produced and used in HPIAC. The derivatized beads bound between 1.5 and 1.85 mg of streptavidin per 2 g

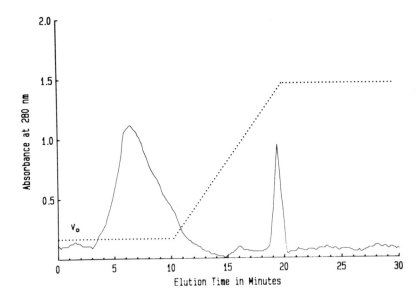

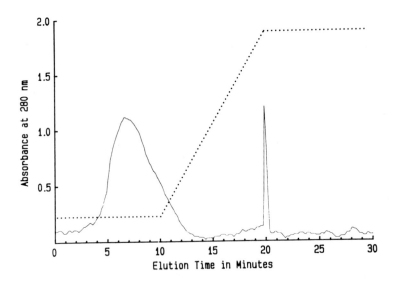

Figure 10-10. Chromatograms of an HPLAC isolation of the B27 antigen isolated from detergent solubilized membranes. The sample, 100 μl, was applied in 0.9% sodium chloride—0.1 M sodium acetate buffer at pH 6.5 and non-adsorbed material was washed off the column before elution started. In (a) the antigen was desorbed by a pH gradient form pH 6.5 to 1.0 by the addition of 0.1 M hydrochloric acid to the initial running buffer. In (b) the antigen was eluted by a chaotropic ion gradient 0 to 2.5 M sodium thiocyanate. Reproduced with permission from Babashak *et al.*[91].

batch of beads and the streptavidin beads bound between 195 and 245 μg of hydrazine biotinylated antibodies. Running the chromatograms at $+4°C$ is preferable to room temperature (both regarding peak performance and bead life time). Also, chaotropic elution is preferable to acid elution since the bead life time increases two-fold.

10.7 References

1. M. Wilchek, T. Miron, J. Kohn, In *Meth. Enzymol. 104*, ed. W. B. Jacoby, Academic Press, New York, 1984, pp. 3–55.
2. W. H. Scouten, *Affinity Chromatography*, John Wiley & Sons, New York, 1981.
3. J.-C. Janson, T. Kristiansen, In *Packings and Stationary Phases in Chromatographic Techniques*; K. K. Unger, ed.; Marcel Dekker: New York, 1989, in press.
4. R. H. Allen, P. W. Majerus, *J. Biol. Chem. 247*, 7709 (1972).
5. D. G. Deutsch, E. T. Mertz, *Science 170*, 1095–1096 (1970).
6. E. A. Bayer, M. Wilchek, *Meth. Biochem. Anal., 26*, 1 (1980).
7. See Data Sheet on Benzamidine-Sepharose 6B from Pharmacia LKB Biotechnology AB, Uppsala, Sweden, for relevant references.
8. P. D. G. Dean, F. A. Middle, C. Longstaff, A. Bannister, J. J. Dembinski, In *Affinity Chromatography and Biological Recognition*; I. M. Chaiken, M. Wilchek, I. Parikh, eds.; Academic Press, Orlando, 1983, 433–443.
9. Affinity Chromatography: Principles and Methods; Handbook published by Pharmacia LKB Biotechnology AB, Uppsala, Sweden.
10. K. Mosbach, In *Adv. in Enzymology*; A. Meister, ed.; John Wiley & Sons, New York, 1978, *46*, 205–278.
11. G. Kopperschläger, H.-J. Böhme, E. Hofmann, In *Adv. in Biochemical Engineering*; A. Fiechter, ed.; Springer Verlag, Berlin, Heidelberg, 1982, *25*, 101–138.
12. R. K. Scopes, *J. Chromatogr. 376*, 131–140 (1986).
13. C. R. Lowe, S. J. Burton, J. C. Pearson, Y. D. Clonis, V. Stead, *J. Chromatogr. 376*, 121–130 (1986).
14. J. Porath, M. Belew, *Trends in Biotechnol. 5*, 225–229 (1987).
15. M. Vuento, A. Vaheri, *Biochem. J. 183*, 331 337 (1979).
16. M. M. Andersson, H. Borg, L. O. Andersson, *Thromb. Res., 5*, 439–452 (1974).
17. J. W. Goding, *Monoclonal Antibodies: Principles and Practice, 2nd ed.*; Academic Press, London, 1986, Chapter 6.
18. H. A. Chase, B. J. Horstmann, S. L. Fowell, *J. Chem. Technol. Biotechnol.* 1989, in press.
19. P. Cuatrecases, M. Wilchek, C. B. Anfinsen, *Proc. Natl. Acad. Sic. USA, 61*, 636 (1968).
20. H. Schott, *Affinity Chromatography, Chromatogr. Sci., Series Volume 27*; Marcel Dekker, New York, 1984.
21. P.-O. Larsson, M. Glad, L. Hansson, M.-O. Månsson, S. Ohlson, K. Mosbach, In *Adv. in Chromatogr.*; C. A. Giddings, ed. Marcel Dekker, New York 1983, *21*, 41–85.
22. K. K. Unger, In *Porous Silica, J. Chromatogr. Library*, Elsevier, Amsterdam, 1979, p. 16.
23. J.-G. Gustavsson, T. Ottosson, K. Wiberg, P. Hedman, In *Proc. 4th European Congress on Biotechnology*. O. M. Neijssel, R. R. van der Meer , K. C. A. Luyben eds. 1987, *2*, 585–588.
24. R. R. Harris, J. J. Rewe, *FEBS-Lett, 29*, 189 (1979).
25. P. Cuatrecasas, *Nature, 228*, 1327 (1970).
26. C. R. L. Lowe, In *Affinity Chromatography*. John Wiley Intern. Publ. New York 1974, p. 28.
27. L. Sundberg, J. Porath, *J. Chromatogr. 90*, 87 (1974).
28. T. Ternyck, S. Avrameas, *FEBS-Lett. 23*, 24 (1972).
29. J. Porath, T. Låås, J.-C. Janson, *J. Chromatogr. 103*, 49–62 (1975).
30. M. Wilchek, T. Oka, Y.J. Topper, *Proc. Nat. Acad. Sci. 72*, 1055–1058 (1975).
31. J. Porath, In *Meth. in Enzymol* B. Jakoby, M. Wilchek eds. Academic Press, New York 1974, *34*, 13–30.
32. R. Axén, J. Porath, S. Ernback, *Nature, 214*, 1302 (1967).
33. R. Axén, S. Ernback, *Eur. J. Biochem. 18*, 351–60 (1971).
34. J. Porath, K. Aspberg, H. Drevin, R. Axén, *J. Chromatogr. 86*, 53 (1973).
35. J. Kohn, M. Wilchek, *Appl. Biochem. Biotechnol. 9*, 285–304 (1984).
36. J. Porath, N. Fornstedt, *J. Chromatogr. 51*, 479 (1970).

37. K. Nilsson, K. Mosbach, In *Meth. in Enzymol.* K. Mosbach, ed. Academic Press, New York 1987, *135*, part B, 65–78.
38. M. Wilchek, T. Miron, *Appl. Biochem. Biotechnol. 11*, 191–193 (1985).
39. M. T. W. Hearn, In *Meth. in Enzymol.* K. Mosbach, ed. Academic Press, New York 1987, *135*, part B, 102–117.
40. G. S. Bethell, J. Ayers, W. S. Hancock, M. T. W. Hearn, *J. Biol. Chem. 254*, 2572 (1979).
41. A. Tengblad, *Biochem. J. 199*, 297 (1981).
42. S. L. Marcus, E. Balbinder, *Anal. Biochem. 48*, 448–459 (1972).
43. H. Anttinen, K. I. Kivirikko, *Biochem. Biophys. Acta, 429*, 750–758 (1976).
44. D. Robinson, N. C. Phillips, B. Winchester, *FEBS-Lett. 53*, 110–112 (1975).
45. P. Cuatrecasas, I. Parikh, *Biochem. 11*, 2291 (1972).
46. J. Brandt, A. Svensson, J. Carlsson, H. Drevin, *J. Solid Phase Biochem. 2*, 105–109 (1977).
47. R. M. K. Dale, D. C. Ward, *Biochem. 14*, 2458–2469 (1975).
48. M. Lynn, In *Enzymology of Immobilized Enzymes Antigens, Antibodies and Peptides.* H. H. Weetall, ed. 1975, *1*, 1.
49. R. Axén, P. Vretblad, J. Porath, *Acta Chem. Scand, 25*, 1129–1132 (1971).
50. I. Ugi, *Angew. Chem, 74*, 9 (1962).
51. L. Goldstein, In *Meth. in Enzymol.*, K. Mosbach, ed., Academic Press, New York 1987, *135*, 90–102.
52. J. L. Guesdon, S. Avrameas, *J. Immunol. Meth. 11*, 129 (1976).
53. J. K. Inman, In *Meth. in Enzymol.* W. B. Jakoby, M. Wilchek, eds. Academic Press, New York 1974, *34*, 30.
54. M. B. Wilson, P. K. Nakane, *J. Immunol. Meth. 12*, 171 (1976).
55. M. Wilchek, R. Laurel, In *Meth. in Enzymol.* W. B. Jakoby, M. Wilchek, eds. Academic Press, New York 1974, *34*, 475.
56. I. Parikh, S. March, P. Cuatrecasas, In *Meth. in Enzymol.* W. B. Jakoby, M. Wilchek, eds. Academic Press, New York 1974, *34*, 77–102.
57. S. Ohlson, L. Hansson, P.-O. Larsson, K. Mosbach *FEBS-Lett. 93*, 5–9 (1978).
58. J. Woodward, H. J. Marquess, C. S. Picker, *Preparative Biochem. 16*, 337–352 (1986).
59. D. S. Secher, D. C. Burke, *Nature, 285*, 446–450 (1980).
60. W. J. Jankowski, W. von Muenchhawson, E. Sulkowski, M. A. Carter, *Biochem. 15*, 5182 (1976).
61. W. M. Moore, C. A. Spilburg, *Biochem. 25*, 5189–5195 (1986).
62. K. Miyata, Y. Yamamoto, M. Ueda, Y. Kawade, K. Matsumoto, I. Kubota, *J. Biochem. 99*, 1681–1688 (1986).
63. L. Kiss, A. Tar, S. Gál, B. L. Toth-Martinez, F. J. Hernádi, *J. Chromatogr. 448*, 109–116 (1988).
64. G. Vlatakis, G. Skarpelis, I. Stratidaki, V. Bouritis, Y. D. Clonis, *Appl. Biochem. Biotechnol. 15*, 201–212 (1987).
65. S. P. J. Brooks, V. D. Bennett, C. H. Suelter, *Anal. Biochem. 164*, 190–198 (1987).
66. L. Miribel, P. Goldschmidt-Clermont, R. M. Galbraith, P. Arnaud, *J. Chromatogr. 363*, 448–455 (1986).
67. Y. Kitagawa, E. Okuhara, E. Shikata, *J. Virol. Meth. 16*, 217–224 (1987).
68. P. T. Swoveland, *J. Virol. Meth. 13*, 333–341 (1986).
69. S. J. Busch, C. R. Duvic, J. L. Ellsworth, J. Ihm, J. A. K. Harmony, *Anal. Biochem. 153*, 178–188 (1986).
70. J. M. F. G. Aerts, W. E. Donker-Koopman, G. J. Murray, J. A. Barranger, J. M. Tager, A. W. Schram, *Anal. Biochem. 154*, 655–663 (1986).
71. T. J. Palker, M. E. Clark, M. G. Sarngadharan, T. J. Matthews, *J. Virol. Meth. 18*, 243–256 (1987).
72. P. B. Seshagiri, P. R. Adiga, *Biochim. Biophys. Acta, 916*, 474–481 (1987).
73. J. C. Zwaagstra, G. D. Armstrong, W.-C. Leung, *J. Virol. Meth., 20*, 21–32 (1988).
74. T. Nakajima, S. Yazawa, T. Kogure, K. Furukawa, *Biochem. Biophys. Acta, 964*, 207–212 (1988).
75. R. L. Easterday, I. M. Easterday, In *Immobilized biochemicals and affinity chromatography*, R. B. Dunlap, ed. Plenum Press, New York, 1974, 123–133.
76. J. Ramwani, R. K. Mishra, *J. Biol. Chem. 261*, 8894–8898 (1986).
77. M. W. Strohsacker, M. D. Minnich, M. A. Clark, R. G. L. Shorr, S. T. Crooke, *J. Chromatogr. 435*, 185–192 (1988).
78. H. Harada, M. Kamei, Y. Tokumoto, S. Yui, F. Koyama, N. Kochibe, T. Endo, A. Kobata, *Anal. Biochem. 164*, 374–381 (1987).
79. R. R. Walters, *Anal. Chem. 55*, 1395–1399 (1983).
80. J. Veser, W. May, *Chromatographia, 22*, No. 7–12 404–406 (1986).

81. U. B. Fredriksson, L. G. Fägerstam, A. W. G. Cole, E. Waldén, Pharmacia LKB Biotechnology AB, Poster presented at Sixth Int. Congr. Immunol. Toronto, Canada, (1986).
82. I. Y. Sahkarov, S. M. Danilov, E. A. Dukhanina, *Biochim. Biophys. Acta, 923,* 143–149 (1987).
83. D. S. Hage, R. R. Walters, *J. Chromatogr., 386,* 37–49 (1987).
84. J. J. Langone, *J. Immunol. Meth., 55,* 277–296 (1982).
85. P. L. Ey, S. J. Prowse, C. R. Jenkin, *Immunochem., 15,* 429–436 (1978).
86. M. P. Chalon, R. W. Milne, J.-P-, Vaerman, *Scand. J. Immunol., 9,* 359–364 (1979).
87. C. L. Villemez, M. A. Russell, P. L. Carlo, *Mol. Immunol., 21,* 993–998 (1984).
88. L. Manil, P. Motté, P. Pernas, F. Troalen, C. Bohuon, D. Bellet, *J. Immunol . Meth., 90,* 25–37 (1986).
89. J. Biewenga, F. Daus, M. L. Modderman, G. M. M. L. Bruin, *Immunol. Comm. 11*(3), 189–200 (1982).
90. D. S. Hage, R. R. Walters, H. W. Hethcote, *Anal. Chem., 58,* 274 (1986).
91. J. V. Babashak, T. M. Phillips, *J. Chromatogr., 444,* 21–28 (1988).
92. R. R. Walters, In *Affinity Chromatography,* P. D. G. Dean, W. S. Johnson and F. A. Middle eds., IRL. Press, Washington, D.C., 1985, p. 25.
93. Heparin-Sepharose® CL-6B For affinity chromatography; Handbook published by Pharmacia LKB Biotechnology AB, Uppsala, Sweden.

11 Affinity Partitioning of Proteins using Aqueous Two-Phase Systems

Göte Johansson
Department of Biochemistry
University of Lund
Chemical Centre
S-221 00 Lund
Sweden

11.1 Introduction

The methods for separation described in this chapter are based on partition of proteins between two water-rich phases. The two phases, in contact with each other, yield a so-called aqueous two-phase system. The water content of the phases is, in general, in the order of 70–95 % and the two-phase systems are obtained by dissolving two polymers in water[1,2]. Another possibility to achieve two water-rich phases is to mix an aqueous solution of a polymer with a suitable salt, e.g. sodium phosphate[1]. In the following, the discussion will be limited to systems based on dextran, polyethylene glycol (PEG), and water.

A protein, included in the two-phase system, is after equilibration (15–60 sec of careful mixing) found more or less in both phases. The relative distribution of the protein between the phases, i.e. its partition, can be described by a partition coefficient, K, defined by Eq. (11-1):

$$K = \frac{C_U}{C_L} \qquad (11\text{-}1)$$

where C_U and C_L are the concentrations of the protein in the upper and lower phase, respectively. Two proteins can be resolved from a mixture by partition in a system in which their K values differ enough. Usually several partition steps are necessary to get an acceptable resolution unless the K values are extreme (e.g. $K_1 < 0.01$ and $K_2 > 100$).

Aqueous two-phase systems can be made selective in their properties of extracting certain proteins. One way to achieve this goal is to localize an affinity ligand (for the target protein) in one of the phases. Ligands used in this way include fatty acids[3], triazine dyes[4,5], coenzymes[6] and more specialized ligands[7,8].

The aqueous two-phase systems were introduced by Albertsson as a tool for separation of a vast number of cell components ranging from proteins to cell organelles[1]. Also it was possible to fractionate various forms of micro-organisms by partition within the two-phase system. In contrast to proteins and nucleic acids the particles are not only recovered in the bulk phases but are also found at the interface between the two phases.

11.2 General Properties

11.2.1 Composition of Two-phase Systems

The composition of the two phases, in a dextran-PEG-water two-phase system, depends both on the percentage of the polymers (usually given in per cent weight by weight), the molecular weights of the polymers and the temperature[2]. The phase composition can be described by a phase diagram (Fig. 11-1) where the points indicating the polymer concentrations for the upper ($\circ$) and lower ($\bullet$) phase are both on the border line (the so-called binodial curve) separating one-phase and two-phase regions. The phase diagrams are useful tools for finding systems with the desired properties.

Systems with total compositions along one of the straight lines (tie-lines) in Fig. 11-1, generated by the composition points of the two phases, will differ in the relative amount of upper and lower phase (but not in their composition). The volume ratio is related to the length of the sections of the tie-line (l_l and l_r in Fig. 11-1) when this is "divided" by the point ($+$) representing the total composition. This relation is given by Eq. (11-2):

$$\frac{V_U}{V_L} = \frac{l_r}{l_l}\frac{d_L}{d_U} \qquad (11\text{-}2)$$

where d_U and d_L are the densities of the upper and lower phase, respectively. Since the densities are nearly equal the expression above can in many cases be approximated by Eq. (11-3)

$$\frac{V_U}{V_L} = \frac{l_r}{l_l} \qquad (11\text{-}3)$$

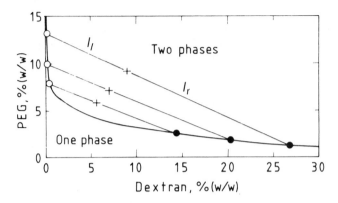

Figure 11-1. Phase diagram for the system dextran (M_w = 500,000), PEG (M_r = 3000–3700), and water at 0°C. The compositions of the upper phases (○) and the lower phases (●) are shown for systems of various total composition (+). Reproduced from Ref. 1 with permission.

The phase diagram (Fig. 11-1) also shows that if the total concentrations of dextran and PEG are both increased the two phases will differ more and more in their compositions. The phases will consequently be increasingly pure in respect to the content of their dominating polymer. If an affinity ligand is attached to one of the polymers it will therefore be increasingly concentrated in one phase the more the total composition is removed from the binodial curve.

11.2.2   Influence of Electrolytes

The pH value and the presence of electrolytes in the system has a pronounced effect on the partition of proteins between the two phases. Salts commonly used in biochemical procedures, such as phosphates, chlorides or Tris-HCl can greatly influence partition[2,9]. The salts themselves have partition coefficients close to one.

The effect of salts on the partition increases with the distance from the binodial curve (increasing percentage of the two polymers) and also depends on the temperature[1,2]. Normal salt concentrations are in the range 25–250 mM.

Stronger effects than those obtainable with salt have been achieved by using charged groups covalently bound to PEG, e.g. sulphonate or trimethylamino groups[10,11]. This is an example of the use of one phase as a liquid ion exchanger. To get effective extractions in this case the content of salt must be kept at a low level, usually well below 10 mM.

11.2.3   Theory for Effect of Salts on Partition of Proteins

The partition of a protein is influenced by the presence of salts and this effect increases with the net charge of the protein. This has led to the idea that a small portion of the ions of the salt, present in excess, form an interfacial double-layer at the interface between the phases. This, in turn, gives rise to an interfacial potential which affects the partition of the charged protein. If the partition coefficient of a non-charged protein ($\text{pH} = Ip$) is K_0, an adjustment of the pH value, which gives the protein a net charge Z, will in most cases influence the partition. The partition coefficient, K, of the protein at this pH value is given by Eq. (11-4)

$$K = K_0 \cdot (K_+/K_-)^{-Z/(m+n)} \tag{11-4}$$

The corresponding logarithmic expression is:

$$\log K - \log K_0 = -Z/(m+n)(\log K_+ - \log K_-) \tag{11-5}$$

where K_+ and K_- are the (hypothetical) partition coefficients of the cation (A^{m+}) and the anion (B^{n-}) of the salt present in excess. These are the partition coefficients the ions should have had if they could partition independent of the electric field caused by the neighbourhood ions. The coefficients for a two-phase system composed of 8% (w/w) dextran 500 and 8% (w/w) PEG 3400 at 20°C are given in Fig. 11-2.

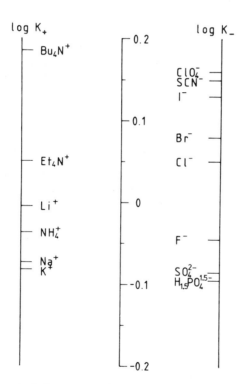

Figure 11-2. The (hypothetical) non-constrainal partition coefficients for a number of cations (K_+) and anions (K_-) in a system containing 8% (w/w) dextran ($M_w = 500,000$), 8% (w/w) PEG ($M_r = 3000$–3700), and 20–25 mM salt at 25°C. The partition coefficient of a salt K_s is, in the case of 1–1 electrolytes, equal to $\sqrt{K_+ \cdot K_-}$, or $\log K_s = \frac{1}{2}(\log K_+ + \log K_-)$. Adapted from Ref. 12, reproduced with permission.

When K_+ is larger than K_- the upper-phase-side of the interface will have a positive charge and it increases the K values for negatively charged proteins while the K values are lowered for positively charged proteins. This is the case when e.g. lithium phosphate buffer is used. If, on the other hand, $K_+ < K_-$ (e.g. with KBr) the charge-dependence is the opposite. As can be seen in Eq. (11-4) the charge-dependent partition, $\log K - \log K_0$ (for a given charge Z) is proportional to $\log K_+ - \log K_-$.

11.2.4   Influence of Molecular Weights of Polymers

Changes in the molecular weights of the phase-forming polymers influence both the K_0 value of a protein and the interfacial potential. A decrease in molecular weight of a polymer changes the non-charge-dependent partition towards the phase rich in this polymer.

11.3   Affinity Partitioning

An affinity ligand (for a protein) can be restricted to one of the phases by attaching it to the main polymer of this phase. Usually only a fraction of the phase-forming polymer is used as ligand carrier.

11.3.1   Preparation of Polymer-Bound Ligand Derivatives

A number of methods for binding of ligands to watersoluble polymers have been described[2,13,14]. Ligands with reactive groups, e.g. reactive textile dyes, can be bound directly, as will be described below. Otherwise reactive groups can be introduced on the polymer, e.g. tresyl-, tosyl-, bromo- or carboxyl groups. The first three groups can, under mild conditions, be replaced with affinity ligands containing amino, mercapto or phenolic groups. Alcohols and amines form ester and amide bonds, respectively, with polymers carrying carboxyl groups by using carbodiimides as water-accepting reagent. By reacting the activated polymers with diamines (in excess) spacers with free amino groups for further binding are introduced. An increasing number of PEG derivatives are now commercially available (e.g. from Sigma Chemical Co). A number of ligands used for affinity partitioning of proteins are given in Table 11-1.

11.3.2   Partition of Ligand-Polymers

Normally not more than 5 per cent of a polymer is used as ligand carrier. When the two constituents of the ligand-polymer derivative are in the molar ratio 1:1 the partition of the polymer-bound ligand is comparable with that of free polymer. Higher degrees of substitution may influence the partition strongly, especially if the ligand carries ionic

Table 11-1. Examples of ligands used for affinity partitioning of proteins.

Ligand	Extracted protein	Ref.
p-Aminobenzamidine	Trypsin	7
Palmitate	Serum albumin	3
Dinitrophenol	S-32 myeloma protein	9
Estradiol	Oxosteroid isomerase	15, 16
Lecithin	Colipase	17
Triazine dyes	Dehydrogenases and kinases	4, 5, 18–22
Nucleotides	Dehydrogenases and kinases	6

Table 11-2. Partition of phosphofructokinase from bakers' yeast in systems containing various concentrations of the two phase-forming polymers dextran ($M_w = 500{,}000$) and PEG ($M_r = 7000$–9000). The systems also contained 50 mM sodium phosphate buffer, pH 7.0, 5 mM mercaptoethanol, and 0.5 mM EDTA. Temperature 0°C. The affinity partitioning effect, log K_{aff}, was caused by exchanging 3% of PEG for Cibracron blue F3G-A PEG (Cb-PEG). Adapted from Ref. 23, produced with permission.

System		Logarithmic partition coefficients		
% w/w dextran	% w/w PEG	log K_{aff}	log $K_{Cb\text{-}PEG}$	log K_{PEG}
4.80	3.20	1.4	0.59	0.24
5.25	3.50	2.0	0.75	0.48
6.00	4.00	2.4	1.02	0.78
6.75	4.50	2.8	1.26	1.00
7.50	5.00	3.4	1.43	1.14
8.25	5.50	3.4	1.63	1.22
9.00	6.00	3.6	1.80	1.27

groups. The partition of the ligand-polymer will be more extreme at higher concentrations of the phase-forming polymers, as has been shown by using Cibacron blue F3G-A bound to PEG, Table 11-2. The partition coefficient, $K_{Cb\text{-}PEG}$, is furthermore larger than the one for PEG itself. The effect of increasing the degree of substitution is illustrated by dextran-bound Procion yellow HE-3G (Table 11-3). Also changes of the molecular weight of the bulk dextran affect the partition of the ligand-dextrans.

11.3.3 Theory for Affinity Partitioning

The effect of the polymer-bound ligand on the partition of a protein can be described by the relative change in the partition coefficient of the protein. If the partition without ligand is given by K^* the partition in the presence of ligand, K, is described by Eq. (11-6):

$$K = K_{aff} \cdot K^* \tag{11-6}$$

Table 11-3. Partition of Procion yellow HE-3G dextran (PrY-dextran) in systems composed of PEG 8000 and dextrans of various molecular weights and 50 mM sodium phosphate buffer, pH 7.9. Temperature, 22°C. Degree of substitution (n) gives the number of dye molecules per dextran molecule. Reproduced by permission from Ref. 24.

Molecular weight of dextran	Composition of system		K of PrY-dextran			
	% w/w dextran	% w/w PEG	$n = 1.3$	$n = 2.3$	$n = 5.3$	$n = 8.3$
40,000	8.00	5.00	0.15	0.30	2.0	17.5
70,000	8.00	4.50	0.23	0.50	4.1	28.2
500,000	5.00	3.80	0.97	1.62	6.2	24.3
2000,000	5.00	4.00	1.03	1.83	7.5	22.5

K_{aff} is the factor quantifying the affinity partitioning effect. K_{aff} increases with concentration of PEG-bound ligand until it reaches a saturation value (Fig. 11-3).

A theoretical model for affinity partitioning has been presented by Flanagan and Barondes[8]. It states that K_{aff}, in systems with excess of polymer-bound ligand is related to $K_{ligand-PEG}$ by Eq. (11-7)

$$K_{aff} = n \cdot K_{ligand-PEG} D_U/D_L \qquad (11-7)$$

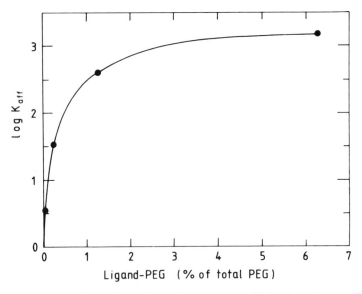

Figure 11-3. Log K_{aff} of phosphofructokinase when an extract of bakers' yeast was partitioned in the system 6% (w/w) dextran ($M_w = 70,000$), 4% (w/w) PEG ($M_r = 35,00\text{–}40,000$), 25 mM sodium phosphate buffer, 5 mM mercaptoethanol, and 0.25 mM EDTA at 0°C, pH 7.0, with increasing concentration of Cibacron blue F3G-A PEG ($M_r = 6500$).

where n is the number of binding sites for ligand on the protein molecule and D_U and D_L are the dissociation constants in upper and lower phase, respectively.

11.3.4   Effects of Salt, Free Ligands and pH Value

The affinity partitioning may be affected by the presence of salt in at least two ways. Firstly the ligand-protein interaction might be weakened (if it is of electrostatic character) or strengthened (if hydrophobic) by increasing the salt concentration. Secondly, if the ligand carries ionized groups the ligand-PEG will act, at low salt concentration, as a liquid ion-exchanger and extract proteins of the opposite net charge. The salt concentration will therefore often be chosen as a compromise to minimize these two effects, with values between 10 and 150 mM. The effects of a number of salts on the partition of phosphofructokinase are shown in Fig. 11-4. Salts may be used to weaken

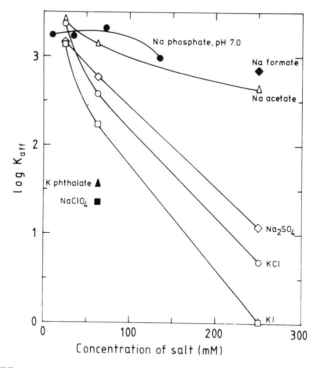

Figure 11-4. Effects of various salts on the affinity partitioning of phosphofructokinase from bakers' yeast. The system contained 7% (w/w) dextran ($M_w = 500,000$), 5% (w/w) PEG ($M_r = 6000$–7500), 10 mM sodium phosphate buffer, pH 7.0, 0.5 mM EDTA, 5 mM mercaptoethanol, and 4 nkat enzyme/g system, together with the salt, at 0°C. The affinity partitioning effect, log K_{aff}, was obtained by replacing 3% of PEG with Cibacron blue F3G-A PEG. Adapted from Ref. 23, reproduced with permission.

unwanted ligand-protein interaction in order to make the affinity extraction more selective.

Temperature also affects the affinity partitioning and this has to be checked for every ligand-protein pair. Other factors, such as pH, may play an important role. When the ligand is a trizine dye the maximal K_{aff} value increases with decreasing pH values[25] but at the same time the selectivity of the extraction is generally reduced.

The molecular weights of dextran and PEG may in some cases affect the K_{aff} values[23] when systems with equal distance from the binodial curves are compared. This

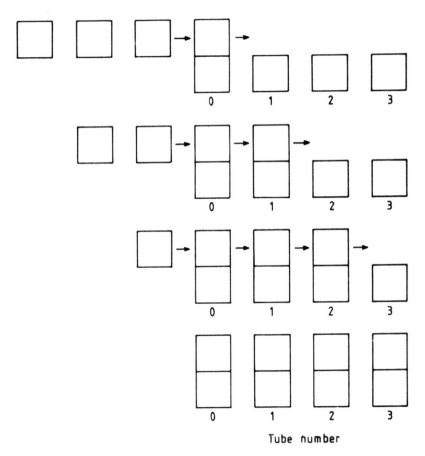

Tube number

Figure 11-5. Principle for counter-current distribution. The sample to be separated is included in a two-phase system (No 0 at the top). The system is equilibrated and when the phases have separated the top phase is transfered to a fresh bottom phase, yielding system No 1. A fresh top phase is also added to the bottom phase of system No 0. After equilibration and settling of the phases the transfer of the top phases is repeated. A counter-current distribution with n transfers will give rise to $n + 1$ systems (No 0—n). Water can be added to the system to obtain single phase fractions which are analysed.

has been demonstrated for phosphofructokinase from bakers' yeast. In several other cases only slight dependence on the molecular weights have been observed.

11.3.5   Counter-Current Distribution of Cell Extracts

The purification obtained by preparative extractions in one or a few steps is suited for isolation of one component from a mixture. Counter-current distribution (CCD) is a procedure where the two phases stepwise are moved over each other. It often allows a number of proteins to be fractionated simultaneously. The principle of CCD is explained in Fig. 11-5. The process can be carried out manually by transfering upper phases of a row of systems which are equilibrated and allowed to separate after each transfer. A large number of transfers makes this operation tedious. Instead ordinary CCD apparatus (according to Craig) can be used, or even better, the apparatus especially constructed for aqueous two-phase sytems[26,27]. After the CCD run water can be added to get one phase. The amount necessary can be determined from the phase diagram. The measurements of protein or enzyme activities along the CCD tubes show distribution curves resembling those of a chromatogram. An example of a manual CCD with 7 transfers is shown in Fig. 11-6. An automatic CCD with 55 transfers is shown in Fig. 11-7.

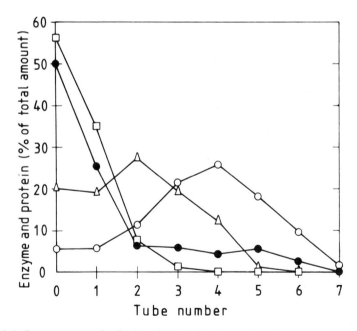

Figure 11-6. Counter-current distribution of extract from bakers' yeast using seven transfers. The sample was loaded in system No 0 and the upper phases have then been moved stepwise to the right. PEG-bound ligand (in the upper phase) was Cibacron blue F3G-A. Composition of system: 7.5% (w/w) dextran ($M_w = 70,000$), 4.2 (w/w) PEG ($M_r = 35,000$–$40,000$), 0.8% (w/w) ligand-PEG ($M_r = 6500$), 12.5 mM potassium phosphate buffer, pH 6.2 at 0°C. ●, Protein; ○, 3-phosphoglycerate kinase; □, hexokinase; and △, glyceraldehydephosphate dehydrogenase.

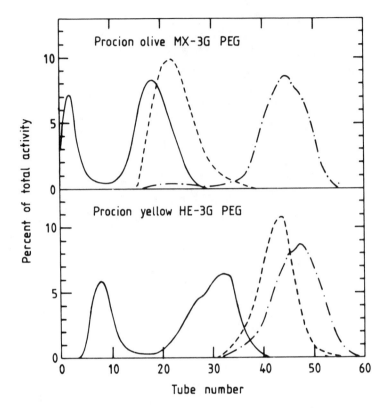

Figure 11-7. Counter-current distribution with 55 transfers of an extract from bakers' yeast. The graphs show the distribution of hexokinase (———), 3-phosphoglycerate kinase (– – –), and phosphofructokinase (– · – · –) when either Procion olive MX-3G or Procion yellow HE-3G are used as PEG-bound affinity ligands. System composition: 7 % dextran 500, 5 % PEG 8000 (including dye-PEG, 1 % of total PEG) and 50 mM sodium phosphate buffer, pH 7.0, 0.2 mM EDTA and 5 mM 2-mecaptoethanol. Temperature 3°C. Adapted from Ref. 28 with permission.

11.4 Practical Procedures

11.4.1 Strategy

The following steps are recommended:

1. Check the partition of the target protein (e.g. an enzyme) and of total protein in one chosen system (without ligand).
2. By varying parameters such as polymer concentration, salt and pH you can see if some purification can be obtained without ligand.
3. Bind a ligand to the polymer occupying the phase into which you want to extract your target protein.

4. Determine the extraction curve by using increasing concentrations of the ligand-polymer.
5. Adjust the system parameters at saturated amounts of ligand-polymer in order to get the most effective purification. Parameters are temperature, polymer concentration, sample concentration, pH and salt (type and concentration).
6. Adjust the volume ratio of your system to allow a reasonable yield of the target protein in the collecting phase. Test if repeated "washings" of this phase with opposite pure phases increases the purification.
7. Carry out the extraction in the scale you find suitable.

11.4.2 Preparation of Phase Systems

The phase systems are prepared from stock solutions of dextran, PEG, salts and buffer. Dextran with molecular weight of 500,000 (dextran 500, 20% solution) and PEG with molcular weight 8000 (40% solution) are normally used. Dry dextran usually contains 4–5% (W/W) of water. Dextran solutions, for which accurate concentrations might be of value, can be analysed by polarimetry using the specific rotation (using sodium light) $[\alpha] = 199$ degree ml dm^{-1} g^{-1}. For partitions of proteins the following system may be tried: 8% dextran 500 and 6% PEG 8000. All the percentage values are in W/W. The systems also contains 25 mmole sodium phosphate buffer per kg (approximately equal to 25 mM).

Systems of 4–20 g weight can be weighed out accurately enough on a top-loaded balance. Also smaller volumes can be handled with a special technique. The preparation of a 4 g system, of the composition suggested above, is given here:

To a 10 ml graduated centrifugation tube is added,

1.6 g 20% dextran 500
0.6 g 40% PEG 8000
0.2 ml 0.5 sodium phosphate buffer, pH 7.5 water to total 3.00 g
1.00 g protein extract is added.

If the protein extract contains buffer or salt this has to be taken into account. The system is thermostated and then carefully mixed by inverting the closed tube for at least 20 sec. After centrifugation (1′ or more at 1000xg) or standing for 10–20 min the phases separate and samples are withdrawn for analysis. The partition coefficients for total protein and for the target protein are determined.

11.4.3 Preparation of Ligand Polymers

The ligand is usually bound to PEG and the target protein will then be extracted into the upper phase.

Also dextran can be used as ligand carrier to extract proteins to the lower phase. Several detailed review articles have summarised the synthetic methods which have been used[2,13,14].

A popular choice of ligands for enzymes[4,5,18–25] and some proteins present in blood plasma[29,30] has been the triazine dyes which are commercially available in activated form (as textile dyes). They react with both PEG and dextran in alkaline water solution.

The PEG derivative is extracted from salt-containing water solution into chloroform and (after evaporation of solvent) purified by treatment of DEAE-cellulose. The dextran derivative is purified by precipitation. The synthetic steps of dye-polymer derivatives using Cibacron blue F3-GA are given below and will apply to most other related dyes. Also other polymers might be derivatized in this way with minor modifications[24,31].

Cibacron blue F3G-A PEG 8000: 100 g of PEG is dissolved in 150 ml of water. 3 g of Cibacron blue F3-A is added under stirring followed by 50 ml 5 M NaOH and 10 g crystallized sodium sulfate. The mixture is mechanically stirred for 7 hours, neutralized with acetic acid and dialyzed overnight to remove salts. The polymer solution (containing unsubstituted PEG, dye-PEG and free dye) is diluted to 1 litre and 100 g of DEAE cellulose (Whatman DE52, neutralized and intensively washed) are added. The mixture is stirred for one hour. Dye-PEG and free dyes both bind to the ion exchanger while unsubstituted PEG stays in solution. The ion exchanger is collected by suction filtration and washed with water. The dye-PEG is then eluted (without suction) with a 2 M potassium chloride solution. The free dyes remain on the ion exchanger. The dye-PEG is extracted from the salt solution by repeated treatments with chloroform (500–800 ml). The pooled chloroform fractions are pooled and dried with anhydrous sodium sulphate and filtered. Finally the solvent is removed by evaporation.

Procion yellow HE-3G dextran 500: 9 g dextran 500 is dissolved in 180 ml water at 70°C and 1 g of Procion yellow HE-3G is added. When the dye has dissolved, 18 ml of a solution containing 1 M NaOH and 0.5 M sodium sulfate is added and the mixture is kept at the same temperature for 45 min. The solution is neutralized with acetic acid and the dextran is precipitated by slow addition, under stirring, of 300 ml ethanol. The precipitate is dissolved in 100 ml of water and again precipitated with 300 ml ethanol. After dissolving the precipitate in 100 ml of water, the remaining free dye is removed by addition of 5 ml 2 M KCl and 30 ml wet DEAE-cellulose (Whatman DE52). The mixture is stirred for 3 hours and the ion exchanger is removed by filtration and the filter cake is washed with 10 ml 0.1 M KCl which is combined with the main filtrate. The treatment with DEAE-cellulose is repeated once and the dye-dextran containing solution is then dialyzed against water. If necessary the solution can be concentrated by evaporation in vacuum.

A very important point for ligand-polymers, in general, is the purity. Also minute amounts of free ligand in the system may considerably decrease the efficiency by competing in binding to the protein. The purity can be checked by thin-layer chromatography[30]. Another useful procedure is to determine the K_{aff} values obtained with the ligand-polymer after it has been further purified (e.g. by recrystallization from methanol, gel chromatography, or ion exchanger).

11.4.4 Protein Measurements

Protein determinations are assayed according to Bradford by using Coomassie Brillant blue G staining in solution[32]. In contrast to other methods the presence of polymers does not have a strong negative influence on the assay. Another possibility is to determine the absorbance at 280 nm. In both cases the phases are diluted 5–100 times before the assay and phases from a protein-free system are used as blanks.

11.5 Applications

11.5.1 Isolation of Phosphofructokinase from Yeast[19]

Baker's yeast (100 g) is ground in a household mixer (with rotating knives) together with 400 g of crushed dry ice (solid CO_2) for 4 min. After evaporation of the carbon dioxide the remains are mixed with 100 ml 50 mM sodium phosphate buffer, pH 7.1, containing EDTA (0.5 mM), 2-mercaptoethanol (5 mM) and phenylmethyl-sulphonyl fluoride (PMSF, 0.5 mM). PEG 8000 is added to a concentration of 4 % (w/w) at 3°C, the mixture is stirred for 15 min and then centrifuged at 500xg for 20 min. The supernatant is collected and the PEG content is increased (from 4%) to 8 % (w/w), incubated and centrifuged as above. The phosphofructokinase (PFK) is enriched in the pellet. Purification 6–7 times relative to protein.

The pellet is dissolved in 15 ml of the buffer used above. The solution is included in a 50 g two-phase system containing (totally) 7.5 % dextran 500, 5 % PEG 8000, 25 mM sodium phosphate buffer (with the additives mentioned above). The system is equilibrated at 3°C by mixing for 30 sec and the phases are allowed to settle, eventually speeded up by centrifugation. The upper phase (containing some proteins with high K values) is removed and replaced with a fresh upper phase where 5 % of the total PEG consists of Cibacron blue F3G-A PEG 8000. The system is equilibrated and the phases are separated. The PFK is now in the upper (blue) phase while bulk proteins mainly stay in the lower phase. The upper phase is "washed" by equilibration with a new fresh lower phase. The collected upper phase is warmed to 25°C and 12 ml of a 40 % solution of potassium phosphates (KH_2PO_4 and K_2HPO_4 in molar ratio 1:1.2) in water are added. A PEG-salt system is formed containing PFK in the lower salt-rich phase. Purification in the affinity partition step: 8–9 times.

Pure PFK is obtained from the salt phase by passing it through a Sephadex G-25 column (using 25 mM buffer with additives as above) to remove salts. The protein fraction is mixed with 1–2 g of DEAE-cellulose for 15 min (at 3°C) and washed on a small filter (or in a column) with the buffer. The PFK is eluted with 200 mM KCl (in the same buffer). As a final step the PFK is purified on a Sepharose 4B-CL column. The total purification is now 142 times. Proteolytic activity (not inhibited by PMSF) is reduced by a factor of 2200.

11.5.2 Isolation of Glucose-6-Phosphate Dehydrogenase

Glucose-6-phosphate dehydrogenase (G6PDH) can be purified from baker's yeast in a similar way[21]. The following modifications are recommended:

1. The precipitation with PEG 8000 is carried out between 6.5 and 12.5 % (w/w).
2. Two-phase partition is carried out in a system containing 10 % dextran 500 and 7 % PEG 8000. For affinity partitioning Procion yellow HE-3G PEG (5 % of total PEG) is used. For the washing with lower phase Cibacron blue F3G-A dextran (2 % of total dextran) may be included to increase the effectivity. More effective than the addition of phosphates (to separate enzyme from dye-PEG) is to add a fresh (dextran) lower phase together with NADP (2.5 g per litre system) and sodium sulphite (0.5 g per liter system). G6PDH is then recovered in the lower phase. Total purification: 43–44 times relative to protein.

3. The lower (G6PDH containing) phase is diluted with 4 volumes of a solution containing 5 mM sodium phosphate buffer, pH 7.5, 75 mM KCl, 2 mM 2-mercaptoethanol, 0.5 mM EDTA and 2 mM magnesium sulfate. The solution is passed through a bed of DEAE-cellulose. When dextran and non-bound proteins have been washed away the enzyme is eluted by increasing the concentration of potassium chloride to 170 mM. Total purification: 330 times.

11.5.3   One-Step Isolation of Lactate Dehydrogenase (LDH) from Swine Muscle[22]

Swine muscle (100 g) is cut in small pieces and homogenized in a household mixer (with rotating knives) with 330 ml ice-cold 40 mM sodium phosphate buffer, pH 7.9, for 5 min. The mixture is centrifuged at 16000xg for 20 min. The supernatant is included in a one kg system of the final composition: 10% (w/w) dextran 500, 7.03% (w/w) PEG 8000, 0.07% Procion yellow HE-3G PEG 8000, and 50 mM sodium phosphate buffer, pH 7.9. The system is equilibrated at 3°C and the phases are separated. The LDH-containing upper phase is washed twice with fresh lower phases from systems with the same composition (but not containing protein). The upper phase is mixed with half its weight of 50% phosphate solution ($NaH_2PO_4.H_2O$ + NaH_2PO_4 is molar ratio 1:1) which has a temperaure of 40°C. The obtained salt phase contains practically pure LDH.

11.6   References

1. P.-Å. Albertsson, *Partition of cell particles and macromolecules* Almqvist & Wiksell, Stockholm, 2 ed. (1971).
2. H. Walter, D. E. Brooks, and D. Fisher *Partitioning in Aqueous Two-Phase Systems: Theory, Methods, Uses, and Applications in Biotechnology*, Academic Press, Orlando, 1985.
3. V. P. Shanbhag, G. Johansson, *Biochem. Biophys. Res. Commun. 61*, 1141–1146 (1974).
4. H. K. Kroner, A. Cordes, A. Schelper, M. Morr, A. F. Bückmann, and M.-R. Kula in *Affinity chromatography and related techniques* (T. C. J. Gribnau, J. Visser, and R. J. F. Nivard eds.). Elsevier, Amsterdam, 1982, pp. 491–501.
5. G. Kopperschläger, G. Lorenz, E. Usbeck, *J. Chromatogr. 259*, 97–105 (1983).
6. M.-R. Kula, G. Johansson, A. F. Bückmann, *Biochem. Soc. Transact. 7*, 1–5 (1979).
7. G. Takerkart, E. Segard, M. Monsigny *FEBS Lett. 42*, 218–220 (1974).
8. S. D. Flanagan, S. H. Barondes, *J. Biol. Chem. 250*, 1484–1489 (1975).
9. H. Walter, G. Johansson, *Anal. Biochem. 155*, 215–242 (1986).
10. G. Johansson, *Biochim. Biophys. Acta 222*, 381–389 (1970).
11. G. Johansson, A. Hartman, P.-Å. Albertsson, *Eur. J. Biochem. 33*, 379–386 (1973).
12. G. Johansson, *Acta Chem. Scand. B-28*, 873–882 (1974).
13. J. M. Harris, *J. Macromol. Sci. C-25*, 325–373 (1985).
14. A. F. Bückmann, M. Morr, G. Johansson, *Makromol. Chem. 182*, 1379–1384 (1981).
15. A. Chaabouni, E. Dellacherie, *J. Chromatogr. 171*, 135–143 (1979).
16. P. Hubert, E. Dellacheire, J. Neel, E.-E. Baulieu, *FEBS Lett. 65*, 169–174 (1976).
17. C. Erlandson-Albertsson, *FEBS Lett. 117*, 295–298 (1980).
18. G. Johansson, M. Andersson, *J. Chromatogr. 303*, 39–51 (1984).
19. G. Kopperschläger, G. Johansson, *Anal. Biochem. 124*, 117–124 (1982).
20. J. Schiemann, G. Kopperschläger, *Plant Scie. Lett. 36*, 205–211 (1985).
21. G. Johansson, M. Joelsson, *Enzyme Microb. Technol. 7*, 629–634 (1985).
22. G. Johansson, M. Joelsson, *Applied Biochem. Biotechnol. 13*, 15–27 (1986).
23. G. Johansson, G. Kopperschläger, P.-Å. Albertsson, *Eur. J. Biochem. 131*, 589–594 (1983).
24. G. Johansson, M. Joelsson, *J. Chromatogr. 393*, 195–208 (1987).
25. G. Johansson, *Methods Enzymol. 104*, 356–364 (1984).

26. P.-Å. Albertsson, *Anal. Biochem. 11*, 121–125 (1965).
27. H.-E. Åkerlund, *J. Biochem. Biophys. Meth. 9*, 133–141 (1984).
28. G. Johansson, M. Andersson, H.-E. Åkerlund, *J. Chromatogr. 298*, 483–493 (1984).
29. G. Birkenmeier, G . Kopperschläger, *Mol. Cell. Biochem. 73*, 99–110 (1987).
30. G. Birkenmeier, G. Kopperschläger, G. Johansson, *Biomed. Chromatogr. 1*, 64–77 (1986).
31. G. Johansson, M. Joelsson, *J. Chromatogr. 411*, 161–166 (1987).
32. M. M. Bradford, *Anal. Biochem. 72*, 248–254 (1976).

III Electrophoresis

12 Electrophoresis in Gels

Torgny Låås

Research & Development
Pharmacia LKB Biotechnology AB
S-751 82 Uppsala
Sweden

12.1 Introduction

The word "electrophoresis" derives from Greek and means "carried by electricity". It was initially used to describe the behaviour of electrically charged colloidal particles in an electric field. The migration of true solutes was originally referred to as "ionophoresis". Eventually electrophoresis became the recognized term for the migration of all kinds of particles in an electrical field[1].

Our interest here is the use of electrophoresis for the separation of proteins. The sample containing the proteins to be separated is placed in an electric field which forces the electrically charged proteins to move. If the experiment is performed in free solution (free electrophoresis) and the proteins have different charge densities they will move at different speeds and can thus be separated. In practice, however, the separation is normally performed not in free solution but in a supporting gel medium. The gel can either act as an "inert" support for the electrophoresis buffer or actively participate in the separation by interacting with the proteins. In the latter case the protein-gel interaction is the actual separation factor while the electrical field merely makes the proteins migrate through the gel.

All these techniques, whatever determines the separation, are collectively called electrophoresis. A prefix indicating the separation principle or the separation medium used is often added (e.g. affinity electrophoresis, starch gel electrophoresis).

Note that nomenclature is not consistent here. The prefix sometimes designates the gel media, sometimes the type of samples that are being separated and sometimes even the characteristics of the electric field. In practice this seldom causes any problem. The meaning is normally obvious from the context. However, there are two very special electrophoretic techniques with independent names. *Isoelectric focusing* (also named electrofocusing) and *isotachophoresis* (also named displacement electrophoresis, steady state electrophoresis, multiphasic zone electrophoresis ...).

Isoelectric focusing (IEF), which separates proteins according to differences in isoelectric point, is of such importance in protein chemistry and has so many special characteristics that it will be dealt with separately (Chapter 13).

Isotachophoresis, on the other hand, has not gained a wide acceptance for protein separations except for the use of its principles for sample concentration by "stacking" in disc-electrophoresis (12.3.3.2) and in the elution of proteins and nucleic acids from gels after electrophoresis (12.3.2).

Today there is such a wealth of techniques available both for preparative and analytical separations that even the experienced worker sometimes finds it difficult to choose the best for a particular situation. A few general statements might therefore be appropriate at this point.

For *preparative purposes* the use of gel electrophoresis is hampered by several factors. First, in large scale applications the joule heat formed causes severe problems. Second, it is difficult to recover the separated fractions without zone contamination and/ or dilution. In view of the efficiency and relative ease of scaling up high resolution chromatographic techniques, chromatography should always by the first choice for standard preparative work.

However, "preparative" may now refer to much smaller amounts than before. Many techniques such as amino acid analysis and sequencing only need ng quantities of protein! In such cases the conventional analytical electrophoretic separations can readily provide the amounts necessary. The problem with preparative gel electrophoresis is then reduced to detection of the substance of interest and transferring it out of the gel for further treatment. With electroelution, the protein can be extracted from the gel and at the same time concentrated (12.3.2). Recovery of proteins after gel electrophoresis is further described in Chapter 16. For pure *analytical purposes* on the other hand, many of the electrophoretic variants have demonstrated outstanding properties. Electrophoresis is the method of choice for gathering information about the composition of a crude sample. Furthermore, approximate estimates of M_r and isoelectric point together with a distribution profile of M_r and pI of contaminating proteins will form a good basis for a purification strategy. Especially, the technique called "titration curve analysis" will provide valuable information for ion exchange chromatography (12.4.9). During a purification procedure it is customary to monitor the progress by electrophoresis after each purification step.

SDS-electrophoresis and IEF are normally the methods of choice for accurate determination of molecular weight and iso-electric point of a protein.

By electrophoresis, large numbers of samples can be compared with ease and accuracy. This is used in clinical chemistry as well as in analysis of isoenzymatic patterns in genetic polymorphism in, for example, forensic medicine. In cell biology, the effects of different manipulations on protein pattern are often analysed by electrophoresis.

By combining two independent separation principles, the separation power of electrophoresis can be drastically increased. Two-dimensional electrophoresis according to the iso-dalt method (Chapter 15) is one example of this. Immunoelectrophoresis and affinity electrophoresis (Chapter 14) are other examples. It should also be pointed out that by combining different detection methods additional separation dimensions may be added (12.2.4).

Electrophoresis in transverse gradients of urea and/or other chaotropic agents can be used for the analysis of protein conformational changes caused by these agents[2].

Eletrophoresis has a number of practical advantages including:

1. Relatively simple and inexpensive equipment
2. High resolution results
3. Multiple sample analysis easy
4. High sensitivity
5. Specific detection easy (immunological, enzymatic etc)
6. Aesthetic appearance of banding pattern.

The merits of electrophoresis for protein analysis have become even more obvious in the last few years with the introduction of commercial instrumentation offering greater speed, convenience, reproducibility and accuracy than before[3,4].

12.2   Basic Concepts

Fundamental to electrophoretical separations is the fact that proteins are electrically charged particles. The charges are derived from amino acids with ionogenic side groups. These are the basic residues arginine, lysine and histidine and the acidic residues glutamic acid and aspartic acid. See also Table 4-2. Tyrosine side chains will also contribute to the protein's overall charge. In addition, the proteins often have associated charged components of non-protein origin such as lipids or carbohydrates. Most of these acids and bases are relatively weak making the overall charge of the protein strongly pH dependent. Most globular proteins are acidic with pI values in the range 4–5[5]. (Fig. 12-1. See also Section 4.3).

12.2.1   Electrophoretic Migration

When exposed to an electrical field, E (volts/cm, V/cm) a force, F, acts upon the protein molecules. The force F depends on the field strength and the net charge, z, of the protein: $F = Ez$. The net charge for a particular protein depends on the pH (Fig. 12-1) but also on the ionic environment. The association of ions as counter ions or by hydrophobic forces (e.g. SDS, see 12.5) will modify the charge of the "naked" protein. As

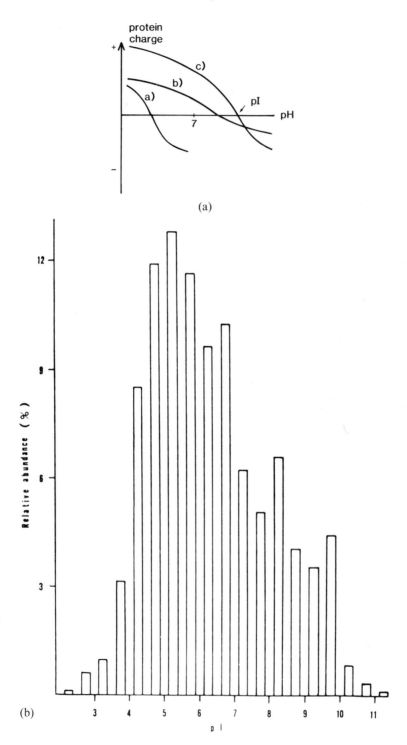

(a)

(b)

soon as the protein starts to move by the action of the electrical force, a counteracting force, F', is manifested due to the "friction" of the medium. F' is directly proportional to the migration speed and hence the protein molecule accelerates until $F = F'$. F' depends on factors such as the size of the molecule, the viscosity of the medium and the properties of the gel medium (if there is a gel).

The migration velocity of a protein in the unit electrical field ($E = 1$ V/cm) under defined conditions (temperature, buffer, pH, gel composition) is called mobility, and is an intrinsic property of each individual protein. We therefore conclude the following relationship between migration velocity (v), mobility (u) and field strength (E):

$$v \text{ (cm/h)} = E \text{ (V/cm)} u \text{ (cm}^2/Vh).$$

Note that the driving force for the migration is set up by the electrical field strength and that the current has no direct effect. However, a higher current will generate more heat which can affect mobility and migration velocity by reducing the viscosity of the medium. (Heat evolvement is proportional to the square of the current according to the formula $P = RI^2$, where P is the effect, I the current and R the resistance).

12.2.2 Gel Media

The earliest electrophoresis experiments were performed in free solution. It was soon realized, however, that the use of anticonvection media offered practical advantages and a number of "water holding" supports such as filter paper, starch, agar, agarose, cellulose, polyacrylamide etc. were introduced. Some of these media actually contributed to the separation by sieving. Of course, the sieving depends on the relative size of the proteins and the pores in the support.

Of all the media tried, two have proved outstanding for protein electrophoresis, polyacrylamide (PAA) and agarose gels. In the concentrations mostly used, PAA can be considered as a sieving gel and agarose a non-sieving gel for proteins. The sieving properties are used to advantage in the different variants of polyacrylamide gel electrophoresis (PAGE).

12.2.2.1 Polyacrylamide Gels

Polyacrylamide gels are prepared by polymerisation of acrylamide in the presence of a bifunctional cross-linking agent[6,7]. Normally this is N',N'-methylene bisacrylamide. This procedure gives a three dimensional network. The pores of the gel can be varied within wide limits. The cross-linking reaction and the properties of the PAA gels obtained under different conditions have been extensively investigated[8,9,10,11,12]. (See also Section 2.2).

Figure 12-1. Proteins as charged particles. Top diagram (a) illustrates the charge dependence of three imaginary proteins, a, b and c. The pH where the net charge of the protein $= 0$ is called the isoelectric point, pI. Proteins with pI in the acidic region are called "acidic" proteins (a) whereas "basic" proteins have basic pI-values (b and c) and "neutral" proteins have pI values close to pH 7. Note that the slope of the curve at pI can be very different. This is especially important in isoelectric focusing (Chapter 13). Bottom diagram (b) demonstrates the distribution of pI values for globular proteins reported up to 1981. Although the number of reported pI values has increased since then, the overall picture, with an predominance of acidic proteins, is without doubt still the same. (Fig. 12-1b reproduced from Ref. 5 with permission of Elsevier Science Publishers B.V. Amsterdam).

PAA gels are conveniently described by the "T" and "C" nomenclature introduced by Hjertén[13].

$$T = \frac{100(A + B)}{V}\,\%(\text{w/v}) \quad \text{and} \quad C = \frac{100B}{A + B}\,\%(\text{w/v})$$

where A and B are the amounts of acrylamide and bisacrylamide repectively expressed in grams. V was defined by Hjertén as the volume of added buffer, but is most often given as the volume of the *gelling solution*. For small values of T the difference between the two designations is small, but at higher T values it is important to specify clearly the exact meaning of T.

Typical values for T in PAGE where the sieving effect of the gel is utilized are 10–20%. C is usually in the order of 3–10%. Highly cross-linked gels, with C values above *ca.* 20–25%, are macroporous with a completely different gel structure as compared to conventional gels[9]. From C of *ca.* 15% the porosity increases with C.

In IEF, where the sieving is an unwanted side effect, a typical gel composition is T5 C3.

12.2.2.2   Agar, Agarose

Agar and agarose are both composed of a gel-forming polysaccharide of alternating galactose and 3,6-anhydrogalactose[14,15,16]. To this gel-forming backbone varying amounts of methyl ether, sulphate and pyruvate group are attached to the sugar residues. There is no principal difference between agar and agarose. The name "agarose" is reserved for the less charged and better defined forms of agar. (See Section 2.2).

Agarose with different specifications suitable for different applictions can be obtained from a number of suppliers.

Since agarose is a non-sieving matrix for proteins, the separation in agarose gels does not benefit from sieving, and suffers from increased diffusion in the open agarose pores. The separations are thus in general not as good as in PAGE unless the pure electrophoresis is combined with some other separation parameter such as in immunoelectrophoresis or affinity electrophoresis.

The only really high resolution electrophoretic technique where agarose may be used with advantage is isoelectric focusing. However, for this purpose, only the purest forms of agarose with extremely low content of charged groups are suitable (12.4.2). Such agaroses are now commercially available.

12.2.3   Electro(endo)osmosis

When applying a voltage over a gel it is frequently observed that the gel shrinks close to one of the electrodes, mostly the anode, and swells at the other end. Sometimes swelling is so large that buffer is extruded from the gel, the gel " sweats". Occasionally the gel shrinkage may cause the gel to completely dry out which, of course, totally ruins the experiment. These effects are all due to the phenomenon called electroosmosis[17,18]. All gel media contain charged, mostly acidic, groups. Polyacrylamide is inevitably deamidated to some extent and agar and agarose always contain sulphate and carboxylic groups. These immobilised negatively charged groups attract positively charged ions from the buffer as counter ions to keep the electroneutrality of the system. These counter ions are not completely immobilized but will occasionally dissociate into solution where they will be carried away in the voltage gradient until they are trapped by the next charged group on the gel matrix. Since the counter ions are normally small, highly hydrated, positively charged cations, the macroscopic effect is a transportation of liquid from anode to cathode. The magnitude of the electroosmotic flow depends primarily on the charge density of the gel, the voltage and the buffer concentration: More immobilized charges mean more counter ions to carry the electroosmotic flow, and a lower buffer concentration forces a higher proportion of the current to be transported by the counter ions. Consequently, as will be discussed in more detail below, the effects of electro-osmosis are especially pronounced in isoelectric focusing with its extremely low ionic strength. In IEF, therefore, only extremely low charged gel media will work properly.

On the other hand, in some systems, as for example in immunoelectrophoresis, a controlled electroosmotic flow may be an advantage (Chapter 15).

In agar and agarose gels electroosmosis can be expressed quantitatively as a relative mobility index, "m_r", where $m_r = -D/(D + A)$. "D" and "A" are the migration distances of uncharged dextran and serum albumin in barbital buffer pH 8.2 with ionic strength 0.05 M [19].

12.2.4 Detection

In the normal case the electrophoretic separation is interrupted while the protein zones are still in the gel and the separation pattern is visualized by some protein staining procedure.

The most frequently used stain is Coomassie (Brilliant) Blue, (CBB), R-250, marketed under trade names such as Serva Blue R, PhastGel Blue R, etc.[20]. Normally the proteins are first precipitated by agents such as trichloracetic acid (*ca.* 10%) and/or formaldehyde. The gel is then exposed to a solution of stain dissolved in dilute acetic acid/ethanol mixture. Finally, excess stain is removed by soaking the gel in acetic acid/ethanol until the background is clear. Several changes of destaining solution may be necessary. Electrophoretic removal of the excess dye by transverse electrophoresis in a destaining apparatus will speed up the process[21].

Staining methods based on the deposition of silver[22,23,24], gold[25,26,27], or iron[25] have been described. Silver staining is claimed to be up to 100 times as sensitive as conventional Coomassie straining, with a detection limit of less than 1 ng of protein. However, by prolonged staining (up to 72 h) at low pH and in the presence of ammonium sulphate Neuhoff *et al* demonstrated that less than 1 ng/mm^2 of protein can be detected even with Coomassie stains[28]. This is almost as sensitive as the silver staining methods.

Ready-to-use kits for silver staining are available. The automatic elecrophoresis machines provide automatic procedures for these staining procedures as well[29,30]. As already mentioned, another dimension of information can be obtained by combining general protein staining with *specific detection methods*. Such methods include staining for lipids[31], carbohydrates[32,33] enzyme activity[34,35,36] and antigenic properties[37,38]. When used in conjunction with total protein staining, a precise localization of the component(s) of interest in the whole protein spectrum is possible.

Of particular interest in this context is the technique often referred to as "Western blotting" although a more descriptive name would be "Protein Blotting". Here the separated proteins in the gel are transferred to a nitrocellulose or nylon membrane by transverse electrophoresis. Having the proteins on a surface instead of in a gel greatly facilitates detection procedures based on immunogenic properties, autoradiography, zymograms etc.[38,39,40]. Protein blotting is further described in Chapter 16. After detection, PAA gels can be stored either wet or dry. Wet gels are preferably soaked in acetic acid (10%) and stored in a sealed plastic bag. Procedures for drying PAA gels have been described[41,42] and there are also gel drying apparatuses available commercially.

12.3   Electrophoretic Techniques

As pointed out in 12.2.2 the highest resolution is obtained in sieving media and polyacrylamide has proved especially useful. The following discussion will therefore be restricted to polyacrylamide gel electrophoresis, PAGE.

12.3.1   Equipment and Mode of Operation

Vertical electrophoresis with sample migration downwards is no doubt still the most common way to run PAGE. The sample is loaded on top of a gel which has its ends in contact with an upper and lower electrolyte. (Fig. 12-2).

The earliest experiments were performed using glass rods in vertical glass tubes, but today gel cassettes containing gels in the form of slabs are more common. The major advantage with slab gels is that a number of samples can be run simultaneously on the same gel, facilitating comparison of the electrophoresis patterns and minimizing the risk of sample confusion. Excellent equipment is available from a number of commercial sources. Some of these allow the use of both rods and slabs. Precast gel slabs are also available from some companies.

During recent years horizontal electrophoresis has become more common. This trend has been accelerated by the introduction of automated equipment for horizontal

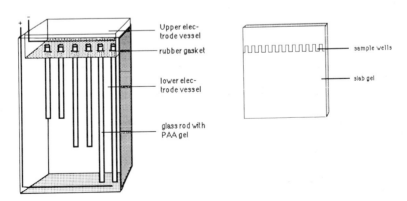

Figure 12-2. Schematic illustration of a typical set-up for vertical electrophoresis. The lower electrolyte buffer (usually anolyte) also serves as coolant for the electrophoresis rods or slabs. Sealing with rubber gaskets prevents the two electrolyte buffers from mixing. Some apparatuses are equipped with facilities for temperature control. Either the buffer can be circulated outside the electrode vessel and cooled or coils with circulating coolant are placed in the bottom buffer vessel.

electrophoresis. This equipment simplifies or eliminates most practical problems such as gel casting, staining and destaining, thus making high performance electrophoresis a convenient and easily accessible technique[3,4].

To a large extent, the choice between vertical and horizontal electrophoresis is a matter of taste. The fundamental principles discussed below are equally applicable to both modes.

In nucleic acid electrophoresis a special form of horizontal electrophoresis, "submarine", is used where the whole gel is immersed in the electrophoesis buffer[43]. So far this technique has not been applied to proteins and is therefore not further dealt with here.

12.3.2 Sample Application and Recovery

Obviously only protein molecules in solution can be analysed by electrophoresis. If the sample is difficult to dissolve, agents such as detergents or urea may prove necessary. Unlike in chromatography, the presence of insoluble material in the sample is not generally critical. In fact, with a concentrating technique such as isoelectric focusing, even whole tissue samples can be directly applied and the extraction performed during the IEF process itself. In some cases, such as in the zone sharpening techniques discussed below, the sample must be transferred into a special buffering environment. This can be done by dialysis or by gel filtration on a small column packed with Sephadex G-25 or G-50.

If it desired to recover the purified protein band, this is easily accomplished by cutting out the band from the gel after localizing it by any of the detection methods available and then simply soaking the crushed gel in buffer. More efficient methods, however, take advantage of the principles of isotachophoresis[44,45,46]. Equipment for electrophoretic extraction is commerically available. This is further described in Chapter 16.

12.3.3 Zone Sharpening Techniques

All protein zones tend to broaden with time due to diffusion. Fortunately this effect can be counteracted by different methods. The importance of these techniques cannot be overestimated. They are fundamental to the high quality results that are regularly obtained today! In addition to the automatic concentration of protein zones at the gel surface, several techniques concentrate the protein zone *before* it enters the gel (sample ionic strength and protein "stacking"), as well as *during* and *after* separation (Pore Gradient PAGE).

12.3.3.1 Sample Ionic Strength

A very simple and straightforward zone sharpening technique is to make sure that the electrical conductance of the sample buffer is lower than that of the running buffer. This will increase the field strength in the sample zone. Proteins will then migrate quickly through the sample zone and into the running buffer where they will slow down and hence become concentrated in the lower field strength[47]. A two to ten times dilution of the buffer will decrease the conductance proportionally with a concomitant proportional increase in field strength. Since migration velocity is directly proportional to field strength (12.2.2) this will result in a two to ten times concentration of the protein zone as it leaves the original sample zone.

12.3.3.2   Protein Stacking and Moving Boundary Electrophoresis

As mentioned in the introduction, this is an application of isotachophoresis. Typical for this very efficient technique for sample concentration is casting of a special, concentrating or "stacking" gel on top of the separation gel. Since the technique utilizes *discontinuities* in buffer composition as well as in PAA concentration, it is frequently referred to as "disc-electrophoresis"[48,49]. The purpose of the stacking gel is simply to hold the buffer required for protein concentration before the protein zone enters the separating gel. The stacking gel is therefore prepared as a low porosity gel with minimal sieving properties.

In the simplest form, the separating gel and the stacking gel both contain a fast moving ion (ex Cl^-) at the start, while the upper electrolyte solution contains a slow moving ion (ex glycine). When the voltage is applied, the fast Cl^- ions leaving the upper part of the stacking gel will be replaced with the slower moving ions entering from the upper buffer reservoir. Since the faster ions migrate first, they will tend to leave the slower ions behind, creating a zone of "ionic vacuum". The low ionic strength in this region causes a high field strength, and when the slower ions enter this region, their migration speed increases so that they catch up with the fast ions. Thus, immediately behind the fast ions, the slower ions are diluted to a concentration giving a field strength that is just right to give both the slower and faster ions the same migration speed. A self-regulating moving boundary is set up. The slower, "trailing" ions cannot pass the faster, "leading" ions. Should they happen to enter the zone with the fast ions they will immediately be passed by the faster ions and caught up by the zone with the other ions of the same kind. Should the faster ions happen to lag behind the boundary into the zone with the slower ions and the higher field strength they will immediately overtake these ions and migrate to the right side of the boundary again.

Suppose now, that the mobility of the fast ion is higher, and that of the slow is lower, than the mobilities of the proteins in the sample. The proteins will then be trapped between the two zones described above and, furthermore, they will be sorted in mobility order according to the same principles as for the buffer ions.

The concentrations of the buffer ions and the sample proteins will be adjusted so that the electrophoretic migration in all zones is the same as in the zone with the leading, fast, electrolyte. The migration velocity and electrolyte concentration in each zone is determined by the Kohlrausch regulating function[50]. We will not go into the mathematical details here. Suffice it to mention that since the protein concentration expressed as charge equivalents/volume must be in the same order of magnitude as that of the leading buffer ion, the concentration effect can be enormous and the sample concentrated to a very narrow zone. Finally, when the proteins enter the separating gel, their mobilities will be reduced. The much smaller trailing ions will continue with unchanged speed and pass the proteins. Due to the sieving effect of the gel the proteins will now separate. (Fig. 12-3).

It is evident that the stacking gel must be long enough for protein stacking to be finished before the moving boundary reaches the separating gel; 5–10 mm is usually sufficient.

Proteins with mobilities outside those of the leading and trailing ions will not be concentrated. Under non-denaturing conditions and with standard buffer systems no

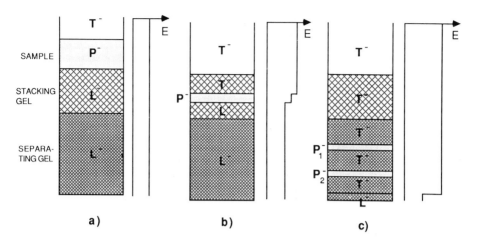

Figure 12-3. Schematic illustration of the principles of disc(ontinuous) electrophoresis.

a) *Starting conditions.* The system is composed of a relatively dense sieving gel as separating gel with a highly porous, low density stacking gel on top. Both these gels contain the fast, leading, ion L^- as anion. The catholyte anion, T^-, has a lower mobility than both the proteins, P^-, and L^-. The protein sample, which can be very dilute, is layered on top of the stacking gel. The cation can be considered the same throughout the system and can be neglected for the sake of argument.

b) *Concentrating phase.* Under the influence of the voltage all ions start to migrate. Given the same field strength in the whole system the L^--ion would leave the other, slower, ions behind, creating an "ionic vacuum". This will not happen, since in this imaginary "ionic vacuum" the field strength would be very high and the slower ions would catch up. The concentrations of the P^- and T^- ions will be adjusted to the concentration that gives exactly the required field strength to make all ionic components migrate at the same speed. In the protein zone this predetermined ionic strength is largely due to the proteins themselves. Since the mass/charge ratio of proteins is high this means that, under suitable conditions, proteins will be highly concentrated. In fact, the different proteins will not only be concentrated but also separated within the protein zone into adjacent zones in order of mobility.

c) *Separating phase.* Eventually the proteins enter the denser, separating gel where their migration velocity decreases because of the sieving interaction with the gel network. The much smaller T^- ion will not be affected by the gel and will now pass the proteins and a new self-regulating moving boundary is set up between the T^- and L^- ions (the "front"). Marker dyes (bromophenol blue etc) will normally migrate in this front since they have a mobility intermediate between L^- and T^-. The proteins will be separated by sieving in the gel matrix.

protein will ever move faster than the leading ion. Many proteins will however move slower than the trailing ion and even in the opposite direction, depending on the pI of the protein and the pH of the electrophoresis buffer. In SDS electrophoresis all proteins have a strong negative charge and will move faster than the trailing ion. Small proteins (M_r less than *ca.* 10,000) may move to the front immediately behind the leading ion (see 12.5.2). By choosing suitable buffers and buffer concentrations the principle of protein stacking can be applied to different pH values for the separation of different types of proteins[51,52].

12.3.3.3 Pore Gradient Gels

In this technique it is not the sample, but the protein zones that are concentrated during the separation. This is accomplished by using a gel with continuously changing acrylamide concentration[53,54,55]. The sample is applied in the low concentration region. When the proteins migrate into a denser gel network the "friction" or sieving effect from the gel will increase. The electrophoretic migration speed will thus decrease and asymptotically reach zero as the protein approaches the gel concentration of its "pore limit", primarily depending on the Stokes radius. Provided the pore limit is reached, gradient gel PAGE is a convenient way to estimate protein molecular weight[56,57,58]. For the best results, gradient PAGE should therefore be performed at a pH where the protein is highly charged, which for the majority of proteins means a basic pH. Buffers with pH values of 8.2–8.6 are standard (Fig. 12-4).

Casting of gradient gels is most conveniently done by filling the cassette from a gradient mixer of the type used in, for example, low pressure ion exchange chromatography. With suitable equipment several gels can be cast simultaneously. The acrylamide concentration can be varied within wide limits, but concentrations below 4% and above 30–40% are not recommended for practical reasons. Detailed instructions for gel casting are provided by several companies. Precast gels with gradients suitable for most applications are available from several manufacturers both for conventional vertical electrophoresis and for automatic electrophoresis equipment.

In gradient PAGE the protein zones are sometimes not as sharp as expected. This may of course depend on the buffer pH, which may have been unsuitable for the particular protein. However, another factor may also conribute. Because of the fact that the concentration of liquid decreases in the gradient, the field strength will increase. The same electrical current must pass through less liquid, leading to higher current density and higher field strength. The slowing down of the front of the zone by the increased sieving effect from the gradient is therefore to some extent counteracted by an increased field strength. Generally, and especially if the proteins are run to pore limit, the concentrating effect predominates.

12.4 Optimizing the Separation

12.4.1 Choice of Buffer pH and Gel Porosity: The Ferguson Plot Analysis

The electrophoretic mobility of a protein in a gel will be determined by the balance between the electrical force and the retarding, "frictional" force. The electrical force is a function of the voltage applied (which is the same for all proteins in the sample) and the protein charge. The retardation force is a function of gel concentration and protein size. The parameters which must be adjusted for optimization are the buffer concentration and pH and gel porosity. In most cases, standard procedures with a buffer pH of 8–9 and a gel concentration of 7–10% in the separating gel will work acceptably for neutral and acidic proteins whereas reversed polarity and a buffer pH of 4–5 will separate basic proteins. If, however, the separation problem is difficult and needs to be optimized the so called Ferguson plot analysis can be used as a powerful tool for establishing the optimal conditions[59].

A conventional Ferguson plot analysis is done as follows: Run the separation at different gel concentrations (T-values) while keeping C and buffer fixed. Measure the

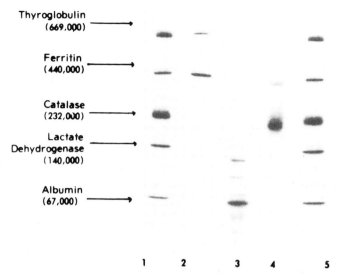

Figure 12-4a. Molecular weight determination by electrophoresis in PAA gradient gels. The picture shows the result of electrophoresis of model proteins on a PAA gel with concentration from 4% to 30% acrylamide (Pharmacia precast gradient gel PAA 4/30). Electrophoresis was run to "pore limit" (16h, 150V) in tris/-borate/EDTA buffer pH 8.4. Samples were applied on the top of the gel.

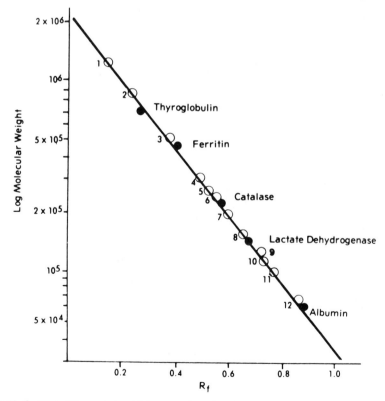

Figure 12-4b. Plot of R_f versus log M_r for a number of globular non-denatured proteins. Notice the excellent linear fit over a wide molecular weight range. (Fig. 12-4a and 12-4b were reproduced from Ref. 57 with permission of Elsevier Science Publishers B.V.).

migration distance of the protein(s) of interest and divide this distance by the migration distance of the marker stain (usually bromophenol blue) to obtain the relative migration value, R_f. Plot the R_f values of the protein to be purified and the major contaminants as a function of T. This is called a "Ferguson plot". Repeat the procedure for different buffers. By analysing the plots, the gel concentration and the buffer that gives the maximal distance between the proteins to be separated can be deduced[60,61,62,63], (Fig. 12-5). The optimal value of T can be mathematically calculated from the data obtained by the Ferguson plot analysis[62]. To simplify the optimization procedure, a system of 19 buffers with different operational pH-values and stacking effects has been described[51].

Another way to perform the Ferguson plot analysis is to run titration curve analysis at different T-concentrations (Section 13.4.8, p. 389). This gives essentially the same information as above with fewer experiments but with less strict control of the buffers. Each gel will provide separation data for the whole pH-spectrum. Measure the migration distances for the relevant proteins at some chosen pH values and make the same plots as for the conventional Ferguson plot. Analyse the plots to find the optimal pH and gel concentration for the separation.

12.4.2 Choice of Buffer Composition and Detergent

Electrophoresis under denaturing conditions in sodium dodecyl sulphate (SDS) is so efficient and well established that it is sometimes adopted too uncritically. Before setting up the electrophoretic analysis it may prove worthwhile to contemplate for a moment which method is most apt for achieving the desired goals.

In all cases where the biological function of the protein can be utilized to gain further relevant information, electrophoresis (or isoelectric focusing) of the native protein, followed by specific detection of enzymatic or other activity etc., will normally be a better

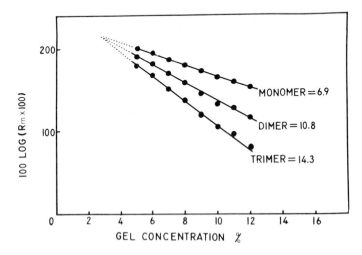

Figure 12-5. Ferguson plot analysis.
Figure 12-5a Size isomers. Ferguson plot of monomer, dimer and trimer of bovine serum albumin. Since all molecules have the same charge density they will asymptotically approach the same mobility at $T = 0$. Better separation at higher gel concentrations.

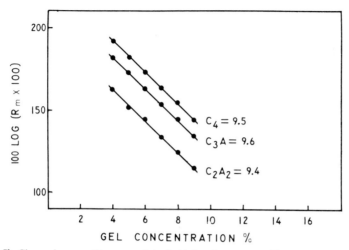

Figure 12-5b Charge isomers. Ferguson plot of aldolase isoenzymes. These proteins all have the same size and are therefore affected by changes in gel concentration in the same way. Separation is in principal not dependent on gel concentration.

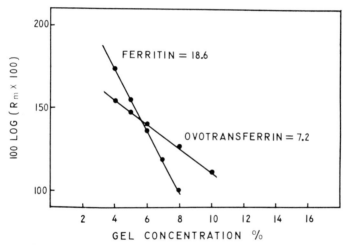

Figure 12-5c Different charge and molecular weight. Ferguson plot of ferritin ($M_R = > 10^6$) and ovotransferrin ($M_R = 87,000$). Ovotransferrin has a low charge and is overrun by ferritin at low gel concentrations. The mobility of the larger ferritin is more dependent on the gel concentration than the mobility of the ovotransferrin. At higher gel concentrations the transferrin is highly retarded because of its size. Note that at $T = 6\%$ and the pH used the two proteins cannot be separated by PAGE. (Fig. 12-5 was reproduced from Ref. 63 with permission of Academic Press).

alternative or a powerful complement to SDS electrophoresis of denatured proteins. Ordinary globular proteins with good solubility in standard buffers can easily be separated in native conformation and with full activity. The separation conditions can be chosen primarily from a pure electrophoretic separation point of view.

Many proteins, however, such as hydrophobic membrane proteins, filamentous proteins etc., with poor solubility in ordinary buffers, pose special problems in electrophoresis. In order to get these proteins in solution for electrophoresis, detergents such as SDS and reducing agents (mercaptoethanol, dithiothreitol etc.) for cleavage of S-S bridges must be used. Mixtures of phenol, acetic acid and urea have been used successfully for the solubilization of membrane proteins[64].

All types of detergents, nonionic-, zwitterionic-, cationic-, and anionic-detergents can be used[65]. In order to keep the protein in its native configuration the mildest possible detergent should first be tried. For the solubilization of membrane proteins without denaturation, Hjelmeland and Chrambach recommend starting with CHAPS (3-((cholamidopropyl)-dimethylammonio)-1-propanesulphonate) and systematically explore the optimal solubilization conditions[65]. (See also p. 8.)

As a last resort, for proteins that are extremely difficult to solubilize, stronger detergents like SDS (see below) can be tried. Some unusually stable proteins retain their activity in SDS and in other cases proteins can regain activity by careful exchange of the stronger detergent with a weaker one[66]. Once the solubilization conditions are found, the electrophoresis can then be performed according to standard procedures described below. Renaturation and the use of different detergents in PAGE has been reviewed[67].

12.5   SDS Gel Electrophoresis and M_r-Determinations

Proteins can be denatured by a variety of agents such as urea and guanidinium hydrochloride as well as by strong positive or negative ionic detergents. Among these, the negatively charged detergent dodecyl sulphate, normally in the form of its sodium salt, has proved extremely useful in connection with electrophoresis separations in acrylamide gels (SDS PAGE)[68,69]. Occasionally, as described below, SDS is used in conjunction with other agents such as urea.

Reasons for the popularity of SDS PAGE can be summarised as follows:

— It separates proteins strictly according to a single molecular parameter, molecular size, and can therefore be used for molecular characterization if the system is calibrated with molecular standards.
— The resolution is generally very high.
— The technique is inexpensive and easy to set up and perform.
— The separation is fast.
— SDS PAGE is applicable to most separation problems since SDS solubilizes most proteins.

The major drawback with the technique is that SDS denatured proteins have lost their biological activities which decreases the possibility of gaining extra information by specific detection methods. Immunogenic detection methods can, however, generally be used since sufficient immunogenic identity is usually inherent in the peptide sequence. It may also be possible, as already mentioned, to renature protein activity[67].

12.5.1   Principles

Fundamental to SDS PAGE is the very strong interaction between the dodecyl sulphate detergent and the protein peptide chain causing the SDS-protein complex to migrate as one well-defined entity. The exact nature of this interaction is not fully elucidated, but it appears that a flexible rod containing about 1.4 g SDS/g protein is formed[70,71].

Some facts about SDS-protein interaction can be summarised as follows:

— The strong solubilizing effect of SDS makes essentially all proteins accessible for electrophoretic analysis including normally insoluble proteins such as filamentous proteins and hydrophobic membrane proteins.
— The SDS-protein interaction is strong enough to make the composition of the SDS-protein complex essentially pH independent.
— All protein-SDS complexes acquire the same conformation and differ only in size.
— All protein-SDS complexes have the same, very high, charge/mass ratio.

Because the charge density and conformation is constant, all SDS-protein complexes migrate with essentially the same velocity in free solution. The separation in SDS PAGE is thus due entirely to the sieving effect of the gel. The high charge density of the SDS-protein complexes gives them a high electrophoretic mobility and short separation times. The various theories describing the sieving effect have recently been reviewed[72]. For practical use it is sufficient to know that there is a direct relationship between log M_r and R_f making determination of protein molecular weight by SDS electrophoresis both a simple and a reliable method[20,69,73,74]. (R_f is defined as the migration distance of the protein relative to that of the tracking dye). By calibrating the system with proteins of known molecular weight, this relationship can be used to determine the molecular weight of unknown proteins, (Fig. 12-4).

SDS PAGE can with advantage be performed in PAA gradient gels. In gradient SDS PAGE the linearity between log M_r and R_f is extended even further as compared to conventional gels[57]. Figure 12-4 shows a typical plot.

The condition most commonly used for SDS PAGE is that described by Laemmli[75] or variants thereof where discontinuous buffers are used for zone sharpening by protein stacking as described above.

The merits of the commonly used buffer systems described by Wyckoff, Rodbard and Chrambach[76], Neville[73], and Laemmli[75] have been compared[77]. The system of Wyckoff *et al* was found to be particularly suitable for smaller proteins/peptides (down to *ca* 6000 in molecular weight) if the SDS concentration was increased from 0.03 to 0.1 %.

Most SDS preparations contain a substantial amount of SDS alkyl analogues. Strangely enough it has been found that the best separations are sometimes obtained not with pure SDS, but with SDS highly contaminated with its alkyl analogues[78,79]. The SDS quality also affects renaturation of proteins[80]. Different qualities of SDS can undoubtedly explain many discrepancies found in SDS electrophoresis[81].

Generally SDS is added to all gels and buffers to concentrations of 0.05–0.1 % (w/v). However, in principle it is only necessary to have SDS in the upper (cathodic) buffer[82]. The dodecyl sulphate molecules will enter the gel before the proteins. Certainly, SDS in the anodic buffer fulfils no purpose unless the buffer is circulated between the two compartments.

12.5.2 Practical Procedures in SDS PAGE

To assure complete denaturation, the sample should be heated to 100°C for 5–10 min in the presence of excess SDS (1–2 %)[69]. If desired, a reducing agent may be included to cleave S-S bridges[73]. 5% mercaptoethanol or dithiothreitol are commonly used. Be careful with plastic tubes. Not all qualities withstand high temperature!

The ionic strength of the sample has little influence on the result. It was found that up to 0.8 M NaCl was well tolerated[83]. Good molecular weight estimates will be obtained even without reduction of intramolecular S-S bonds. Unreduced protein migrated only $ca.$ 14% slower[84]. Normally sucrose or glycerol is added to increase sample density and a small amount of bromophenol blue (from a 1% stock solution) is added as tracking dye. This facilitates sample application and the position of the tracking dye can also be used for R_f calculations.

The amount of sample that is required depends primarily on the sensitivity of the detection method used. With conventional Coomassie detection with comparatively low sensitivity, 1–5 μg protein per band will be adequate whereas with high sensitive immunological, enzymatic or silver staining methods ng quantities or less are sufficient. Optimal conditions have to be worked out for each application.

The SDS-protein complexes are extremely soluble and do not precipitate or take up stain easily. The staining procedures must therefore be designed so that the SDS-protein complexes are dissolved and the SDS removed from the gel before the actual staining.

The most efficient removal of SDS is via transverse electrophoresis in high concentrations (45%) of ethanol or isopropanol before staining[85]. Soaking the gel in 12.5% trichloroacetic acid will precipitate the proteins and remove SDS by dilution at the same time[84].

12.5.3 M_r Determinations

If molecular weight calibration proteins are run on the same gel as the sample, the protein pattern obtained after SDS electrophoresis will give a good survey of the molecular weight distribution of proteins. Molecular weights of individual proteins can be calculated with good accuracy from log M_r versus R_f plots. Kits with calibration proteins can be obtained from several manufacturers. If the sample is run in both reduced and unreduced form the number of covalently linked subunits can be derived.

Molecular weights down to about 10–14,000 daltons can be estimated fairly accurately using standard procedures. Smaller proteins and peptides will not migrate strictly according to molecular size or migrate in the buffer front and not separate at all. Also, these small peptide-SDS complexes are of similar size to the SDS micelles[62] and all peptides below a certain size ($ca.$ 10,000) will concentrate in the SDS-micelle front. By adding 8 M urea to the system (0.1% SDS, 0.1 M tris-phosphate pH 6.8, $T = 12.5\%$), Swank and Munkres succeeded in separating peptides down to $ca.$ 1000 in M_r, essentially according to molecular weight[86]. By running the buffer system of Wyckoff et $al.$[74] in a PAA gradient (10.2–30.2%), proteins and peptides ranging from $ca.$ 1500–100,000 in molecular weight can be separated in the same gel[87].

Some proteins do not behave normally in SDS PAGE. Glycoproteins do not bind SDS to the same extent as "normal" proteins and M_r estimations must be made with much care. Molecular weights may easily be overestimated by as much as 30%[88,89]. With extremely positively charged proteins such as histones, it seems that the charges of SDS are balanced to a significant degree by the proteins' own charges, which will make M_r determinations difficult[90]. Very acidic proteins such as pepsin, papain and glucose oxidase were found to bind very small amounts of SDS[91].

12.6 Applications

12.6.1 Polyacrylamide Gradient Gel Electrophoresis of Concanavalin A

Different preparations of the lectin Concanavalin A were analysed by polyacryl-amide gradient gel electrophoresis (Gradient PAGE)[92]. With gradient PAGE an extra component in addition to the major band was observed in all the preparations. This band was not seen in conventional PAGE in homogeneous gel. Preparations with different metal ion composition did not give significantly different electrophoretic patterns (Fig. 12-6) in contrast to preparations from different commercial sources (Fig. 12-7).

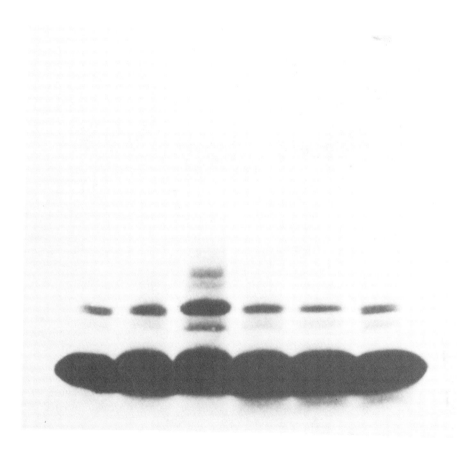

Figure 12-6. Gradient PAGE of Con A samples containing different amounts of metal. Samples were applied on top. From left to right the samples had the following metal ion concentrations (in %): 1. 0.25 Mn^{2+}, 0.23 Ca^{2+}; 2. 0.16 Mn^{2+}. 0.18 Ca^{2+}; 3. 0.0007 Mn^{2+}, 0.001 Ca^{2+}; 4. 0.003 Mn^{2+}, 0.009 Ca^{2+}; 5. 0.0007 Mn^{2+}, 0.003 Ca^{2+}; 6. 0.17 Mn^{2+}, 0.21 Ca^{2+}. (Reproduced from Ref. 92 with permission of Elsevier Science Publishers).

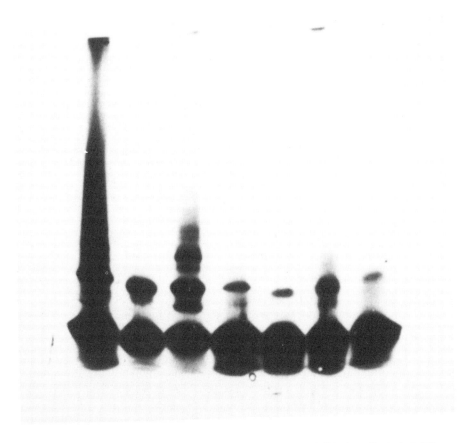

Figure 12.7. Gradient PAGE of commercial Con A preparations: 1. Calbiochem Grade A; 2. Miles-Yeda recrystalized twice; 4–6. Different batches from Pharmacia; 7. Sigma Grade III. (Lane 3 is the pattern from *Canavalia rosea* lectin prepared by affinity chromatography on Sephadex). (Reproduced from Ref. 92 with permission of Elsevier Science Publishers).

Experimental procedure

Electrophoresis was performed in Pharmacia equipment for vertical electrophoresis (GE 2/4) in premade gradient gels (Pharmacia PAA 4/30 gels) with monomer concentration from 4% to 30%. Before application of the sample, the gels were equilibriated by preelectrophoresis in the running buffer, 0.02 M acetate pH 4.0.

The sample contained 10–30 mg/ml protein and 10% sucrose (to facilitate sample application) and 5–15 μl was applied. The protein concentration was estimated spectrophotometrically.

Electrophoresis was performed for 4 h at 40 mA.

After electrophoresis, the gels were stained by soaking in 0.5% Amido Black in 7% acetic acid for 10–30 min. Finally the gels were electrophoretically destained in 7% acetic acid (36 V, 30 min.).

12.6.2   Analysing Urine Proteins by Micro-SDS-PAGE on PhastSystem

SDS-PAGE of urine has been found to be a versatile tool in the differential diagnosis of proteinuric diseases. The micro scale allows the analysis to be performed fast and on small sample amounts[93]. In this case, electrophoresis was performed horizontally in the Pharmacia PhastSystem. The ready to use premade gel slabs are supported on one side by a plastic foil. The gel is laid on the cooling plate with the plastic foil downwards. A film of water between the cooling plate and the plastic foil assures good contact. The whole system is microprocessor controlled. The temperature of the cooling plate is regulated by a Peltier element. Electrophoresis is performed according to preset values of voltage, volthours and time. Sample application is performed automatically after the preset volthours, Fig. 12-8.

Experimental procedure

Protein concentration in the sample was 1 mg/ml. When necessary the urine samples were diluted. SDS was added to a final concentration of 1% w/v.

Electrophoresis was run in the Pharmacia PhastSystem according to manufacturer's instructions on 8–25% PhastGels. Eight 1 μl samples were run on each gel at a constant temperature of 16°C for *ca.* 30 min. Staining was done automatically with Coomassie Blue in the development unit as described in the PhastSystem manual. Staining and destaining took less than 1 h.

12.6.3   Separation of Small Proteins and Peptides by SDS Gradient-PAGE.

By using the buffer system of Wyckoff *et al.*[76] in a high concentration PAA gradient system, Bothe, Simonis and von Döhren were able to separate proteins over a broad size range, from 100,000 down to *ca.* 1,500 without urea (Figs. 12-9 and 12-10). The limiting factor in the analysis of small proteins was the problem with fixing and staining the smallest proteins[87].

Experimental

Gels with a linear acrylamide concentration gradient were cast by mixing a low concentration solution (Solution I) and a high concentration solution (Solution II) in a conventional chromatography gradient mixer while filling the gel cassette from the bottom with a peristaltic pump. Gel dimensions were $140 \times 160 \times 1.2$ mm. The gel solution was overlayered with n-butanol to obtain a flat upper surface.

Solution I: $T = 10$, $C = 2$; 0.11 M ammediol; 0.047 M HCl; 0.03% ammonium persulphate; 0.1% SDS; 0.005% TEMED.

Solution II: $T = 30$, $C = 0.66$; 0.11 M ammediol; 0.047 M HCl; 0.015% ammonium persulphate; 0.1% SDS; 0.005% TEMED.

The monomers are added from a stock solution and diluted with water accordingly. Before adding SDS, ammonium persulphate and TEMED the solutions were carefully degassed (vacuum). It is recommended that the ammonium persulphate solution is prepared daily.

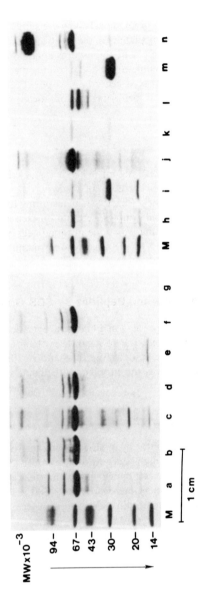

Figure 12-8. SDS-PAGE of urine (a–m) and serum (n) samples from patients with different proteinuric diseases in 8–25 % gradient slab gels. M: LMW markers (Pharmacia); proteinuric patterns: (a) glomerular with partial micromolecular; (b), (d) and (f) glomerular; (c) glomerulo-tubular; (e) and (h) tubular; (g) physiological; (i) light chain monomer with tubular; (j) tubular with Ig; (k) light chain monomer; (l) light chain dimer with tubular; (m) dark band of light chain monomer and weak band of dimer; (n) serum from the same patient as (m) (multiple myeloma, IgG/kappa). Sample migration was from the top of the figure.

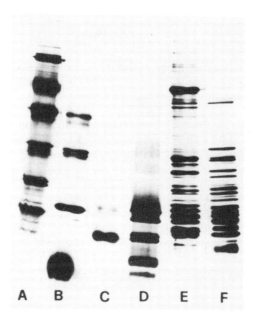

Figure 12-9. Separation of protein standards by SDS PAGE in a 10.2 to 30.2 % linear gradient gel. Samples were run downwards. The channels contain (A) Phosphorylase b (94,000), Bovine serum albumin (67,000), Ovalbumin (43,000), Carbonic anhydrase (30,000), Soybean trypsin inhibitor (21,100), L-lactalbumin (14,400) (= Pharmacia LMW molecular weight markers) (5.7 μg); (B) lysozyme (14,388) and bacitracin (1,500), 0.5 μg of each; (C) aprotinin, monomer (6,500) and dimer (1 μg) and insulin B (2 μg); (D) cyanogen bromide cleavage of myoglobin (7.5 μg) (= Pharmacia peptide molcular weight standards, molecular weights 17,200; 14,630; 8,235; 6,383; 2,556; 1,695); (E) E. coli 30S ribosomal proteins (8,369–61,159) and (F) E. coli 50 S ribosomal proteins (5,381–29,730). (Reproduced from Ref. 87 with permission of Academic Press).

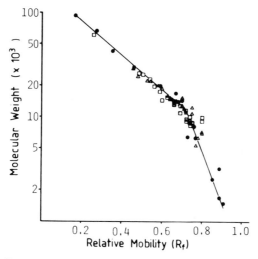

Figure 12-10. Plot of log molecular weight versus relative mobility, R_f for proteins and peptides of known molecular weights when run according to the method described. The relative mobility was measured in relation to the mobility of the tracking dye Bromo Phenol Blue. Note the inflection at molcular weights of *ca.* 10,000—12,000. (Reproduced from Ref. 87 with permission of Academic Press).

371

The pH of these solutions will be 8.96. Polymerization was allowed to take place overnight at room temperature. Because of the ammonium persulphate gradient, the polymerization proceeded from top to bottom, minimizing the risk for gradient disturbance by heat convection.

Samples were prepared by mixing the peptide solutions with an equal volume of double concentration buffer containing glycerol (40%), bromophenol blue (0.01%), SDS (2%) and 2-mercaptoethanol (1%). This solution was heated in a boiling water bath for 5 minutes to assure complete denaturation.

Electrophoresis was performed with the discontinuous buffer system of Bury[76] in a home-built vertical electrophoresis apparatus. The experiment was conducted at 30 mA until the tracking dye reached the bottom of the gel (*ca.* 3 h).

Gels were silver stained according to the method of Oakely *et al.*[94] with some modifications: Gels were immediately put in a fixing solution of 20% trichloroacetic acid (TCA) for about 1 h to prevent loss of smaller peptides and then put in 7.5% acetic acid overnight before the glutaraldehyde step. Staining was interrupted by placing the gels in 7.5% acetic acid.

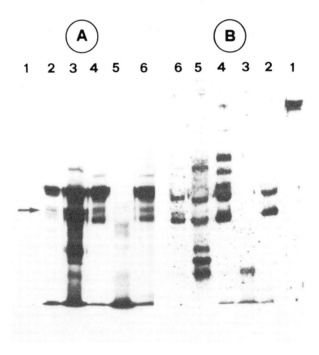

Figure 12-11. Separation of samples from human bronchoalveolar lavages by SDS electrophoresis. The samples were applied on top of the gel. (A) Separation gel stained with Coomassie Blue (total protein). (B) Nitrocellulose blot stained for fibronectin and fibronectin-derived peptides. Lane 1, Pure human fibronectin; lane 3 non-smoker control; lane 2,4 and 6 smokers; lane 5 acute emphysema (?). The protease inhibitor alpha-1-antitrypsin is indicated by the arrow (a set of bands).

12.6.4   Studying Fibronectin in Lungs of Smokers and Non-Smokers by Electrophoresis and Immunoblotting

Samples from lungs of smokers and non-smokers were analysed by SDS PAGE. After electrophoresis the proteins were transferred to nitrocellulose membranes and stained for total protein as well as specifically for fibronectin, Fig. 12-11.

Increased levels of proteolytic fragments from fibronectin in the samples from smokers are clearly visible. It is suggested that the increased proteolytic activity is a consequence of lower concentrations of the protease inhibitor alpha-1-antitrypsin in smokers[166].

Experimental

To prevent aggregation of the fibronectin, SDS as well as a high concentration of reducing agent was necessary. Samples were dissociated in 1 M Tris-HCl buffer pH 6.8 containing 4 % SDS and 3 % dithitreitol.

Electrophoresis was conducted in a 0.75 mm T7.5 C2 PAA gel in tris-glycine buffer pH 8.3, 0.5 % SDS and 0.1 % mercaptoethanol.

50 μl samples were applied and electrophoresis continued for *ca.* 4 h at 200 V until the Bromophenol Blue tracking dye left the gel. Immediately after electrophoresis half of the gel was fixed and stained with Coomassie Blue. The proteins in the other half were electrophoretically transferred to nitrocellulose membranes by transverse electrophoresis in a low ionic strength buffer (3.7 mM tris, 0.75 mM veronal, 7.5 mM glycine; pH 8.9) containing 0.5 % SDS and 1 % mercaptoethanol. Transfer was conducted for 6 h at 10 V/cm.

Immunological detection of fibronectin and fibronectin peptides was done by an indirect immunoperoxidase technique. First, the membrane was soaked in buffer (PBS-Tween = 50 mM phosphate, 0.9 % NaCl, 0.05 % Tween-20 pH 7.2) The buffer was changed several times to remove SDS and mercaptoethanol. Then the membrane was treated with rabbit anti fibronectin 1/200 in PBS-Tween for 2 h at 37°C followed by repetitive washes with buffer to remove unspecifically bound reagents. To allow specific detection, the membrane was treated with peroxidase conjugated goat and rabbit IgG 1/4000 in PBS-Tween.

Finally the position of the fibronectin and fibronectin peptides was visualized by incubation in 0.05 % 3-amino-9-ethylcarbazol, 0.001 % H_2O_2 in 50 mM acetate buffer pH 5. The brickred colour developed within the first minutes. The reaction was stopped with PBS-Tween. Tween-20 was used as blocking agent. Proteins commonly used for this purpose (e.g. Haemoglobin, serum albumin, gelatin) all contain fibronectin in amounts that will give background problems.

12.7   References

1. J. Th. G. Overbeek, and Bijsterbosch, In *Electrokinetic Separation Methods*; P. G. Righetti, C. J. van Oss, and J. W. Vanderhoff, eds. Elsevier/North Holland 1979; pp. 1–32.
2. D. P. Goldenberg, T. E. Creighton, *Anal. Biochem. 138*, 1–18 (1984).
3. J. Brewer, E. Grund, P. Hagerlid, I. Olsson, J. Lizana, In *Electrophoresis '86* M. J. Dunn, ed. VCH Verlagsgesellschaft Meinheim BRD, 1986, pp. 226–229.
4. I. Olsson, U.-B. Axiö-Fredriksson, M. Degerman, B. Olsson, *Electrophoresis 9, 1*, 16–22 (1988).
5. E. Gianazza, P. G. Righetti, *J. Chromatogr. 193*, 1–8 (1980).

6. S. Raymond, L. S. Weintraub, *Science 130*, 711–713 (1959).
7. L. Ornstein, B. J. Davies, *Disc Electrophoresis*; Distillation Products Div., Eastman Kodak Co., Rochester 1962, New York.
8. A. B. Bosisio, C. Loeherlein, R. S. Snyder, P. G. Righetti, *J. Chromatogr. 189*, 317–330 (1980).
9. P. G. Righetti, B. C. W. Brost, R. S. Snyder, *J. Bichem. Biophys. Meth. 4*, 347–363 (1981).
10. C. Gelfi, P. G. Righetti, *Electrophoresis 2*, 213–219 (1981).
11. C. Gelfi, P. G. Righetti, *Electrophoresis 2*, 220–228 (1981).
12. P. G. Righetti, C. Gelfi, A. B. Bosisio, *Electrophoresis 2*, 291–295 (1981).
13. S. Hjertén, *Arch. Biochem. Biophys. Suppl. 1*, 147–151 (1962).
14. S. Arnott, F. A. Scott, W. E. Dea, C. M. Moorhouse, D. A. Rees, *J. Mol. Biol. 90*, 269–284 (1974).
15. M. Duckworth, W. Yaphe, *Carbohyd. Res. 16*, 189–197 (1971).
16. P. Serwer, *Electrophoresis 4*, 375–382 (1983).
17. T. W. Nee, *J. Chromatogr. 105*, 231–249 (1975).
18. P. Serwer, S. J. Hayes, *Electrophoresis 3*, 80–85 (1982).
19. R. J. Wieme, *Agar gel electrophoresis*; Elsevier Amsterdam, 1965, pp. 110–113.
20. K. Weber, M. Osborn, *J. Biol. Chem. 244*, 4406–4412 (1969).
21. C. Schwabe, *Anal. Biochem. 17*, 201–209 (1966).
22. R. C. Switzer, C. R. Merril, S. Shifrin, *Anal. Biochem. 98*, 231–237 (1979).
23. C. M. Merril, In *Electrophoresis '86*; M. J. Dunn, ed. VCH Verlagsgesellschaft Meinheim BRD, 1986, pp. 273–290.
24. H. Blum, H. Beier, H. J. Gross, *Electrophoresis 8*, 93–99 (1987).
25. M. Moeremans, G. Daneels, M. de Raeymaeker, J. de Mey, In *Electrophoresis '86*; M. J. Dunn ed.; VCH Verlagsgesellschaft Meinheim BRD, 1986, 328–329.
26. R. Rohringer, D. W. Holden, *Anal. Biochem. 144*, 118–127 (1985).
27. J. B. Hunter, S. M. Hunter, *Anal. Biohem. 164*, 430–433 (1987).
28. V. Neuhoff, R. Stamm, H. Eibl, *Electrophoresis 6*, 427–448 (1985).
29. I. Olsson, R. Wheeler, C. Johansson, B. Ekström, N. Stafström, R. Bikhabhai, G. Jacobson *Electrophoresis 9, 1*, 22–27 (1988).
30. J. Heukeshoven, R. Dernick, *Electrophoresis 9, 1*, 28–32 (1988).
31. Ö. Gaal, G. A. Medgyesi, L. Vereczkey, In *Electrophoresis in the Separation of Biological Macromolecules*; John Wiley & Sons, 1980, pp. 327–335.
32. A. H. Wardi, G. A. Michos, *Anal. Biochem. 49*, 607–609 (1972).
33. G. Dubray, G. Bezard, *Anal. Biochem. 325–329 (1982)*.
34. C. J. Smyth, T. Wadström. *Anal. Biochem. 65*, 137–152 (1975).
35. H. Harris, D. A. Hopkinson, *Handbook of Enzyme Electrophoresis in Human Genetics*, 1976 North Holland Publishing Company, Amsterdam.
36. A. Kinzkofer, B. J. Radola, *Electrophoresis 4*, 408–417 (1983).
37. A. M. Johnson, *Annals. Clin. and Lab. Sci. Vol. 8, No 3*. 195–200 (1978).
38. U. Beisiegel, *Electrophoresis 7*, 1–18 (1986).
39. H. Towbin, T. Staehlin, J. Gordon, *Proc. Natl. Acad. Sci. U.S.A. 76*, 4350–4354 (1979).
40. W. N. Burnette, *Anal. Biochem. 112*, 195–203 (1982).
41. K. Wallevik, J. C. Jensenius *J. Biochem. Biophys. Meth. 6*, 17–21 (1982).
42. B. B. Samal, *Anal. Biochem. 163*, 42–44 (1987).
43. P. G. Sealy, E. M. Southern, In *Gel Electrophoresis of Nucleic Acids: A Practical Approach*; D. Rickwood, B. D. Hames, eds.; 1982, pp. 39–76.
44. L. G. Öfverstedt, K. Hammarström, N. Balgobin, S. Hjertén, U. Pettersson, J. Chattopadyaya, *J. Biochim. Biophys. Acta 782*, 120–126 (1984).
45. S. Hjertén, Q. Z. Liu, S. L. Zhao, *J. Biochem. Biophys. Meth. 7*, 101–113 (1983).
46. B. An der Lan, R. Horuk, J. V. Sullivan, A. Chrambach, *Electrophoresis 4*, 335–337 (1983).
47. S. Hjertén, S. Jerstedt, A. Tiselius, *Anal. Biochem. 11*, 219–223 (1965).
48. L. Ornstein, *Ann. N. Y. Acad. Sci. 121*, 321–349 (1964).
49. B. J. Davies, *Ann. N. Y. Acad. Sci. 121*, 404–427 (1964).
50. F. Kohlrausch, *Ann. Phys. Chem. 62*, 209–239 (1897).
51. A. Chrambach, T. M. Jovin, *Electrophoresis 4*, 190–204 (1983).
52. T. M. Jovin, *Biochemistry 12, 5*, 871–898 (1973).
53. G. G. Slater, *Fed. Proc. 24*, 225 (1965).
54. J. Margolis, K. G. Kenrick, *Biochem. Biophys. Res. Commun. 27, 1*, 68–73 (1967).
55. G. G. Slater, *Anal. Biochem. 24*, 215–217 (1968).
56. P. Lambin, J. M. Fine, *Anal. Biochem. 98*, 160–168 (1979).
57. M. Lasky, In *Electrophoresis '78*; Catsimpoolas ed.; Elsevier North Holland, 1978, pp. 195–209.

58. G. M. Rothe, H. Purkhanbaba, *Electrophoresis 3*, 33–42 (1982).
59. K. A. Ferguson, *Metabolism 13*, No 10, Part 2, 985–1002 (1964).
60. A. Chrambach, *Molecular & Cellular Biochemistry, 29, 1*, 23–46 (1980).
61. A. Chrambach, *J. Chromatogr. 320*, 1–14 (1985).
62. D. Rodbard, A. Chrambach, G. H. Weiss, In *Electrophoresis and isoelectric focusing in Polyacrylamide Gel*; R. C. Allen, H. R. Maurer, eds.; W. de Gruyter, Berlin-New York, 1974, pp. 62–104.
63. J. L. Hedrick, A. J. Smith, *Arch. Biochem. Biophys. 126*, 155–164 (1968).
64. S. R. Gallagher, R. T. Leonard, *Anal. Biochem. 162*, 350–357 (1987).
65. L. M. Hjelmeland, A. Chrambach, *Meth. Enzymol.* B. Jacoby, ed.; pp 305–318 (1984).
66. S. Hjertén, *Biochim. Biophys. Acta. 736*, 130–136 (1983).
67. G. M. Rothe, W. D. Maurer, In *Gel Electrophoresis of proteins*, M. J. Dunn, ed.; Wright Bristol, 1986, pp. 108–112.
68. J. V. Maizel, Jr. *Science 151*, 988–990 (1966).
69. A. L. Shapiro, E. Vinuela, J. V. Maizel, Jr. *Bichem. Biophys. Res. Commun. 28, 5*, 815–820 (1967).
70. J. A. Reynolds, C. Tanford, *J. Biol. Chem. 245, 19*, 5161–5165 (1970).
71. T. Takagi, K. Tsuji, K. Shirahama, *J. Biochem. 77*, 939–947 (1975).
72. G. M. Rothe, W. D. Maurer, In *Gel Electrophoresis of proteins*; M. J. Dunn, ed.; Wright Bristol, 1986, pp. 37–140.
73. D. M. Neville, Jr. *J. Biol. Chem. 246, 20*, 6328–6334 (1971).
74. A. K. Dunker, R. R. Rueckert, *J. Biol. Chem. 244, 18*, 5074–5080 (1969).
75. U. K. Laemmli, *Nature 227*, 680–685 (1970).
76. M. Wyckoff, D. Rodbard, A. Chrambach, *Anal. Biochem. 78*, 459–482 (1977).
77. A. F. Bury, *J. Chromatogr. 213*, 491–500 (1981).
77. M. M. Margulies, H. L. Tiffany, *Anal. Biochem. 136*, 309–313 (1984).
79. H. D. Matheka, P.-J. Enzmann, H. L. Bachrach, B. Migi, *Anal. Biochem. 81*, 9–17 (1977).
80. D. Best, P. J. Warr, K. Gull, *Anal. Biochem. 114*, 281–284 (1981).
81. H. Anwar, P. A. Lambert, M. R. W. Brown, *Biochim. Biophys. Acta. 761*, 119–125 (1983).
82. J. T. Stoklosa, H. W. Latz, *Biochem. Biophys. Res. Commun. 58*, 74–79 (1974).
83. Y. P. See, P. M. Olley, G. Jackowski, *Electrophoresis 6*, 382–387 (1985).
84. A. Chrambach, *Anal. Biochem. 20*, 150–154 (1967).
85. H. M. Phillips, *Anal. Biochem. 117*, 398–401 (1981).
86. R. T. Swank, K. D. Munkres, *Anal. Biochem. 39*, 462–477 (1971).
87. D. Bothe, M. Simonis, H. von Döhren, *Anal. Biochem. 151*, 49–54 (1985).
88. J. P. Segrest, R. L. Jackson, E. P. Andrews, V. T. Marchesi, *Biochem. Biophys. Res. Commun. 44*, 390–395 (1971).
89. H. Glossman, D. M. Neville, Jr. *J. Biol. Chem. 246, 20*, 6339–6346 (1971).
90. S. Panyim, R. Chalkley, *Arch. Biochem. Biophys. 130*, 854–857 (1969).
91. C. A. Nelson, *J. Biol. Chem. 246*, 3895–3901 (1971).
92. B. Karlstam, *J. Chromatogr. 211*, 233–238 (1981).
93. K. Cheong (Kim), N. Arold, M. Weber, V. Neuhoff, In *Electrophoresis '86*; M. J. Dunn, ed.; Walter de Greyter & Co., Berlin- New York, 1986, pp. 334–337.
94. B. R. Oakley, D. R. Kirsch, and N. R. Morris, *Anal. Biochem. 105*, 361–363 (1980).
95. J. V. Castell, I. Guillén, A. Móntoya, Mª. J. Gómez-L, In *Electrophoresis '86*; M. J. Dunn, ed.; Walter de Greyter & Co., Berlin- New York, 1986, pp. 330–333.

13 ISOELECTRIC FOCUSING IN GELS

Torgny Låås

Research & Development
Pharmacia LKB Biotechnology AB
S-751 82 Uppsala
Sweden

13.1 Introduction

Isoelectric focusing (IEF) can be described as electrophoresis in a pH gradient set up between a cathode and an anode with the cathode at a higher pH than the anode. Proteins, being amphoteric species will be positively charged at pH values below their pI and negatively charged above. This means that wherever a protein is in the pH gradient it will migrate towards its pI. (Fig. 13–1.)

Ikeda and Suzuki were the first to apply the principles of isoelectric focusing in their isolation of glutamic acid from plant protein hydrolysates[1] while Williams and Waterman were the first to describe the basic principles behind today's IEF clearly[2]. The first separations resembling IEF as we know it today were performed by Kolin in the 1950's in sucrose gradients composed by careful layering of buffers with different pH values on top of each other[3].

The major obstacle at that time was the stabilization of the pH gradient. It was not until Svensson (later named Rilbe) and Vesterberg managed to synthesise "carrier ampholytes" with properties to create and stabilize the pH gradient that IEF could be developed into everyman's tool as we know it today[4]. The most essential property of good carrier ampholytes is buffer capacity at the isoelectric point. Svensson and Vesterberg called their pH gradients "natural" since they developed automatically during electrophoresis, in contrast to "artificial" pH gradients created in advance.

The use of artificial pH gradients was revived by the commercialisation of Immobiline. With Immobiline™ the pH gradient is literally immobilised as part of the gel matrix[5]. Many of the problems connected with conventional IEF are circumvented with the Immobiline gradients while some new problems arise, as will be discussed below. The two systems therefore complement each other.

13.2   Theory of IEF

Under the influence of the electrical force the pH gradient will be established by the carrier ampholytes, and the protein species focused at their isoelectric points as described in Fig. 13-1. The focusing effect of the electrical force is counteracted by diffusion which is directly proportional to the protein concentration gradient in the zone. Eventually, a steady state is established where the electrokinetic transport of protein into the zone is exactly balanced by the diffusion out of the zone, Fig. 13-1.

From the factors that regulate the widths of the protein zones and the distance between the zones, Svensson and Vesterberg derived an equation for the resolution of two similar proteins[6], based on the following assumptions:

1. Straight and continuous pH gradients, dpH/dx.
2. Constant field strength, E.
3. The two different proteins have the same diffusion coefficient, D.
4. The electrophoretic mobility change with pH, du/dpH, is constant and the same for both proteins.
5. Two closely spaced proteins are considered separated when the position of their peak maxima differs by 3 standard deviations or more.

The minimum difference in pI, (ΔpI), for two proteins to be resolved is then

$$\Delta pI = -3 \sqrt{\frac{D(dpH/dx)}{E(du/dpH)}} \tag{13-1}$$

From the equation it is seen that reducing the *diffusion*, D, would increase resolution. With a given separation, the only way to accomplish this is to increase the viscosity of the medium. Inert non-charged substances such as sucrose, glycerol etc. may be added or the experiment can be performed in a sieving media such as a high concentration PAA gel.

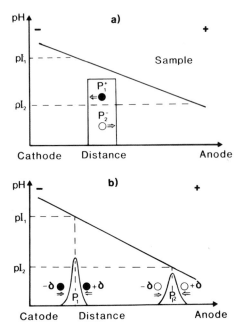

Figure 13-1. Principles of isoelectric focusing. **a)** Schematic illustration of a sample with two proteins P_1 and P_2 placed in the centre of a pH gradient. P_1 with the more alkaline pI is positively charged at the application spot and the acidic P_2 is negatively charged. Under the influence of the electric field the two proteins start to migrate as indicated by the arrows. **b)** As the proteins approach their pI, they gradually become less and less charged. The proteins will thus concentrate at the position where pH = pI. However, the proteins cannot concentrate in an indefinitely concentrated zone. Zone widening by diffusion is inevitable. Any protein molecule diffusing away from pI will acquire a net charge and be transferred back to pI again by electrophoresis. A balance will be set up between electrophoretical accumulation at pI and diffusion.

Increased viscosity will however also affect the mobility of the proteins. Not only will this make the whole experiment slower, but also counteract the resolution by decreasing the factor du/dpH in the equation. Therefore, increasing the viscosity is not generally a successful way to improve the resolution although it may explain why there is a clear tendency for better resolution in sieving PAA gels than in the more porous agarose gels.

Since the diffusion coefficient is inversely related to molcular size it follows that larger proteins will tend to focus better than smaller ones, other things being equal. IEF of peptides is particularly difficult although with suitable precautions good results can often be obtained[7].

The shallower the *gradient, dpH/dx*, (low value of dpH/dx), the further apart will two proteins be and hence better separated. Note however that the factor only applies as the square root (as in fact all other factors as well). There are however some drawbacks with the use of extremely shallow gradients: Long focusing times since proteins must migrate a relatively long distance close to the pI with very low charge; Only the limited number of proteins with pI values within the narrow pH interval can be analysed simultaneously; The carrier ampholytes may not manage to maintain a completely smooth pH gradient.

This is not applicable to IEF with Immobiline. IEF in very shallow gradients is in fact the greatest advantage with Immobiline IEF.

Higher *field strength*, *E*, will not only increase the resolution, the experimental time is also reduced. Too high field strength may give heat problems if the cooling is inefficient, especially when focusing in the very basic or acid pH regions or if carrier ampholytes with unevenly distributed conductance are used. Recommendations by manufacturers are generally safe.

The approach should always be to work with the highest "safe" field strength[8].

The higher the *pH dependence of the mobility*, (du/dpH), the better the focusing. A high electrophoretic mobility close to the pI will efficiently transfer diffused protein molecules back to pI. This is essentially an intrinsic factor of the protein that cannot be manipulated. The effect of modifying mobility by affecting the viscosity will be counter-acted by the effect of viscosity on diffusion as already discussed and the overall effect difficult to predict. A high value of du/dpH results from the presence of many groups with pK_a-values close to the pI. Statistically this is more likely to be the case for a larger than for a smaller protein. Both du/dpH and diffusion thus favour the focusing of large proteins and the influence of these factors explains the difficulties of focusing small proteins and peptides to sharp zones.

Good indications of the prospects for IEF will be obtained by titration curve analysis (p. 389). Proteins with shallow profiles close to their pI are less likely to focus well than the ones that cross the pI at a steep angle.

13.3   Formation of Natural pH Gradients

13.3.1   Carrier Ampholytes and pH Gradient Formation

The formation of a pH gradient is schematically illustrated in Fig. 13-2. Hydrogen ions form at the anode and hydroxyl ions at the cathode in the electrode reactions. This results in regions of low and high pH near the anode and cathode respectively and steep pH gradients as one moves into the bulk solution. An amphoteric species with a pI lower than the average pH in the system will concentrate in the steep gradient close to the anode. A substance with good buffering capacity at its pI will create a pH plateau around its pI. Given a sufficient number of such substances with evenly distributed pI values, their corresponding plateaux will overlap, resulting in a continuous pH gradient. The amphoteric substances that form and stabilize the pH gradient are collectively called "carrier ampholytes".

The most essential property for a good carrier ampholyte molecule is a good buffering capacity at its isoelectric point[9,10]. This requires many pK values close to the isoelectric point for each molecular species, making most naturally occurring ampholytic substances, especially most naturally occurring amino acids, useless as carrier ampho-lytes. Svensson realized that the only way to produce suitable carrier ampholytes was to synthesize substances with the required properties. It was not until the first synthetic carrier ampholytes were successfully prepared that isoelectric focusing could be devel-oped into the practically useful technique of today[4].

The established pH gradient is maintained by hundreds or thousands of carrier ampholyte molecules lined up in order of pI with partially overlapping distributions. Since there are no other ionic species in the system, each carrier ampholyte must act as

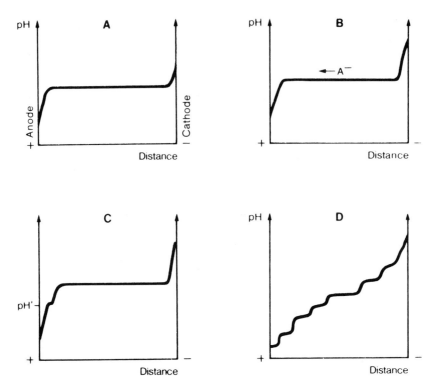

Figure 13-2. Formation of the pH gradient by carrier ampholytes. **A** Imagine a system with electrolysis of water. Electrode reactions will produce H^+ at the anode and OH^- at the cathode causing the pH to change close to the electrodes. **B** Let us introduce the amphoteric substance A in the system. If A has an acidic pI it is negatively charged in the center of the system. Under the influence of the electric field, A will migrate towards the anode. **C** A will accummulate at its pI in the (steep) pH gradient close to the anode just as described for the proteins in Fig. 13-1. Suppose further that A has a good buffering capacity at its pI. A will then tend to create a pH plateau at that position. **D** Given a number of species similar to A but with different pI values, a stepwise pH gradient will be formed. If the number of different "A-molecules" is large enough, (thousands as in commercial carrier ampholyte mixtures) the different steps are smoothed out to a continuous even pH gradient with the different carrier ampholytes extending through the whole system, each focused at its pI.

counter ion to other carrier ampholytes. Consequently, each position in the pH gradient will have a unique chemical composition. Electrical conductance and buffer capacity will therefore vary over the pH gradient. Regions with low buffer capacity are more prone to distortion. In preparative experiments with high protein loads, buffering capacity from the proteins may affect the pH gradient.

Local heating will occur in the regions with the highest field strength (lowest conductance) and these regions will determine how high an overall voltage can be used[11]. Consequently, other regions with lower field strength will not be focused at optimal conditions. Optimal conditions over the whole pH gradient thus requires even field strength (conductance) and buffering capacity across the gradient, Fig. 13-3.

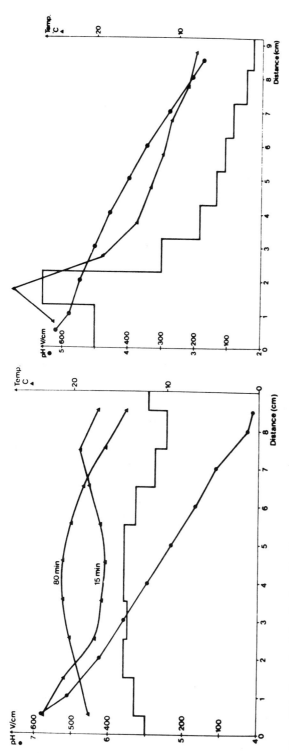

Figure 13-3. Example of distribution of buffer capacity and field strength over the pH gradient. The ideal carrier ampholyte mixture has a completely even distribution of buffer capacity and field strength over the pH gradient to assure even focusing and even heat evolution. Pharmalyte 4-6.5 comes close to this ideal whereas Pharmalyte 2.5-5 has much more uneven properties. With this pH interval it is inevitable that the high conductance contribution from the protons at the acid end of the pH gradient reduces the field strength in this region. The consequence of this is poor focusing in the acidic region because of the low field strength and high temperature in the other end of the gradient with risk of over heating. (Reproduced from Ref. 11 with permission from Pergamon Press):

381

13.3.2   pH Gradient Engineering

A large number of carrier ampholyte mixtures are available giving different pH gradients. Many can also be obtained in precast gels ready to use. The optimal pH gradient will depend on the purpose of the experiment. For screening purposes, a broadrange interval (pH 3–10 or similar) should be used. A narrow pH range interval is useful for careful pI determinations or when analysing proteins with very similar pI points. Generally, one should not use a narrower gradient than necessary because the shallower gradient will lead to longer focusing times and more diffuse bands. When choosing pH gradient one should be aware that the interval stated by the manufacturer can only be an approximation. The exact gradient obtained depends on many factors such as choice of electrolyte solutions, gradient medium (PAA or agarose), focusing time etc.

Despite the large number of pH intervals available, there may be occasions where none of them fits perfectly. In such cases one can either choose to work with Immobiline or use "pH gradient engineering" in any of the following variants:

1. Extend a given pH interval by adding carrier ampholytes covering the adjacent or a partly overlapping region. Extension into the extreme pH ends can be accomplished by adding acidic[12] or basic compounds[13].
2. Expand a certain pH area by adding an amphoteric substance, "spacer", such as an amino acid. The spacer should be a "bad" ampholyte so that it does not focus too well[14,15,16,17].
3. Extend a certain pH range by manipulating the thickness of the gel. The gradient will be shallower in areas with thinner gel.[18,19].
4. Manipulating the carrier ampholyte concentration will also affect the steepness of the final gradient. Areas with lower concentration will give shallower gradients[18].

The different methods can be combined, as was demonstrated by Gill[19].

Generally IEF will give a true representation of the isoelectric spectrum of the sample. However, IEF of immunoglobulins in standard carrier ampholyte mixtures results in distinct bands in the otherwise continuous smear of immunoglobulin molecules. This was shown to depend on heterogeneity in the carrier ampholyte distribution[20]. For a more truthful representation of the distribution of pI points in an immunoglobulin sample, the best results were obtained in mixtures of different carrier ampholyte preparations. A mixture of three different Pharmalyte intervals to maximize the number of carrer ampholytes in the interesting region was found to give the best results[20].

13.4   Experimental Techniques

13.4.1   Equipment for Preparative and Analytical IEF

Traditionally *preparative IEF* has been performed in a vertical, cooled glass column stabilized towards heat convection by a density gradient of sucrose, glycerol or similar substance[21]. The many drawbacks of this technique can be circumvented by performing the separation in a horizontal bed of Sephadex[22,23]. Isoelectrically precipitated protein will not move in the Sephadex and disturb other regions. More efficient cooling allows higher voltages to be used. Although being easy in principle, the technique requires some practice to get the right slurry density when preparing the bed, and in addition a new separation problem is created, the elution of the protein out of the gel. Furthermore the consumption of expensive carrier ampholytes is high.

Since both techniques have severe drawbacks, the first choice when preparative separation of protein according to pI is required should be chromatography by chromatofocusing (Chapter 5). Some proteins tend, however, to form complexes or aggregates at the low ionic strength used in chromatofocusing and in such cases preparative IEF may still be the best alternative.

Analytical IEF is performed exclusively in gels of either PAA or agarose. PAGIF can be performed vertically in gel rods or horizontally in thin gel layers. PAGIF in gel rods is hardly used today except as the first step in 2-D electrophoresis (Chapter 15) and will therefore not be dealt with further here. Horizontal slab gel IEF is the technique of choice when high resolution, high reproducibility and high quality results are required. It is ideal for screening and comparisons of many samples run side by side. In comparison with agarose, PAA gels tends to give sharper bands, fewer problems with electroosmosis and gradient drift (see below) and generally more reproducible results. Precast PAGIF slabs ready to use are commercially available for both conventional and Immobiline horizontal IEF as well as for the modern electrophoresis machines.

As already mentioned, IEF in agarose requires special quality agarose, and even then focusing does not always proceed as troublefree as with PAA, especially at extreme pH values. Even high quality agaroses contain some sulphate and carboxylic groups[24]. At the extremely low ionic strength conditions used in IEF, these groups will generate electroosmosis with concomitant problems. Only the purest forms of agarose are suitable for IEF[25,26]. To circumvent the problem the agarose can be "charge balanced" by careful addition of positively charged groups[27]. In comparison with PAA, agarose offers the following advantages:

1. Easy gel casting with non-toxic substances
2. Suitable for specific detection methods such as immunoprinting, immunoblotting and certain enzymatic staining procedures,
3. Better focusing of large proteins due to the larger pores in the agarose gel.

13.4.2 Casting of Gels

PAA gels down to *ca.* 0.5–1 mm in thickness are most easily cast in gel cassettes composed of two glass plates sealed with a rubber gasket. The gel composition is usually T5C3 or similar. The gelling mixture of acrylamide monomers, carrier ampholytes and TEMED is prepared and carefully degassed by vacuum. Usually the casting mixture also contains 10% of sorbitol, glycerol or similar substance. The addition of an electrophoretically inert substance of this kind increases the osmotic pressure of the system, thereby minimizing the formation of ripples on the gel surface during IEF. Immediately before filling the cassette, the polymerization reaction is initiated by the addition of a freshly prepared solution of ammonium persulphate. The amount of persulphate should be kept low, since excess ionic components from the polymerization mixture tend to give disturbances (wavy bands, pH gradient modifications). The low concentration of persulphate makes degassing critical since oxygen is an effective inhibitor of polymerization. It should also be realised that the persulphate decomposes in solution.

It is customary to prepare a stock solution of acrylamide and store it over some ion exchange material (Mixed Bed in H^+ and OH^- form). Detailed procedures are provided by carrier ampholyte manufacturers.

For casting *ultra-thin PAA gels*, either the "flap technique"[28,29] or the "sliding technique" is recommended[30]. When preparing *agarose gels*, the agarose is first dissolved in boiling water to the desired concentration (commonly *ca.* 1%). Addition of a neutral substance such as sorbitol to 10% will improve the physical properties of the gel, giving less ripple formation during IEF and less risk of water extrusion. After cooling of the solution to 70–80°C the carrier ampholytes are added and the solution carefully mixed. The gel is then cast by simply pouring the viscous solution on the gel support. The surface tension will prevent the solution from overflowing. As gel support a glass plate can be used or preferably a specially treated plastic foil (GelBond™). A few drops of water between the plastic foil and the flat support (glass plate) pressed to a thin film will assure that the plastic foil stays flat during gel casting[27].

13.4.3 Sample

The sample should have as low an ionic strength as possible. Too high a salt concentration in the sample will result in curved protein bands (see 13.7). It is seldom necessary to concentrate the sample, due both to the concentrating effect of the IEF process itself and to the extremely sensitive detection procedures now available. If necessary, concentration is most conveniently done in an Amicon concentration cell.

Crude extracts may be very viscous due to the presence of nucleic acids. Treatment of the sample with DNAse will help. An even simpler alternative that sometimes works is to apply the sample close to the anode. The negatively charged nucleic acid will then quickly separate from the proteins which can then focus normally. Normally the sample is applied with the use of a plastic mask with cut out holes or with the use of filter paper pieces soaked in sample. Plastic masks work excellently on PAA gels with their sticky surface. On agarose gels, the samples have a tendency to penetrate under the plastic. The use of filter paper pieces soaked in sample is therefore generally a safer method for sample application on agarose gels. If the sample is very crude, this method also has the advantage that precipitated proteins or other solid material in the sample will stay in the filter paper and can be removed from the gel. High quality paper without unspecific adsorption of protein, or other undesired side effects, must be used. Such papers are available from companies providing electrophoresis equipment.

A good guide to the best sample application spot can be obtained by titration curve analysis (13.4.8). If possible, apply the protein from the side with the highest electrophoretic mobility. Any regions with pH instability will also be revealed by this analysis.

One should also try not to apply the sample directly over the area of greatest interest since there will always be some disturbances from precipitated protein etc. on the spot of sample application.

One advantage of IEF is the ability to apply very crude samples. Soluble proteins may be analysed by applying pieces of tissue directly on the gel surface[31,32,33,34]. Proteins will diffuse out of the tissue and into the gel where they will focus.

In preparative IEF in Sephadex[®], the sample is usually mixed with the carrier ampholytes and slurry and applied as a sample zone with the aid of a special sample application tray[23].

Within reasonable limits, the sample concentration is of less importance for the IEF process than the total amount of each protein. The optimal amount will depend on a number of factors. With Coomassie staining of total protein after focusing in gels of *ca.* 10 cm length 10–50 μg protein or 1–5 μg per protein band will generally give a good result. In shorter gels with steeper gradients the bands will be narrower and smaller amounts will give a suitable staining intensity[35]. With silver staining the amounts should be reduced about 50 to 100 times.

The optimum amount of sample depends on the purpose of the experiment. If maximum sharpness is required for accurate pI determination, the smallest detectable amount should be used. If, on the other hand, the whole protein pattern is to be screened or small amounts of impurities are to be looked for, larger amounts must be used. When specific detection methods such as immunochemical or enzymatic detections are being used, much smaller amounts can be detected. With radioactively labelled antibodies down to picogram amounts of protein can be detected[36].

13.4.4 Controlling Experimental Parameters, The "Steady State"

At the beginning of the experiment when the voltage is applied, the majority of the carrier ampholytes find themselves in a pH far away from their pI values. They are therefore highly charged and the electrical conductivity of the system is good. As the pH gradient forms and the carrier ampholytes concentrate at their pI points, their electrical charges and consequently also the electrical conductivity of the whole system approaches zero. The establishment of the pH gradient can therefore be followed by monitoring the change of the electrical properties of the system. Generally IEF experiments are performed at constant power (wattage). The pH gradient developed can then be followed by monitoring either the increase in voltage towards an asymptotic steady state value or the decreasing current (Fig. 13-4).

However, the carrier ampholytes move much faster than the proteins, so focusing must continue for some time after the establishment of the pH gradient. How long this is depends on the electrophoretic mobilities of the proteins in the sample. The time required to reach the pI is a function of the protein mobility over the pH values it has to pass through and the voltage. For a given experimental set up (gel size, type and concentration

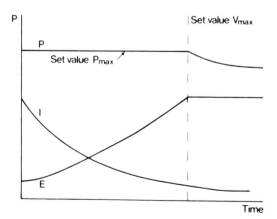

Figure 13-4. Voltage (E), power (P) and current (I) during a typical IEF experiment run initially at constant power and changed to constant voltage close to the end of the experiment.

and sample application spot), the necessary extent of the experiment is most accurately expressed in volt-hours (Vh). It was found that, with the same experimental set-up, the same degree of focusing was reached after the same number of volthours independent of the voltage used and the source of the carrier ampholytes[37,38]. With a focusing distance of *ca.* 10 cm, the fastest proteins were focused after *ca.* 1500 Vh, whereas the slower ones needed at least *ca.* 3000 Vh[37].

Once the pH gradient is established and the proteins focused at their isoelectric points a "steady state" is reached which ideally should persist indefinitely. However, in practice a true steady state is never established but the system gradually changes, mainly through the gradient drift, and eventually deteriorates. Consequently, for each experiment there is an optimal duration of the focusing. The carrier ampholyte manufacturers give guidelines on how long to extend the experiments under different conditions.

If the aim of the experiment is to determine an unknown isoelectric point, the optimal focusing time has to be carefully determined in each case. Since the proteins have different mobilities they will need different times to reach their isoelectric points. When the position of the protein band is the same whether it is applied at the acidic or the basic side of the gel, the protein is with certainty focussed at its true pI. Finding the protein band at the same pH after focusing for different volthours is also a good indication that the steady state has been reached.

The Vh are registered most accurately and conveniently with a volthour registering device. Parameters that must be specified for the Vh integration to be meaningful and the experiment to be reproducible are: Gel type and composition, temperature (preferably that in the gel, but at least the cooling conditions should be given), carrier ampholyte used, electrode distance, point of sample application and additives if any. Gel width and thickness and carrier ampholyte concentration are theoretically not needed to define the steady state conditions as these factors will be compensated for in the Vh integration, but it is recommended that they are given anyway. The carrier ampholyte concentration especially can affect other aspects of the IEF experiment as for example pH gradient stability or protein solubility.

13.4.5   Measuring pH and Determination of pI

To obtain reproducible and relevant results for the comparison of different samples, or for an approximate estimation of the pI spectrum of the sample, an approximate pH gradient determination may be adequate. To determine a proper pI value, one must first assure that the protein really is focussed at its pI. Focusing for 20–50% longer than the minimum number of Vh will assure this (see above). Then the pH must be measured accurately which is not a trivial task. In IEF it is especially important that consideration is given to *temperature* and *solvent composition*. Glass electrodes used for pH measurements are normally calibrated with standard buffers at carefully specified compositions. These can either be prepared in the laboratory according to formulae given by the National Bureau of Standards[39] or bought premade ready to use. The theory of pH determination is quite complicated. Here we will only discuss factors of special relevance in connection with IEF.

The phenomenon called "*liquid junction potential*" (the potential drop from glass to liquid) enters into the output of any pH reading. Any factor that affects the chemical potential of the liquid will also affect the liquid junction potential. Consequently, for accurate pH readings, the conditions must be as similar as possible to those used at the calibration. Factors of particular importance are stirring, temperature and solvent composition. pH readings in solutions with significantly different composition to the standard buffers must be interpreted with great care[40,41,42,43,44]. For example, pH readings of solutions containing high concentrations of urea should be referred to as "apparent" pH or properly corrected[43].

Measuring high pH values accurately is particularly difficult. At high pH, carbon dioxide in the air is rapidly absorbed. Standard buffers should therefore be prepared fresh from CO_2 free water and discarded after use. Exposure to air of the gel surface of a basic pH gradient will lower the pH. The pH measurement should therefore be taken as soon as possible[45].

pK_a values, and consequently pI values, are temperature dependent. This is especially prominent at basic pH[46]. The temperature of interest is obviously that in the gel where the focusing takes place, which will necessarily be significantly higher than that of the coolant. If carrier ampholytes with an uneven conductivity profile are used, the temperature will be different at different places in the gradient[8,11] (Fig. 13-3). By the use of, for example, a thermocouple thermometer it is relatively easy to measure the temperature of the gel at steady state towards the end of the experiment. To avoid complicated and not very exact temperature compensations, it is advisable to calibrate the pH electrode at the temperature at which the reading is to be taken.

A much easier way to determine the pH gradient accurately is to use pI marker proteins with carefully determined pI points and the guides presented in the manuals from some pI protein marker manufacturers[47]. By using these guides, the experiment can be arranged so that the protein of interest focuses at a given temperature, and its pI can be determined by comparison with the positions of the pI marker proteins. For the sake of convenience, and since many pK values are reported at room temperature, it has been recommended that pI determinations are made at 24–25°C[8].

13.4.6   IEF at Acid and Basic pH:es

The electrical conductance of the system should be even over the whole pH gradient for optimal performance. At low pH (below pH 3–3.5) this condition cannot be fulfilled due to the high concentration of protons. The electrical field strength at these pH values is thus very low, making focusing of acid proteins at their pI very slow and uncertain.

Another problem with focusing acid proteins is related to the nature of the proteins. Acid proteins usually contain large amounts of acidic carbohydrates making them very difficult to precipitate with standard procedures. Also, the binding of common stains, such as Coomassie, depends to a large extent on ionic forces between the negatively charged stain and the proteins. At the acid pH used for the staining, "normal" proteins are highly positively charged in opposition to the acidic proteins. Very acidic proteins therefore stain poorly with standard procedures.

The extreme alkaline end of the pH scale also offers special problems that one must be aware of:

1. The stability of the PAA gel is not very high at basic pH values. The amido bond is sensitive to alkaline hydrolysis giving rise to negative charges on the PAA gel matrix causing electroosmosis and gradient drift. PAA gels with basic carrier ampholytes should therefore not be stored for more than a few days in the refrigerator.
2. Carbon dioxide in the system will dissolve at the alkaline end of the gradient lowering the pH in that region. It will then migrate as bicarbonate ions towards the acid region

where it will evaporate as CO_2. The result is a continuous acidification of the alkaline end of the pH gradient. In addition, the bands tend to be tilted since the gel surface is more affected by the CO_2 then the bottom of the gel. The migration of the pH gradient also makes protein bands fuzzier and broader than normal.

3. Agarose gel tend to give problems with gradient drift at high pH[47]. Generally PAA gels give better performance. In a closed system, such as a gel rod for use in 2-D electrophoresis, carbon dioxide in the system cannot exaporate but will displace carrier ampholytes and concentrate in a pH plateau around pH 5-7. This region will thus contain few carrier ampholytes, making pH measurements unreliable.

The remedy of CO_2 problems is to exclude CO_2 (and HCO_3^- ions) from the system as far as possible. Use degassed or nitrogen saturated water when preparing the buffers. This is especially important with the basic buffers and electrode solutions since HCO_3^- in these buffers cannot be removed by degassing. In flat bed focusing the equipment can be flushed with nitrogen or argon and in rod focusing the cathodic buffer should be prepared as free as possible of carbon dioxide. CO_2 traps in the form of 1M NaOH in a vessel or sponge close to the gel will also improve the result. As already mentioned the pH readings should be taken as soon as possible to minimize the risk of pH being reduced by CO_2. The frequently used procedure of soaking gel pieces in water before pH measurement is a very unreliable method, especially at alkaline pH values and in the carbon dioxide pK-region at pH 5-7.

If required, the pH span may be extended by addition of basic amino acids such as lysine and arginine[13].

13.4.7   Additives: IEF in Urea

In principle it should be possible to use all non-charged, or zero net charged, additives without problem. This is also generally the case. The most commonly used additives are detergents and urea. Zwitterionic and non-ionic detergents will not interfere with the focusing process provided they are pure. Unfortunately this is not always the case. In such cases the charged contaminants must be removed from the detergents by for example passage over a zwitterionic ion exchanger in H^+ and OH^- form.

The use of urea is accompanied by other problems as urea decomposes to ammonium isocyanate which is very reactive towards primary amines in the proteins (carbamylation)[48]. This reaction transfers the positive amino group into a negative group and lowers the pI.

Fortunately the rearrangement of urea to isocyanate and ammonia is a slow reaction at temperatures at and below room temperature. In the cold, urea solutions can be kept safely for at least a week but for important experiments use freshly prepared solutions of high quality urea and/or purify the solution over a zwitterionic ion exchanger as described for detergents. Also note that by prefocusing before sample application the isocyanate will migrate out of the gel.

The use of high concentrations of urea in agarose IEF is particularly difficult. The high temperatues needed to disolve the agarose will enhance the isocyanate formation of the urea and in addition high concentrations of urea will impair the formation of agarose gels. These problems can be handled by suitable modification of the gel casting procedure and by using a 2% agarose gel[49]. Special precautions must be taken when measuring pH in gels containing urea[43].

13.4.8 Titration Curve Analysis

By using a combination of IEF and electrophoresis it is possible to prepare a physically visible titration curve of a protein. A square slab gel for horizontal PAGIF with broad range carrier ampholytes (pH 3–10 or similar) is prepared. Normally the gel casting mould is prepared so that a sample application well extanding almost across the whole gel is formed. An IEF experiment is run without sample just enough to assure that a stable pH gradient is formed. The sample well should be along the pH gradient so that pH is acidic at one end and basic at the other. The gel is then rotated 90° and the well filled with sample. The sample is now electrophoresed across the pH gradient. Since a gel with low porosity is used, the electrophoretic mobility is essentially a function of protein charge and the pattern for each protein will correspond to its titration curve[50,51] (13.8.4).

Titration curve analysis can be very useful in the prediction of purification strategies[52,53] (Chapter 4) and can also be used for Ferguson plot analysis (12.4.1).

13.5 IEF in Immobilised pH Gradients. Immobiline™

The general theory and most of the factors discussed above for conventional IEF are applicable also to IEF in Immobiline gradients. The presentation below will therefore concentrate on factors of special relevance to IEF in Immobiline.

13.5.1 Principles and pH Gradient Generation

The only principal difference between conventional IEF in carrier ampholytes as described above, and IEF in Immobiline™, is the way the pH gradient is formed and maintained. In 1982 a new technique for creating and maintaining the pH gradient was made available by the introduction of Immobiline[5]. Immobiline is the trade mark for special derivatives of acrylamide containing amines or carboxylic acid functions. Immobiline is used to form immobilised artificial pH gradients in PAA gels. At present there are seven different types, four basic (pK values 6.2, 7.0, 8.5 and 9.3) and three acidic ones (pK values 3.6, 4.4 and 4.6), of such chemicals available.

When casting an Immobiline gel, two solutions of gel monomers are made. To one of the solutions is added, in addition, Immobiline titrated to the acid-end pH and to the other solution Immobiline titrated to the alkaline-end pH. These solutions are then mixed to form a continuous gradient during the filling of the gel cassette. The practical procedure is quite analogous to the casting of PAA gradient gels but with a Gel Bond PAG plastic support included in the cassette. The two Immobiline solutions will titrate each other, and, if properly chosen, form a straight pH gradient. To stabilise the gradient, the botton solution (usually the most acidic) is made denser with glycerol (25%w/v). Since the Immobiline molecules contain the acrylic double bond they will be covalently incorporated in the gel network during the polymerization process and hence the pH gradient will be covalently immobilised. When polymerisation is complete, the cassette is opened and the gel on the plastic support is used for flat bed IEF as in the conventional case[54]. The introduction of charges in the gel does not give problems with electro-osmosis during IEF since the gel *net charge* is zero at all pH values. (Compare charge-balanced agarose.)

Before use, the gel must be carefully washed in water to remove residual reagents (at least 1 h for a 0.5 mm thick gel). Care must also be taken to assure that the gel is swollen to the same degree after washing as during casting[55,56]. Narrow range pH gradients are prepared by choosing an Immobiline with a pK within the same desired pH range as the buffering Immobiline. This Immobiline is then titrated with non-buffering Immobiline types having pK values outside the pH range. In this way gradients spanning up to about 1 pH unit can be prepared[5]. In order to make longer pH gradients, more complicated procedures must be used[57]. Detailed instructions on how to prepare Immobiline gradients are obtainable from the manufacturer as well as from the literature[58,59].

Obviously Immobiline IEF can only be performed in PAA gels. It is not recommended to deviate from the gel casting conditions suggested by the manufacturer since this will affect the exact degree of incorporation of Immobiline in the PAA gel and hence the pH gradient. Urea and neutral detergents are well tolerated in Immobiline systems[60].

13.5.2   Characteristics of Immobiline Gels

Immobiline gels have many attractive advantages over conventional IEF gels:

— By definition gradient drift can not take place
— Well defined chemicals give better control of pH gradients
— Ultra flat pH gradients can easily be prepared (down to 0.01 pH unit/cm) with corresponding increase in resolution
— Problems such as wavy bands etc. related to inadequate carrier ampholyte quality are eliminated.
— High loading capacity
— Less sensitive to salts and buffers in the sample

The separation power was clearly demonstrated in the very first paper[5]. It was estimated that bands with pI values differing as little as 0.001 pH unit could be resolved. However, limitations in the technique as compared to conventional IEF soon became obvious: The gel casting is quite cumbersome (This has been solved at least partly with the advent of commercially available premade Immobiline plates); Long preruns[61] and long separation times (overnight runs are standard); Very difficult to measure pH in the gels, in fact it seems to be impossible without the addition of carrier ampholytes to the system[61]; Lower detection limits for protein than in conventional IEF due to loss of sample material[62]; Limited stability of Immobiline causing variation of pH gradients depending on age of Immobiline[63]. The reason for the longer separation times is probably related to interactions between the proteins and the gel matrix. There is strong evidence to suggest that this is primarily due to hydrophobic interaction between the proteins and basic Immobiline molecules[64]. Ionic interactions may also contribute. Consequently, a fraction of the proteins is more or less tightly "bound" to the matrix and only the proteins in solution can actually migrate. This is particularly obvious for extremely charged substances such as histones and nucleic acids which do not move at all in Immobiline gradients[64].

By adding carrier ampholytes to Immobiline systems (Hybrid IEF) it seems that it is possible to take advantage of the best properties from both IEF-systems[62,65,66,67], the only obvious drawback being the relatively cumbersome gel preparation. Immobilised pH gradients can be used with advantage in 2-D electrophoresis. With an automatic electrophoresis machine working with small dimensions, silver stained 2-D Immobiline gels can be produced in only 3.5 h[68].

13.6 Optimizing the IEF System

13.6.1 Choosing IEF Type: Conventional/Immobiline

The optimal system and experimental conditions will of course depend on the separation problem, the type of sample and the purpose of the separation.

For preparative separations, flat bed IEF in Sephadex is generally to be recommended (p. 382). Choose the narrowest possible pH gradient and follow the recommendations of the carrier ampholyte manufacturer. To separate proteins differing by less than *ca.* 0.05–0.1 pH units, IEF in Immobiline gradients should be tried. Proteins can be extracted from the gel by isotachophoresis by any of the suggested techniques[69,70,71,72,73].

In analytical IEF there are essentially three choices:

1. Conventional IEF in agarose,
2. Conventional IEF in PAA and
3. IEF in Immobiline (in PAA gel).

If the protein to be analysed has a very high molecular weight (higher than 150,000 to 200,000) agarose should be the first choice. In the majority of cases, however, IEF in PAA is to be preferred.

In IEF in PAA gel (PAGIF) the choice stands between conventional IEF with carrier ampholytes and IEF in Immobiline gels. It is the author's opinion that the first choice should always be conventional IEF, because of its higher speed and convenience. Staining and pH measurements are also more straight-forward in conventional IEF, which is important when pI values are to be recorded. The preparation of Immobiline pH gradients covering more than *ca.* one pH unit is also quite cumbersome. Neither is the use of Immobiline ready-made plates as straight-forward as using ready-made plates for conventional IEF since the Immobiline plates have to be swelled and washed before use. However, if the problem requires separation of proteins with very small differences in pI and pH gradient engineering does not work satisfactorily, Immobiline IEF (or Hybrid IEF) will be the natural choice. Immobiline IEF should be also chosen in cases where the iso-pH lines must be absolutely straight, for example, when it is important to detect very slight differences in pI in different samples.

Generally it is advisable to use precast gels as much as possible to save time and effort and ensure a high degree of reproducibility.

13.6.2 Gel Length and Thickness

Intuitively it seems advantageous to use long gels in order for the proteins to be spaced apart as much as possible. In practise, however, the use of very long gels is accompanied with a number of drawbacks:

1. The longer the gel, the longer the path for the proteins to migrate. Eventually some proteins may not be able to reach their pI during the lifetime of the pH gradient. At least, the duration of the experiment will be inconveniently long.

2. For practical reasons, very long gels cannot be run at the same field strengths as shorter ones, thus the bands will be less sharp in the longer gel. The actual resolution is thus only marginally increased with very long gels.
3. For a given pH span, the gradient will be shallower in the long gel. The bands will thus not focus as sharply as in the shorter one, reducing the increase in resolution otherwise expected from the longer gel.

After investigating the effect of gel length on overall performance, it was concluded that there is generally little to gain by using gels longer than *ca.* 10 cm and, in fact, considerably shorter gels can be used advantageously in many cases[35,74]. This fact has been exploited in the modern electrophoresis machines working with miniaturized gels[75,76].

The trend in recent years has been to decrease the thickness of the gels used. Thin gels offer advantages such as more efficient cooling, apparently sharper bands (The effect of tilted bands is less severe), fast solvent penetration in staining and destaining and lower consumption of carrier ampholytes. Too thin a gel will on the other hand be accompanied by a number of practical problems: Small nominal differences in gel thickness will become relatively large with concomitant effects on field strength, heat production and gradient distortions. Also because of the small amounts of carrier ampholytes, the whole system becomes more sensitive to all kinds of disturbances, for example from carbon dioxide and evaporation. The optimal gel thickness seems to be 0.2—1 mm for most applications.

13.6.3   Running Conditions: Voltage and power

As discussed above, the sharpest bands and best resolution are obtained with maximum voltage. The limits are set only by the ability of the system to control the heat produced, which is directly proportional to the voltage. The following points should be considered for optimum performance:

1. Use an experimental set up with good cooling capacity. For analytical purposes this is thin or ultrathin (0.2-0.5 mm) layer PAGIF on an apparatus with a metal cooling plate.
2. Use the carrier ampholytes with the most even conductivity profile. In regions with lower conductivity, the voltage and heat production will be higher. These areas will dictate the overall voltage that can be applied. The focusing in the rest of the gradient will suffer from a focusing voltage less than optimum.
3. Do not use more carrier ampholytes and/or Immobiline than needed to stabilize the pH gradient. Excess carrier ampholytes will increase the current and the heat produced, making it impossible to use high voltage.
4. Run the experiment at constant power. This will ensure the maximum voltage during the whole separation, which minimizes separation time and maximizes resolution at the final stage of the experiment.

13.7   Trouble Shooting

A good result in slab IEF requires that the iso-pH lines extend straight over the plate. Uneven gradients give rise to wavy protein bands. With conventional IEF using carrier ampholytes, a number of factors affect the straightness of the iso-pH lines. The most important is the chemical composition of the system, especially the *carrier ampholytes* used. Some carrier ampholyte mixtures are more easily disturbed than others. This depends mostly on the buffer capacity distribution during the pH gradient development. Most disturbances appear during the development of the gradient. Straighter iso-pH lines are, therefore, generally achieved if the pH gradient is allowed to form ("prefocusing") before the sample is applied. Most often the disturbances appear from the anode as "acid fingers", possibly due to the great mobility of the protons when they manage to break through the buffering barrier of the carrier ampholytes.

Uneven *contact between the electrodes* and the gel will affect the gradient. Make sure that the electrodes are straight and in good condition and that the strips are evenly wet and in good contact with the gel. The electrodes must be applied to the strips with a certain pressure.

Sample composition is also very important. Bow-shaped bands generally occur because of excess salt in the sample. The effect of buffering salts is especially strong. Neutral salts, for example NaCl, or buffering salts applied at a non-buffering pH can generally be tolerated up to *ca.* 0.5 M [74]. With reasonable amounts of buffering salts, an improved result may be obtained by changing the sample application site to a pH where the buffering capacity of the salt is smaller. If neither this nor prefocusing is sufficient, the salts have to be removed before sample application. Standard methods such as gel filtration or dialysis can be used. The *ionic strength of the gel casting mixture* should be kept as low as possible. The persulphate concentration, especially, should be kept to a minimum since sulphate emanating from the persulphate is covalently attached to the ends of the PAA chains during the initiation of the polymerisation reaction. Low concentration of persulphate makes exhaustive degassing a necessity, since oxygen inhibits the polymerisation.

The choice of *electrode solutions* will also strongly affect the pH gradient. A suitable electrode solution will compensate for disturbances due, for example, to uneven electrode contact. The choice of electrode solution will also affect the exact span of the pH gradient.

The phenomenon of *gradient drift* is the major reason why conventional IEF is not quite the steady state technique it theoretically should be. The major cause of gradient drift is electroosmosis, although other factors such as the absorption of carbon dioxide at high pH and evaporation at the acidic pH or the different mobilities of the positive and negative forms of the carrier ampholyte molecules may play a part [77]. The gradient drift is mostly towards the cathode ("cathodic drift") reflecting a negatively charged matrix. Sometimes the opposite can be seen, and sometimes both directions of drift may be manifested in the same experiment, reflecting a pH depending net charge of the gel matrix. The physical manifestation of gradient drift in flat bed IEF is liquid transportation from one electrode towards the other. In severe cases liquid droplets are extruded from the gel and/or the gel dries out at the other side, eventually to complete dryness. To eliminate the problem, a gel matrix with lower charge content must be used. Change to higher quality chemicals or try to deionize the monomers as described (p. 384). Be sure that high pH gels are used fresh since the acrylamide deamidates at high pH. It seems that gradient drift can never be completely eliminated with conventional IEF, but it is seldom a problem in practice when using PAA gels or IEF grades of agarose.

Proteins are generally poorly soluble at their isoelectric point. Isoelectric precipitation was often a serious problem when focusing in vertical density gradient stabilized systems. In horizontal systems, both preparative and analytical, isoelectric precipitation is in fact seldom a problem.

Should there be precipitation, urea or detergents may be added to increase the solubility.

13.8   Applications

13.8.1   Analysing Unconcentrated Cerebrospinal Fluid by Agarose IEF and Silver Staining

The protein composition of cerebrospinal fluid is of great interest in the diagnosis of neuronal diseases such as multiple sclerosis (MS). The protein content is, however, too low to allow conventional staining method to be applied without previous concentration of the sample. Silver staining, sensitive enough for this purpose, is not generally compatible with agarose. The method presented here circumvents the problems and makes silver staining possible also in agarose[78]. The essential feature, according to the author, is the inclusion of 5% glycerol in all the solutions. Figure 13-5 demonstrates the increased sensitivity of the silver stain method as compared to Coomassie Blue. Typical CSF patterns from healthy individuals and individuals suffering from MS are shown in Fig. 13-6.

Experimental

Gels containing 0.75% agarose (special IEF grade, Isolab USA) and carrier ampholytes 3–10 (Isolab) were cast as 100×85 mm, 0.8 mm thick layers on plastic foil (GelBond, FMC USA) Fifteen μl sample (CSF or pI markers from Pharmacia) was applied in all cases. IEF was conducted over a 70 mm distance at a final voltage of 250–270 V/cm.

Silver staining: Immediately after IEF the gels were immersed in a fixing solution containing 15% trichloroacetic acid, 5%w/v $ZnSO_4$, and 5% v/v glycerol in distilled water for not more than 10 min. The gels were then washed in 500 ml of 5% glycerol solution for 1 h and in fresh glycerol solution for an additional hour. Gels were completely dried under an infrared lamp at a 30 cm distance for *ca.* 45 min. For silver staining, gels were placed in a clean glass or plastic container. Two solutions were made up:

Solution I: 5% w/v sodium carbonate.
Solution II: 0.2% ammonium nitrate, 0.2% silver nitrate, 1% tungstosilic acid, and 0.52% formaldehyde (mixed in the given order).

Solution II (68 ml) was added to solution I (32 ml) under vigorous stirring and immediately poured on the dried gel. The gel was incubated under gentle agitation until the silver solution turned grey (about 4–5 min) when the gel was quickly rinsed with distilled water, freed from surface stain by careful wiping with moistened Kimwipe™. The gel was put in silver staining solution again and the procedure repeated. Finally, staining was terminated by placing the gel in 1% acetic acid. The gel was soaked in distilled water for 10 min and dried. The pattern is said to be stable indefinitely if protected from light.

13.8.2   Direct IEF of Raw Muscle Tissue without Previous Extraction

Because of the concentrating effect of the IEF process itself, the demands on the composition of the sample are very modest. This may be illustrated by "direct tissue IEF" as described by Saravis[31,32,33] or "spread tissue IEF" (STIF)[34]. In STIF the sample is

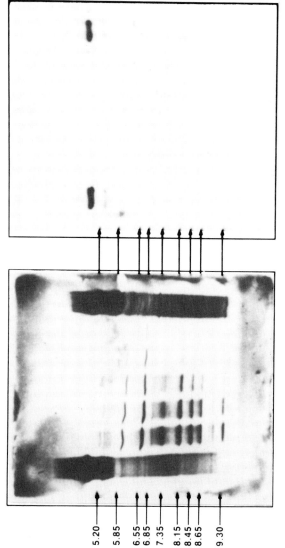

pH

5.20
5.85
6.55
6.85
7.35
8.15
8.45
8.65
9.30

Figure 13-5. A silver stained gel (left) and an identical gel stained with Coomassie Brilliant Blue G-250 (right) after IEF on Agarose. The samples are from left to right: serum diluted 200 times, pI markers with total protein 1.95 µg, 0.97 µg, 0.4 9 µg, 0.10 µg, cerebrospinal fluid from patient with multiple sclerosis diluted 2 times. The anode is at the top. (Reproduced from Ref. 78).

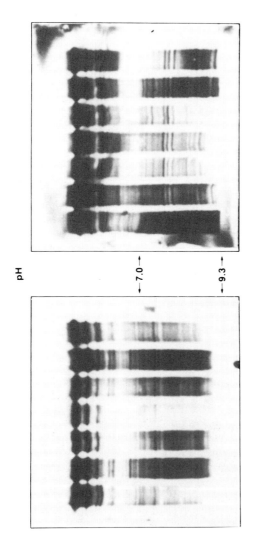

pH

← 7.0 →

← 9.3 →

Figure 13-6. Two silver-stained gels of unconcentrated CSF after IEF on agarose. On the gel to the left samples 2,3,5,6 and on the gel to the right samples 1,2,6,7, are from patients with clinically documented multiple sclerosis. All other samples are normal. The anode is at the top. (Reproduced from Ref. 78).

applied as a thin coat of tissue on a filter-paper applicator. No preparation of the tissue whatsoever is necessary before sample application. Only the contact time has to be optimized to get good results. STIF has been used for species determination of raw meat which is of great interest in food control Fig. 13-7.

Experimental

Agarose gels were prepared by adding 0.9 ml each of Ampholine 4–6 and Ampholine 6–8 to a solution (70°C) of 0.3 g Agarose IEF, 3.6 g sorbitol and 27.0 ml deionized water. The mixture was poured in a 1.0 mm thick gel casting frame with GelBond plastic foil as gel support and left to gel at room temperature for 20 min followed by 2 h at 4°C.

The gels were placed on the cooling plate set at 10°C and briefly blotted with Whatman filter paper to remove excess liquid. Electrolyte wicks were soaked in 0.05 M H_2SO_4 (anode) and 1 M NaOH (cathode). The samples were applied 4 cm from the cathode wick using Whatman nr 1 filter paper 5 × 10 mm. The sample tissue (20–25 mg fresh defrosted tissue) was spread on the filter paper.

After sample application, IEF was performed for 8 min with settings 700 V, 10 W. After 8 min the sample applicators were removed and focusing continued. After an additional 45 min the settings were adjusted to 100 V, 10 W and focusing continued for another 60 min. Total volthours applied were 1500. After IEF, the pH gradient was measured with a surface electrode. Gels were stained according to the instructions for Agarose IEF with Coomassie Blue R-250.

All electrophoresis chemicals and equipment were from Pharmacia LKB.

13.8.3   Analysis of Serum Gc-Globulin Phenotypes by IEF and Immunofixation

Phenotypic characterization of human Gc-globulin is of interest in, for instance, forensic science and clinical medicine. IEF of serum in PhastSystem™ in combination

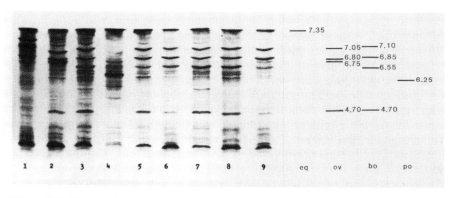

Figure 13-7. The use of spread tissue isoelectric focusing (STIF) for identification of raw muscle tissue. The right hand part illustrates schematically the bands used for identification while the left part shows typical patterns obtained with (1) equine species, (2) ovine species, (3), (5)–(9) bovine species and (4) porcine species. pI values of band at 25°C indicated to the right. (Reproduced from Ref. 34).

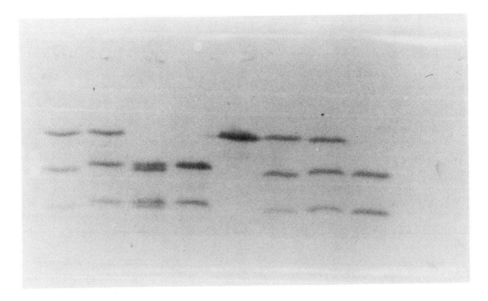

2-1F 2-1S 1S-1F 1S-1S 2-2 2-1F 2-1S 1S-1S

Figure 13-8. Different phenotypes of human Gc-globulin separated by IEF in PhastSystem and analysed by immunofixation. The whole procedure was finished within 1 h. The identities of the different phenotypes are marked in the figure. Cathode at the top. (Reproduced from photograph kindly provided by B. Olsson and J. Olsson, Pharmacia LKB, Uppsala).

with immunologically based detection methods is a fast, reliable and convenient method for this analysis as demonstrated by Fig. 13-8 originally presented at the FEBS Meeting in Ljubljana 1987.

Experimental

IEF was performed in Pharmacia PhastSystem with gels containing Pharmalyte 4.9–5.2. Gels used in PhastSystem are 50×43 mm and the system is completely microprocessor-controlled with automatic switching between the different settings. The following running program was used:

Before sample application the gels were prerun with settings 1000 V, 1.5 mA, 3.5 W for 75 Vh. The samples (0.5 µl human serum) were applied automatically with the sample applicator and IEF was started with settings 200 V, 2 mA, 3.5 W and run for 15 Vh when the settings were changed to 2000V, 2.5 mA and 3.5 W and focusing continued for an additional 510 Vh. Total focusing time was *ca.* 30 min and the gel was thermostated at 15°C.

For immunofixation, cellulose acetate membranes (37×45 mm) were soaked in 250 µl of a solution containing antiserum towards human Gc-globulins (DAKO) diluted 1:3 with 0.05M phosphate buffer pH 7.5, 4% polyethyleneglycol 6000. Immediately after IEF, the cellulose membranes were placed on the gel surface for 1 min. To remove excess non-precipitated proteins, the membranes were washed 3×5 min in 0.05 M phosphate buffer pH 7.5, 0.9% NaCl. The remaining immunoprecipitates on the membranes were stained with 0.05% PhastGel Blue R in 30% methanol, 10% acetic acid. Membranes were destained in 30% methanol, 10% acetic acid and preserved in 10% glycerol in 10% acetic acid.

13.8.4   Electrophoretic Titration of Beef Lactic Dehydrogenase in Agarose-Sephadex and Polyacrylamide

Lactic dehydrogenase is an enzyme composed of four subunits with a native molecular weight of *ca.* 140,000. The subunits exist in two different types H and M. This allows for five different isoenzyme variants, H_4, H_3M, H_2M_2, HM_3, and M_4. Electrophoretic titration curves were successfully produced in both agarose-Sephadex and polyacrylamide[52] (Fig. 13-9), the principal differences being that the bands tended to be sharper in poly-acrylamide and that the electrophoretic mobility was higher in agarose-Sephadex. In this case where all proteins of interest have the same size, this is of no importance since the relative positions of the bands will not be affected. When analysing proteins with greatly differing sizes, however, a highly porous gel such as agarose or agarose-Sephadex is to be preferred.

In this model system the proteins were stained with Coomassie. The same pattern can be obtained from a more complex pattern by specific staining (zymogram, immunological etc).

Experimental

The gels were made in a specially produced casting frame with a trough for the sample well. Polyacrylamide gels were prepared from a stock solution of 10% acrylamide

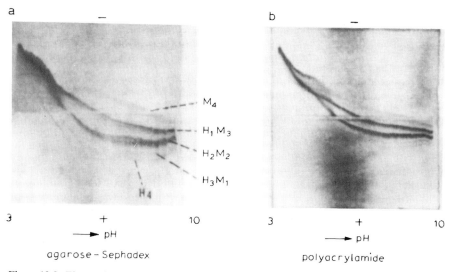

Figure 13-9. Electrophoretic titration curve analysis of beef lactate dehydrogenase isoenzymes in agarose-Sephadex (left) and polyacrylamide (right). Note: (1) The electrophoretic mobility is higher in agarose than in polyacrylamide. The difference would be less pronounced with a smaller protein. (2) The different slopes of the curves where they cross the sample application site (their pI points) indicate how well the isoenzymes are expected to perform in IEF. M_4 would be the least suitable for IEF analysis. (Reproduced from Ref. 52 with permission of Elsevier Science Publishers).

and 3% bisacrylamide deionized with Amberlite MB-3. Gels were cast on glass plates treated with Silane 174 to assure good adherence between gel and glass. To 45 ml gelling solution, sufficient for two gels, was added 22.5 ml acrylamide stock solution, 12 ml 50% glycerol and 3 ml Pharmalyte 3-10 and water to 45 ml. The final gel composition was T5C3. The polymerization was initiated by the addition of 100 μl ammonium persulphate (60 mg/ml). The solution was quickly injected into the casting mould and allowed to polymerize for 90 min at room temperature. The gels were stored at 40°C in a moist chamber after removal from the gel cassette.

Agarose-Sephadex gels for titration curve analysis were cast in the same gel casting frame but on GelBond plastic foil instead of glass plates. To a final volume of 45 ml was added 0.45 g Agarose IEF (Pharmacia), 0.75 g Sephadex G-200 (Pharmacia), and 4.5 g sorbitol and water to 45 ml. The solution was boiled and cooled to 70°C before 3 ml Pharmalyte 3-10 was added. The solution was injected into the casting frame and allowed to gel at room temperature for ca. 45 min when the gels were removed and stored as the PAA gels.

For electrophoresis a Pharmacia Flat Bed Apparatus was used with coolant temperature of 12°C. For the agarose-Sephadex gels the anode strip was soaked in 1 M phosphoric acid and for the PAA gels in 0.04 M aspartic acid. The cathode strip for both gels was soaked in 1 M sodium hydroxide. For pH gradient generation, the gels were electrophoresed with the sample application trough perpendicular to the electrodes. The agarose gel was run for 750 Vh at 15W and the PAA gel 750 Vh at 15 W.

For electrophoresis the gels were rotated a quarter turn, the wicks soaked again in their respective solutions and the sample was applied in the trough. Sample amounts were 50 μl containing ca. 2.5 mg/ml. Agarose gels were run at 1000 V for 100 volthours and PAA gels at 1000 V for 150 Vh.

After electrophoresis the gels were fixed for 30 min in 10% trichloroacetic acid.

13.8.5 Improved Resolution by IEF in Immobiline Gels

The enormous resolving power of IEF in shallow pH gradients stabilized by Immobiline is demonstrated in Fig. 13-10. From the results it was estimated that it is possible to separate proteins with pI values differing as little as 0.001 pH unit[5].

Experimental

Gel casting was made with the LKB 2117-901 gradient gel kit. Gels were 250 × 110 × 0.5 mm and contained slots (5 × 2 × 0.2 mm) for sample application. The glass surface of the cassette was treated with Repelsilane to avoid sticking of the gel to the glass after gel casting. Acrylamide and bisacrylamide were added from stock solutions so as to give a final gel composition of $T = 5$, $C = 3$. Immobiline 4.6 was added to a final concentration of 5×10^{-3} M and titrated with Immobiline 9.3 to the desired end values for the different pH gradient slopes. The acidic Immobiline solution was always used as the heavy solution and contained 20%(v/v) glycerol. Immediately before filling the cassettes, 5 μl TEMED and 5 μl 40% persulphate were added to each chamber as polymerization catalysts.

The gels were polymerized for 1 h at 50°C.

After polymerization the gels were removed from the cassettes and soaked in distilled water for half an hour to remove excess reagents. Before use, the gels were drained to their initial weight.

The IEF was performed overnight with electrodes directly on the gel surface with settings: 10 W, 2500 V and 25 mA.

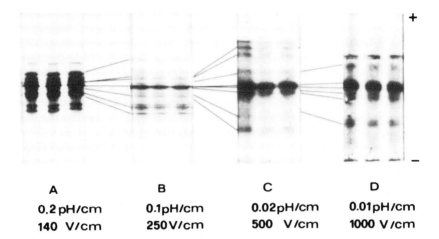

A

0.2 pH/cm

140 V/cm

B

0.1pH/cm

250V/cm

C

0.02pH/cm

500 V/cm

D

0.01pH/cm

1000 V/cm

Figure 13-10. Isoelectric focusing in immobilised pH gradient as compared to conventional IEF with carrier ampholytes. Strip A is ovalbumin focused in a narrow pH gradient of Ampholine and B–D show the same sample in Immobiline gradients of different slopes. Sample amounts per track from left to right 40, 20 and 20 μg. Note the absolutely straight bands in the Immobiline gels and the increased resolution. Also note the band broadening which inevitably accompanies the increased band spacing. It is not practically possible to compensate completely the band-broadening effect by the shallower gradient in extreme cases such as C and D with increased field strength. (Reproduced from Ref. 5 with permission of Elsevier Science Publishers).

The achieve the necessary band sharpening in the experiments with the shallowest gradients (C and D), the electrodes had to be moved closer to the sample (5 resp 2.5 cm) and the final focusing prolonged for an additional 3 hours.

Staining and destaining were performed with Coomassie Blue in the conventional way.

13.9   References

1. K. Ikeda, S. Suzuki, *US Patent 10 5891* (1912).
2. R. R. Williams, R. E. Waterman, *Proc. Soc. Exp. Biol. N.Y. 27,* 56–59 (1929).
3. A. Kolin, *J. Chem. Phys. 22,* 1628–1629 (1954).
4. O. Vesterberg, *Acta Chem. Scand. 23,* 2653–2665 (1969).
5. B. Bjellqvist, K. Ek, P. G. Righetti, E. Gianazza, A. Görg, R. Westermeier, W. Postel, *J. Biochem. Biophys. Meth. 6,* 317–339 (1982).
6. O. Vesterberg, H. Svensson, *Acta Chem. Scand. 20,* 820–834 (1966).
7. T. Bibring, J. Baxandall, *Anal. Biochem. 85,* 1, 1–15 (1978).
8. T. Låås, I. Olsson, L. Söderberg, *Anal. Biochem. 101* 449–461 (1980).
9. H. Svensson, *Acta Chem. Scand. 16,* 456–466 (1962).
10. H. Svensson, In *Protides Biol. Fluids* H. Peeters, ed. Pergamon Press, 1967, Vol. 15, pp. 515–522.
11. T. Låås, I. Olsson L. Söderberg, In *Protides of Biological Fluids* Vol. 27. H. Peeters, ed. 1979, pp. 683–686.
12. U.-H. Stenman, R. Gräsbeck, *Scand. J. Clin. Lab. Invest. Suppl. 116:26,* (1971).
13. G. Yao-Jun, R. Bishop, *J. Chromatogr. 234,* 459–462 (1982).
14. M. L. Caspers, Y. Posey, R. K. Brown, *Anal. Biochem. 79,* 166–180 (1977).
15. P. Gill, J. G. Sutton, *Electrophoresis 6,* 23–26 (1985).
16. P. Gill, *Electrophoresis 6,* 282–286 (1985).
17. J. Lizana, I. Olsson, *J. Clin. Chem. Clin. Biochem. 22,* 545–549 (1984).
18. T. Låås, I. Olsson, *Anal. Biochem. 114,* 167–172 (1981).
19. P. Gill, *Electrophoresis 6,* 552–555 (1985).

20. I. Olsson, T. Låås, In *Electrophoresis '82*; D. Stathakos, ed.; Walter de Greyter, 1982, pp. 157-167.
21. H. Svensson, *Arch. Biochem. Biophys. Suppl. 1*, 132-136 (1962).
22. B. J. Radola, *Biochim. Biophys. Acta* 181-195 (1974).
23. L. Wahlström, R. Björkman, In *Protides of Biological Fluids 27*, H. Peeters, ed. Pergamon Press, Oxford, 1979, pp. 703-706.
24. M. Duckworth, W. Yaphe, *Carbohvd. Res. 16*, 189-197 (1971).
25. C. A. Saravis, N. Zamcheck, *J. Immunol. Methods 29*, 91-96 (1979).
26. A. Rosén, K. Ek, P. Åman, *J. Immunol. Methods 28*, 1-11 (1979).
27. In *Agarose IEF, a supporting matrix for isoelectric focusing*. Pharmacia LKB Uppsala Sweden.
28. A. Görg, W. Postel, R. Westermeier, *Anal. Biochem. 89*, 60-70 (1978).
29. B. J. Radola, *Electrophoresis 1*, 43-56 (1980).
30. H. Garoff, W. Ansorge, *Anal. Biochem. 115*, 450-457 (1981).
31. C. A. Saravis, M. O'Brien, N. Zamcheck, *J. Immunol. Methods 29*, 97-100 (1979).
32. B. J. Thompson, A. H. M. Burghes, M. J. Dunn, V. Dubowitz, *Electrophoresis 2*, 251-258 (1981).
33. B. J. Thompson, M. J. Dunn, A. H. M. Burghes, V. Dubowitz, *Electrophoresis 3*, 307-314 (1982).
34. A. L. D'Andrea, V. M. T. Leedham, *Electrophoresis 6*, 468-469 (1985).
35. T. Låås, I. Olsson, In *Electrophoresis '81* R. C. Allen, P. Arnaud, eds. Walter de Greyter & Co Berlin, N. Y. 1981, pp. 191-203.
36. J. M. Gershoni, G. E. Palade, *Anal. Biochem. 131*, 1-15 (1983).
37. T. Låås, I. Olsson *Electrophoresis 2*, 235-239 (1981).
38. In *Isoelectric Focusing Principles & Methods* Pharmacia Fine Chemicals, Uppsala, Sweden 1982 pp. 18-23.
39. R. G. Bates, Determination of pH, 2nd ed, Wiley-Interscience New York 1973.
40. W. J. Gelsema, C. L. de Ligny, *J. Chromatogr. 130*, 41-50 (1977).
41. W. J. Gelsema, C. L. de Ligny, N. G. van der Veen, *J. Chromatogr. 140*, 149-155 (1977).
42. W. J. Gelsema, C. L. de Ligny, N. G. van der Veen, *J. Chromatogr. 154*, 161-174 (1978).
43. W. J. Gelsema, C. L. de Ligny, N. G. van der Veen, *J . Chromatogr. 171*, 171-181 (1979).
44. J. A. Illingworth, *Biochem. J. 195*, 259-262 (1981).
45. H. Delincee, B. J. Radola, *Anal. Biochem. 90*, 609-623 (1978).
46. S. Fredriksson, *J. Chromatogr. 151*, 347-355 (1978).
47. In *Isoelectric Focusing Principles & Methods*, Pharmacia Fine Chemicals Uppsala, Sweden, 1982, pp. 125-129.
48. G. R. Stark, W. H. Stein, S. Moore, *J. Biol. Chem. 235, 11*, 3177-3181 (1960).
49. I. Olsson, T. Låås. *J. Chromatogr. 215*, 373-378 (1981).
50. A. Rosengren, B. Bjellqvist, V. Gasparic, In *Electrophocusing and isotachophoresis* B. J. Radola, D. Graesslin, eds. Walter de Greyter, 1977, pp. 165-171.
51. P. G. Righetti, E. Gianazza, In *Electrophoresis '81* R. C. Allen, P. Arnaud, eds. Walter de Greyter & Co., Berlin-New York 1981, pp 655-665.
52. L. A. Haff, L. Fägerstam, A. R. Barry, *J. Chromatogr. 266*, 409-425 (1983).
53. U.-B. Axiö-Fredriksson, R. Bhikhabai, J. Brewer, L. G. Fägerstam, H. Lindblom, P. Steffner, In *Protides. Biol. Fluids*, Vol. 34, H. Peeters, ed. Pergamon Press, 1986, pp. 735-739.
54. P. G. Righetti, *J. Chromatogr. 300*, 165-223 (1984).
55. C. Gelfi, P. G. Righetti, *Electrophoresis 5*, 257-262 (1984).
56. K. Altland, A. Banzhoff, R. Hackler, U. Rossman, *Electrophoresis 6*, 379-381 (1984).
57. E. Gianazza, F. Celentano, G. Dossi, B. Bjellqvist, P. G. Righetti, *Electrophoresis 5*, 88-97 (1984).
58. P. G. Righetti, K. Ek, B. Bjellqvist, *J. Chromatogr. 291*, 31-42 (1984).
59. LKB Application Notes 321, 324.
60. E. Gianazza, G. Artoni, P. G. Righetti, *Electrophoresis 4*, 321-326 (1983).
61. C. Gelfi, A. Morelli, E. Rovida, P. G. Righetti, *J. Biochem. Biophys. Meth.* 113-124 (1986).
62. K. Altland, U. Rossman, *Electrophoresis 6*, 314-325 (1985).
63. P. Pietta, E. Pocaterra, A. Fiorino, E. Gianazza, P. G. Righetti, *Electrophoresis 6*, 162-170 (1985).
64. T. Rabilloud, C. Gelfi, M. L. Bissi, P. G. Righetti, *Electrophoresis 8*, 305-312 (1987).
65. M. A. Rimpiläinen, P. G. Righetti, *Electrophoresis 6*, 419-422 (1985).
66. J. Fawcett, A. Chrambach, In *Protides Biol. Fluids* Vol. 33, H. Peeters, ed. Pergamon Press, 1985, 439-442.
67. A. Görg, W. Postel, J. Weser, S. Gunther, J. R. Strahler, S. M. Hanash, L. Somerlot, *Electrophoresis 8*, 45-51 (1987).

68. A. Görg, W. Postel, S. Gunther, C. Friedrich, *Electrophoresis* 9, 57–59 (1988).
69. L. G. Öfverstedt, K. Hammarström, N. Balgobin, S. Hjertén, U. Pettersson, J. Chattopadyaya, *Biochim. Biophys. Acta 782*, 120–126 (1984).
70. S. Hjertén, Q. Z. Liu, S. L. Zhao, *J. Biochem. Biophys. Meth. 7*, 101–113 (1983).
71. B. An der Lan, R. Horuk, J. V. Sullivan, A. Chrambach, *Electrophoresis 4*, 335–337 (1983).
72. P. G. Righetti, A. Morelli, C. Gelfi, R. Westermeier, *J. Biochem. Biophys. Meth. 13*, 151–159 (1986).
73. R. Charlionet, C. Morcamp, R. Sesboue, J. P. Martin, *J. Chromatogr. 205*, 355–366 (1981).
74. Separation News, Pharmacia Fine Chemicals 8 (1979).
75. J. Brewer, E. Grund, P. Hagerlid, I. Olsson, J. Lizana, In *Electrophoresis '86* M. J. Dunn, ed. VCH Verlags-gesellschaft Meinheim BRD, 1986, pp. 226–229.
76. I. Olsson, U.-B. Axiö-Fredriksson, M. Degerman, B. Olsson, *Electrophoresis 9*, 16–22 (1988).
77. H. Rilbe, In *Electrofocusing and Isotachophoresis* B. J. Radola, D. Graesslin eds, 1977, pp. 35–40.
78. J. A. Black, *Electrophoresis 6*, 27–29 (1985).

14 Immunoelectrophoresis

Jorge A. Lizana

Scientific & Technical Service
Pharmacia LKB Biotechnology AB
S-751 82 Uppsala
Sweden

Abbreviations

CA   = Cellulose acetate
TRICINE = N-tris (hydroxymethyl) methyl glycine

I	= Ionic strength
IE	= Immunoelectrophoresis
IEF	= Isoelectric focusing
TRIS	= 2-amino-2 (hydroxymethyl) propane-1,3-diol
CIE	= Crossed immunoelectrophoresis
CRIE	= Crossed radioimmunoelectrophoresis
SDS	= Sodium dodecyl sulphate
DOC	= Deoxycholate
PEG	= Polythylene glycol 6000
FITC	= Fluorescein isothiocyanate
FPLC	= Fast protein liquid chromatography
RAST	= Radioimmunosorbent test
M_r	= Molecular weight
ELP	= Electrophoresis

14.1 Introduction

The use of methods based on the specific reaction between an antigen (protein) and its antibody is more than 80 years old. The formation of specific immunoprecipitates allows several visualisation assays not possible with ordinary protein electrophoresis methods. Proteins that migrate very close to each other or even overlap in gel electrophoresis or isoelectric focusing will normally give different precipitation lines and thus can be distinguished from each other. The possible similarities or identities of one or several proteins, e.g. different forms of aggregates can be discovered, since the immuno-precipitin lines will fuse or cross. The pioneer work of Oudin in the 40's, extended and modified by Ouchterlony and Elek, set the basis for the use of methods employing free diffusion of the antigens and antibodies into an agar/agarose matrix. However, these immunodiffusion methods were very slow and this prompted several groups of re-searchers to develop a new generation of assays using an electric field as the motive force, giving better separations and faster results.

The introduction of immunoelectrophoresis by Grabar and Williams[1] in 1953 was a great technical advance and opened the way for the study of complex mixtures of proteins (antigens) in biological fluids. In this technique, the proteins are first separated by electrophoresis in agar gels. Subsequently the immunoprecipitin reaction is performed by diffusion of the antigens and antibodies on the same gel. Two serious disadvantages limit the value of this technique: the impairment of electrophoretic resolution due to the long immunodiffusion step (usually overnight migration), and the non-quantitative nature of the technique. Laurell and his group[2], later developed several qualitative and quantita-tive methods based on electrophoresis of proteins through agarose gels containing antibodies. Extension and modifications of these methods have resulted in other types of closely related immunoelectrophoretic techniques (Fig. 14-1) which are being used by an increasing number of laboratories in many areas of the biomedical sciences. Most of these immunoelectrophoretic methods have been compiled in some very useful publications[2–7].

Parallel to the development of immunoelectrophoresis, the associated instrumenta-tion, such as the electrophoresis apparatus, power supplies and related accessories, have also been developed and improved to give reliable long-term operation, proper cooling, high power output and low running costs. Agarose gels and antibodies have also been improved. Agaroses with specifications tailored to the various immunoelectrophoretic methods are currently available, and a growing range of antibody preparations with different qualities are available from commercial sources[40].

This chapter describes immunoelectrophoretic methods commonly used as tools to follow the progress of protein (antigens) purification schemes, to check the purity of the fractioned materials, and to quantitate specific proteins.

14.2 Immunochemical Principles

14.2.1 The Precipitin Reaction

The specific reaction between an antigen and its corresponding antibody, known as the precipitin reaction, constitute the basis for the development of the array of methods available today for researchers in biomedical sciences.

The addition of increasing amounts of an antigen to a fixed amount of antibody produces antigen-antibody complexes which grow in size until they form a precipitate. This region of increasing precipitation is known as the "antibody excess" zone. Above this zone, an equivalence between the antigen and antibody concentrations is obtained, "the equivalence zone", and the amount of precipitate shows no further increase. Further addition of antigen gives rise to a solubilization of the immunoprecipitates and formation of soluble complexes, "antigen excess" zone[5,16,56].

The precipitin reaction can be performed in free solution or in agarose gels where the antigen-antibody complexes can be seen as opaque precipitates. The precipitin reaction is temperature and pH dependent. Immunoprecipitation is usually performed at pH between 7 and 9; the immunoprecipitates are soluble below pH 4.5 and above pH 10. Temperatures between 4 and 37°C can be used[4].

14.2.2 Antigens

Antigen is any foreign substance that, when introduced parenterally into an animal, will elicit an immune response with production of antibodies. The antibodies formed are generally found in plasma and should react specifically with the antigen used to induce its formation. Actually the whole antigen is not recognized as foreign of "non-self" but only parts of the antigen molecule. These parts are called "antigenic determinants" or epitopes and are defined as the elementary structure which can combine with only one antibody molecule (or one antibody site). Ovalbumin (M_r 43,000), when injected into a rabbit has 10 antigenic determinants and thyroglobulin (M_r 60,000) as many as 40.

Of great importance for a successful immune response, is the accessibility of the antigenic determinants on the surface of the antigen molecule. This is closely related to the molecular size, complexity and rigidity of the antigen[5,16,56]. Other important aspects in the elicitation of the immune response by the antigen are the dosage, immunization site and use of adjuvants to increase the persistence of the antigen in the host[4,16,39].

14.2.3 Antibodies

Antibody is a plasma protein belonging to the family called immunoglobulins. All immunoglobulins independently of their specificity can be described on the basis of a common structure of four polypeptides: two identical heavy chains and two identical light chains with constant and variable regions. In human the immunoglobulins can be divided into five major classes with indicated name of the heavy chain: IgG (γ), IgA (α), IgM (μ), IgD (δ) and IgE (η). There are two types of light chains κ and λ. Each immunoglobulin molecule contains two light chains. Individual molecules may possess either κ or λ chains, but never both. Each immunoglobulin molecule has two antigen binding sites, which are identical. The most abundant immunoglobulin in serum is IgG, its concentration being 10 to several hundred times higher than the other immunoglobulins. IgG is the immunoglobulin of choice in immunoelectrophoresis.

It is now well established that the antigenic determinants of an antigen are recognized by B-lymphocytes carrying specific receptors. This results in a division and differentiation process and the lymphoid cells start to produce an antibody with specificity for the antigenic determinant recognized by the receptor-carrying B-lymphocyte first encountered by the antigen. All antibodies produced by this cell line thus have the same chemical structure. This is called a monoclonal antibody.

When an antigen—which usually contains a number of antigenic determinants—is introduced into a host, the host responds with a number of antibodies against the various determinants. This is called a polyclonal antibody response. The antibodies made against one antigen are thus a population of immunoglobulins with the same specificity towards that antigen, "monospecific", but directed to the various antigenic determinants of that antigen. Besides, two antibodies binding to the same determinant may also have different affinities, i.e. they bind with different affinity constants. Monospecific, polyclonal antibodies with high affinity are preferred for immunoelectrophoresis of specific proteins.

When two (or more) antigens are injected into an animal, "polyspecific" antibodies are obtained; that is, a population of immunoglobulins composed of subpopulations with specificity for one antigen or the other. Again, within each subpopulation there are immunoglobulin molecules against the various determinants and with different affinities.

Polyclonal antibodies are required for immunoprecipitation since the antigen-antibody network can only be formed through the binding of several antibody molecules to different binding sites on the antigen. Such a three-dimensional complex is impossible between antigens and monoclonal antibodies, unless the antigen itself is an aggregate of identical subunits.

The lymphoid cell lines producing antibodies specific for antigenic determinants (monoclonal antibodies) have a very short life when cultured *in vitro*. In 1975 Köhler and Milstein[64] reported that these cells can be immortalized by fusion with myeloma cell lines. The resulting hybrid cells, "hybridomas", can be cloned and maintained indefinitely. At any time hybridoma samples can be grown in culture or injected into animals for large scale production of monoclonal antibody[63].

The choice of polyclonal antibodies for the various immunochemical techniques is disussed in section 14.4.3.

14.3 Choice of Immunoelectrophoretic Method

There is a growing trend to replace the traditional immunoelectrophoresis method[1] by methods based upon electrophoresis of proteins in agarose gels containing antibodies[2]. Figure 14-1 shows the gel arrangements most used in immunoelectrophoretic techniques. For researchers engaged in protein fractionation and isolation, the choice of the method is influenced by the capacity, ease of use and speed of the technique, as well as by the availability of antibodies. When dealing with eluates from chromatographic separations, usually the first choice is a semi-quantitative screening method such as fused-rocket IE. For further peak characterization, crossed-IE or one of its modifications is often used. Quantitative determinations are best obtained by rocket-IE.

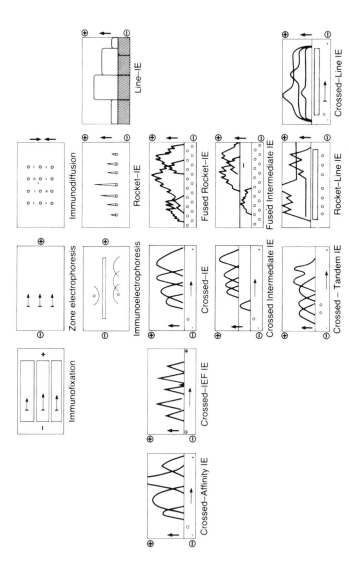

Figure 14.1 Gel arrangements of the various immunoelectrophoretic and immunodiffusion techniques. Zone electrophoresis can be combined with immunoprecipitation by placing a CA membrane containing specific antibodies onto the gel (immunofixation), or by diffusion of the antibodies into the gel (immunoelectrophoresis). Immunoelectrophoresis can be speeded up by using electrophoresis in both dimensions: CIE. The resolution of CIE can be improved by addition of carrier ampholytes (IEF), lectins or affinity media into the first dimension gel. The incorporation of an intermediate gel containing antibodies or affinity media increases the flexibility of CIE. Immunodiffusion can be speeded up by electrophoresis of the samples into an antibody-containing gel: rocket-IE. If instead of sample wells, a trench (line) is used, larger samples can be applied: crossed line-IE. When the sample wells are closely placed, and the samples are allowed to diffuse before electrophoresis, fused rockets are obtained. The incorporation of an intermediate gel with specific antibodies will trap the corresponding proteins and fused rockets will be obtained in the intermediate gel. The combination of line-IE with rocket-IE increases the discriminating power of rocket-IE.

14.3.1 Rocket Immunoelectrophoresis

Rocket-IE, also called electroimmunoassay[2], has become a very popular technique for the quantification of specific proteins and is widely used in routine analysis[6].

A protein migrating electrophoretically through a gel containing specific antibodies to that protein will form antigen-antibody complexes. These complexes will migrate further into the gel until they reach the point of equivalence between the antigen and antibody concentration. Here the antigen-antibody complexes precipitate and no further migration occurs. The result is a rocket-shaped precipitate, the area of which is proportional to the concentration of antigen (protein) in the sample. It is important to run the experiment at a pH in which the antibodies are at their pI's and do not migrate (see section 14.4.1, p. 415). Recommendations for improving the curve fitting in rocket-IE[8] and a computer program to better describe the antigen concentration/peak height relationship, have been proposed[9]. Recently, a very useful compilation of artifacts encountered in the practise of immunoelectrophoresis has been reported[17].

Rocket-IE can be performed at high voltage 10 V/cm ("fast rockets") for about 2-4 hours or at low voltage 2 V/cm ("slow rockets") for about 12-18 hours. The choice is often a question of available time and sufficient cooling capacity. With the modern instrumentation available there is a trend to run fast rocket-IE. Besides, the lateral diffusion is reduced and the precision improved (Fig. 14-2).

Proteins which migrate very slowly (e.g., IgG, IgA) will form cigar-shaped anodic-cathodic precipitates around the application well, the total area of these precipitates must be used for quantitation. In such cases the precision of the assay is impaired. Carbamylation of the samples lowers the pI of the protein and increases its anodic mobility. Another possibility is to carbamylate the antibodies[10]. True rockets are thus obtained and the precision of the assay improved.

14.3.2 Fused-Rocket Immunoelectrophoresis

This technique is a modification of rocket-IE and a very useful tool to follow the purfication of a protein during the various fractionation steps[2-4]. Peaks that appear to be symmetrical in a chromatogram are often shown to be composed of several components by this technique. (Fig. 14-3).

Samples of the fractions from column chromatography are applied in a row of closely placed wells and allowed to diffuse into the surrounding antibody-free agarose gel for 30-60 minutes. Thereafter samples are electrophoresed into an antibody-containing agarose gel. The result of the first diffusion step is that the rocket-shaped precipitates of the different samples, formed during the second dimension, are "fused" together and elution profiles for each individual protein are thus obtained in a single run[11].

In addition to its analytical discrimination, fused rocket-IE also gives a semiquantitative estimation of all the proteins present in the different peaks, provided that there are enough high-titre antibodies against each of them. A modification called fused rocket-IE with an intermediate gel[12] (Fig. 14-1) solves the identification problem of the pattern obtained with common fused rockets, but it is limited by the availability of monospecific antibodies (which are incorporated into the intermediate gel) to the protein under purification.

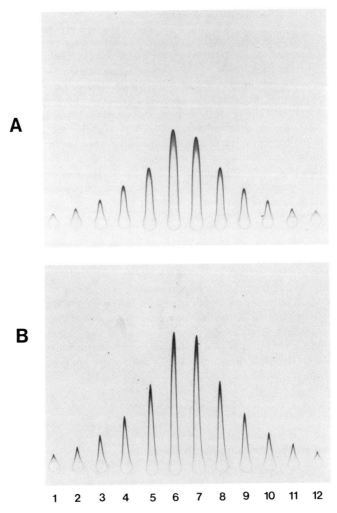

Figure 14-2. Rocket immunoelectrophoresis of human serum albumin. Agarose A (1 %) gel containing anti-albumin (7 μl/cm^2 of gel). Tris-barbital buffer (pH 8.6, I = 0.020). (A) low voltage rockets (2 V/cm) run for 18 hours; (B) high voltage rockets (10 V/cm) run for 3 hours. Wells 1–6: albumin calibration solutions 5.6 to 120 mg/1; wells 7–12 are duplicates of wells 1–6. Sample volume: 5 μl.

14.3.3 Crossed Immunoelectrophoresis

In crossed immunoelectrophoresis (CIE) the proteins are first separated by zone electrophoresis in agarose gel followed by a second electrophoretic run, perpendicular to the first dimension, into an antibody-containing agarose gel[2]. During the second dimension antigen-antibody complexes for each antigen-antibody system present are formed. Due to antigen excess, these complexes migrate further into the antibody-containing gel until all the antigen is used up. Precipitation zones are thus formed (Fig. 14-4). For a given concentration of an antibody component the area of the individual

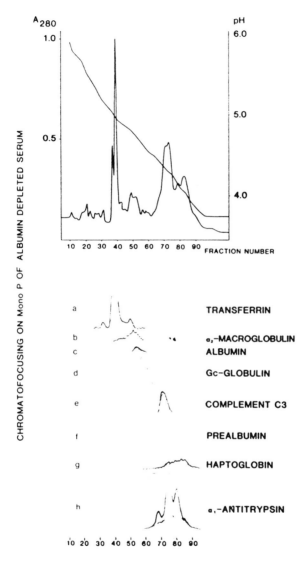

Figure 14-3. FPLC chromatofocusing of serum proteins in the pH interval 6.0-3.8, followed by fused rocket-IE. Aliquots (6 μl) of the fractions were analysed by fused rocket-IE using monospecific antisera against: transferrin (a), α_2-macroglobulin (b), albumin (c), Gc-globulin (d), complement C3 (e), prealbumin (f) haptoglobin (g) and α_1-antitrypsin (h). Reproduced from Ref. 11 with the kind permissions of the authors and editors.

percipitation zone (or peak) is proportional to the concentration of the corresponding antigen. CIE is therefore a powerful combination of qualitative electrophoresis and semi-quantitative immunoelectrophoresis.

With this technique it is possible to analyse complex mixtures of proteins; e.g., proteins in human plasma and other body fluids. The resolution is very high compared to

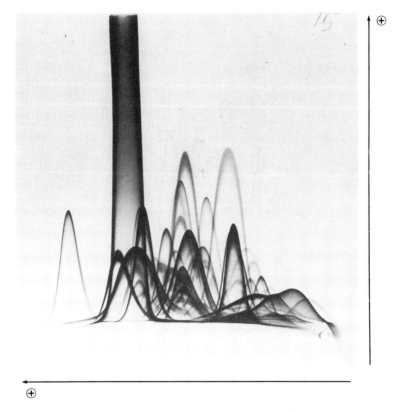

Figure 14-4. Crossed immunoelectrophoresis of human serum. First dimension: Agarose A (1 %) gel run for 90 min at 10 V/cm. Barbital/barbiturate buffer (pH 8.6, I = 0.075). Sample: 3 μl of fresh serum. Second dimension: Agarose C (1 %) containing antibodies against whole human serum (9 ul/cm² of gel). Run overnight at 3 V/cm. Tris-barbital pH 8.6, I = 0.020.

traditional IE. The sensitivity is also high and can be increased by increasing the sample size to 20–30 μl in the first dimension gel and by decreasing the amount of antiserum used in the second dimension gel[4]. The information obtained is mainly semiquantitative, but the Clarke and Freeman modification[14] should give quantitative results. CIE is very common in biomedical research, but is too complicated for routine diagnosis as it requires experience in the evaluation of the results. In some areas it is being replaced by immunofixation[15]. Other areas of use of CIE are the estimation of subpopulations of heterogeneous antigenic compounds (e.g. demonstration of polyclonal, oligoclonal and monoclonal IgG mixtures); studies of association and dissociation phenomena (e.g., complex formation between Protein A and IgG); studies of genetic variants of proteins; and classification of antigens as immunologically identical, non-identical or partially identical[4,16,48].

A number of modifications of this method and hybrids with other techniques have been developed. These include; CIE with intermediate gel, crossed IEF-IE, crossed affinity-IE, crossed line-IE and tandem-CIE (Fig. 14-1)[2-4,16]. Some of these techniques are described below.

14.3.4 Crossed Electrofocusing Immunoelectrophoresis

In crossed IEF-IE the proteins are first separated by isoelectric fosucing (IEF) in polyacrylamide or agarose IEF gels containing carrier ampholytes, followed by electrophoresis in antibody-containing agarose gels[4,18,19]. The resolution obtained by IEF is much superior to that obtained with agarose zone electrophoresis, and similarly higher resolution can be expected with crossed IEF-IE (compare Figs. 14-4 and 14-5).

The wider acceptance of the technique has been delayed by technical problems caused by the electro-osmosis of the different supporting media in the IEF (polyacrylamide) and electrophoresis (agarose) steps. The "laying-on" technique[18] was proposed to overcome such problems and involves simply applying a lane of the IEF gel, containing the focused proteins, on top of the second antibody-agarose gel. With the introduction of Agarose IEF, a charge-balanced agarose specially designed for isoelectric focusing, agarose can now be used as the supporting gel in both dimensions, eliminating the gel junction problems caused by electro-osmosis[19-21].

The technique is very straightforward. IEF gels should be cast on plastic films. After running IEF for 1–2 hours at 15 W, 1500 V (2000–3000 Vh), the portions of the IEF gel that were under the electrode strips are removed since they contain very acid or basic electrode solutions and will produce disturbances in the second dimension antibody-containing agarose gel. A lane containing the focused proteins is excised and placed on the second dimension gel. Tris-barbital buffer pH 8.6, I = 0.050 is recommended for the second dimension. The high ionic strength is needed to buffer the carrier ampholytes contained in the lane of the IEF gel. Fast electrophoresis at 10 V/cm for 3–4 hours is recommended if the resolution of the sharply focused protein band is to be retained during the second dimension run[19]. A typical crossed IEF-IE experiment using Agarose IEF is shown in Fig. 14-5.

Crossed IEF-IE is very well suited for checking the purity of isolated proteins, and studies of hereditary polymorphism, microheterogeneity, fragmentation of proteins and demonstration of abnormal fractions[18-23].

14.3.5 Crossed Affinity Immunoelectrophoresis

The combination of affinity chromatography and immunoelctrophoresis constitute a general technique for the study of ligand-protein interactions during electrophoresis[30,31]. Successful applications of this principle include the use of free and immobilized lectins to study glycoproteins[28,29], and hydrophobic interaction media combined with CIE for the study of amphiphilic proteins[34].

Characterization and purification by interaction with lectins is a common procedure for many glycoproteins. Two general approaches can be used to study lectin-glycoprotein interactions during electrophoresis: (a) The free or immobilized lectin can be included in the first dimension gel of CIE or, (b) The free or immobilized lecin can be included in an intermediate gel between the first and second dimension of CIE[27-31]. The presence of a lectin (free or immobilized) in the first dimension gel or in the intermediate gel does not inhibit or disturb the formation of immunoprecipitates of other proteins present in the sample.

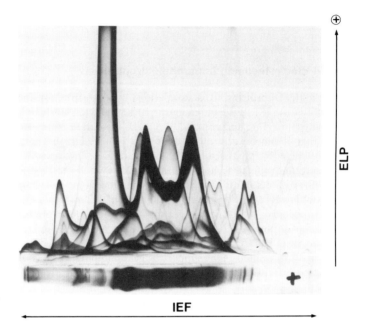

Figure 14-5. Crossed electrofocusing immunoelectrophoresis of human serum. First dimension: Agarose IEF (1 %) gel containing Pharmalyte pH 2.5–5 + 4–6.5 (1:1), run at 15 W, 1500 V for 2500 Volthours at 10°C. Sample: 3 μl of fresh serum. Second dimension: Agarose C (1 %) containing antibodies against whole human serum (9 μl/cm²) run for 4 hours at 10 V/cm. Tris-barbital buffer pH 8.6, I = 0.050. Reproduced from Ref. 19 with the kind permissions of the authors and editors.

Concanavalin A (Con A) has been the lectin most widely used, though other lectins, e.g., Lentil lectin, Wheat Germ lectin, Soybean lectin, *Ulex europeus* lectin and Peanut agglutinin have also been tried[30,31]. The finding that a minor form of alpha-fetoprotein, which does not react with Con A and may be helpful in the laboratory diagnosis of neutral tube defects, has increased the interest in this technique in clinical situations[32,33].

When lectins are included in the first dimension gel of crossed immunoelectrophoresis. The biospecific interactions between the lectin and the carbohydrate moiety of the glycoproteins present in the sample produce a retardation, partial or total, of the electrophoretic mobility of the glycoproteins. The second dimension run into an agarose gel containing antibodies will reveal the heterogeneity of the carbohydrate part by the appearance of several peaks for a single protein. Control runs without the lectin in the first dimension gel are required for comparison of the electrophoretic pattern.

Introduction of the lectin in an intermediate gel, after the sample proteins have been separated in the first dimension agarose gel, produces disappearance, diminishing or even splitting of immunoprecipitates in the second dimension gel. Again, control runs without the lectin are required for comparison of the patterns[27–30].

Crossed hydrophobic interaction immunoelectrophoresis is a related technique in which Phenyl- or Octyl-Sepharose is incorporated into the first dimension gel or in an intermediate gel. The amphiphilic proteins present in the sample are trapped or retarded by the gel media and this is seen by the shape and characteristics of the immunoprecipitates formed in the second dimension[4,34].

14.3.6 Crossed Radioimmunoelectrophoresis

Crossed radioimmunoelectrophoresis (CRIE) permits direct indentification of individual allergens in complex allergen extracts without previous separation[7]. In this respect the technique is unique. CRIE is also a very valuable tool for the control of allergen extracts. The technique is based upon binding of specific IgE to immunoprecipitates in a crossed immunoelectrophoresis plate after incubation with serum samples from patients with IgE-mediated allergy. The IgE bound to the allergen is demonstrated by using anti-IgE labelled with ^{125}I and subsequent autoradiography[24]. The high resolving power of the CRIE method is useful for establishing the allergenic composition of various allergen extracts. Furthermore, the amount of specific IgE-binding to the precipitates may be semi-quantified using a CRIE reference system[25]. The often used CRIE method requires, however, a very long exposure time.

Recently, a procedure for performing CRIE with significantly reduced exposure times has been developed[26]. The rapid method has been effected by using a highly sensitive X-ray film, X-ray intensifying screen and by exposing at low temperature ($-70°C$) in a Kodac X-Omatic cassette. Exposure at $-70°C$ appears to be critical to obtain optimal radiostaining. Furthermore, the isotape concentration has been doubled without any loss of specificity or reproducibility. The rapid procedure makes it possible to obtain a CRIE screening within a day and a complete CRIE classification in 5 days. A CRIE experiment of *A. fumigatus* antigens[27] is shown in Fig. 14-6.

14.4 General Techniques

14.4.1 Choice of the Buffer

Although an optimal buffer can be found for each system, most workers use buffers based on barbituric acid (barbital) because the resolution obtained is sufficient for most of the immunoelectrophoretic techniques.

Tris-barbital buffer, pH 8.6, ionic strength (I) 0.100

5,5-diethylbarbituric acid	224 g
Tris	443 g
Calcium lactate	20 g
Sodium azide	10 g
Distilled water to	10 l

This stock solution can be stored at 4°C for several weeks.

The pH of the buffer should be above the pI of the proteins under study. At pH 8.6 most of the serum proteins have a net negative charge and migrate towards the anode. The antibodies should not migrate during the electrophoretic run, i.e. the pH of the buffer should be similar to their pI. The pI of rabbit immunoglobulins is around 8.6.

The majority of the immunoelectrophoretic techniques use an ionic strength of 0.020. Dilute 200 ml of the stock solution to one litre with distilled water. If other ionic strengths are required, dilute the stock solution accordingly. High ionic strengths give sharper bands and improved resolution, but a lot of heat develops, making it necessary to use an apparatus with high cooling capacity to avoid drying/burning of the gel. The

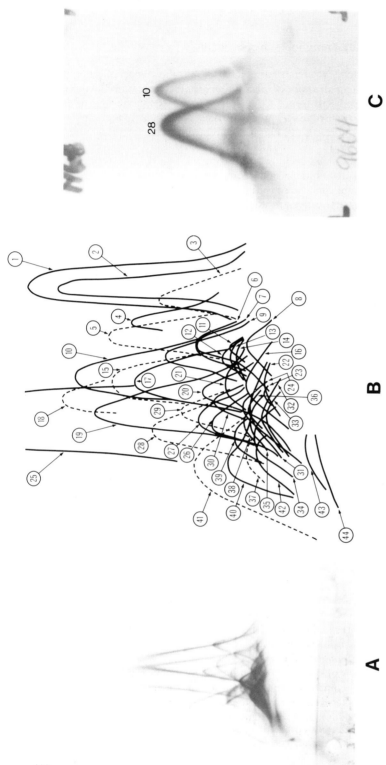

Figure 14-6. (A) Crossed immunoelectrophoresis of *A. fumigatus* antigens against hyperimmune rabbit antisera. Sample: 0.2 mg of mycelial extract. First dimension: Agarose A (1%) gel run for 35 min at 10 V/cm (15°C). Tris/tricine buffer pH 8.6, I = 0.025. Second dimension: Agarose A (1%) containing rabbit antiserum (10 μl/cm²) run at 2V/cm for 18 hours. (B) Schematic drawing of the immunoprecipitates arbitrarily numbered 1–44 from the anode: ——— denotes sharp precipitates seen after CBB staining and ---------- denotes precipitates only visible after radiostaining. (C) X-ray film from CRIE showing an allergenic patient serum containing high amounts of immunoprecipitates 10 and 28. Immunoprecipitates 24, 30 and 40 are also present. Reproduced from Ref. 27 with the kind permissions of the authors and editors.

buffer above is not recommended for agarose zone electrophoresis of plasma proteins. The barbital/Na-barbital buffer (pH 8.6) of ionic strength 0.075 as suggested by Jeppsson et al.[35] should be used since it gives sharper bands and better resolution. Barbiturate has potential as a drug of abuse and in recent years government regulations have made the handling of this substance difficult. A non-barbital buffer using Tris-tricine has been proposed and should give similar results to Tris-barbiturate[36].

Tris-tricine buffer, pH 8.6, ionic strength 0.100

Tricine	226 g
Tris	443 g
Calcium lactate	20 g
Sodium azide	10 g
Distilled water to	10 l

This stock solution can be stored at 4°C for several weeks. To obtain an ionic strength of 0.020 dilute accordingly with distilled water.

14.4.2 Choice of the Gel Matrix

Immunoelectrophoresis can be performed in polyacrylamide, starch, cellulose acetate, fibrin-agar and agar/agarose gels (reviewed in Ref. 16). Agarose gel and cellulose acetate membranes are the two matrices most used today. Cellulose membranes are mainly used in immunofixation following zone electrophoresis or isoelectric focusing for identification and phenotyping of proteins[15,37,38]. Over the years agarose has emerged as the matrix of choice for immunoelectrophoretic techniques. In the majority of applications it is important to use an agarose that has a low ionized group content to maintain a suitable degree of electroosmosis (see 12.2.3, p. 354) and to avoid any risk of adsorption of materials by the gel.

For most electrophoretic techniques, agarose with an electroosmosis of 0.13 (relative mobility, $M_r = -0.13$) should give good results. Electroosmosis can even be beneficial as a mechanism which may increase the separation between particular proteins during electrophoresis, and is essential for successful counterflow immunoelectrophoresis (see Ref. 16 for review). Recommendations concerning which type of agarose to use for the various immunoelectrophoretic techniques are given in Table 14-1. Of great importance is the batch to batch reproducibility of the electroosmosis of the selected agarose, to avoid shifts in the migration of the proteins and confusion in the interpretation of the results.

Agarose will form a gel in a 0.1% solution, but concentrations of around 1% are commonly used to give a much stronger gel which is easier to handle. A 1% gel has a molecular exclusion limit for spherical particles well above 50 million M_r and allows free diffusion of antigens, antibodies and immune complexes. A 1% solution of agarose a in Tris-barbital buffer, pH 8.6, ionic strength 0.020 is prepared as follows: weigh out the appropriate amount of agarose and dissolve in the buffer by gentle heating. Ensure that all the agarose has dissolved by boiling the solution a few moments. A 1% solution can be stored at 4°C for several weeks and liquefied repeatedly by heat. If antibodies are to be included in the gel, cool the solution to about 55–57°C, add the antisera, mix and pour on the plate. If lectins or other proteins are to be included, proceed as with antibodies. Agarose gels can be cast on glass plates or on the hydrophilic side of plastic polyester

Table 14-1. Selection of agarose for different techniques.

Technique	Agarose type*		
	A	B	C
Zone electrophoresis	●	○	○
Counter immunoelectrophoresis		●	
Zone immunoelectrophoresis	●		
Immunoelectrophoresis	●	●	
Rocket immunoelectrophoresis	●	○	○
Fused rockets	●	○	○
Line immunoelectrophoresis	●	○	
Crossed immunoelectrophoresis	●		●
Crossed electrofocusing IE	○		●
Crossed affinity IE	●		○
Double immunodiffusion	●	○	
Radial immunodiffusion	●	○	

Recommended type: ●
Used by some authors in modified procedures: ○

Agarose A: medium electroosmosis ($M_r = -0.13 \pm 0.01$)
Agarose B: high electroosmosis   ($M_r = -0.25 \pm 0.01$)
Agarose C: low electroosmosis   ($M_r = -0.02 \pm 0.01$)
Agarose IEF: charge-balanced for isoelectric focusing

* Pharmacia LKB Biotechnology AB, Uppsala, Sweden. Other suppliers offer similar agaroses e.g. Litex AS, Copenhagen, Denmark; FMC Marine Colloids, Maine, USA.

sheets (GelBond™). The agarose solution can be poured directly on a plate fitted with a gel casting frame (Fig. 14-7). Casting between glass plates with a U-shaped frame is done only when an absolutely even gel thickness is desired.

The sizes vary very much depending on the particular technique and on different authors, but there is a trend to use small gels. Recommended thickness is 1 mm. Thicker gels will tend to develop a temperature gradient during the running and give skewed bands.

14.4.3   Choice of the Antibodies

Although whole unfractionated antisera can always be used in immunoelectrophoretic techniques, these preparations have many disadvantages: the high level of non-antibody proteins present in a crude preparation is such that a longer washing time will be required to obtain a clear background. Crude preparations usually become turbid on storage due to the precipitation of denatured lipoproteins. They also contain excess antigen and soluble complexes of antibody and antigen used to absorb the antiserum. Besides, there is always the uncertainty of interaction(s) between the non-antibody proteins of the antiserum and proteins in the sample, which can cause artifacts.

Figure 14-7. Instruments and accessories for immunoelectrophoresis. (1) Power supply; (2) electrophoresis apparatus; (3) cooling plate with printed pattern for easy centering of gels; (4) glass plates and plastic film; (5) levelling table and clamps; (6) gel casting frame; (7) gasket for casting gels between glasses; (8) punching template and holder; (9) punches; (10) sample application foil for agarose electrophoresis; (11) electrode wicks; (12) voltage probe; (13) spirit level; (14) agaroses; (15) antibodies.

It is recommended to use antibody preparations that contain only the immunoglobulin fraction. They are more stable than crude preparations and the titre is better controlled. Besides, the protein content is much lower and the tedious washing steps can be considerably shortened or even eliminated. Crude antisera preparations can be fractionated by salting out with ammonium sulphate (250 g/l rabbit serum), buffer exchange with a small column of Sephadex G-25 (PD-10) and chromatography on DEAE Sephadex A-50 or DEAE Sepharose CL-4B at pH 5.0 in sodium acetate/acetic acid buffer (I = 0.05). After another buffer exchange with 0.1 M NaCl/15 mM sodium azide, the gamma-globulin fraction obtained can be stored for years at 4°C. The loss in activity is around 5% per year. Storage at −20°C reduces the loss in activity even further[39].

In most instances the choice of species in which antibodies are raised may reasonably be made on the basis of what is available and the volume of antiserum required. Rabbits have been shown to produce antibodies with optimal precipitating activity and are widely used in immunoelectrophoretic techniques. Rabbits are easy to handle and by pooling the serum from several animals the individual immune reaction, which can give atypical antibodies, is eliminated. Goat, sheep and swine antisera are also equally useful. An extensive list of suppliers of crude and fractionated antibody preparations can be found in Ref. 40.

14.4.4 Choice of the Electrophoretic Conditions

A suitable DC power supply for immunoelectrophoresis should deliver stabilized voltage up to 500 V (400 mA). Most of the experiments are run at constant voltage. Slow electrophoresis at 2–5 V/cm requires 80 to 130 V and overnight run. Fast electrophoresis at 10–20 V/cm requires 350 to 400 V and 2 to 3 hours to run; a good cooling capacity is critical. Ultra rapid electrophoresis with potential gradients of up to 62 V/cm and running times down to 9–25 minutes have been reported (see Chapter 23 in Ref. 4).

The electrophoresis apparatus (Fig. 14-7) consists of a box with two removable buffer vessels and a cooling plate. Each buffer vessel contains a built-in electrode. A printed pattern on the cooling plate facilitates the centering of the gel plates and application of the samples. A black background allows easy visualization of the white immunoprecipitates.

For most applications tap water (10–18°C) circulating at a flow rate of 1.5–2 l/min will be sufficient. Ethanol or ethylene glycol (20% v/v) solutions may also be used to cool the system. Recommended maximum flow of coolant is 4 l/min. Too intensive cooling may cause water condensation on the gel surface and flooding of the immunoprecipitates. Insufficient cooling will increase evaporation from the gel and result in gel drying and/or burning. The lid seals the apparatus from the surrounding air and protects the operator from any electric hazard. The pairs of holes on either side of the centre-line of the lid are for insertion of the voltage probe for measurement of the field strength (V/cm) across the gel during the course of an experiment.

The best electrophoresis wicks are those made of highly absorbent paper protected on one side by a film of polyethylene which prevents the evaporation of buffer from the wicks. The wicks should be cut to the exact width of the gel. The plastic surface should be uppermost. If wicks with plastic film are not available, two layers of Whatman No.3 filter paper (or one layer of Whatman No.17) can be used. Ensure that there is good contact between the wicks and the gel, otherwise artifacts due to uneven distribution of the electric field can be obtained.

The necessary instrumentation for running immunoelectrophoresis and accessories for gel and punching are shown in Fig. 14-7.

14.4.5   Staining and Destaining

14.4.5.1   Coomassie Brilliant Blue

After the electrophoretic step the immunoelectrophoresis plates do not need fixation since the antigens are already fixed in the immunoprecipitates. The plates are covered with several layers of filter paper and a clean glass plate, and on top of this a weight of about 1 kilo is placed. Leave to press for 15 minutes. Carefully remove the wet papers. The plate is now reduced to a thin film which is easy to handle and allows easy washing, staning and destaining.

Place the thin film in a tray containing NaCl (9 g/l) to wash away unreacted proteins (from sample and antiserum). Depending on the purity of the antiserum used, the washing time can be shortened from overnight to less than 2 hours. Rinse the plate with distilled water (30 seconds). Press again for 15 minutes and dry the gel in hot air.

Staining solutions: Stock solution of 5 g of Coomassie Brilliant Blue R-250 (CBB), or Page Blue 83, in 1 litre of 96% ethanol. The alcoholic stock solution is stable for weeks. To an aliquot of this solution an equal volume of 20% acetic acid is added immediately before use. Filtering is not necessary. The alcoholic stain with acetic acid results in precipitation of the dye and should be discarded after a few days of use. Staining of the immunoprecipitates is completed in 15–30 minutes. Staining for longer times (overnight) will not increase the uptake of the dye by the immunoprecipitates, it will only increase the background staining of the gel[41]. CBB is 3–5 times more sensitive than Amido Black 10B and should be the dye of choice for immunoprecipitates.

Destaining solutions: After staining, the immunoplates are rinsed in water (10–20 seconds) to remove excess dye and placed in a tray containing the destaining solution: ethanol (96%) 450 ml, acetic acid 100 ml and distilled water 450 ml. Destain until clear background is obtained (10–15 minutes). The destaining solution can be reused after filtering through activated charcoal.

Staining and destaining can be accelerated by a factor of two or more if all the steps above are carried out at 40–50°C. Working under a fume hood is recommended.

14.4.5.2 Silver Staining

In recent years, various authors have described silver staining of proteins and immunoprecipitates in agarose gels[58–61]. In general, all these silver staining methods are 20 to 40 times more sensitive that those using CBB, and therefore very useful for the detection of small amounts of proteins. For instance McLachlan and Burns[60] have reported a sensitivity of 0.9 ng/mm^2 in human IgG-antiIgG precipitates, about 20 times higher than immunoprecipitates stained with CBB.

After electrophoresis the immunoplate is washed with saline and dried to a thin film as described in the section 14.4.5.1 above. Staining is carried out as follows:

Prepare two solutions. Solution A, containing 5 g of Na_2CO_3 in 100 ml of distilled water. Solution B, containing 0.4 g NH_4NO_3, 0.4 g $AgNO_3$, 1.0 g silicotungstic acid, and 2.8 ml of a 370 g/L solution of formaldehyde in 200 ml of distilled water. Add solution B to solution A with constant stirring, in a volume ratio of 2:1, just before staining and pour the mixture into the staining tray containing the dried immunoplate. Agitate the tray and when the colour of mixture changes from white to gray after about 3 minutes, remove the gel and immerse it in water. Meanwhile, freshly prepare another mixture of stain and again stain the gel with it. When the gel pattern shows the desired intensity (usually after 2 to 3 minutes), stop the reaction by immersing the gel in a 5% acetic acid solution for 15 minutes. Finally, rinse the gel in water for 5 minutes and dry it.

14.4.6 Enhancing the Sensitivity of Immunoelectrophoresis

14.4.6.1 Nonionic Polymers Solutions

Neutral polymers (dextrans, polyethylene glycols, polyvinyl-pyrrolidones) have been shown to enhance the antigen-antibody reaction. This enhancement has been explained in terms of a steric exclusion of the antigen-antibody complexes from the domain of the polysaccharide (for review see Ref. 42). The use of this effect in immuno-electrophoresis was first reported in 1968[43]. Dextran T-10 and polyethylene glycol 6,000 (PEG), in concentrations of 2–4%, are at present the most used polymers for enhancement of the immune reaction.

The polymers are usually included in the gel by addition to the agarose solution before casting. Immunoelectrophoresis is then performed as usual. The result is a dramatic sharpening of the immunoprecipitin lines; weak immunoprecipitates become more apparent. An alternative immersion technique has also been reported[44]. The agarose plate with the immunoprecipitates is immersed in 4% Dextran T-40 or PEG 6000 solutions for 60 minutes. Thereafter the gels are washed, pressed and stained as usual. Somewhat better results are obtained with the inclusion technique as compared to the immersion technique.

14.4.6.2   Enzyme-Labelled Antibodies

The visualization of weak immunoprecipitates can be enhanced with enzyme-labelled antibodies. The increase in sensitivity is about 50 times compared to Coomassie Brilliant Blue staining.

After electrophoresis the immunoplates are incubated with a second antibody to which an enzyme has been coupled. The second antibody is specific for the antibody fixed on the immunoprecipitate. Development of the enzyme colour reaction will yield stained immunoprecipitates[45,46]. Do not use labelled antibodies as the first antibody since the introduction of an enzyme or any other group to the antibody molecule will cause a shift in the isoelectric point. New pH conditions must then be found to prevent the migration of the labelled-antibody contained in the agarose plates.

Several enzymes can be used to label the antibody: peroxidase, β-galactosidase, glucose oxidas, alkaline phosphatase, etc. The detailed methodology of coupling enzymes to antibodies has been reviewed by Avrameas et al.[47]. For labelling, either the purified IgG fraction or the affinity purified antibody can be used. Better results were obtained when labelled F_{ab} antibody fragments, rather than the whole antibody molecule were used[47]. An extensive list of suppliers of enzyme-labelled antibodies can be found in Ref. 40.

14.4.6.3   Fluorescent-Labelled Antibodies

A fluorescent label can be used in the same way as an enzyme label for the amplification of immunoprecipitates (see above). In this case the immunoprocipitates are visualized by UV illumination. The most used fluorochrome is fluorescein isothiocyanate (FITC). Conjugation of proteins by FITC is a well standardized technique which does not require too much bench work[48]. Low FITC concentrations (around 10 μg/mg of protein) are used with antibodies and fragments of antibodies to avoid hyperconjugation and loss of antibody activity. Conjugation is carried out at pH 9 for 18 hours. Unreacted FITC is separated from FITC-antibodies by gel filtration on Sephadex G-25. Conjugated and unconjugated antibody molecules can be further separated by ion exchange chromatography[16]. A list of suppliers of fluorescent antibodies can be found in Ref. 40.

An alternative is to incubate the immunoprecipitates with FITC-Protein A. Protein A will bind to the F_c region of the IgG in the immunoprecipitate[49].

14.4.6.4   Radio-Labelled Antibodies

The immunoprecipitates can be made visible by radiolabelling the antigen or antibody. Radioiodine (^{125}I) is the most used tracer to label antibodies: the procedure is well-known and standardized[48]. The gel arrangements for radioimmunoelectrophoresis are similar to those in Fig. 14-1. After electrophoresis, the plates are washed, dried and subjected to autoradiography. In one application the detection limit for alpha-fetoprotein as measured by rocket radio-IE was as low as 0.3 μg; an increase in sensitivity of 50 to 100 times as compared to CBB staining[50].

Another approach is to incubate the "cold" immunoprecipitates with a second radio-labelled antibody. The main application of this approach is crossed-radioimmuno-electrophoresis (CRIE) for the identification of major and minor allergens by reaction of the immunoprecipitate with a patient's specific IgE followed by reaction with ^{125}I-anti IgE and autoradiography (for review see Ref. 7).

14.4.7   Carbamylation of Proteins

Carbamylation is used in immunoelectrophoresis to increase the mobility of proteins that migrate very slowly (see section 14.3.1, p. 409).

Mix equal volumes of the protein sample (antigens) and standards with freshly prepared KCNO (2 M). Incubate the mixture at 45°C for 30 minutes (alternatively, leave overnight at room temperature). Cool to 10–15°C in a water bath. Apply the samples and run the rocket-IE as usual. Carbamylation of the *antibodies* contained in the agarose antibody plate can be performed in the same way as described above. In this case the electrophoresis must be performed at pH 5.0 instead of pH 8.6. Recommended buffer is TEMED 0.01 M, acetic acid 0.02 M (pH 5.0). The carbamylated antibodies will not migrate during electrophoresis because carbamylation has lowered their pIs to about 5. The sample antigens will migrate towards the cathode (the reverse side compared to using pH 8.6).

14.4.8   Use of Detergents in Immunoelectrophoresis

Nonionic detergents solubilize the proteins, usually without denaturation. This means that the antigenic determinants of the protein remain available for interaction with the antibodies. A number of nonionic detergents (Table 14-2), at concentrations of 0.05–1.0% (w/v), have been shown to have no effect on the antigen-antibody

Table 14-2. Effect of detergents on the immune reaction (data from Ref. 52).

Detergent type	concentration (%) w/v	inhibition (extent %)
Ionic		
SDS	0.05	YES (60%)
	0.10	YES (80%)
	0.20	YES (90%)
DOC	0.10	YES (60%)
	0.25	YES (80%)
Nonionic		
Triton X-100	0.05–1.0	NO
Nonidet NP-40		NO
Tween 20		NO
Tween 80		NO
Brij 58		NO
Berol EMU-043		NO
Lubrol WX		NO

If in the reaction medium there is a combination of SDS or DOC and another nonionic detergent, the inhibition by SDS or DOC is reduced. Thus, membrane proteins can be solubilized in SDS or DOC and run in agarose gels containing nonionic detergents.

reaction[51−53]. The detergent is used at a concentration below the critical micellar concentration and is incorporated into the agarose gel before casting. Thereafter, IE is carried out as normal. It is not necessary to add the nonionic detergent to the buffer.

Ionic detergents (anionic and cationic) unfold the tertiary structure of proteins and very often denature the protein they solubilize. Therefore it is not surprising that in most cases the antigen-antibody reaction is inhibited when these detergents are present in the medium[52,62] (Table 14-2). However, in some cases direct immunochemical identification of proteins is possible in gels containing SDS[55,56].

Sodium dodecyl sulphate (SDS) is very often used to solubilize membrane proteins and its presence during electrophoresis is necessary to avoid aggregation of the solubilized proteins. However, when SDS is present in concentrations above 0.05 % (w/v) it binds antigens or antibodies and charges their electrophoretic mobility, spoiling the resolution of the separation. Besides, the antigen-antibody reaction is inhibited. Distorted immunoprecipitates can be obtained[51,54]. If in the reaction medium there is a mixture of SDS and a nonionic detergent, the inhibition of the antigen-antibody reaction by SDS is reduced. This is possibly due to the formation of mixed micelles between SDS and the nonionic detergent resulting in a decrease of the free SDS monomers and/or to a competitive binding of the SDS and nonionic detergent for the proteins[52,53]. Anderton and Torpe[55] have reviewed the methods available for removing SDS from polyacrylamide gels prior to immunochemical analysis.

Zwitterionic detergents e.g., Empigen BB and the sulfobetaine series, can be used directly in connection with immunochemical analysis since they do not interfere with the immune reactions[54,56].

Chaotrophic agents such as urea can be used to solubilize membrane antigens. Urea inhibits the immune reaction[9]. However, concentrations of 1.5 M in conjunction with nonionic detergents have been successfully used in immunoelectrophoresis[51].

14.5   Applications

14.5.1   FPLC™ chromatofocusing fractionation of human serum proteins assessed by fused rocket immunoelectrophoresis.

Human serum (1 ml) was first depleted of albumin by affinity chromatography on a column (115 × 10 mm) packed with Blue Sepharose CL-6B and equilibrated in 50 mM Tris/HCl with 100 mM NaCl, pH 7.0 (Fig. 14-3, top). The albumin-depleted material was then equilibrated in the chromatofocusing start buffer (Bis/Tris 25 mM, pH 6.3) and applied to a MonoP column mounted in an FPLC System. The eluent was Polybuffer 74 (diluted 1:10 in HCl) pH 4.0. Flow rate 1 ml/min. Fractions of 0.4 ml were collected. Aliquots (6 μl) of the fractions were analyzed by fused rocket-IE (Fig. 14-3, bottom). Gels (120 × 240 × 1 mm) were cast with 26 ml of 1 % agarose A in Tris-barbital buffer (pH 8.6, I = 0.020) and 300 μl of the monospecific antiserum towards (a) transferrin, (b) α_2-macroglobulin, (c) albumin, (d) Gc-globulin, (e) completement C3, (f) pre-albumin, (g) haptoglobin and (h) α_1-antritryspin. In each gel 80 samples were run (fractions No. 10 to 90). The gels were run at 5 V/cm overnight and later dried, stained and destained as described in section 14.4.5.1. The fractions with the selected proteins were pooled and submitted to further fractionation by chromatofocusing in other pH ranges[11].

14.5.2 Microheterogeneity of α_2-Macroglobulin Studied by Crossed Electrofocusing Immunoelectrophoresis.

Serum α_2-macroglobulin (M_r 820,000) is an acute phase reactant and shows proteinase inhibitor activity which controls the clotting and fibrinolytic system. There is also evidence that this protein shows genetic polymorphism[57]. The use of crossed IEF-IE allows study of the microheterogeneity of this serum protein (Fig. 14-8).

An aliquot of fresh human serum (10 μl) was run on an Agarose IEF gel containing Pharmalyte pH 4–6.5 and 2.5–5 (1:1 ratio) for 2500 Vh. IEF was run at constant power (15 W) with maximum set voltage and current of 1500 V and 25 mA. Cooling water at 10°C was circulated through the apparatus. After IEF, a lane containing the focused proteins was cut out and placed on top of a second dimension gel (1% Agarose C) containing anti-α_2-macroglobulin (2 μl/cm^2). Electrophoresis was run for 4 hours at 10 V/cm in buffer Tris-barbital pH 8.6, I = 0.05. Four components with pIs between 4.9 and 5.3 were found. This approach can be further extended to study the variations and interrelations of these components of α_2-macroglobulin in normal subjects and patients with various diseases.

14.5.3 Antigenic Variability of Different Strains of *A. fumigatus* Studied by Crossed Immuno- and Crossed Radio Immunoelectrophoresis.

The antigenic and allergenic activity of different strains of fungi *A. fumigatus* can be examined by combination of CIE, CRIE, and RAST-inhibition (Fig. 14-6). CIE was performed as described in section 14.3.3. Samples of 10 μl of crude *A. fumigatus* extract (20 mg dry weight/ml) were run in the first dimension at 10 V/cm for 35 minutes. The

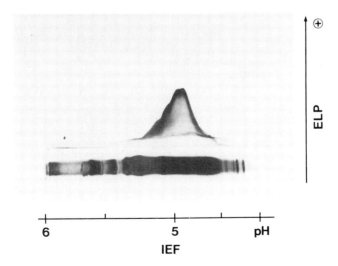

Figure 14-8. Crossed IEF-IE of α_2-macroglobulin. First dimension: Agarose IEF-(1%) gel containing Pharmalyte pH 2.5–5 + 4–6.5 (1:1) run at 15 W, 1500 V for 2500 Vh (10°C). A lane (5 mm width) containing the focused proteins was cut out and laid on the second dimension gel. Sample: 10 μl of fresh serum. Second dimensions: Agarose C (1%) contaning anti α_2-macroglobulin (90 μl/cm^2 of gel) run as described under Fig. 14-5. Reproduced from Ref. 19 with the kind permissions of the authors and editors.

second dimension electrophoresis was performed at 2 V/cm for 18 hours on an Agarose A(1 %) gel containing rabbit antiserum against *A. fumigatus* (10 μl/cm^2). The immunoplate was then stained with CBB (see p. 420). Radiostaining was performed by incubation of the immunoplate with the patient's serum for 18 hours at 22°C followed by incubation with ^{125}I- anti-IgE solution (72000 cpm/ml) for 18 hours at 22°C. After washing and drying, the plate was placed on X-ray photographic film for rapid autoradiographic exposure in cassette at low temperature (-70°C) as described in Ref. 26.

Fortyfour different antigens were demonstrated in mycelial extracts of *A. fumigatus* by crossed-IE stained with CBB and radiostaining (Fig. 14-6A and B, respectively). Eighteen different allergens were identified using CRIE: two major allergents (antigens No. 10 and 40 in Fig. 14-6B), ten intermediate allergens and six minor allergens[27]. In the next step, individual serum samples of patients with fungal allergy were screened for reactivity to *A. fumigatus* antigens. In Fig. 14-6C the X-ray film from a CRIE experiment shows a patient with high levels of IgE antibodies agaist antigens No. 10 and No. 28. Three other antigens besides showed a weak IgE response[27].

Using this approach it is possible to identify individual allergents in extracts and to obtain the specific pattern of IgE response for each patient.

14.6   References

1. P. Grabar, C. A. Willians, *Biochem. Biophys. Acta 10*, 193–202 (1953).
2. Electrophoretic and electro-immuno-chemical analysis of proteins. C.-B., Laurell, ed., *Scand. J. Clin. Lab. Invest.* 1972, 29 (suppl).
3. Quantitative immunoelectrophoresis: New developments and applications. N. H. Axelsen, ed. *Scand. J. Immunol.* 1975, 4(suppl. 2).
4. Handbook of immunoprecipitation-in-gel techniques. N. H. Axelsen, ed., *Scand. J. Immunol.* 1983, 17(suppl. 10).
5. Ö. Ouchterlony, L.-Å. Nilsson, In *Handbook of experimental immunology*. D. M. Wier, ed., Blackwell Scientific Publications, Oxford, 1978, Vol. 1, Chapter 19.
6. R. Verbruggen *Clin. Chem. 21*, 5–43 (1975).
7. H. Løwenstein, *Prog. Allergy 25*, 1–62 (1978).
8. U. W. Mueller, J. M. Potter, *Anal. Biochem. 100*, 51–53 (1979).
9. H. P. Princé, D. Burnett, D. B. Ramsden, *J. Chromatog. 143*, 321–323 (1977).
10. B. Weeke, *Scand. J. Clin. Lab. Invest. 21*, 351–354 (1968).
11. L. G. Fägerstam, J. Lizana, U.-B. Axiö-Fredriksson, L. Wahlström, *J. Chromatog. 226*, 523–526 (1983).
12. P. J. Svendsen, N. H. Axelsen, *J. Immunol. Methods 1*, 169–173 (1972).
13. I. Brandslund, H. C. Siersted, S.-E. Svehag, B. Teisner, *J. Immunol. Methods 44*, 63–69 (1981).
14. H. G. M. Clarke, T. Freeman, *Protides Biol. Fluids 14*, 503–509 (1966).
15. R. J. Ritchie, R. M. Smith, *Clin. Chem. 22*, 497–499 (1976).
16. J. Clausen, *Immunochemical techniques for the identification and estimation of macromolecules.* Elsevier/North Holland Biomedical Press, Amsterdam, 1981.
17. O. J. Bjerrum, *Electrophoresis 6*, 209–226 (1985).
18. J. Söderholm, C. J. Smyth, T. Wadström, *Scand. J. Immunol. 4.*(suppl. 2), 107–113 (1975).
19. J. Lizana, I. Olsson, In *XI International Congress of Clinical Chemistry.* E. Kaiser, F. Gabl, M. M. Müller, M. Bayer, eds., Walter de Gruyter & Co., Berlin 1982, pp. 1225–1233.
20. J. Lizana, I. Olsson, A. Savill, *Clin. Chem. 28*, 1569 (1982).
21. M. Thyman, *Human Hered. 31*, 214–221 (1981).
22. P. D. Eckersall, J. A. Beeley, *Electrophoresis 1*, 62–67 (1980).
23. Å. Siden, *J. Neurol. 217*, 103–109 (1977).
24. B. Weeke, H. Löwenstein, *Scand. J. Immunol. 2*(suppl. 1), 149–154 (1973).
25. L. Aukrust, K. Aas, *J. Scand. J. Immunol. 6*, 1093–1099. (1977).
26. T. Uhlin, R. Einarsson, *Int. Arch. Allergy. Appl. Immunol. 70*, 213–219 (1983).
27. I. Wallenbeck, L. Aukrust, R. Einarsson, *Int. Arch. Allergy. Appl. Immunol. 73*, 166–172 (1984).

28. T. C. Bøg-Hansen, *Anal. Biochem. 56*, 480-488 (1973).
29. P. Owen, J. D. Oppnheim, M. S. Nachbar, R. E. Kessler, *Anal. Biochem. 80*, 446-457 (1977).
30. *Lectins-Biology, Biochemistry, Clinical Biochemistry.* T. C. Bøg-Hansen, ed., Walter de Gruyter & Co., Berlin, 1980, Vol. 1.
31. V. Horejsi, *Anal. Biochem. 112*, 1-8 (1981).
32. K. Toftager-Larsen, E. Kjaersgaard, J.-Chr. Jacobsen, B. Nørgaard-Pedersen, *Clin. Chem. 26*, 1656-1659 (1980).
33. J. Breborowicz, A. Mackiewicz, D. Breborowicz, *Scand. J. Immunol. 14*, 15-20 (1981).
34. O. J. Bjerrum, *Anal. Biochem. 90*, 331-348 (1978).
35. J. O. Jeppsson, C.-B. Laurell, B. Franzén, *Clin. Chem. 25*, 629-638 (1979).
36. J. F. Monthony, E. G. Wallace, D. M. Allen, *Clin. Chem. 24*, 1825-1828 (1978).
37. A. M. Johnson, *Annals Clin. Lab. Sci. 8*, 195-216 (1978).
38. J. Lizana, A. Savill, I. Olsson, In *Electrophoresis '81.* R. C. Allen, P. Arnaud, eds. Walter de Gruyter & Co., Berlin, pp. 550-561 (1981).
39. R. J. Mayer, J. H. Walker *Immunochemical methods in the Biological Sciences: Enzymes and proteins.* Academic Press, New York, 1980.
40. W. D. Linscott, *Linscott's directory of Immunological and Biological reagents.* Mill Valley, California, USA, 4th Edition, 1986.
41. V. Neuhoff, R. Stamm, H. Eibl, *Electrophoresis 6*, 427-448 (1985).
42. K. Hellsing, In *Automated immunoanalysis.* R. F. Ritchie, ed., Marcel Dekker Inc., New York, 1977, pp. 67-112.
43. M. Céska, F. Grossmüller, *Experientia 24*, 391-392 (1968).
44. N. St. G. Hyslop, D. G. Cochrane, *J. Immunol. Methods 6*, 99-107 (1974).
45. J.-L. Guesdon, S. Avrameas, *Immunochemistry 11*, 595-598 (1974).
46. S. Avrameas, J. Uriel, *Comp. Rend. Acad. Sci. Paris* Serie D *262*, 2543-2545 (1966).
47. S. Avrameas, T. Ternynch, J.-L. Guesdon *Scand. J. Immunol 8*(suppl. 7) 7-23 (1978).
48. G. D. Johnson, E. J. Holborow, J. Dorling, In *Handbook of Experimental Immunology.* D. M. Weir ed., Blackwell Scientific Publications, Oxford, 1978, Vol. 1, Chapter 15.
49. J. W. Goding, *J. Immunol. Methods 20*, 241-253 (1978).
50. B. Nørgaard-Pedersen, K. Toftager-Larsen, N. H. Axelsen, *Scand. J. Immunol 17*(suppl. 10) 259-263 (1983).
51. O. J. Bjerrum, P. Lundahl, *Scand. J. Immunol. 2*(suppl. 1), 139-146 (1973).
52. G. Dimitriadis, *Anal. Biochem. 98*, 445-451 (1979).
53. O. J. Bjerrum, J. H. Gerlach, T. C. Bøg-Hansen, J. B. Hertz, *Electrophoresis 3*, 89-92 (1982).
54. L. M. Hjelmeland, A. Chrambach, *Electrophoresis 2*, 1-11 (1981).
55. B. H. Anderton, R. C. Thorpe, *Immunol. Today 1*, 122-127 (1980).
56. K. Burridge, In *Methods in Enzymology.* S. P. Collowick, N. O. Kaplan, eds., Academic Press, New York, Vol. 50, 1978, pp. 54-64.
57. A. Rosén, K. Ek, P. Åman, *J. Immunol. Methods 28*, 1-11 (1979).
58. L. Kerenyi, F. Gallyas, *Clin. Chim. Acta 38*, 465-467 (1972).
59. E. W. Willoughby, A. Lambert *Anal. Biochem. 130*, 353-358 (1984).
60. R. McLachlan, D. Burns, In *Electrophoresis '84*, V. Neuhoff, ed., Verlag-Chemie, Wienheim, pp. 324-327 (1984).
61. O. Vestergerg, B. Gramstrup-Christensen *Electrophoresis 5*, 282-285 (1984).
62. L. D. Lee, H. P. Baden, *J. Immunol. Methods 18*, 381-385 (1977).
63. C. Milstein, In *Antibodies as a tool. The applications of immunochemistry.* J. J. Marchalonis, G. W. Warr, eds., John Wiley & Sons Ltd., Chichester, (1982).
64. G. Köhler, C. Milstein, *Nature 256*, 495-497 (1975).

15 Protein Mapping by Two-Dimensional Polyacrylamide Gel Electrophoresis (O'Farrell Technique)

Karl-Erik Johansson

The National Veterinary Institute
Box 7073
S-750 07 Uppsala
Sweden

15.1   Introduction

Many problems in biological and medical research involve analysis and comparison of protein compositions of a great number of complex samples. High resolution is a prerequisite for successful analysis of complex protein mixtures and is especially important when the proteins of interest are present in relatively low concentrations. To avoid low or irreproducible yields and time-consuming purification work, one should be able to perform the analysis of each sample under identical conditions and in as few steps as possible. Conventional polyacrylamide gel electrophoresis (PAGE) or any other one-dimensional separation method will not always give the required resolution when the samples contain more than approximately 100 components. The development of two-dimensional (2-D) PAGE by O'Farrell has revolutionized the field of protein analysis[1] and this technique has proved to be one of the most powerful methods for protein mapping of cells or sub-cellular fractions. The extremely high resolution capacity has been achieved by combining two methods with high resolving power and which separate proteins according to independent parameters. The two methods are isoelectric focusing (IEF), by which proteins are separated according to charge, and PAGE in the presence of sodium dodecyl sulphate (SDS-PAGE), by which proteins are separated according to size. Thus, IEF and SDS-PAGE are combined into a two-dimensional technique, which is often termed the O'Farrell technique to distinguish it from 2-D PAGE based on other parameters. In this chapter, only the O'Farrell technique will be discussed.

Many samples can be analyzed simultaneously by 2-D PAGE provided that the right equipment is used and the running conditions can be standardized. Whole cell extracts are often analyzed and this requires that the solubility of membrane and ribosomal proteins and the like be enhanced by addition of urea and a neutral detergent to the samples and to the IEF gel. It is possible to resolve at least 100 components by either IEF or SDS-PAGE. The theoretical resolution capacity of a 2-D technique, in which these two methods are combined, should therefore be 10,000 components, provided they are evenly distributed over the separation ranges and present at about the same concentration. However, in practice it has not been possible to detect more than 1500-2000 protein spots on one gel plate. IEF can be used to separate proteins with a charge difference of only one unit. For instance, the ribosomal protein L12 can be separated from L7, which differ only in that L7 is acetylated on the α-amino group while L12 is not[1]. Thus, samples to be analyzed by 2-D PAGE should always be handled in a way that minimizes the risk of introducing charge modifications. If charge modification is introduced, the corresponding protein will appear as multiple spots in the IEF dimension.

There is no technique for protein analysis available today which can compete with 2-D PAGE with respect to the resolution capacity. However, 2-D PAGE is a complicated method, particularly if it has to be used in concert with computer aided pattern analysis (CAPA), and should not be chosen if the problem can be solved with simpler methods. Thus, 2-D PAGE is not a suitable method, if it is going to be used only a few times per year in the laboratory. It has to be used continuously for optimal results.

It is tempting to compare 2-D PAGE with another 2-D electrophoresis method, crossed immunoelectrophoresis (CIE), which is treated in Chapter 14 of this volume. An immunoprecipitate in CIE is not likely to contain more than one antigen, and a spot in 2-D PAGE probably contains one protein only. Even if two proteins happen to have the same isoelectric point (pI) and molecular weight, they tend to displace each other in the gel and appear as discrete spots[2], at least if they are present in similar concentrations. After introduction of silver staining methods, both 1-D and 2-D PAGE are more sensitive than CIE, particularly for proteins of low immunogenicity. However, if a good antiserum is available, CIE is a simple method for very accurate quantifications. CIE has to be performed under non-denaturing conditions and antigens can therefore be identified by zymogram techniques. After the introduction of protein blotting techniques (see Chapter 16), proteins can also be identified after SDS-PAGE by an immune assay. CIE has a resolution capacity of about 50 components, which means that 2-D PAGE has a much greater resolving power. However, one should keep in mind that CIE is better suited for quantification than 2-D PAGE since the area subtended by an immunoprecipitation arc is easily determined (cf Chapter 14). The characteristics of different immunochemical analysis methods are further discussed in Chapter 16.

The article on 2-D PAGE by O'Farrell, which was published in 1975[1], has been cited several thousand times and the literature on 2-D PAGE is continuously (and so far also exponentially) growing. There are many excellent reviews[3,4], manuals[5,6], books[7,8], application publications[7,10], and proceedings from symposia[11,12]. Three scientific events have further contributed to the utility of 2-D PAGE. First, the introduction of blotting techniques[13], which can be used for specific identification of protein spots[14]. Second, the development of CAPA[15], which can facilitate storage and comparison of the information in 2-D gels. Third, the use of silver staining for detection of proteins in polyacrylamide gels[15].

2-D PAGE requires a lot of experimental manipulations and the different steps will be discussed below, but for those who are going to set up the technique, the book edited by Celis and Bravo[7] and the book written by Dunbar[8] are highly recommended.

15.2   Preparation of Samples

The sample preparation is very important for the reproducibility as well as the result of a 2-D PAGE. For instance, incomplete unfolding of a peptide chain can result in multiple spots in the SDS-PAGE direction, and incomplete disaggregation of protein complexes might give multiple spots or streaking in both directions, which is typical, as an example, for the outer membrane OmpA of *Escherichia coli*.[17] This protein is difficult to unfold (cf. Fig. 15-5C). Unfolding and disaggregation is more efficient at high temperature, but charge modifications may also be introduced[3,18]. However, multiple spots due to post-translation modification has to be distinguished from artefactual multiple spot formation. For instance, partial phosphorylation and glycosylation will introduce charge differences in the protein, resulting in multiple spots in the IEF direction.

Instead of analyzing whole cell extracts by 2-D PAGE, it is sometimes valuable to pre-fractionate the material into membranes and cytoplasmic components, or other subcellular fractions[4,19]. Proteins which are present in low concentrations can be enriched to a level where it is possible to detect them by 2-D PAGE[19]. The 2-D protein patterns will be less complicated after pre-fractionation of the material and identification of protein spots might be easier. Figure 15-5 shows that a pH gradient, which does not give optimal resolution of one subcellular fraction (the cytoplasmic and periplasmic proteins) might well give very good resolution of another subcellular fraction (outer membrane proteins). Thus, by pre-fractionation procedures, it could be easier to select the pH gradient and the gel composition (or gradient), which give optimal resolution, Furthermore, identification of protein spots can be facilitated.

15.2.1   Radiolabelling

2-D PAGE can be used to establish complete protein maps of very complex samples such as whole cell extracts, cytoplasmic proteins, membranes or other subcellular fractions. Such samples are likely to contain several hundred components, which are present at various concentrations. Conventional Coomassie blue staining, is, therefore, in general not adequate for 2-D PAGE. More sensitive detection methods, such as silver staining or radiolabelling in combination with audoradiography, have to be used.

If the material under study originates from cells which can be grown in culture it is convenient to label the proteins internally to facilitate their detection. Bacteria, for instance, can be grown in the presence of [^{35}S]sulphate[1,20] or a mixture of [^{14}C]-labelled amino acids[1]. Nutritional auxotrophs for amino acids are often available or can be constructed and such strains of bacteria can be grown under limiting concentrations of the radiolabelled amino acid. *E. coli* proteins are often labelled by cultivation of the cells in the presence of [^{35}S]methionine because biosynthesis of methionine is depressed if it is present in the growth medium and it can be obtained with high specific activity. Transformed cells, for which many amino acids are essential, are also often grown in the presence of [^{35}S]methionine[21,22]. By optimizing the cultivation conditions, proteins with high specific activity can be obtained[21].

There is a great interest in using 2-D PAGE for determination of the protein composition of clinical material[9,10] in which the proteins cannot easily be labelled internally. Such material can instead be detected by silver staining (see below) or external labelling and autoradiography. Proteins of high specific activity can be prepared by labelling with ^{125}I by the lactoperoxidase-mediated reaction[23]. Iodination is a practical method because it is possible to label surface components of intact cells specifically, and thereby distinguish between membrane and cytoplasmic proteins. However, iodination has been shown to introduce charge heterogeneity[24], which may cause artefacts, or low reproducibility in the IEF step. Reductive methylation of amino groups with [3H]sodium borohydride is preferable because charge heterogeneities are not introduced[24].

15.2.2 Solubilization

15.2.2.1 Conditions

The reproducibility of 2-D PAGE is highly dependent upon the conditions used for solubilization and disaggregation of the proteins in the sample. It is essential to find conditions which:

1. Solubilize the proteins in the sample quantitatively.
2. Disaggregate protein complexes and unfold the peptide chains completely.
3. Are compatible with IEF.

In general, SDS cannot be used for solubilization, at least not at high concentration, since it modifies the charge of the proteins and interferes with the formation of the pH gradient in the IEF step. A non-ionic detergent is, therefore, used in combination with a high concentration of urea and a reducing agent in the sample buffer. Two commonly used neutral detergents are Nonidet P-40[1] and Triton X-100[25], which are polyoxyethylene p-t-octyl phenols. They both have hydrophilic polyoxyethylene head groups, but of different chain lengths, and hydrophobic t-octyl-phenyl groups[26]. These neutral detergents are not sufficiently efficient to solubilize all kind of proteins or to disaggregate protein complexes, and 9 M urea is also used in the sample buffer. A reducing agent, such as 2-mercaptoethanol or dithiothreitol, must be added to disrupt disulphide bonds, allowing a more complete disaggregation of the polypeptide chains.

There are proteins (e.g. some membrane proteins, ribosomal proteins, and histones), which resist solubilization even under reducing conditions in the presence of urea and a neutral detergent. For analysis of such proteins, it is possible to add small amounts of SDS to the sample buffer without decreasing the resolution in the IEF step too much[18]. This solubilization method has been used for analysis of the outer membrane proteins OmpA[17] and TraT[29] from *E. coli* by 2-D PAGE. These proteins are both difficult to solubilize and disaggregate under normal conditions. Protein complexes have to be completely disaggregated, and polypeptide chains have to be completely unfolded in the sample buffer; otherwise one protein may give rise to several spots in the final 2-D pattern.

Presence of a reducing agent in the gel buffer will result in two or three lines in the molecular weight range 55,000–70,000.[27] These lines are visible upon Coomassie blue or silver staining but not in autoradiograms. It has been suggested that these bands originate from contaminations with skin proteins, which attain that size upon reduction[28]. Figure 15-5B shows that these artefactual lines can easily be distinguished from protein spots in a 2-D gel.

Ribosomal proteins are often difficult to analyze by the O'Farrell technique, and these proteins are therefore analyzed by other 2-D PAGE methods based on separation at different pH and in the presence of urea.[30]

15.2.2.2   Protein Concentration

The protein concentration in the sample must be high in order to detect minor components, to avoid diluting the sample buffer more than about 10%, and to keep the sample volume small. Whole cell extracts or membrane extracts can be prepared by treating the cell or membrane pellet with sample buffer. Soluble proteins often have to be concentrated by e.g. ultrafiltration or precipitation with cold 10% trichloroacetic acid and washing of the precipitate with acetone[31]. Lyophilization should be avoided, since charge modification due to oxidation and deamidation can occur[1].

15.2.2.3   Interfering Substances

High salt concentration in the sample must be avoided unless the ampholytes are covalently linked to the polyacylamide gel in the IEF step (see below). Samples solubilized or suspended in e.g. 2–5 mM EDTA (pH 7.5) can be treated with sample buffer and analyzed by 2-D PAGE. Even small amounts of particulate or unsolubilized material in the sample will cause streaking in the IEF dimension. The sample should be clarified by centrifugation in a microcentrifuge at $10,000 \times g$ for 5–10 min after treatment with the sample buffer. Presence of too much nucleic acid can disturb the IEF, and the sample should be treated with DNase[1] (and RNase[25]) before analysis if the effect of nucleic acid is too severe. Lipids have also been reported to interfere with the IEF step in 2-D PAGE and can be removed by extraction with ethanol[25]. Specific proteins can decrease the resolution in 2-D PAGE, especially if they are very abundant. For instance, spectrin and actin have been shown to disturb the 2-D pattern of erythrocyte membranes[32].

15.3   Isoelectric Focusing

IEF is treated in detail in Chapter 13 and this paragraph will therefore mainly be focused on factors which are important for 2-D PAGE. Briefly, proteins are separated according to charge by electrophoresis in a pH gradient where they will migrate to the position corresponding to their pI. This principle will result in very sharp protein bands because electrophoresis is continuously counteracting diffusion.

15.3.1   Orders of Dimension

It is most common to run the IEF before the SDS-PAGE, because the sample load can be larger on an IEF gel than on an SDS-gel. Furthermore, if SDS-PAGE is peformed first, most of the SDS will have to be removed from the gel slice prior to analysis by IEF. Removal of SDS will also result in protein losses from the SDS-gel. Despite these difficulties, SDS-PAGE has been used in the first dimension of 2-D PAGE to overcome solubility problems (for a review, see Ref. 3). When IEF is used in the first dimension, it has to be performed under the same conditions used for solubilization, i.e. in the presence of a neutral detergent and a high concentration of urea, to avoid precipitation of proteins in the gel.

15.3.2 Chemicals

15.3.2.1 Carrier Ampholytes

So-called carrier ampholytes are included in the IEF gel to create the pH gradient. Carrier ampholytes are now available from several companies, all of which use different procedures for their production. There seem to be difficulties involved in production of these ampholytes, since variations between different batches have been reported. Such batch variations could be very troublesome when comparing large series of 2-D gels where different batches of carrier ampholytes have been used. Pharmacia has developed a special ampholyte (2-D Pharmalyte, pH range 3–10), which is produced with negligible batch variations. However, it should be remembered that a linear pH gradient is in many cases not very useful, and the actual pH range in a system containing a high concentration of urea can be difficult to predict[33]. Protein patterns are not identical when carrier ampholytes from different manufacturers are used, and the different ampholytes give better resolution in different pH regions[3]. For optimal resolution of a certain sample, ampholytes of different pH regions often have to be mixed, and the optimal proportion, which gives the most even distribution of protein spots over the whole gel, has to be determined[34]. In general, microbial proteins seem to be more acidic[1,25] than proteins from mammalian cells[35], and this must be considered when designing the pH gradient.

15.3.2.2 Monomers and Gel Composition

Acrylamide and N,N'-methylenebisacrylamide (bis) have to be of high quality because presence of acrylic acid contaminants might ruin the pH gradient. The monomer solution should therefore be freshly prepared and if necessary de-ionized by treatment with Amberlite MB1[36]. In the first dimension, proteins should be separated according to charge only, and the pore size of the IEF gel should be large enough to avoid a molecular sieving effect. A gel composition of $T = 4\%$; $C = 5\%$ is often used. For a definition of T and C, see p. 354 and Ref. 37. Proteins of high molecular weight will also be slightly retarded in gels of this composition.

15.3.2.3 Other Chemicals

The same detergent that is used for solubilization of the sample should be included in the IEF gel, as well as urea of high concentration (8–9 M). Urea must be of high quality (ultra pure) because presence of isocyanate, which is formed by decomposition of urea, can result in carbamylation of proteins and thereby cause charge heterogeneity[1]. A solution of urea can also be deionized with Amberlite MB1. Dust particles should be removed from solutions by filtration through a small pore size (0.2–0.5 μm) filter.

15.3.3 Gel Shape

IEF can be performed in rod gels or slab gels, but rod gels have mainly been used in connection with 2-D PAGE. The inner diameter of the tube used for rod gel IEF should not exceed 2–3 mm, and the length is usually 12–15 cm. Any electrophoesis apparatus which can accommodate this tube size can be used. Figure 15-1A shows an apparatus for

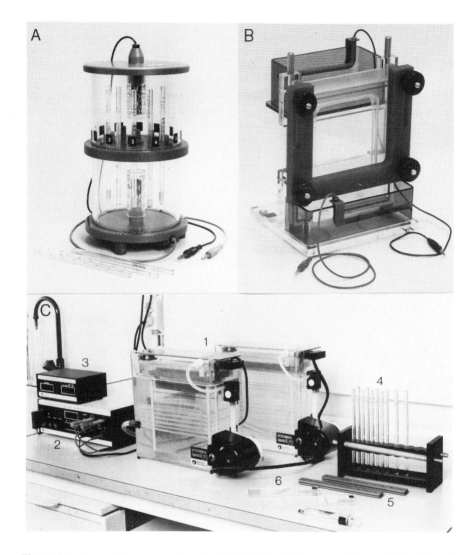

Figure 15-1. Some of the equipment used in 2-D PAGE, which has been built at the Workshop of Uppsala Biomedical Center, Box 570, S-751 23 Uppsala, Sweden. A, Apparatus for IEF in glass tubes. B, Apparatus for SDS-PAGE (the gel cassette is also shown in Fig. 15-4B). C. Equipment for 2-PAGE is commercially available from many different companies. Two large or four intermediate size gels can be analyzed simultaneously in the GE-2/4 LS apparatus from Pharmacia. 1, Electrophoresis tanks. 2, Power supply. 3, Volthour integrator. 4, Tubes for IEF. 5, IEF tube gaskets. 6, IEF gel transfer tools. (c.f. Fig. 15-4A).

IEF in rod-shaped gels. Rod gels are easy to cast and to handle when they are transfered to the SDS-gel for the running in the second dimension. However, resolution can be improved, especially for basic proteins, by using a thin slab gel for IEF[38]. Such a gel cannot stretch since it has been bound to a plastic support, and the pH gradient is therefore more stable[38].

15.3.4   Sample Application

In the original description[1], the sample was applied at the cathodic side of the (rod-shaped) gel after a short prerun. Presence of 2-mercaptoethanol in the sample buffer will disturb the pH gradient and it is therefore better to apply the sample at the anodic side of the IEF gel[39]. It also seems preferable to apply samples containing nucleic acids on the anodic side[40]. An overlay solution, containing a lower concentration of urea than the sample buffer, can be used to protect the sample from the extreme pH of the electrolyte[1].

15.3.5   Running Conditions

When rod gels of the above size are used, the current should not exceed 1 mA per gel, which means that the IEF can initially be performed at approximately 300 V. The voltage should then be continously increased to reach 800–1000 V for the last couple of hours. The running conditions are often expressed as the number of volthours (Vh) but it is better to use Vh/cm^2 (Ref. 3), since the length of the gel will also effect the time it takes to reach steady-state. In general, steady-state will be reached in 30–40 Vh/cm^2 (Ref. 41). High voltage should be used at the end of the running to assure efficient focusing.

A non-equilibrium pH gradient electrophoresis (NEPHGE) system has been developed to improve resolution of basic proteins[42]. By running the IEF for only 10–15 Vh/cm^2, steady-state will never be attained, but resolution of basic proteins is profoundly improved[42]. Thus, to obtain a complete 2-D protein pattern of a sample containing certain basic proteins, both ordinary IEF and the NEPHGE system have to be used in the first dimension. It is very important to standardize the conditions used for the NEPHGE system, to get reproducible results. It should also be remembered that the size of a protein will influence its final position in the NEPHGE system, and the two parameters (charge and size) used for separation by ordinary 2-D PAGE will therefore not be truly independent.

15.3.6   Estimation of the pH Gradient

The pH gradient in the IEF gel can be estimated by running standard proteins with known pI in parallel with or on the sample gel. Carbamylated standard proteins have been used for determination of the pH gradient[33]. Carbamylated standard proteins, which can be added to the sample, are available from Pharmacia as Carbamylyte[TM 43]. However, standards cannot be used to determine the pH gradient in the NEPHGE system. Another method is to run an extra gel, which is sliced into 0.5 cm pieces after completed IEF. Each slice is then eluted over night with 0.2 ml of degassed H_2O and the pHs of the fractions are determined by using a small electrode. A drawback to this method is that the pH gradient of the gel which was actually used for the second dimension cannot be checked.

Figure 15-2 shows the pH gradients, determined after elution of parallel gels, obtained by using ampholytes of different pH regions and in different proportions. Some of these gradients were used for IEF of cytoplasmic and periplasmic proteins from *E. coli*

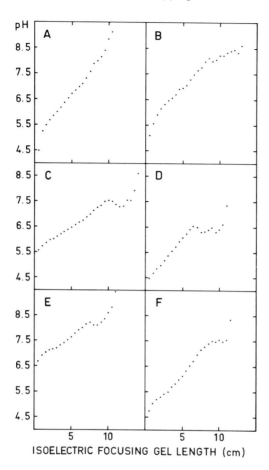

Figure 15-2. The pH gradients obtained in 12 cm rod shaped gels with Pharmalyte[ⁱᵉ] cocktails of different pH ranges and proportions. This figure illustrates how the pH gradient can be modified by changing the pH ranges and the proportions of the carrier ampholytes. The corresponding protein patterns are shown in Fig. 15-3. A, (3–10) + (4–6.5); (1:1). B, (3–10). C, (3–10) + (5–8); (1:4). D, (2.5–5) + (5–8); (1:1). E, (5–8) + (8–10.5); (1:1). F, (2.5–5) + (5–8) + (8–10.5); (2:1:1).

and the final 2-D protein maps are shown in Fig. 15-3 and Fig. 15-5A. Some of the acid proteins are lost if the pH is not low enough at the anodic side (cf. Fig. 15-2C and Fig. 15-3A). If the pH gradient is too steep in the acid region, the resolution in the IEF direction will be poor (cf. Fig. 15-2B and Fig. 15-3B). The gradient shown in Fig. 15-2F gives good resolution of cytoplasmic *E. coli* proteins in the IEF direction, see Fig. 15-3C. Furthermore, the acid proteins L7 and L12 with pIs of 4.7 and 4.85 have not been lost and they are resolved.

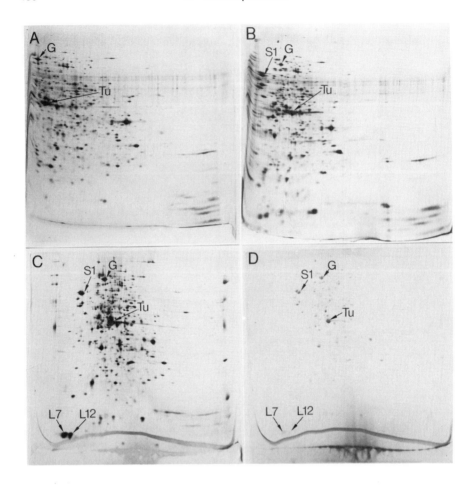

Figure 15-3. Analysis of cytoplasmic and periplasmic proteins from *E. coli* by 2-D PAGE with different pH gradients in the first dimension (cf. Fig. 15-2). The ribosomal proteins S1, L7, and L12 as well as the elongation factors G and Tu have been tentatively identified by comparison with published 2-D protein maps[20]. The gels are always shown with the low pH side to the left and with low molecular weights at the bottom. The protein patterns were developed by silver staining except in D. The following cocktails of Pharmalyte were used. A, (3–10) + (5–8); (1:4). B, (3–10). C, (2.5–5) + (5–8) + (8–10.5); (2:1:1). D, The same gel as in Fig. 15-3C, but stained with Coomassie blue. (A–C reproduced from Ref. 33 with permission from Pergamon Books, Ltd. Oxford).

15.3.7   Transfer of the IEF Gel to the SDS Gel

15.3.7.1   Equilibration

If all IEF gels (in tubes) cannot be simultaneously processed, they may be frozen (preferably at −70°C), although optimal resolution is obtained when the second dimension is run as soon as possible. The gels are extruded from the tubes by introducing

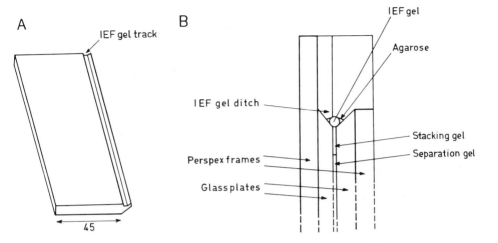

Figure 15-4. Schematic drawings of some important details of the equipment for 2-D PAGE. The number refer to length in mm. A, The transfer tool for IEF gels. B, Cross section of an assembled gel cassette for 2-D PAGE.

water between the gel and the glass by means of a needle and a syring. The gel should then be transferred to a test tube with gel buffer contaning Tris, SDS, 2-mercaptoethanol, and glycerol (pH 6.8)[1]. The gel should be "equilibrated" for 10–30 min under slow shaking. The equilibration will unfortunately cause losses of low molecular weight proteins, in particular, since the pores of the IEF gel are large. It is therefore important to keep the equilibration time constant. Some authors prefer to omit the equilibration step, although it may cause some streaking in the second dimension. If bromophenol blue is added to the gel buffer it is easy to follow the electrophoresis in the second dimension.

It has been found that treatment of the IEF gel with a gel buffer containing 55% methanol can improve resolution[44].

15.3.7.2 Gel Transfer

After equilibration, the gel is placed in the shallow track of a special tool for the transfer of gels (see Fig. 15-4A). Agarose (0.3%) of low electroendosmosis is then poured into the "ditch" of the electrophoresis apparatus (see Fig. 15-4B), in which the SDS gel has previously been cast. By tilting the transfer tool over the ditch, the IEF gel can now be introduced into the SDS-PAGE apparatus without stretching[45]. It is very important to avoid stretching of the IEF gel, to facilitate later comparison of the 2-D protein patterns.

15.4   Sodium Dodecyl Sulphate Polyacrylamide Gel Electrophoresis

SDS-PAGE will only be described briefly since it is treated in detail in Chapter 12, p. 364. An apparatus which is convenient for 2-D PAGE and the corresponding gel cassette are shown in Fig. 15-1B and Fig. 15-4B, respectively. In principle, proteins are separated according to size in a molecular sieving gel because they attain the same negative surface charge per unit area in the presence of SDS[46]. There are, however, exceptions, such as glycoproteins, membrane proteins, and proteins which are extremely charged. A new model of the interactions between SDS and proteins has recently been proposed[47]. It is necessary to use a discontinuous buffer system for optimal resolution in 2-D PAGE. Discontinuous buffer systems have high resolving capacities because the protein will be eluted from the IEF gel and concentrated (stacked) in the so-called stacking gel by isotachophoresis[48]. The buffer system of Laemmli[49] was used in the original paper on 2-D PAGE[1], but others can be used as well. A complete experimental setup (Pharmacia) for 2-D PAGE is shown in Fig. 15-1C.

15.4.1   Chemicals

The quality of acrylamide and bis does not seem to be as critical as in IEF. Most stock solutions can be stored at 4°C for several months (except potassium persulphate), but the stacking buffer, which usually has a pH of 6.1–6.8, should regularly be checked for microbial growth. All stock solutions used in gel casting should be filtered through a small pore size (0.2–0.5 μm) filter to remove particles, which is especially important if the gel is to be silver stained.

15.4.2   Buffer Systems

15.4.2.1   The Maizel System[31]

This buffer system is often referred to as the Dobberstein system[50], particularly when used in combination with a gradient gel, but the original method was described by Maizel[31]. The system has a very high resolving capacity and the buffers are easy to prepare since the pH is not as critical as for some of the other buffer systems. One running buffer, containing Tris-glycine and SDS, is used and two different Tris-HCl buffers are used in the stacking and in the separation gel, respectively. The buffer system, referred to as the Laemmli system[49], is also a modified Maizel system.

15.4.2.2   The Neville System[51]

The buffers for this system are a little more complicated to prepare because different buffers are used as catholyte and anolyte, and the pH of the buffers is critical. However, this system gives sharp bands even for many glycoproteins and hence a better resolution in the final 2-D pattern. The Neville system can always be recommended for molecular weight determinations when glycoproteins are used as reference proteins.

15.4.2.3   Other Systems

Combinations of the Maizel and the Neville systems have been used, but more interesting is a recently introduced gel and buffer system, which resolves proteins in the molecular weight range 2,500–90,000 [52]. Another system has been described[53] which can resolve proteins with molecular weights of up to 8×10^6. These systems might be very useful for 2-D PAGE of proteins with extreme molecular weights.

15.4.3 Gel Systems

15.4.3.1 Homogeneous Gels

Homogeneous gels can be used at different concentrations, and a convenient way to express the gel composition has been introduced by Hjertén[37]. The defintions of the total concentration T and the degree of cross-linking C are also given on p. 354. A gel composition of $T = 10\%$; $C = 1\%$ will give good resolution in the molecular weight range 20,000–200,000 and $T = 15\%$; $C = 1\%$ is useful for the molecular weight range 10,000–100,000.

15.4.3.2 Gradient Gels

If the proteins in a certain sample are not evenly distributed over the molecular weight range, it is impossible to have good resolution over the whole area of a 2-D gel with a homogeneous gel. For such samples it could be valuable to utilize a gradient gel for the second dimension. Linear gradients are often not sufficient, but O'Farrell found e.g. that exponential gradient gels gave good resolution of *E. coli* cell proteins[1]. It is not easy to cast and process a great number of 2-D gels, particularly if a gradient is used for the second dimension. This is, however, necessary when several samples are to be compared by 2-D PAGE. Anderson and Anderson have therefore designed an apparatus by which it is possible to cast and run several gels simultaneously, which is termed the Iso-dalt system[54,55]. The Iso-dalt system is to be used in combination with their computerized 2-D gel analysis program termed Tycho[15].

15.4.4 Gel Size

The size of the final 2-D gel will of course influence the resolution. Large gels will give better resolution but are more difficult to handle. The area of the gel is often within the range 15 × 18 cm to 25 × 35 cm and the thickness mostly 0.5–2 mm. Very small gels (5 × 6 cm) have been used for 2-D analysis in microscale[56], but very large gels have been shown to give the best resolution[38,57]. For routine or screening purposes and for monitoring purification experiments, however, the resolution obtained by 2-D PAGE with small gels, for instance the PhastSystem™ (Pharmacia LKB), might be adequate[58].

15.4.5 Molecular Weight Determination

It is easier to calibrate a 2-D PAGE with respect to molecular weight than pI. Standard molecular weight markers can be used for calibration in two different ways. If the sample proteins are to be detected by autoradiography, the markers can be mixed with the gel buffer or the agarose solution[5]. This procedure will result in calibration lines on the gel, which has the advantage of automatic compensation for non-uniform migration over the gel. Non-uniform migration is common in 2-D PAGE as judged from the appearance of the bromophenol blue front, particularly if the second dimension is run at too high voltage, see Fig. 15-3C.

Wells for a reference sample can also be introduced in the agarose (in the gel ditch) beside the IEF gel by using small slot formers of perspex. The molecular weight markers can then be electrophoresed along the edges of the gel. See e.g. Fig. 15-5 where low and high molecular weight standard proteins have been used.

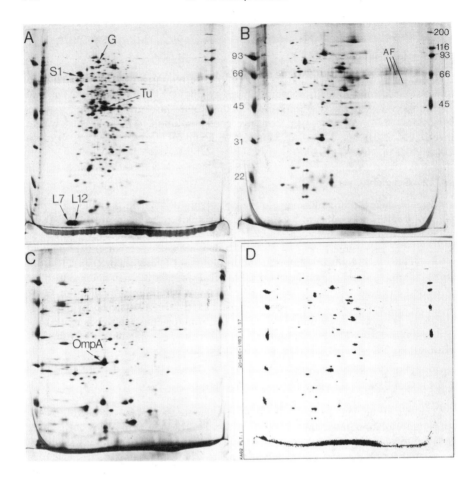

Figure 15-5. Analysis of different cell fractions from *E. coli* by 2-D PAGE. The membrane fractions were obtained by density gradient centrifugation of total membranes prepared by sonication of lysozyme-EDTA spheroplasts[66]. The IEF was performed with a Pharmalyte cocktail (2.5-5) + (5-8); (1:1), cf. Fig. 15-2D. The protein patterns were developed by silver staining. A, Cytoplasmic and periplasmic proteins. B, Cytoplasmic membranes. The artefact lines are indicated with AF. C, Outer membranes. The position of the OmpA protein is shown. D, A computer digitized image of the silver stained gel shown in B. A photographic negative (5 × 5 cm) was scanned in an Optronics P-1000 rotating drum scanner with a resolution of 50 μm. Graphic representation was obtained with a Printronics 600 line printer connected to a VAX 11/750 minicomputer.

15.5   Detection of Protein

15.5.1   Coomassie Blue Staining

The most common method to stain ordinary SDS gels is to use Coomassie blue. However, using Coomassie blue alone is in general not adequate for 2-D gels, since Coomassie blue does not detect spots containing less than about 0.5 μg of protein. The amounts of individual components in e.g. a cell extract can differ by several orders of magnitude. Therefore, Coomassie blue staining can be combined with a method of higher sensitivity (silver staining or autoradiograhy) to cover a broad range of protein concentrations. This procedure will result in good resolution of both major and minor components (cf. Fig. 15-3C and D as well as Fig. 15-7A and C). If the gel is to be stained with Coomassie blue, the proteins have to be fixed in aqueous methanol or trichloro-acetic acid. The latter is the more efficient, which is important for proteins of low molecular weight[59].

15.5.2 Silver Staining

Silver staining was first developed for histological purposes, but was also found to be useful for staining of protein bands after electrophoresis. The early method was troublesome and irreproducible, but after several modifications, the procedure has been very much improved. The reproducibility is good, background staining low, and the sensitivity at least 100 fold better than Coomassie blue staining. The different silver staining procedures have recently been reviewed[4,16]. The chemistry behind the silver staining process is not quite clear, but silver ions or diamine complexes of silver are probably bound to negatively charged groups on the proteins or to free aldehyde groups introduced by fixation with glutardialdehyde. After extensive rinsing of the gel to remove excess silver, the pattern is developed by reduction of bound silver to its metal form with formaldehyde. Many modifications of the original silver staining procedure have been published, and one of them, which works satisfactorily for 2-D gels, has been used for silver staining of gels shown in this chapter[60]. It is possible to detect 1–10 ng of protein in a spot.

If silver staining is to be combined with fluorography, it should be kept in mind that silver will quench low-energy beta particles, which could be particularly troublesome for tritium[61].

15.5.3 Radiolabelling

15.5.3.1 Autoradiography

Proteins can be detected by direct autoradiography if they have been labelled with a γ-emitter (e.g. ^{125}I or ^{59}Fe) or a hard β-emitter (e.g. ^{32}P). The gel is first dried and then placed in close contact with an X-ray film and exposed for a time period, which is dependent upon the specific activity of the protein and the energy of the emitted radiation. The sensitivity can be increased by using an X-ray intensifying screen. 2-D gels can be dried onto filter paper in special slab gel dryers, which are commerically available. A simple but very efficient technique has been introduced by Wallevik et al.,[62] see below.

Direct autoradiography can also be used if the proteins have been labelled with a β-emitter of intermediate energy level (e.g. ^{35}S and ^{14}C) and O'Farrell showed that it is possible to detect a protein constituting as little as 10^{-4}–10^{-5}% of the total protein, corresponding to 1–10 ng[1]. The sensitivity can be increased about 10-fold by using fluorography (see below). The 2-D protein patterns shown in Fig. 15-6 were obtained by direct autoradiography of gels, on which [^{35}S]-labelled proteins had been analyzed.

Figure 15-6. 2-D PAGE of whole cell extracts of different strains of *E. coli*. The cells were grown in the presence of [³⁵S]methionine and the protein patterns were developed by autoradiography. IEF was performed with Ampholines® of the pH ranges (3.5–10) + (5–7); (7:3). A, A wild type strain with intact tryptophan (trp) operon, in which e.g. tryptophan synthetase α and β (the trpA and trpB proteins) are expressed. B, A mutant strain in which the trp operon was deleted[37]. By courtesy of Leif Kirsebom, Department of Microbiology, Uppsala University, Uppsala, Sweden.

15.5.3.2   Fluorography

Low-energy β-particles (from e.g. ^{3}H) cannot be detected by direct autoradiography. A more sensitive technique, termed fluorography has been developed for such radiation[63]. In this method a fluor, e.g. 2,5-diphenyl-oxazole (PPO), is incorporated into the gel before exposure. When it is excited by β-particles, PPO will emit light which can be used to expose a blue-sensitive X-ray film. The sensitivity can be further increased by pre-exposing the X-ray film with a very short light flash[64]. PPO is not water soluble and the gel has to be soaked in a solution of PPO in dimethylsulphoxide (DMSO). When the gel is then transferred to water, PPO precipitates and the PPO-impregnated gel can now be dried and used in fluorography. Drying can be done with a slab gel dryer or between sheets of cellophane[62] as described below. The water-soluble fluor sodium salicylate can be used as substitute for PPO with only a slight decrease in resolution[65]. Sodium salicylate is less expensive and less toxic than PPO. Another advantage is that DMSO, which is also toxic, does not have to be used.

15.6   Drying of Polyacrylamide Gels

Apparatus for the drying of polyacrylamide gels at elevated temperature and under vacuum are available from many companies dealing with electrophoresis equipment. The gels easily crack in these apparatus if they are not completely dry when the vacuum tubing is disconnected. Gels can be dried surprisingly fast without using vacuum or high temperature. In the method of Wallevik et al.,[62] the gel is dried between two sheets of wet cellophane held together by a pair of plastic frames. After a few hours in the air stream of a fume hood or in front of a fan of some kind, the gel is dry and can easily be used for autoradiography or fluorography if a fluor was incorporated into the gel. These dried gels can also be stored as permanent transparencies. It is very convenient to have the gel as a transparency, because it can be placed on other gel transparencies, autoradiograms, or blotting media (see below) for accurate manual comparisons. However, computerized comparisons are preferably done on autoradiograms or photos of wet gels, which are less distorted (see 15.8).

15.7   Identification of Protein Spots

A 2-D protein map is of course only useful when the protein(s) of interest can be identified, or if differences in the 2-D patterns can be correlated to certain properties of the material from which the samples originated. Fortunately there are several methods available for identificaton, based on different properties of the proteins.

15.7.1   Comparison with Published 2-D Maps

Identification by comparison with published maps can so far only be of limited use since so few extensive maps have been published. In the case of E. coli K-12 proteins, however, 157 of the more than 1100 spots seen on 2-D gels have been identified[20,67], and a gene-protein index is under construction[68]. A human protein index is also under construction[69], and protein maps of several cell types have been published[9,10]. If a published protein map is to be used for identification, it is essential to use conditions for the 2-D PAGE identical to those used by the author, and it could still be impossible to identify minor components unambiguously. A coordinate system with some easily identified major components facilitates the use of a 2-D protein map[68] and the comparison can be both fast and accurate if a suitable computer with a matching and stretching program is available.[70]

15.7.2   Co-Electrophoresis

If the protein to be identified is available in pure form or can be purified to a reasonable degree, its position in the 2-D map can be determined by co-electrophoresis. By manipulating the proportion protein/cell fraction in the sample, it should be possible to identify the spot of which the intensity has been influenced. An example is shown in Fig. 15-7, where membranes from *Acholeplasma laidlawii* have been analyzed by 2-D PAGE (Fig. 15-7A) as well as a mixture of membranes and the purified membrane protein T_{4b} (Fig. 15-7B). The spot corresponding to protein T_{4b} can easily be identified by comparison. Identification can be facilitated by using an integrating 2-D film scanner. Many spots in the *E. coli* K-12 protein map have been identified by co-electrophoresis[68].

15.7.3   Specific Intrinsic Labelling

Proteins which carry a prosthetic group can be identified by radiolabelling of this group in vivo. The resulting 2-D protein map, obtained by staining, is then compared with an autoradiography of the gel. Alternatively, a dual isotope method can be used[4]. Phosphoproteins have been identified in 2-D protein patterns by intrinsic labelling with [^{32}P]phosphate[25], and it would also be possible to identify flavoproteins by labelling in vivo with [^{14}C]riboflavin[71]. Iron-containing proteins can be identified by labelling with ^{59}Fe, as was done for identification of immunoprecipitates in crossed immuno-electrophoresis[71,72]. A limitation with this method is that some prosthetic groups are lost under the conditions used for SDS-PAGE. Furthermore, there are often several proteins from one kind of cell, which carry the same prosthetic group, e.g. flavoproteins.

15.7.4   Genetic Approaches

15.7.4.1   Mutations

Genetic approaches can be used for bacteria, especially *E. coli*, from which more than 1000 genes have been identified[73]. By comparing cell extracts from wild type cells and a strain carrying a known mutation, it should be possible to identify the corresponding gene product[68]. A nonsense mutation for instance, would change the position of a spot to lower molecular weight, since the corresponding protein will become truncated. A missense mutation will change the position in the IEF direction if the charge of the new amino acid is different from the original one. Insertions and deletions can in principle give new spots anywhere in the 2-D protein pattern, because new stop or start codons might have been introduced. An example is shown in Fig. 15-6, where a wild type strain has been compared with a strain from which the tryptophan (trp) operon has been deleted[40]. The positions of the two gene products of the trp operon are indicated in Fig. 15-6A. The corresponding spots in Fig. 15-6B are lacking.

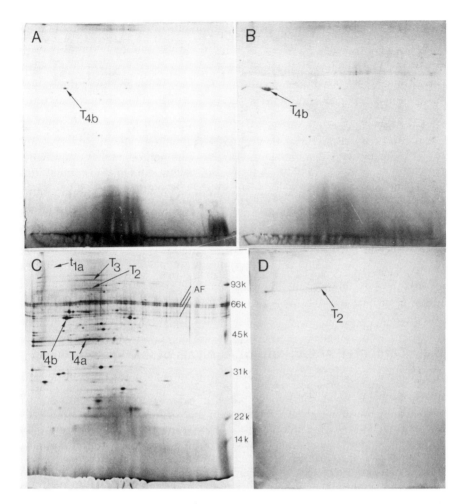

Figure 15-7. Analysis of *Acholeplasma laidlawii* membrane proteins by 2-D PAGE. IEF was performed with the Pharmalyte cocktail (2.5–5) + (5–8) + (8–10.5); (2:1:1). A, Analysis of membranes and detection by Coomassie staining. B, Co-electrophoresis of total membrane proteins and the membrane protein T_{4b}. C, Analysis of membranes and detection by silver staining. The indicated spots were identified by co-electrophoresis or electroblotting with an immune assay (see D). D, Electrophoretic blotting (see Chapter 16) of a 2-D gel on which membrane proteins were analyzed. The protein T_2 was detected by producing an immune overlay with anti-T_2 and using horse-radish peroxidase-conjugated secondary antibodies.

15.7.4.2   Plasmids

A gene library can be used for identification of protein spots. The Clarke-Carbon library[74], in which pieces of the entire *E. coli* genome have been inserted into a ColE1 vector, has been used[68]. These ColE1 derivatives are introduced in minicells[75] or maxicells[76], where only plasmid-encoded proteins are expressed. Minicells do not contain chromosomes, and in maxicells, the chromosomes have been specifically inactivated by irradiation with ultraviolet light. The cells are then incubated in the presence of [^{35}S]methionine and cell extracts are analyzed by 2-D PAGE and autoradiography.

15.7.5   Physiological Approaches

Inducible and derepressable proteins can be identified by changing the growth conditions of the cells and analyzing cell extracts from the different cultures by 2-D PAGE[68]. For instance, the λ-receptor[77] of E. coli membranes, which is involved in transport of maltose, is induced by the presence of maltose in the growth medium. Thus, by comparison of 2-D protein maps of cells grown in the presence and in the absence of maltose, it should be possible to identify the λ-receptor.

15.7.6   Blotting Techniques

Blotting techniques can be used in combination with all kinds of gel electrophoresis. It is often referred to as the third dimension of 2-D PAGE. Protein blotting is described in detail in Chapter 16.

15.8   Computer Aided Pattern Analysis of 2-D Protein Maps

The amount of information in a 2-D PAGE of a complex protein sample is immense, but manual comparison is still possible, provided that only small series of experiments are performed. However, computer aided pattern analysis (CAPA) will save a lot of time and tedious work or even evaluate the results of large experimental series in a way which would not have been possible to do manually[15]. With a suitable program, the computer can determine whether a certain protein map is identical, with respect to relative spot positions, to a standard protein map. Quantitative differences between spots corresponding to identical proteins can also be determined and all data can easily be stored and retrieved later. CAPA of 2-D gels is a large and complicated field and can, therefore, only be briefly outlined in this chapter.

15.8.1   Hardware

Different systems for CAPA of 2-D PAGE are used and have been reviewed[4]. One of the most well-known is the Tycho system[15], in which a rotating drum scanner connected to a minicomputer is used to digitize the autoradiographic films or photographic negatives of the 2-D gels. The minicomputer is a PDP 11/60 or a VAX 11/780 with one or two array processors, in which most of the arithmetic computation is done, and a pair of 300 M-byte disc units for the storage of data. For interactive work, the images can be displayed on a color video display with a trackball controlling two cursors. A large flatbed plotter is used for graphic representation of the results.

A computer digitized 2-D gel is shown in Fig. 15-5D. As can be seen, background staining has been reduced at the expense of sensitivity.

15.8.2 Software

CAPA of 2-D PAGE is complicated because of the large variability of patterns from different electrophoretic experiments. This variability is due to:

1. Difficulties in obtaining reproducible pH gradients.
2. Variations between different batches of ampholytes.
3. Swelling and stretching of the IEF gel during equilibration and transfer.
4. Difficulties in producing SDS gels with exactly the same properties if they are not cast simultaneously.
5. Swelling and deformation of the 2-D gel during staining and drying.

The variability will result in 2-D gel patterns, which are similar within a small area of the gel but not identical. CAPA of 2-D gels is therefore more complicated than for example pattern analysis in astronomy because it must be possible to compensate for the pattern variability by using a matching and stretching program.

Programs which can compensate for the variability between different 2-D gels by matching and stretching have been constructed[4], and the one used in the Tycho system[15] has the following properties[78]. Gel patterns are overlayed onto one another in such a way that corresponding spots come in registration over the whole gel. A composite gel pattern is constructed by overlaying the gel patterns from several experiments. The composite gel pattern is used as reference gel, with which all other gels are compared. This implies that all protein spots that have a counterpart in the reference pattern can be matched. The user of the system has to select a small number (approximately 10) of easily identified spots, which can be used as reference spots in the stretching procedure. Then the program runs automatically, background noise is reduced by filtering, all spots are matched and identified, the amount of protein in a spot is determined as integrated density, and the information is stored. Optimal resolution and quantification of overlapping spots is obtained by using a two-dimensional least-squares Gaussian fitting approach.

15.9 Applications

Analysis of proteins by 2-D PAGE is a rapidly expanding field, and it is impossible to cover all aspects and all kind of applications from analysis of microbial proteins[1] to sweat proteins from humans[79] in one chapter. Only a few examples, which should illustrate the potentials of this method, have therefore been selected.

15.9.1 The Human Protein Index[69]

The number of proteins in all human cell types and during all stages of differentiation and maturation has been estimated to be between 30,000 and 50,000[15]. To be able to make complete protein maps of cell extracts from different tissues, where the number of proteins in one cell type could be in the range 3,000 to 8,000, techniques with very high resolving capacity, like 2-D PAGE, have to be used. Furthermore, CAPA has to be used to acquire, handle, and store the enormous amount of data for rapid retrieval. It would be very useful for clinical purposes to have a complete Human Protein Index (HPI)[69], where all known proteins, appearing as spots in 2-D PAGE, would be listed. The HPI would be associated with a data base containing all available information about the proteins, e.g. functions, physico-chemical data, diseases which can be correlated to changes in the positions on the 2-D map, and relevant references. The usefulness of a HPI has been vividly described by Anderson and Anderson[69], and it has been shown to be feasible to construct the HPI as well as the associated data base. Furthermore, in the near future, several laboratories will have computer facilities for storage and retrieval of data from 2-D PAGE[19].

15.9.2   The Gene-Protein Index of *E. coil* K-12[68]

E. coli K-12 is the most extensively studied organism and more than 1,000 genes have now been characterized and mapped[73]. A gene-protein index of *E. coli* K-12, which would be very useful for biochemical and genetic studies, is under construction by Neidhardt[68]. The goal of this project is to have all known genes and proteins of *E. coli* cross-referenced as well as all chimeric plasmids known to contain a defined segment of the genom. The authors did not earlier use CAPA for evaluation of the experiments, which shows that 2-D PAGE can be very useful even in laboratories without computer facilities.

Many *E. coli* membrane proteins have been characterized by CIE as antigens by Owen[71]. Immunoprecipitates can be excised and analyzed by 2-D PAGE and it would be useful to have cross-references also to the antigens in an *E. coli* gene-protein index.

To give a general idea of a typical experiment, the different steps for an analysis of *E. coli* whole cell proteins are given below:

1. Dissolve freshly prepared *E. coli* cells (concentrated by centrifugation) corresponding to 100–200 μg of protein in 20–40 μl sample buffer containing Triton X-100, urea, 2-mercaptoethanol and ampholytes at room temperature[1].
2. Clarify the sample by centrifugation (10 min at 10,000–15,000 xg)
3. Run IEF in a rodshaped gel (2 × 150 mm) for 8,000–10,000 Vh. Do not exceed 1 mA per gel and use constant power or voltage. As an example use Pharmalyte: (2.5–5) + (5–8) + (3–10) in the proportions 2:1:1[33]. Run a gel without sample in parallel for estimation of the pH gradient.
4. Remove the gel from the glass tube by introducing water between the gel and the glass with a syringe and a needle[45]. Soak the gel in 5 ml of a suitable sample buffer for SDS-PAGE during 15 min.
5. Cast an SDS-gel beforehand. One possibility is to use a homogeneous gel concentration of $T = 15\%$; $C = 1\%$ in the Neville buffer system[51].
6. Transfer the IEF gel by using a special transfer tool[45] to the ditch of the gel cassette and use 0.3% hot agarose solution as glue.
7. Run the second dimension for approximately 500 Vh.
8. Interrupt the electrophoresis and stain the gel by the silverstaining procedure[59] or use it for a blotting experiment (see Chapter 16).

15.9.3   2-D PAGE of Mycoplasmas

Mycoplasmas are prokaryotic parasites, which lack a cell wall and have, therefore, been used for membrane studies[81]. *A. laidlawii* was the first mycoplasma, from which cell proteins were analyzed by 2-D PAGE[25]. Approximately 320 protein spots were detected, and by the lactoperoxidase-mediated iodination it was shown that 40 of the 140 membrane associated proteins were exposed on the exterior face of the membrane. Other mycoplasmas have also been studied, particularly the human pathogen *Mycoplasma pneumoniae*[81], and proteins associated with virulence and haemadsorption have been identified after analysis of several mutant strains by 2-D PAGE[82].

Figure 15-7C shows a 2-D map of *A. laidlawii* membrane proteins. The designations of the proteins have earlier been introduced[83] and the proteins T_3 and T_{4b} were identified by co-electrophoresis (cf. Fig. 15-7A and B). The proteins t_{1a}, T_2, and T_{4a} were identified by electrophoretic blotting and immune assay (cf. Fig. 15-7C and D). Monospecific antisera were raised against the proteins t_{1a}, T_2, and T_{4a} by immunization with immunoprecipitates obtained by line immunoelectrophoresis of pre-fractionated material[84]. All proteins which have been identified in the 2-D map except T_{4b} exhibit streaking in the IEF direction (see Fig. 15-7C).

15.10 Future Prospects

2-D PAGE has been improved since it was first introduced and by small modifications almost any kind of protein sample can now be analyzed. However, further improvements can be expected and the method has to be technically simplified before it will become a routine method, e.g. in hospitals. One problem, which has not been satisfactorily resolved, is the pH gradient instability due to cathodic drift, which is particularly pronounced in rod shaped gels. The new ampholytes, which are covalently linked to the gel (Immobiline[R])[85], will probably be more used in future for 2-D PAGE. The extended pH intervals, which can now be generated with Immunobilines, are also useful in certain applications of 2-D PAGE[86]. Casting of IEF gels with Immobilines is unfortunately more complicated than casting ordinary IEF gels, because the pH gradient has to be generated before polymerization and, therfore, has to be stabilized by another gradient e.g. in urea. Thus, it would be useful to be able to reduce the amount of urea by using more efficient detergents than Triton X-100 or Nonidet P-40, but which still have to be compatible with IEF.

CAPA of 2-D PAGE will probably be more commonly used in future and standardization from the beginning is therefore important[19]. Standardized terminology, ampholyte mixtures, gel compositions, and computer systems will facilitate comparison of results and exchange of information between laboratories. The evaluation is very important when 2-D PAGE is used for screening of a great number of samples. The software for the QUEST (QUantitative Electrophoresis, STandardized) system for CAPA of 2-D gels developed by Garrels[87] is commercially available (Protein Databases Inc., U.S.A). Complete systems for automatic scanning and evaluation of 2-D gels, like Visage[TM] (BioImage, U.S.A.), are also commerically available. The most sophisticated versions are, however, very expensive.

2-D PAGE was introduced by O'Farrell in 1975.[1] The outstanding resolution capacity of the technique has been illustrated in a very large number of articles and after development of silver staining, protein electroblotting and CAPA, the potentials of the method seem to be almost unlimited. If 2-D PAGE is combined with blotting[13] and monoclonal antibodies[88] are used as probe in the immune assay, you will have a technique with an extremely high resolution capacity and the highest possible specificity.

15.11   Acknowledgements

I am grateful to Kurt Nordström for stimulating discussions on *E. coli* proteins, Leif Kirsebom and Torgny Låås for helpful advice and valuable discussions concerning 2-D PAGE, and Lars-Göran Öfverstedt for producing the digitized gel image shown in Fig. 15-5D. In addition, I would like to thank Nils Juntti for a generous gift of horse-radish peroxidase-conjugated antibodies, Hans Pettersson, Per Krantz and Göte Eriksson at the BMC-Workshop for construction of the electrophoresis equipment, Olle Lavemark and Bengt Ekberg for taking the photos, Carina Bohlin for typing the manuscript, Mia Blomqvist for the drawing of Fig. 15-4A, and Kristina Mead for linguistic revision.

Financial support has been obtained from the Foundations of O.E. and Edla Johansson, Carl Trygger, and Helge Ax:son Johnson.

15.12   References

1. P. H. O'Farrell, *J. Biol. Chem. 250*, 4007 (1975).
2. E. M. McConkey, *Anal. Biochem. 96*, 39 (1979).
3. M. J. Dunn, A. H. M. Burghes, *Electrophoresis 4*, 97 (1983).
4. M. J. Dunn, A. H. M. Burghes, *Electrophoresis 4*, 173 (1983).
5. C. Sidman, In *Immunological Methods*; I. Lefkovits and B. Pernis, eds., Academic Press,: New York, 1981, Vol. 2, Chapter 3, pp. 57–74.
6. J. Sinclair, D. Rickwood, In *Gel Electrophoresis of Proteins: a Practical Approach*, B. D. Hames, and D. Rickwood, eds.; IRL Press: Oxford, 1981, Chapter 5, pp. 189–218.
7. *Two-Dimensional Gel Electrophoresis of Proteins: Methods and Applications*; J. E. Celis, R. Bravo, eds, Academic Press, New York, 1984.
8. B. S. Dunbar, *Two-Dimensional Electrophoresis and Immunological Techniques*; Plenum: New York, 1987.
9. J. S. King, ed. *Clin. Chem.* Special issue: Two-Dimensional Gel Electrophoresis *28*, pp. 737–1092 (1982).
10 J. S. King, ed. *Clin. Chem.* Special issue: Two-Dimensional Electrophoresis *30*, pp. 1897–2195 (1984).
11. *Electrophoresis '86*, M. J. Dunn, VCH Verlagsgesellschaft: Weinheim, (1986).
12. *Recent Progresses in Two-Dimensional Electrophoresis*; M.-M. Galteau, G. Siest, eds; Presses Universitaries de Nancy: Nancy, 1986.
13. E. M. Southern, *J. Mol. Biol. 98*, 503 (1975).
14. K. Gooderham, In *Techniques in Molecular Biology*, J. Walker and W. Gaastra, eds.: Crom Helm, London, 1983, Chapter 3, pp. 49–61.
15. N. L. Anderson, J. Taylor, A. E. Scandora, B. P. Coulter, N. G. Anderson *Clin Chem. 27*, 1807 (1981).
16. R. C. Allen, *Trends Anal. Chem. 2*, 206 (1983).
17. V. Henning, I. Sonntag, I. Hindennach, *Eur. J. Biochem. 92*, 491 (1978).
18. G. F.-L. Ames, K. Nikaido, *Biochemistry 15*, 616 (1976).
19. B. F. C. Clark, *Nature 292*, 491 (1981).
20. P. L. Bloch, T. A. Phillips, F. C. Neidhardt, *J. Bacteriol. 141*, 1409 (1980).
21. J. E. Celis, R. Bravo, *Trends Biochem. Sci. 6*, 197 (1981).
22. M.-M. Portier, B. Croizat, F. Gros, *FEBS Lett. 146*, 283 (1982).
23. J. J. Langone, *Methods Enzymol. 70*, 221 (1980).
24. O. Kuhn, F. H. Wilt, *Anal. Biochem. 105* , 274 (1980)
25. D. B. Archer, A. W. Rodwell, E. S. Rodwell, *Biochim. Biophys. Acta 513*, 268 (1978).
26. A. Helenius, K. Simons, *Biochim. Biophys. Acta 415*, 29 (1975).
27. B. Tasheva, G. Dessev, *Anal. Biochem. 129*, 98 (1983).
28. D. Ochs, *Anal. Biochem. 135*, 470 (1983).
29. P. A. Manning, L. Beutin, M. Achtman *J. Bacteriol. 142*, 285 (1980).
30. G. A. Howard, R. R. Traut, *FEBS Lett. 29*, 177 (1973).

31. J. V. Maizel, Jr. In *Methods in Virology*, K. Maramorosch, H. Koprowski, eds.; Academic Press, New York, 1971, Vol. 5, Chapter 5, pp 179–246.
32. B. B. Rosenblum, S. M. Hanasch, N. Yew, J. V. Neel, *Clin. Chem. 28*, 925 (1982).
33. K.-E. Johansson, *Protides Biol. Fluids. 33*, 475 (1985).
34. S. L. Tollaksen, J. J. Edwards, N. G. Anderson, *Electrophoresis 2*, 155 (1981).
35. P. H. O'Farrell, P. Z. O'Farrell, *Methods Cell Biol. 16*, 407 (1977).
36. A. H. M. Burghes, M. J. Dunn, H. E. Statham, V. Dubowitz, *Electrophoresis 3*, 177 (1982).
37. S. Hjertén, *Arch. Biochem. Biophys. Suppl. 1*, 147 (1962).
38. A. H. M. Burghes, M. J. Dunn, V. Dubowitz, *Electrophoresis 3*, 354 (1982).
39. P. G. Righetti, G. Tudor, E. Gianazza, *J. Biochem. Biophys. Methods 6*, 219 (1982).
40. L. Kirsebom, Ph.D. Thesis, University of Uppsala, (1985).
41. G. R. Finlayson, A. Chrambach, *Anal. Biochem. 40*, 292 (1971).
42. P. O'Farrell, M. M. Goodman, P. H O'Farrell, *Cell 12*, 1133 (1977).
43. J. Lizana, K.-E. Johansson, *Protides Biol. Fluids 34*, 741 (1986).
44. W. E. Wright, *Separation Science and Technology 17*, 1689 (1982–83).
45. K.-E. Johansson, In *Handbook of Immunoblotting of Proteins*; O. J. Bjerrum, N. H. H. Heegaard, eds., CRC Press: Boca Raton, 1988, Chapter 3, pp. 31–50.
46. G. A. Banker, C. W. Cotman *J. Biol. Chem. 247*, 5856 (1972).
47. P. Lundahl, E. Greijer, M. Sandberg, S. Cardell, K.-O. Eriksson, *Biochim. Biophys. Acta 20* (1986).
48. L. M. Hjelmeland, A. Chrambach, *Electrophoresis 3*, 9 (1982).
49. U. K. Laemmli, *Nature 227*, 680 (1970).
50. G. Blobel, B. Dobberstein, *J. Cell Biol. 67*, 835 (1975).
51. D. M. Neville, Jr. *J. Biol. Chem. 246*, 6328 (1971).
52. B. L. Anderson, R. W. Berry, A. Telser, *Anal. Biochem. 132*, 365 (1983).
53. B. A. Perret, R. Felix, M. Furlan, E. A. Beck, *Anal. Biochem. 131*, 46 (1983).
54. N. L. Anderson, N. G. Anderson, *Anal. Biochem. 85*, 331 (1978).
55. N. L. Anderson, N. G. Anderson, *Anal. Biochem. 85*, 341 (1978).
56. W. Beltle, S. Zimmerman, M. Bopp, K.-F. Murach, *Electrophoresis 4*, 143 (1983).
57. B. F. Johnson, *Anal. Biochem. 127*, 235 (1982).
58. C. Jägersten, *Abstract of Paper*, Scand Elpho 87: Uppsala, (1987).
59. M. D. Frey , B. J. Radola, *Electrophoresis 3*, 27 (1982).
60. P. Tunón, K.-E. Johansson, *J. Biochem. Biophys. Methods 9*, 171 (1984).
61. C. R. Merril, D. Goldman, M. L. van Keuren, *Electrophoresis 3*, 17 (1982).
62. K. Wallevik, J. C. Jensenius, I. Andersen, A. M. Pulsen, *J. Biochem. Biophys. Methods 6*, 17 (1982).
63. W. M. Bonner, R. A. Laskey, *Eur. J. Biochem. 46*, 83 (1974).
64. R. A. Laskey, *Methods Enzymol 65*, 363 (1980).
65. J. P. Chamberlain, *Anal. Biochem. 98*, 132 (1979).
66. M. J. Osborn, R. Munson, *Methods Enzymol 31A*, (1974).
67. T. A. Phillips, P. L. Bloch, F. C. Neidhardt, *J. Bacteriol. 144*, 1024 (1980).
68. F. C. Neidhardt, V. Vaughn, T. A. Philips, P. L. Bloch, *Microbiol. Rev. 47*, 231 (1983).
69. N. G. Anderson, L. Anderson, *Clin. Chem. 28*, 739 (1980).
70. J. Taylor, N. L. Anderson, A. E. Scandora, Jr., K. E. Willard, N. G. Anderson, *Clin. Chem. 28*, 861 (1980).
71. P. Owen, In *Electroimmunochemical Analysis of Membrane Proteins*, O. J. Bjerrum, ed., Elsevier: Amsterdam, 1983, Chapter 19, pp. 347–373.
72. C. Jägersten, L. Odelstad, K.-E. Johansson, *FEBS Letts. 144*, 130 (1982).
73. B. J. Bachman, *Microbiol. Rev. 47*, 180 (1983).
74. L. Clarke, J. Carbon, *Cell 9*, 91 (1976).
75. J. Reeve, *Methods Enzymol 68*, 493 (1979).
76. A. Sancar, A. M. Hack, W. D. Rupp, *J. Bacteriol. 137*, 692 (1979).
77. L. Randall-Hazelbauer, M. Schwartz, *J. Bacteriol. 116*, 1436 (1973).
78. J. Taylor, N. L. Anderson, N. G. Anderson, In *Electrophoresis '81. Advanced Methods. Biochemical and Clinical Applications*, R. C. Allen and P. Arnaud, eds. Walter de Gruyter: New York 1981, pp. 383–400.
79. R. W. Rubin, N. S. Penneys, *Anal. Biochem. 131*, 520 (1983).
80. S. Razin, *Microbiol. Rev. 42*, 414 (1978).
81. E. J. Hansen, R. M. Wilson, J. B. Baseman, *Infect. Immun. 24*, 468 (1979).
82. D. C. Krause, D. K. Leith, R. M. Wilson, J. B. Baseman, *Infect Immun. 35*, 809 (1982).

83. K.-E. Johansson, *Ann. Clin. Res. 14*, 278 (1982).
84. K.-E. Johansson, In *Electroimmunochemical Analysis of Membrane Proteins*, O. J. Bjerrum, ed.; Elsevier: Amsterdam, 1983, Chapter 18, pp. 321–346.
85. P. G. Righetti, E. Gianazza, B. Bjellqvist, *J. Biochem. Biophys. Methods 8*, 89 (1983).
86. P. G. Rigetti, E. Gianazza, In *Recent Progresses in Two-Dimensional Electrophoresis*, M.-M. Galteau, G. Siest, eds.; Presses Universitaires de Nancy, 1986, pp. 11–19.
87. J. I. Garrles, J. T. Farrar, C. B. Burwell, IV, In *Two-Dimensional Gel Electrophoresis of Proteins: Methods and Applications*, J. E. Celis, R. Bravo, eds.; Academic Press, 1984, pp. 37–91.
88. G. Köhler, C. Milstein, *Nature (London) 256*, 495 (1975).

16 Protein Recovery and Blotting Techniques

Karl-Erik Johansson

The National Veterinary Institute
Box 7073
S-750 07 Uppsala
Sweden

16.1 Introduction

It is impossible to define the difference between analytical and preparative electrophoresis. Experiments, which were reagarded as analytical some years ago, are not necessarily so today. Technical developments have made it possible to carry out advanced experiments with extremely small amounts of protein. It is for instance possible to perform amino acid analysis on 25–50 pmoles and at least partial sequencing (25–30 steps) on 50–100 pmoles of protein. To raise monospecific antisera, only a few micrograms of a pure protein is needed. An ordinary polyacylamide gel electrophoresis in the presence of sodium dodecyl sulphate (SDS-PAGE) might well be preparative, because of these advances, and rapid and efficient methods for elution of proteins from polyacylamide gels are badly needed.

Either one of two principles may be utilized to elute protein bands from a polyacrylamide gel for further analysis. The first method involves eluting the material from the gel into a solution (using various procedures) and performing the next experiment in a test tube. This is referred to as protein recovery. The second approach is to transfer the protein pattern to an immobilizing medium. This technique is referred to as protein blotting. Further analysis is then carried on with the proteins immobilized to the new medium. The two principles can also be combined, in which case proteins are first transferred from the gel to an immobilizing medium and then eluted into solution.

16.2   Protein Recovery

Methods are available for the recovery of proteins from preparative polyacylamide gel electrophoresis (PAGE) performed under native conditions by continuous elution during electrophoresis[1]. The proteins are transported into an elution chamber, which is packed with some porous material and connected to the bottom of the column. The protein bands are then eluted from this elution chamber by using a peristaltic pump. The proteins become rather diluted by this procedure and it is difficult to balance elutions speed and voltage. Another recently introduced method, the Elfe™ system, (Genofit, Switzerland) uses a flow of liquid, which can be created by electro-osmosis, for elution. Electro-osmosis (see p. 354) has earlier been used for transportation of proteins in agarose gels, as in counter immunoelectrophoresis[2]. In the Elfe system, the proteins are separated in a rod shaped gel with a capillary in the centre. The buffer stream is created by a charged membrane at the bottom of the column and a glass fibre membrane between the gel and the charged membrane. When the protein zones leave the gel they are trapped in the buffer stream due to electro-osmosis at the charged membrane. The liquid flow continues through the glass fibre membrane and further into the capillary to which a tubing is connected. No pump is needed.

The resolution of a preparative PAGE is never as good as in an analytical experiment, and the amount of protein which can be purified in an analytical PAGE is often adequate for many experiments. Therefore, it is essential to be able to elute proteins from slices of polyacrylamide gels originating from analytical PAGE experiments by rapid and efficient procedures. Proteins can be eluted from a homogeneous gel by diffusion or by electrophoresis.

16.2.1   Elution by Diffusion

The gel piece containing the protein band of interest is cut into small pieces or homogenized, and placed in a test tube with a suitable buffer[3]. The procedure is time consuming, the yield for high molecular weight (M_r) proteins is low and the reproducibility is not very good. Also, the protein preparation will contain polyacrylamide contaminants. Thus, elution by diffusion can hardly be recommended for proteins. It may be used for elution of oligonucleotides from agarose gels, although electroelution is preferable for these substances as well (see below).

16.2.2   Elution by Electrophoresis

Many different devices have been constructed for electrophoretic elution of proteins from polyacylamide gels, and some of them are commercially available. The Elfe system (described above) can be used for electroelution of proteins from gel slices by casting only a stacking gel in the electrophoresis tube, and applying the gel slices on top of this stacking gel. Another possibility is to place the gel slices on top of a small column of Sephadex® and use displacement electrophoresis (isotachophoresis) to force the proteins out into the Sephadex bed[4]. The column is then connected to a peristaltic pump and eluted. A small syringe (2 ml), or a prepacked PD-10 Sephadex G-25 column may also be used in combination with a displacement electrophoresis buffer system[5].

Displacement (moving boundary) electrophoresis has mainly been used for the separation of low M_r molecules for the stacking of proteins before entering the separation

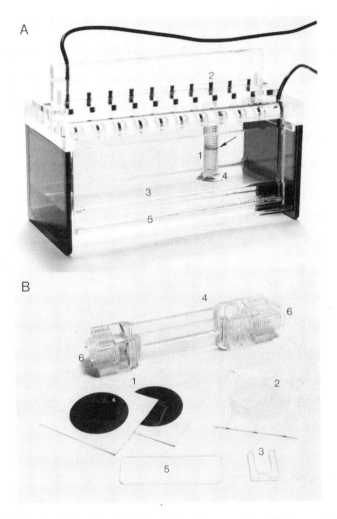

Figure 16-1. Devices for electrophoretic elution of proteins or nucleic acid fragments from poly-acrylamide gels. A, an apparatus built by the Workshop of Uppsala Biomedical Center. The syringe packed with Sephadex (1) is connected to a cathode (2) and the anolyte (3). A dialysis membrane (4) is fixed at the tip of the syringe with an O-ring, which is not visible. Ten positions for syringes are available. The 10 cathodes are mounted in the lid and connected in parallel. The position of the marker dye is denoted with an arrow. B, the Biotrap™ or S & S Elutrap™ (in U.S.A. and Canada). Membranes (1 and 2) are inserted with U-shaped spacers (3) into the cell (4). A special tool (5) is used to loosen and retighten the pressure screws (6). The chambers are then loaded with buffer and gel pieces. Electroelution is performed in an apparatus for horizontal agarose gel electrophoresis[7].

gel in SDS-PAGE (see p. 358), but it has also proved very useful for elution of proteins[4,5] (or nucleic acid fragments[6]) from gels and for protein blotting (see below). When displacement electrophoresis is used for elution, the protein becomes concentrated in a very narrow zone in the Sephadex column and the concentration continues to be high after elution, since the column is short[5]. An apparatus which can be used for electroelutions of 10 gel slices simultaneously is shown in Fig. 16-1A. The electrophoresis chamber, which can be loaded with up to 10 disposable 2 ml syringes, has been built at the Workshop of Uppsala Biomedical Center (Box 570, S-751 23 Uppsala, Sweden). Glass fibre membranes (Whatman 934-AH) are punched out with the aid of one of the syringes and inserted into the bottom of the syringes to be used for electroelution. They are packed with Sephadex G-50 equilibrated with anolyte (= leading buffer) and sealed at the tip with a dialysis tubing, which can be fixed with an O-ring. As anolyte, 20 mM HCl + 40 mM Tris can be used. The syringes are then filled up with the catholyte (= terminating buffer), which can be 100 mM β-alanin, and loaded with the gel slices and a suitable marker dye (bromophenol blue). Electroelution is performed at 100–200 V for approximately 30 min or until the dye has formed a sharp band 0.5–1.0 mm above the glass fibre membrane in the syringe. The syringe is then disconnected from the electrophoresis cell, the dialysis membrane is removed, and the material is eluted by gravity with about 500 μl of the catholyte. This system is used frequently for the elution of nuleic acid fragments[6] but can also be used for proteins[4,5]. The recovery is in general about 90%.

In another device, the so-called Biotrap™ from Schleicher & Schuell, (See Fig. 16-1B) the gel slices are placed between two protein-permeable membranes[7]. The device is placed in an apparatus for horizontal gel electrophoresis. The proteins are then eluted by electrophoresis into a second chamber, which is closed by a third membrane. The exclusion limit for this membrane is about 5000 for charged macromolecules in an electric field. The protein solution is easily recovered with a Pasteur pipette from above, since the system is open. A recovery of 80–97%, depending upon the M_r of the protein, has been reported[7]. The volume of the second chamber is small (a few hundred μl) and the protein will not become too diluted. The Biotrap system has also been successfully used for electroelution of DNA fragments and oligonucleotides[8].

A combination of a salt barrier and a sucrose gradient has been used for recovery of protein zones from polyacylamide gels[9]. Electrophoretic elution against a highly conductive salt barrier is used in the elution system UEA from IBI (U.S.A.), which is primarily intended for nucleic acid fragments.

16.3   Protein Blotting

Blotting was originally developed by Southern for the transfer of DNA-fragments, separated by agarose gel electrophoresis from the gel, to a nitrocellulose (NC) membrane[10]. This technique is, therefore, often termed Southern blotting although it is preferable to use the term DNA blotting[11] to avoid confusion with RNA blotting (Nortern blotting)[12] and protein blotting (Western blotting)[13]. The general purpose of a blotting experiment is to transfer electrophoretically separated macromolecules from a gel to a new medium (termed transfer medium, blotting medium or immobilizing medium), where the electrophoresis pattern is preserved and where the macromolecules are much more easily accessible for reaction with other macromolecules (probes). Another purpose in the case of protein blotting is to renature the proteins after electrophoresis under denaturing conditions (e.g. SDS-PAGE) to allow detection based on specific affinity. Protein blotting has become a major tool in biochemistry and is today a routine method in many protein research laboratories. The usefulness of blotting techniques in protein chemistry has been illustrated in many review articles[11,14–17], methodology chapters[18–21] and a handbook[22].

16.3.1 Principles

NC membranes are often used as transfer media in protein blotting. Proteins are assumed to bind to these membranes by hydrophobic and ionic interactions[19]. The transfer of proteins to NC membranes can be achieved by diffusion, capillary force, vacuum techniques or electrophoresis. The last technique, which is often termed electrophoretic blotting (or electroblotting), is by far the most rapid and efficient method for protein blotting[11]. The principle steps for an electrophoretic blotting experiment of proteins is shown in Fig. 16-2. The polyacrylamide gel is first equilibrated in a pre-transfer solution (with the same composition as the transfer buffer) for the removal of SDS. Often, the transfer buffer contains 20% (v/v) methanol, which increases binding of proteins to NC, increases the efficiency by which SDS is removed from the proteins and counteracts the tendency of the gel to swell during the electrotransfer. The proteins are at least partly renatured upon the removal of SDS during equilibration and electrotransfer. After equilibration, the gel is packed into a blotting sandwich, shown in Fig. 16-3 and immersed in the buffer tank. After electrotransfer, the immobilizing medium has to be treated with a solution of an inert protein or a neutral detergent (Tween 20) to block free binding sites of the membrane. This is called blocking or quenching. Commonly used

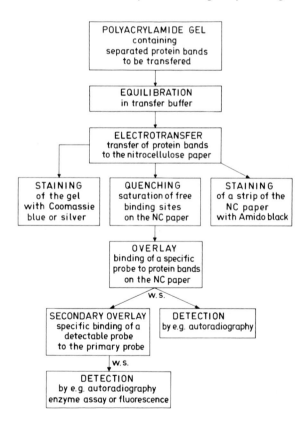

Figure 16-2. The principal steps in electrophoretic protein blotting. Not shown are some washing steps (w.s.) to remove excess of probe after production of the overlays.

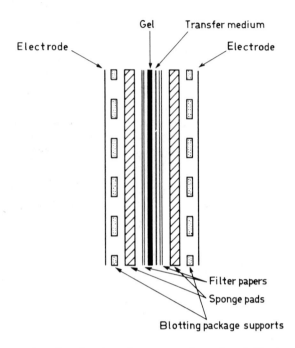

Figure 16-3. Cross section of a gel package for wet state electrophoretic blotting. All components should be in close contact and immersed in the buffer tank. See also Fig. 16-4A.

proteins are albumin, skim milk powder, foetal calf serum or gelatin. This treatment will prevent the probe from binding to unspecific binding sites on the membrane. The next step in the blotting experiment is to identify a specific protein by producing an overlay with a probe, which is further discussed under 16.3.4.

Equipment for electrophoretic blotting (see Fig. 16-4) is commercially available from many different companies, and can also be constructed easily[23]. A power supply, which can generate fairly high current (at least 0.5 A), is also needed. Two different systems have been used for electrophoretic blotting: the conventional "wet-state" electrophoretic blotting (Fig. 16-4A) and semi-dry electroblotting (Fig. 16-4B)[24]. Semi-dry electroblotting is faster than conventional electrophoretic blotting. It can be performed under lower voltage, and only small amounts of buffer are needed, since thick filter papers are used as buffer reservoirs. A very homogeneous electric field is obtained with graphite electrodes which are in close contact with the blotting package. When platinum wires are used, as in wet-state electrophoretic blotting, inhomogeneities in the electric field have been observed[15]. Furthermore, a displacement electrophoresis system is created by the discontinous buffers used in semi-dry electroblotting, which results in efficient electroelution. However, continuous buffer systems can also be utilized[25].

16.3.2 Transfer Media

NC membranes are the most commonly used transfer media in protein blotting. One problem in most blotting experiments is that the replica does not reflect the actual protein composition of the sample. This is due to the fact that certain proteins

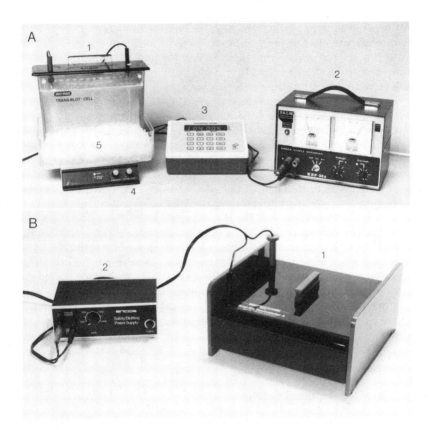

Figure 16-4. Commercially available equipment for electrophoretic blotting. A, the Trans-Blot™ Cell (1) from Bio-Rad. The figure shows an experimental setup for a double replica blotting[35] with the power supply (2) and the electrotransfer controller (3) from PALM, Uppsala, Sweden. A magnetic stirrer (4) and an ice bath (5) are used to avoid high temperature in the blotting cell. B, equipment for semi-dry electroblotting[24] from Ancos, Denmark, with the blotting cell (1) and the low voltage power supply (2).

(particularly those of high M_r) are not quantitatively eluted from the gel during the electrotransfer. Furthermore, proteins of low M_r have a tendency to penetrate the membrane without getting stuck during electrotransfer. The former problem can be overcome by omitting methanol in the transfer buffer, but this increases the tendency for low M_r proteins to penetrate the membrane. One can also choose another medium, for instance nylon membranes, which are claimed to have a 6-fold higher binding capacity[26].

Neutral and postively charged nylon membranes are available, and both are suitable for protein blotting. Membranes of smaller pore size can be used to increase binding of low M_r proteins. NC membranes can be stained for total protein by conventional Amido Black staining, but nylon membranes have to be stained by more sophisticated techniques, for instance the ImmunoGold procedure in combination with a polyspecific antiserum[27]. Nylon membranes also require prolonged treatment with a blocking solution. Another way to increase the binding capacity is to use transfer media to which reactive groups are covalently bound[23]. Two varieties are diazobenzyl-oxymethyl

(DBM) paper and diazophenylthioether (DPT) paper. These methods are less popular due to cumbersome handling procedures. The papers have to be activated by diazotization prior to use. The method can, however, be very useful for low molecular weight proteins which are not easily bound by conventional procedures.

DEAE paper is used for preparative purpose, since proteins can easily be eluted after electrotransfer by increasing the ionic strength. DEAE paper is, however, fragile and difficult to handle. Hydrophobic polyvinylidene difluoride-based (PVDF) membranes have recently been introduced[28]. These membranes are mechanically stable, have a high binding capacity, and can be stained for protein. The ultimate membrane has not yet been invented, but the field is highly competitive and new products are constantly being introduced on the market. Of interest is the activated (polybrene-coated) glass fibre membrane, which can be used directly in gas-phase sequencing[29]. When this membrane is used in electrophoretic blotting experiments, the protein bands or spots are localized by staining with a fluorescent dye. The spots are then excised and used directly in the sequencer. It has proved possible to sequence 5-150 pmoles of protein using this procedure[29]. It was recently shown also that PVDF membranes can be used in the gas-phase sequencer[30], and these membranes are much cheaper than the activated glass fibre membranes. It was shown that 7-250 pmoles of protein could be spotted or electrotransferred on to the PVDF membrane and sequenced directly after detection with Coomassie blue[30].

Good manuals about blotting are available from many of the companies producing immobilizing media, e.g. Bio-Rad (U.S.A.), Schleicher & Schuell (F.R.G.), Amersham (U.K.), Du Pont (U.S.A.), Millipore (U.S.A.), Pall (U.S.A.) and Janssen (Belgium).

16.3.3 Multiple Replica Techniques

In certain applications the possibility of preparing two or more identical replicas from one gel, as for instance when several probes (antibodies) are available, has been used.[31] This applies to both one- and two-dimensional gels. It is also possible to use one of the replicas for protein staining to localize different spots and the other(s) for probing, or sequencing from an activated glass membrane[29] or a PVDF membrane[30], which would be particularly valuable for two-dimensional gels. It is possible in principle to make M_r determinations by plasma desorption mass spectrometry (PDMS) directly on an NC membrane from a blotting experiment[32], and it would also be useful here to have two blots, one for localization of the spots and the other for the PDMS analysis.

Two approaches have been used to produce several replicas from one gel by electrotransfer[26,33]. In one approach[26], the gel is first overloaded with protein and in the blotting step a stack of up to 10 transfer media is applied to one side of the gel. If the amount of protein exceeds the binding capacity of one membrane, material will be transported through the first one and bind to the next. The binding capacity of NC-membranes can be decreased by omitting methanol to obtain as many replicas as possible. The second approach[33] is to perform the electrophoretic blotting with one membrane, but to interrupt and exchange the membrane long before the gel has been depleted of protein. This can be repeated several times to obtain many replicas.

Both methods have a serious drawback: the replicas will not be identical, and the more replicas that are produced, the greater will be the difference between the first and the last one. Two identical replicas can be produced by using diffusion instead of electrophoresis and simply applying one membrane on each side of the gel[34]. The process is, however, slow and the transfer efficiency is low, particularly for proteins of high M_r. Two identical replicas can be produced from one gel by the DREB (Double REplica Blotting) technique[31,35]. Here, oscillating electrotransfer is utilized and by increasing the efficient electrotransfer time in each direction, two identical replicas can be produced. The electrotransfer time is in principle doubled for each cycle, which secures even distribution of the proteins on to the two transfer media.

16.3.4   Probes and Detection Methods

When the transfer medium is incubated with a probe, an "overlay" is produced (see Fig. 16-2). The probe can be detectable or non-detectable. A non-detectable probe requires that a secondary overlay with a detectable probe be produced. The most common probe in protein blotting is a mono-specific antibody. The antibody can be polyclonal, monoclonal, or an Fab fragment, as it does not have to be precipitating. The protein in an electrophoresis pattern for which the antibody has an affinity can after transfer and production of an immune overlay be detected by autoradiography if the antibody has been radiolabelled with for example ^{125}I. The indirect sandwich technique, in which different antibody probes are detected with the same radiolabelled secondary antibody (raised in another species against the primary antibody), can also be used. Protein A from *Staphylococcus aureus*, which has high affinity for the Fc part of an immunoglobulin molecule, has proved to be very useful for detection of immunocomplexes in immune overlays[36]. Thus, $[^{125}I]$-labelled protein A can react with immunocomplexes in the same way as a secondary antibody[13], and it is possible to detect 0.1 ng of protein by this procedure[37]. However, if monoclonals are the primary antibodies, it should be kept in mind that protein A does not bind to all subclasses of immunoglobulins[38].

Enzyme immunoassay is often utilized for detection of immunocomplexes[39]. However, in contrast to enzyme-linked immunosorbent assay (ELISA), a substrate (or colour developing reagent) which forms an insoluble product has to be used[21]. In this system, the secondary antibodies are often conjugated with horse-radish peroxidase (HRP). The immunocomplex is then detected by incubating the nitrocellulose membrane in the substrate solution, which is H_2O_2, and a suitable hydrogen donor (= colour developing reagent) such as o-dianisidine, 3,3-diaminobenzidine tetrahydrochloride, 3-amino-9-ethyl-carbazole (AEC) or 4-chloro-1-naphthol (4-CN). AEC and 4-CN are commonly used and they give very sensitive systems where 0.05 ng of protein can be detected as a brownish or a blue precipitate, respectively. It should be kept in mind that some of the HRP colour developing reagents are known or suspected carcinogens and should be handled accordingly. There are, however, no indications so far that 4-CN is carcinogenic. Alkaline phosphatase conjugated secondary antibodies are also used to detect immunocomplexes and with 5-bromo-4-chloro-indoxyl phosphate (BCIP) as substrate; a high sensitivity is obtained. Figure 15-7D shows the result of a blotting experiment from a 2-D PAGE of *Acholeplasma laidlawii* membranes, where the probe was a monospecific antibody against the membrane protein T_2. Anti-T_2 was then detected by using a HRP conjugated secondary antibody and developing with 4-CN.

Protein bands or spots can be identified in protein electrophoresis patterns by probes other than antibodies. For instance, DNA binding proteins can be detected by using a nick-translated $[^{32}P]$ DNA-probe[34], glycoproteins can be detected by $[^{125}I]$ lectins[40], and different receptors can be detected by suitable radiolabelled ligands or inhibitors as probes[11,41].

16.3.5 Comparison of Immunochemical Detection Methods

Protein blotting with antibodies as probes is often called immunoblotting[22]. This is a comparatively new immunochemical method and it could therefore be worthwhile to compare it with other protein analytical techniques based on antibodies. The methods and their characteristics are listed in Table 16-1. The following methods are compared. Nephelometry[42], where immunoprecipitation occurs in solution and the amount of precipitate is determined by light scattering. This technique can be made fully automatic and the investment of work per sample is therefore small. In ELISA[21,43] the immune reaction takes place on a polystyrene surface, and the amount of antibodies (or antigen) is determined by an immune assay similar to the one in immunoblotting. Immunodiffusion, rocket immunoelectrophoresis (RIE) and crossed immunoelectrophoresis (CIE) are further discussed in Chapter 14.

Table 16-1, indicates if precipitating antibodies are required or if non-precipitating antibodies (like monoclonals) suffice, as well as the relative amount of antibodies consumed during analysis and whether the experiment is performed under native (N) or denaturing (D) conditions. In some of the techniques, the antigens have to be soluble or solubilized (s), and in others they can be particulate (p) as well. The sensitivities and the specificities are compared. Here low specificity is taken to mean that the specificity is dependent on the specificity of the antibodies. In other words, if monoclonals are used in ELISA, the specificity is of course very high, but in immunoblotting and CIE the specificity is independent of the specificity of the antibodies, since immune reactions with the separated individual antigens are studied.

Monoclonal antibodies[44] have a few properties which must be considered during immunoblotting experiments[45,46]. Since monoclonals are directed against only one epitope of an antigen, they are extremely specific. If antigenic variations are expected, the specificity may even be too high. The sensitivity with monoclonals is lower than with polyclonals, since only one antibody molecule can be bound per antigen. Moreover, some monoclonals are more susceptible to denaturation than the average antibody. Finally, since monoclonals are derived from one epitope only, the risk is greater that the target epitope is conformational, and may not have been renatured during the blotting experiment.

16.3.6 Applications

Protein blotting has proved to be an extremely versatile tool in protein chemistry, and no other immunochemical characterization method (like CIE) has ever gained the same popularity as immunoblotting. Protein blotting is used not only for analysis and characterization of antigens, but also for characterization of antisera, and as a diagnostic tool.

16.3.6.1 Characterization of Antigens

Immunological cross-reactions between porcine mycoplasmas have been studied by immunoblotting, and a common cross-reacting antigen has been identified[47]. Figure 16-5 shows an immunoblotting experiment with an ordinary SDS-gel on which different porcine mycoplasmal antigens were separated. This figure illustrates a typical electrophoretic blotting experiment of proteins as described below. Note here that the assays used in blotting experiments are often more sensitive than staining of gels with

Table 16-1. Comparison of methods of protein analysis based on immunochemical assays.

Method	Ab[1]	Amount of Ab	Analysis[2] conditions	Ag[3]	Sensitivity	Specificity	Quantitative	Investment of work
Nephelometry	p	Intermediate	N	s	Intermediate	Low[5]	Yes	Small
ELISA	p/np	Small	N	s/p	High	Low[5]	Yes	Small
Dot-immunobinding	p/np	Small	N	s/p	High	Low[5]	Yes	Small
Immunoblotting	p/np	Small	D/N	s	High	High	No	Large
Immunodiffusion[4]	p	Small	N	s	Low	Low[5]	No	Intermediate
RIE	p	Large	N	s	Intermediate	Low[5]	Yes	Intermediate
CIE	p	Large	N	s	Intermediate	High	Yes	Intermediate

[1] Antibodies (Ab) may be precipitating (p) or non-precipitating (np).
[2] Native (N) or denaturing (D).
[3] The antigen (Ag) may be soluble (s) or particulate (p).
[4] Single radial immunodiffusion.
[5] The specificity is dependent on the specificity of the antibodies.

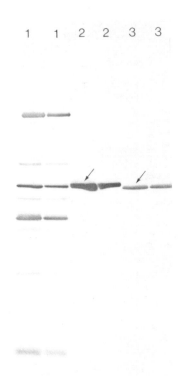

Figure 16-5. The nitrocellulose membrane from an immunoblotting experiment performed in the Trans-Blot Cell, where the following samples were applied from left: 1, *Mycoplasma flocculare*; 2, *M. hyopneunomie* and 3, *M. hyorhinis*. Migration direction was from the top to the bottom in the electrophoresis. Swine antiserum against *M. flocculare* was used as primary overlay and HRP conjugated anti-swine immunoglobulins as secondary overlay. The homologous reaction (lane 1) can easily be distinguished from the heterologous. The main cross-reacting antigen is denoted with an arrow. Reproduced from Bölske and Johansson, manuscript in preparation.

Coomassie blue. It is in general adequate to apply 1/10-1/2 of the amount of protein that is applied in ordinary SDS-PAGE.

1. After completion of the electrophoresis, the gel was soaked in transfer buffer (25 mM Tris and 192 mM glycine dissolved in 20% aqueous methanol) for 30 min.
2. The blotting package was then assembled with an NC membrane (see Fig. 16-3) and electrotransfer was performed for 3 hours at 100 V, which gave a current of 300–400 mA.
3. After electrotransfer, free binding sites on the NC membrane were blocked by incubation for 1 hour in 1.0% gelatin dissolved in phosphate buffered saline (PBS) at room temperature.
4. The NC-membrane was incubated for 4 hours under agitation at room temperature with the primary antibodies (hyperimmune serum from a pig) diluted 100-fold in 0.1% gelatin dissolved in PBS.

5. After washing (3 × 5 min) in PBS × 0.1 % gelatin, the NC membrane was incubated for 2 hours with secondary antibodies (HRP conjugated anti-swine immunoglobulins from DAKOPATTS, Denmark) diluted 1000-fold in the above buffer.
6. After washing (as above), the NC membrane was incubated for 5–20 min in the colour developing solution, which was prepared from freshly mixed solutions A and B: Solution A: 60 mg 4-CN were dissolved in 20 ml cold methanol. Solution B: 60 μl H_2O_2 (30 %) were added to 100 ml of PBS at room temperature.
7. The colour development was interrupted by rinsing of the NC membrane with water.

16.3.6.2 Characterization of Antisera

If standard antigens are separated by SDS-PAGE and transferred to an immobilizing medium, different antisera (e.g. monoclonals) can be screened by blotting with strips cut from the immobilizing medium. Antigens from pathogenic micro-organisms are utilized to produce strips for screening of animal or human sera. For instance, swine sera has been screened for presence of antibodies against pathogenic mycoplasmas by immunoblotting[48]. A more well-known example these days is the confirmatory test for presence of antibodies against human immunodeficiency virus (HIV) in patient sera[49]. A serum which is positive in ELISA is further analyzed by blotting against SDS-PAGE separated HIV antigens, and if a serum shows positive signals against antigens of certain M_r, it is regarded as positive.

16.4 Acknowledgements

I am grateful to Professor Bengt Hurvell for valuable support, Irja Johansson and Dr Gösta Eggertsen for helpful discussions on protein blotting and Dr Gunnar Johansson and Dr Andreas Clad for valuable discussions on protein recovery. I would like to thank Marie-Louise Strandberg for skilful technical assistance, Gösta Forsling for making the drawings, Bengt Ekberg for taking the photos, Carina Bohlin for typing the manuscript and Kristina Mead for linguistic revision. Financial support has been obtained from the board of the National Veterinary Institute.

16.5 References

1. S. Hjertén, S. Jerstedt, A. Tiselius, *Anal. Biochem. 27*, 108 (1969).
2. A. T. Andrews, *Electrophoresis: Theory, Techniques, and Biochemical and Clinical Applications*; Clarendon Press: Oxford (1986).
3. C. Bernabeu, F. P. Conde, D. Vazquez, *Anal. Biochem. 84*, 97 (1977).
4. L. Öfverstedt. J. Sundelin. G. Johansson, *Anal. Biochem. 134*, 361 (1983).
5. L.-G. Öfverstedt, G. Johansson, G. Fröman, S. Hjertén, *Electrophoresis 2*, 168 (1981).
6. L.-G. Öfverstedt, K. Hammarström, N. Balgobin, S. Hjertén, U. Petterson, J. Chattopadhyaya, *Biochem. Biophys. Acta 782*, 120 (1984).
7. E. Jacobs, A. Clad, *Anal. Biochem. 154*, 583 (1986).
8. U. Göbel, R. Maas, A. Clad, *J. Biochem. Biophys. Methods 14*, 245 (1987).
9. S. Hjertén, In *Methodological Developments in Biochemistry*, Vol. 2, E. Reid, ed.; Longman: London, 1977, pp. 39–47.
10. E. M. Southern, *J. Mol. Biol. 98*, 503 (1975).
11. J. M. Gershoni, G. E. Palade, *Anal. Biochem. 131*, 1 (1983).
12. J. C. Alwine, D. J. Kemp, B. A. Parker, J. Reiser, J. Renart, G. R. Stark, G. M. Wahl, *Methods Enzymol. 68*, 220 (1979).

13. W. N. Burnett, *Anal. Biochem. 112*, 195 (1981).
14. H. Towbin, J. Gordon, *J. Immunol. Methods. 72*, 313 (1984).
15. J. M. Gershoni, *Trends Biochem. Sci. 10*, 103 (1985).
16. G. Bers, D. Garfin, *BioTechniques 3*, 276 (1985).
17. U. Beisiegel, *Electrophoresis 7*, 1 (1986).
18. A. Haid, M. Suissa, *Methods Enzymol. 96*, 192 (1983).
19. K. Gooderham, In *Techniques in Molecular Biology*, J. Walker, W. Gasstra, eds.; Croom Helm: London 1983, pp. 49–61.
20. J. Symington, In *Two-Dimensional Gel Electrophoresis of Proteins, Methods and Applictions*, J. Celis, R. Bravo, eds.; Academic Press: Orlando 1984, pp. 127–168.
21. P. Tijssen, In *Laboratory Techniques in Biochemistry and Molecular Biology, Vol. 15. Practice and Theory of Enzyme Immunoassays*, R. H. Burdon, P. H. Van Knippenberg, eds.; Elsevier: Amsterdam 1985, pp. 423–447.
22. *Handbook of Immunoblotting of Proteins, Vol. 1 and 2*, O. J. Bjerrum, N. H. H. Heegard, eds. CRC Press: Boca Raton 1988.
23. M. Bittner, P. Kupferer, C. F. Morris, *Anal. Biochem. 102*, 459 (1980).
24. J. Kyhse-Andersen, *J. Biochem. Biophys. Methods 10*, 203 (1984).
25. O. J. Bjerrum, C. Schafer-Nielsen, In *Electrophoresis 86*, M. J. Dunn, ed., VCH Verlagsgesellschaft: Weinheim 1986, pp. 315–328.
26. J. M. Gershoni, G. E. Palade, *Anal. Biochem. 124*, 396 (1982).
27. M. Moeremans, G. Daneels, A. Van Dijck, G. Langanger, J. De Mey, *J. Immunol. Methods 74*, 353 (1984).
28. M. G. Pluskal, M. B. Przekop, M. R. Kavonian, C. Vecoli, D. A. Hicks, *BioTechniques 4*, 272 (1986).
29. R. H. Aebersold, D. B. Teplow, L. E. Hood, S. B. H. Kent, *J. Biol. Chem. 261*, 4229 (1986).
30. P. Matsudaira, *J. Biol. Chem. 262*, 10035 (1987).
31. K-E. Johansson, *Electrophoresis 8*, 379 (1987).
32. G. P. Jonsson, A. B. Hedin, P. L. Håkansson, B. U. R. Sundquist, G. S. Säve, P. F. Nielsen, P. Roepstorff, K.-E. Johansson, I. Kamensky, M. S. L. Lindberg, *Anal. Chem. 58*, 1084 (1986).
33. R. P. Legocki, D. P. S. Verma, *Anal. Biochem. 111*, 385 (1981).
34. B. Bowen, J. Steinberg, U. K. Laemmli, H. Weintraub, *Nucleic Acid Res. 8*, 1 (1980).
35. K.-E. Johansson, *J. Biochem. Biophys. Methods 13*, 197 (1986).
36. A. Forsgren, V. Ghetie, R. Lindmark, J. Sjöquist, In *Staphylococci and Staphylococci Disease, Vol. 2, The Organism in Vivo and in Vitro*, C. S. F. Easmon and C. Adlam, eds.; Academic Press: New York 1983, pp. 229–480.
37. R. T. M. J. Vaessen, J. Kreike, G. S. P. Groot, *FEBS Lett. 124*, 193 (1981).
38. R. Lindmark, K. Thorén-Tolling, J. Sjöquist, *J. Immunol. Methods 62*, 1 (1983).
39. H. Towbin, T. Staehelin, J. Gordon, *Proc. Natl. Acad, Sci. U.S.A. 76*, 4350 (1979).
40. H. A. Erlich, J. R. Levinson, S. N. Cohen, H. O. McDevitt, *J. Biol. Chem. 254*, 12240 (1979).
41. K. Kahita, S. Giddings, M. A. Permutt, *Proc. Nat. Acad. Sci. U.S.A. 79*, 2803 (1982).
42. R. F. Ritchie, *Protides Biol. Fluids 21*, 569 (1974).
43. E. Engvall, P. Perlman, *J. Immunol. 109*, 129 (1972).
44. G. Köhler, C. Milstein, *Nature (London) 256*, 495 (1975).
45. K-E. Johansson, In *Handbook of Immunoblotting of Proteins*, O. J. Bjerrum, N. H. H. Heegaard, eds.; CRC Press: Boca Raton 1988, Chapter 6.3.2, pp. 159–165.
46. E. Kurstak, *Enzyme Immunodiagnosis*. Orlando: Academic Press 1986, pp. 77–85.
47. G. Bölske, K.-E. Johansson, *Curr. Microbiol. 15*, 233 (1987).
48. T. F. Young, R. F. Ross, *Am. J. Vet. Res. 48*, 651 (1987).
49. E. Lauritzen, B. Ø. Lindhardt, In *Handbook of Immunoblotting of Proteins*, O. J. Bjerrum, N. H. H. Heegaard, eds.; CRC Press: Boca Raton 1988, Chapter 9.1.2, pp. 117–129.

17 Capillary Electrophoretic Separations

Wolfgang Thormann* and Millicent A. Firestone

Center for Separation Science
University of Arizona
Tucson, AZ 85721, USA

LIST OF ABBREVIATIONS

CE	capillary electrophoresis
CZE	capillary zone electrophoresis
CIEF	capillary isoelectric focusing
CITP	capillary isotachophoresis
CFE	continuous flow electrophoresis
DZE	discontinuous zone electrophoresis
FFF	field flow fractionation
EFFF	electrical field flow fractionation
EHFFF	electrical hyperlayer field flow fractionation
EPC	electropolarization chromatography
HFFF	hyperlayer (focusing) field flow fractionation
HPLC	high performance liquid chromatography
IEF	isoelectric focusing
ITP	isotachophoresis
MECC	micellar electrokinetic capillary chromatography
ZE	zone electrophoresis
SDS	sodium dodecyl sulphate
CHAPSO	3[(3-cholamidopropyl)-dimethylammonio]-2-hydroxy-1-propanesulfonate
HPMC	hydroxypropylmethyl cellulose
PEG	polyethylene glycol
PVA	polyvinyl alcohol
LPA	linear polyacrylamide
Tris	tris(hydroxymethyl)aminomethane
Bistris	bis(2-hydroxyethyl)imino-tris(hydroxymethyl)methane
Ammediol	2-amino-2-methyl-1,3-propanediol

* Present address: Department of Clinical Pharmacology, University of Bern, Murtenstrasse 35, CH-3010 Bern, Switzerland.

Summary

Electrophoretic separations in narrow bore plastic tubes, glass capillaries and thin fluid films between parallel plates have received considerable attention within the last two decades. In this period a number of instrumental approaches for capillary zone electrophoresis, capillary isotachophoresis, capillary isoelectric focusing, micellar electrokinetic capillary chromatography, continuous flow electrophoresis and electrical field flow fractionation or electropolarization chromatography have emerged. High separation efficiencies have been demonstrated in conjunction with high-resolution on-column sample detection. Capillary electrophoresis is complementing other existing capillary separation techniques, such as gas chromatography, liquid chromatography and supercritical fluid chromatography, as well as gel electrophoresis. The principle of capillary electrophoresis, its instrumentation and potential is reviewed. In addition, a special section has been included describing electrophoresis performed in narrow bore Teflon tubes.

17.1 Introduction—Capillary Electrophoretic Techniques

Electrophoresis is the premier analytical separation method for biological compounds such as proteins and polynucleotides despite recent applications of HPLC to these biopolymers. Unfortunately, unlike chromatographic methods, classical electrophoresis, that is electrophoresis conducted in solid support media such as gels, is rather slow, laborious, relatively irreproducible, difficult to quantify, and not really adaptable to automation. For example, although an actual gel electrophoretic separation can be achieved within minutes, completion of the entire experiment requires several hours. Obviously, a more instrumental approach is needed. One possible solution has been the exploration and development of capillary electrophoresis (CE). In the last two decades, CE techniques have received considerable attention, having demonstrated highly efficient separations (Fig. 17-1)[1-5]. CE methods should be regarded as complementary or as

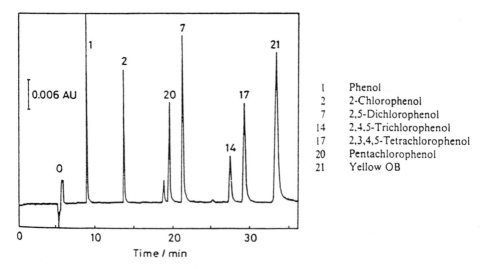

Figure 17-1. MECC of phenolic compounds (from Ref. 18).

attractive alternatives to other capillary separation techniques, such as gas chromatography, liquid chromatography, supercritical fluid chromatography and field flow fractionation. The advantages of CE are high resolution, speed, easy automation and adaptation for micropreparative work. More importantly, CE techniques can exploit numerous separation principles, making them flexible and easily applied to a variety of separation problems. For example, CE has been used to separate a broad spectrum of species and compounds ranging from small molecules (inorganic and organic ions) to large molecules and particles (proteins, cells).

The various CE techniques can be categorized according to their mode of operation (Fig. 17-2; Table 17-1). In the conventional configuration the electric field is applied parallel to the capillary axis (A)[1-45]. Electrophoretic separations of this nature have successfully been conducted in quiescent or flowing solutions with narrow bore plastic tubes[8-10,12,13,26,27,31,34,45], glass or fused silica capillaries[6,7,14-25,28-30,35-37,42-44] and in rectangular troughs[3,11,38-41]. The specific techniques comprising this group are capillary zone electrophoresis (CZE), discontinuous CZE (DCZE), capillary isotachophoresis (CITP), capillary isoelectric focusing (CIEF) and micellar electrokinetic capillary chromatography (MECC). The differentiating features of these methods lie in the

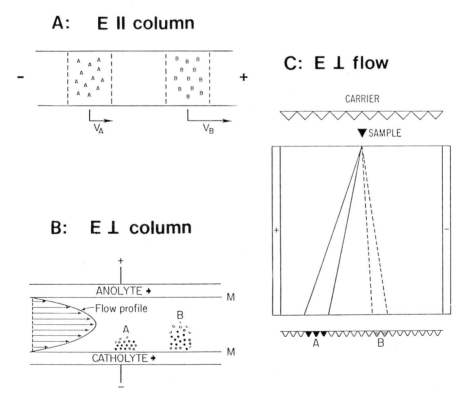

Figure 17-2. Schematic representation of the separation of two components by electrokinetic capillary methods. A = electric field parallel to column axis (conventional CE); B = electric field perpendicular to column axis and flow (EFFF); C = CFE with electric field perpendicular to buffer flow. M = membrane.

Table 17-1. Electrokinetic separation methods in Capillaries

(A) 1. Capillaries with minimal electro-osmosis

free solution
Methods: ZE, ITP, IEF, DZE
Materials: plastic, surface treated glass and fused silica
Option: hydrodynamic flow
Instruments: LKB Tachophor, Shimadzu, CS-Analyzer, Microphore 1000

gel filled
Methods: ZE, DZE
Materials: glass, fused silica
Option: entrapment of complexing agent
Instruments: Microphore 1000

(A) 2. Capillaries with electro-osmotic flow

Methods: ZE, DZE, MECC
Materials: fused silica, glass
Option: micelles
Instruments: Microphore 1000

(B) Electrical (hyperlayer) field flow fractionation

Methods: ZE, IEF
Materials: plastic (ribbonlike channels, hollow fibre)
Instruments: research stage

(C) Analytical continuous flow electrophoresis

Methods: ZE, ITP, IEF
Instruments: Hirschmann ACE 710, Elphor VaP22

---

initial and boundary conditions applied, which determine the character of the migrating sample zones (vide infra). In the second category, the electric field is applied perpendicular to the capillary axis, and the separation of charged solutes is based upon differential interaction with the electric field plus coupling with the mobile phase profile (B). These separations are typically performed in thin ribbonlike channels[46–49] or in hollow ultrafiltration fibres[50–52]. The techniques representing this class are electrical field flow fractionation (EFFF), electropolarization chromatography (EPC) and electrical hyperlayer field flow fractionation (EHFFF). In the third category, the electric field is also oriented perpendicular to the direction of flow, but parallel to the flow profile of a thin fluid film flowing between two parallel plates (C). This represents the configuration of continuous flow electrophoresis (CFE) which can be operated in various modes, including ZE, ITP and IEF[53,54].

CZE[1,2,13–37] is conducted in a continuous buffer where the samples are the only discontinuities present. Under the influence of the electric field, sample zones migrate without exhibiting any steady state behaviour and thus their shape and position continuously change with time. In this technique, separation is based upon differences in net mobility (Fig. 17-2A). Conversely, CITP[1,2,6–12] is performed in a discontinuous buffer system, the so-called leading and terminating electrolytes. The sample components are introduced in small quantities at the interface between the two electrolytes. The establishment of ITP zones requires the net mobility of samples to be intermediate to the mobilities of the buffers. Under the influence of an applied electric field, the sample components separate according to their net mobilities by forming a pattern of consecutive zones between the leader and the terminator. The system attains a migrating steady state in which all components have the same velocity (hence the prefix isotacho). Ideally enough sample is applied to produce zones with constant composition, whose length are proportional to the amount present. DZE[2] is conceptually a composite of CZE and CITP. As in CITP the sample is introduced at the interface between two buffers. However, the terminating constituent's mobility is slightly larger than the mobilities of the sample components. It therefore penetrates the sample zones from the rear side upon current flow. After a sharpening at the buffer transition the sample components separate zone electrophoretically within the terminating electrolyte. In CIEF[1,2,38-45], sample components are sorted according to their isoelectric point in an equilibrium gradient. In its most common form, ampholytes are focused in a pH gradient. Typically, a mixture of carrier ampholytes, having different pI values, is exposed to an electric field in a convection free medium. Eventually, a pH gradient is formed where the most acidic component has condensed in the most anodal position of the capillary, the most basic at the cathode and constituents of intermediate pIs in between. Proteins and other amphoteric compounds can be separated in this gradient provided their isoelectric points are sufficiently different. A stationary steady state is attained.

MECC[18,21] is an electrokinetic method allowing the separation of neutral and charged molecules. In this technique two distinct phases are used, an aqueous and a micellar, or pseudo-stationary, phase. These two phases are established by employing buffers containing surfactants, which are added above their critical micellar concentration. Electrophoresis takes place in open tubular capillaries with electroosmotic flow. Nonionic solutes partition between the two phases and elute with zone velocities between the two phases. Separation is of chromatographic nature. For ionic solutes which are also differentially distributed between the two phases, separation is of chromatographic and electrophoretic nature. EFFF[46-52] uses the coupling of an electric field with a mobile phase flow profile to achieve differential migration under flowing conditions (Fig. 17-2B). The electric field, applied perpendicular to the flow direction, causes the solute molecules to accumulate in a layer of distinct thickness near one channel wall. The solute is then transported by flow along the channel at a rate determined by the mean thickness of the layer. In electrical hyperlayer FFF (EHFFF)[66] separation is based upon partitioning by the concomitant presence of a pH equilibrium gradient and an electric field, as in isoelectric focusing (IEF). CFE[53,54] is conducted in a thin film of fluid flowing between two parallel plates. Both the carrier buffers and the sample are continuously introduced at one end of the electrophoresis chamber and are fractionated by an outlet array at the other end (Fig. 17-2C). CFE is considered a CE technique since fluid stabilization is achieved by a small gap (typically 0.5 mm or less) between the two parallel plates. Athough CFE is mainly a preparative methodology for simultaneous separation and fractionation under mild operating conditions, it is also used for analytical purposes.

A fundamental requirement for the application of electrophoresis is the presence of fully or partly charged solutes. CE has displayed considerable versatility, having been applied successfully to numerous analyses and separations. CE, unlike chromatography, has proved to be an excellent method for very small molecules like metal ions and amino acids, and particles like eukaryotic cells. CE is also well suited to the analysis and isolation of molecules found in complex matrices, such as those encountered in the biotechnology industry, particularly proteins, peptides, enzymes, oligonucleotides, as well as small biomolecules such as antibiotics[4]. Although CE is primarily conducted in aqueous solutions there have been reports of the successful use of non-aqueous[55] and mixed[56] media. In addition to performing CE in free fluid it can also be carried out in gel-filled capillaries[28,29,42]. A recent study used the gel as a sieving matrix to characterize proteins according to molecular weight by SDS electrophoresis[22]. One approach employed gels for the entrapment of complexing species, thereby achieving separation as in affinity electrophoresis. One example is the use of beta-cyclodextrin to separate chiral compounds[57]. As already mentioned above, CE has been expanded to analyze neutral compounds by MECC[18]. Another approach comprised the solvoprotic association of neutral compounds with tetraalkylammonium ions followed by their electrophoretic separation[58]. Applications of analytical CFE are mainly concerned with the separation and characterization of cells, cell organelles and cell membrane systems. CFE permits quick determination of electrophoretic mobility distribution of cells or proteins. It can be used as an aid in understanding the phenomena which take place on the cell surface, such as the effect of drugs, antigens and mitogens[53].

17.2   Conventional Capillary-Type Electrophoretic Instrumentation

The capillary cross section used in conventional CE with the applied electric field parallel to the capillary axis are depicted in Fig. 17-3. Cell configurations used in particle (cell, micro) electrophoresis[59] in which the mobilities of the particles are measured without achieving their separation are included in this graph for purposes of completeness. Early CE investigations were carried out in either glass or quartz glass capillaries of 0.1 to 1 mm ID[6,7]. A second generation comprised instruments with narrow bore plastic tubes (e.g. polytetrafluoroethylene (Teflon), fluorinated ethylene-propylene (FEP) of 0.2 to 0.8 mm ID[8–10,12,13]) or thin ribbonlike channels of rectangular cross sections (0.1 to 0.5 mm height) having various plastic and glass walls[3,11,38–41]. A number of commercial instruments equipped with capillaries of circular cross section emerged from these developments. These are the Tachophor 2127 of LKB (Bromma, Sweden), the Tachophor Delta of Itaba (Jarfalla, Sweden), the IP-1A, IP-2A and IP-3A Analyzers of Shimadzu (Tokyo, Japan) and the CS ITP Analyzer of VVZ PJT (Spisska Nova Ves, Czechoslovakia). More recent approaches concentrate on the use of open-tubular, very narrow bore glass or fused silica capillaries of 0.02 to 0.10 mm ID[14–25]. These capillaries permit rapid removal of ohmic heat, application of high voltages (300 V/cm), realization of short runs and exhibit a strong electro-osmotic flow if no special wall coatings are applied[15,60]. Thin-walled polymer coated fused silica has two inherent advantages over glass. It is extremely flexible, permitting easy manipulation, and is transparent to UV light. The first commercial instrument based on this approach was recently introduced (Microphore 1000, Microphoretic Systems, Sunnyvale CA, USA). Only a few CE reports are found in which gel-filled capillaries were used[22,28,29,42]. A few reports describe procedures for fractionation employing microsyringes[61], elution onto a strip of cellulose acetate[62] and, for use with open tubular capillaries of very small IDs, an array of collection cones of 25 μl volume each[17]. The second approach was commercialized by LKB, Sweden, under the trade name 'Tachofrac'.

The design variables for electrokinetic capillary analyzers are presented in Table 17-2 and the basic arrangement used in conventional CE is depicted schematically in Fig.

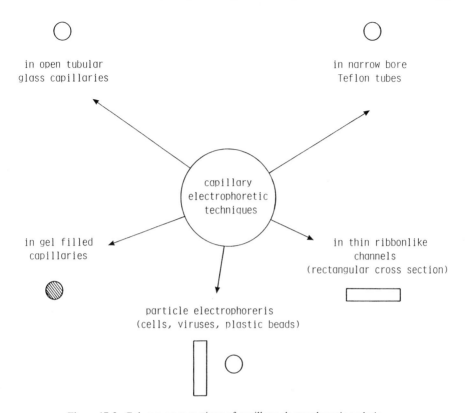

Figure 17-3. Column cross sections of capillary electrophoretic techniques.

17-4. It features a capillary tube mounted between two electrode compartments (*E*). The sample is injected in small quantities into the beginning of the separation column (*S*) and detected at the column end (*D*). The power supply and a method for buffer introduction (e.g. pump) complete the basic arrangement (Fig. 17-4a). Introduction of the buffers and the sample, data manipulation (i.e. collection, storage and evaluation) and fraction collection (Fig. 17-4b) are ideally controlled by a computer. The computer system also allows the establishment of data dictionaries as in other analytical procedures, such as in GC-MS. In systems lacking electro-osmotic flow pulse free electrolyte pumps can be used for the application of counter or co-flow of electrolyte. This is useful for (i) changing the column volume without changing the physical capillary length in ZE and ITP, (ii) decreasing analysis time, again in ITP or ZE, (iii) monitoring the zone spectrum in IEF, as well as (iv) fractionating and (v) detecting outside of the capillary. The last two features (Fig. 17-4b) do not require a pump for the open tubular configuration with electro-osmotic flow.

In principle many of the detection methods developed for HPLC can also be employed for CE (Table 17-3), with spectroscopic techniques (absorption, fluorescence) being the most popular. In addition, unique principles can be used which rely on the electric current flow through the separation capillary, such as the monitoring of potential gradient, conductivity and temperature in ITP[8]. Most electrical sensors read with microelectrodes which have direct contact with the solution in the capillary, whereas

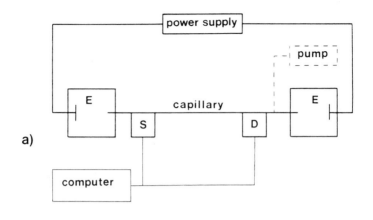

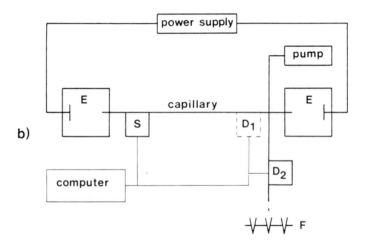

Figure 17-4. Schematic representation of conventional CE setup with on-column detection (a) as well as off-column sample monitoring and fractionation (b). *E*: electrode compartments; *S*: sample inlet system; *D*, *D1*: on-column detector; *D2*: off-column detector; *F*: fraction collector.

most optical measurements are taken through the capillary walls. The advantage of electrical sensing is high resolution and the disadvantage is the possibility of troublesome alterations of the tiny electrode surfaces. Spectroscopic measurements through quartz capillaries are usually trouble free. In the case of plastic tubes (Teflon, FEP), however, absorption of the capillary walls can be significant, especially at wavelengths <254 nm. Therefore a moulded detection block with optical fibres conducting the light directly into and out of the liquid has been constructed (LKB 2127-140, Bromma, Sweden, see also Ref. 32), simplifying the monitoring of solutes (peptides) at wavelengths as low as 206 nm. In CE tiny zone volumes must be measured accurately. In any case, the detection limit is

Table 17-2. Design variables for electrokinetic capillary analyzer

(1) modes
  — zone electrophoresis, discontinuous zone electrophoresis
  — isotachophoresis
  — isoelectric focusing
  — micellar electrokinetic chromatography

(2) separation capillary
  — material (fused silica, glass, Teflon or other plastic)
  — length (10 to 100 cm depending on material and mode)
  — cross section (circular, rectangular)
  — gel filled/free solution
  — surface modification of capillary walls

(3) sample injection
  — electrokinetic
  — hydrodynamic flow
  — syringe

(4) sample detection
  — on-line
  — off-line
  — off-line (hyphenated)
  — single/multiple/array/scan sensors
  — pre-or post-column derivatizations

(5) data analysis/automation
  — microcomputer system
  — auto sampler
  — fraction collector

(6) peripheral components
  — electrode compartments
  — power supply (current/voltage regulated)
  — electrolyte pumps

Table 17-3. Detection methods for electrokinetic capillary analyses

Detection method	on-column	off-column
temperature	×	×
conductivity	×	×
electric field	×	×
UV-VIS absorbance	×	×
fluorescense	×	×
polarimetry-interferometry	×	×
circular dichroism	×	×
radiometry	×	×
amperometry	×	×
pH		×
MS		×

determined by both operational and equipment parameters. The technology for the design of detector cells with nanolitre to microlitre volumes in conjunction with capillary ID's > 0.1 mm is well established and led to the construction of many excellent sensors. The fabrication of microprobes of smaller dimensions, however, is much more demanding and is an active research endeavour in many laboratories, much the same as for open tubular capillary chromatography[5]. In most arrangements, one or two on-column sensors placed at the capillary end are used and monitoring occurs while the electrophoretic zone structures migrate across their location. The advent of versatile and inexpensive microprocessor systems also permits simple control of sensor arrays, thereby providing increased analytical information through the repeated detection of the whole separation pattern along the capillary or the comparison of electropherograms from at least three completely different locations of the separation trough. Thus the CapScan apparatus[1,3,38–41] has been developed featuring a linear array of sensing microelectrodes as one wall of a rectangular (about 0.4 × 1 mm) capillary, a scanning device for the repetitive monitoring of the potential difference between adjacent detection electrodes thereby visualizing electric field profiles as a function of time.

CZE is probably the only electrophoretic methodology which benefits from electro-osmosis in that an electro-osmotic flow along the capillary permits the simultaneous separation and analysis of cations and anions[14]. At sufficiently high buffer pH the entire pattern migrates toward the cathode because the electro-osmotic velocity becomes larger than the electrophoretic migration in opposite direction. Electro-osmotic flow is necessary for MECC and advantageous for e.g. amperometic[36] and MS[37] off-line detection in CZE, but needs to be minimized for CITP and CIEF. This is typically accomplished by coating the capillary wall with a polymer, such as linear polyacrylamide[60]. For CITP only a small number of electrolyte systems permit separations in either migration direction for which sensors for both sample trains have to be present[63]. CIEF requires either an array or scan detector otherwise the sample pattern must be mobilized following focusing[38,41–45]. Mobilization has been achieved by pumping the entire IEF zone structure past a stationary UV monitor with the voltage applied to avoid a decay of the pattern. Electrophoretic elution, accomplished by replacing the acid at the anode by a base (or the base at the cathode by an acid) after focusing, was also investigated[42–45]. This second possibility was reported to be more universal because of its applicability to gel systems.

Using column ID's between 10 and 100 μm makes the customary microlitre syringe for sample injection obsolete. The sample is typically introduced by hydrodynamic flow (suction) or by an electrokinetic technique. For the latter approach, the electrode compartment at the high voltage end is replaced with a reservoir containing the sample. Application of high voltage for a short amount of time allows the insertion of a small amount of sample into the capillary. The sample reservoir is then exchanged by the buffer compartment before further application of current. An autosampler for injection by the two methodologies has recently been described[16]. As an alternative method, electrophoretic sample splitting has been investigated[33].

17.3   Electrophoresis in Narrow Bore Teflon Tubes

It is of current interest to explore free fluid ZE and IEF as automated instrumental approaches to electrophoresis. The vast majority of these studies are conducted in capillaries of glass or fused silica which permit effective heat removal and fabrication of capillaries with very small inner diameters (10–100 μ). Difficulties arise, however, when high molecular weight compounds, such as proteins, are used, since they interact unfavourably with the column walls[23]. In a recent CZE study, gel-filled fused silica capillaries were employed in order to circumvent this problem, and provided excellent resolution when separating proteins according to molecular weight[22]. While this approach is characterized by high resolution, the tedium of introducing a gel matrix into a small bore capillary makes it less attractive than procedures performed in simple free solution. Alternatively, the use of narrow bore PTFE tubes for free solution ZE[13,26,27,31,34] and IEF[40,45] has only been sparsely explored, although they have been employed successfully for ITP analyses for about two decades[8–12]. One advantage PTFE offers is its rather low surface charge and therefore, under an applied electric field, electro-osmosis is minimal. Conversely, PTFE suffers from several disadvantages, low heat dissipation and difficulty in manufacturing tubes having appropriate ID's (smaller than 100 μm). In addition, modification of Teflon's surface properties is difficult. A systematic study of the applicability of Teflon capillaries for the CZE and CIEF analysis of compounds ranging from small molecular weight (e.g. dyes and peptides) to large molecular weight (proteins) was recently undertaken in our laboratory. The Tachophor 2127 capillary analyzer (LKB, Bromma, Sweden), featuring both on-line conductivity and absorbance detection at the end of a 0.5 mm ID PTFE column was used for CZE[26,31,34] and a modification thereof for CIEF[45]. The impact of the addition of polymers, such as PVA, LPA, PEG and HPMC, and solubilizing agents, such as urea, CHAPSO and SDS, to the buffers was also explored. Some of the additives considered in this work severely hamper UV absorbance detection, as is shown with the spectra in Fig. 17-5.

Figure 17-6 represents CZE pherograms of an equimolar dye mixture composed of amaranth red (1) and fluorescein (2). The carrier buffer employed for this study was 10 mM HCl and Bistris (pH 6.02), selected for its low UV absorption, thereby reducing background. In this buffer system, the chloride ion has a higher mobility than any of the sample constituents. This produces a sharp front boundary. As can be seen in panel A, highly asymmetric sample zones are obtained with 5.0 nmol samples in both the UV and conductivity signals. Lesser amounts of sample, e.g. 0.50 nmol (B), yield much more symmetrical peaks. In panel C, corresponding to 0.050 nmol sample, the UV signal is almost completely symmetrical. With this amount of sample, the electric field change has become unmeasurable with the Tachophor detector. Here the carrier electrolyte determines the physical properties such as conductivity and pH within the column. Under such conditions diffusional dispersion is primarily responsible for the zone shape. Figure 17-7 depicts the results obtained with a 5-component model mixture comprising amaranth red (1), acetate (2), bromphenol blue (3), glutamate (4) and fluorescein (5). All five components can be monitored via their conductivity change if enough sample is present (A,B), whereas the three dyes are readily detected via their UV absorbance in all cases (A–C). The conductivity detector was used at its highest sensitivity, where a conductivity change of the order of 1 μS/cm can be resolved.

Another important observation supporting the impact of sample-carrier buffer interaction is the retention time behaviour. Table 17-4 shows the experimental data for the single component sample, fluorescein, in the carrier buffer 10 mM HCl/Bistris (pH 6.0) without fluid stabilizing agent and in the presence of 1 % HPMC. The effect of 1 % HPMC on dispersion of fluorescein is minimal. The most apparent difference appears to be a reduction in the magnitude of the UV response of the dye when the polymer is added. All retention times were taken from the resulting UV signals at 254 nm.

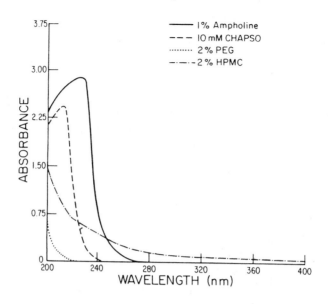

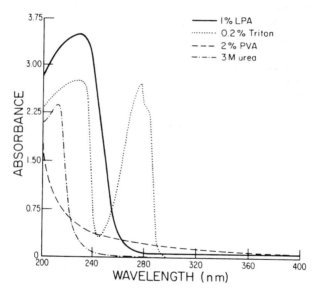

Figure 17-5. Absorption spectra of additives for free fluid electrophoresis. Solutions of 10 mM SDS and 3 mM urea did not exhibit any noticeable absorption within the given scale and wavelength range.

This data clearly indicates that the retention time is affected by the amount of sample injected. Specifically, with larger amounts of sample, the retention times decrease significantly, which is indicative of the presence of conductivity changes across the migrating sample. For very dilute samples the retention time essentially becomes invariant. From the UV tracings collected the plate height was calculated to decrease

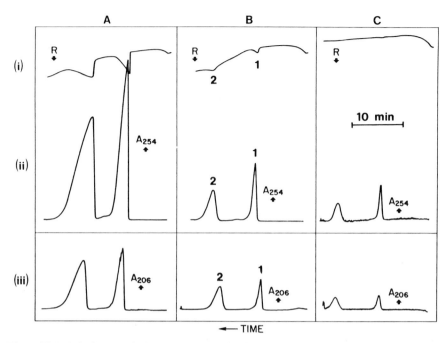

Figure 17-6. Anionic CZE of a 2 component sample using the Tachophor with a capillary of 22.2 cm length. The applied current was a constant 200 μA and the carrier buffer was 10 mM HCl, Bistris (pH 6.01). The injected sample volume was 0.5 μl in all cases. Detection of amaranth red (1) and fluorescein (2) occurred via their conductivity change expressed by an increase in resistance **R** (i), as well as by absorbance at 254 nm (ii) and at 206 nm (iii). Sampled amount of each component: 5 nmol (A), 0.5 nmol (B) and 0.05 nmol (C).

Table 17-4. Retention characteristics of fluorescein in 10 mM HCl/Bistris at pH 6.0 as detected at 254 nm*

		without fluid stabilizing agent				with 1% HPMC	
sample concentration [mM]	sample amount [pmol]	time** [cm]	$W_{1/2}$** [cm]	N***	H*** [μm]	N***	H*** [μm]
10.0	5000	6.95	1.65	98	2700	154	1600
6.0	3000	7.60	1.40	163	1610	288	870
3.0	1500	8.30	1.00	382	688	696	360
1.0	500	8.45	0.85	548	480	2106	120
0.5	250	8.70	0.79	672	391	4351	57
0.25	125	8.90	0.80	686	383	****	****
0.050	25	9.00	0.69	941	279	****	****
0.025	12.5	9.00	0.55	1481	178	****	****

* Experimental parameters: injected sample volume: 0.5 μl; current: 200 μA; temperature: 19°C; distance (L) between location of sample injection and UV monitor: 26.3 cm.
** expressed as chart paper length (1 cm corresponds to 2 minutes).
*** The number of theoretical plates (N) was evaluated using $N = 5.53 \, (t/W_{1/2})^2$ where $W_{1/2}$ is width at half height and t is the retention time; the plate height (H) equals L/N where L is the capillary length.
**** beyond detection limit because of HPMC background absorption.

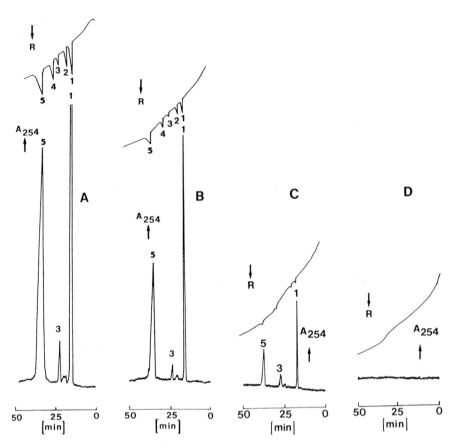

Figure 17-7. Anionic CZE of 5 component model mixture using the Tachophor with a capillary of 44.4 cm length and 0.5 mm ID. The applied current was a constant 150 μA and the carrier electrolyte was 10 mM HCl, 20 mM histidine (pH 6.1; conductivity 55.9 mS/cm). The injected sample volumes (sample quantities) in A to D were 2 μl (2 nmol each), 0.5 μl (500 pmol each), 0.1 μl (100 pmol each) and none respectively. Tracings of the conductivity (R) and absorbance measurements at 254 nm are shown.

from about 2 mm to 100 μm for the 0.5 mm ID Teflon capillary and the 200 μA current employed in this study. Very similar results were obtained for the cationic analysis of L-His-L-Phe in a 5 mM HCl/beta-alanine buffer (pH = 3.94)[34]. The results presented here clearly show that the plate theory constitutes a limiting case which is asymptotically approached through dilution of the sample. As a rule of thumb, and in agreement with the findings of (i) Mikkers et al.[13] using an approximate model and (ii) Thormann et al.[26] using computer simulation, a concentration ratio between sample and carrier component with the same sign < 0.01 needs to be present in order to achieve a relative migration of the solute which does not vary appreciably along the column.

The anionic separation of Hb and A1b in 10 mM NaH_2PO_4 and Tris buffer (pH 8.6) are presented in Fig. 17-8. The distributions of these proteins, having widely spread pIs, are rather broad without the presence of any additives. Addition of 1 % HPMC to the buffer (Fig. 17-8B) produces somewhat sharper peaks, the impact of which is not only

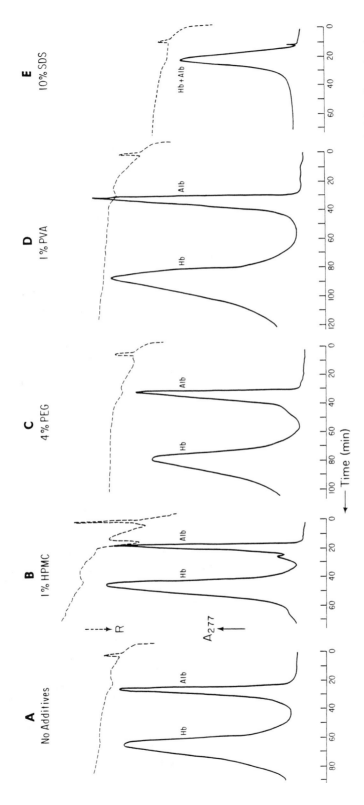

Figure 17-8. Anionic CZE separation of Hb and Alb in 10 mM NaH$_2$PO$_4$ and Tris buffer at pH 8.6. The applied current was a constant 300 μA. The injected sample volume was 0.5 μl with protein quantities of 14 μg (Hb) and 18 μg (Alb). A capillary of 25 cm length was used. Additives are as indicated.

483

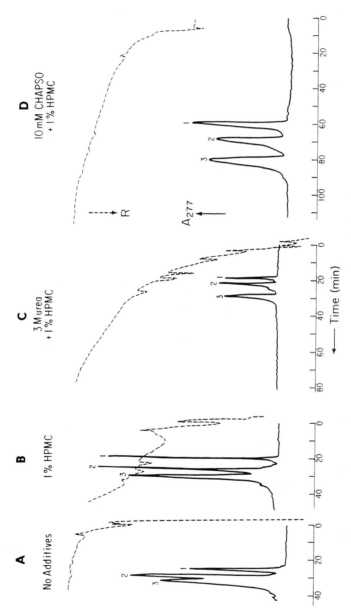

Figure 17-9. Cationic CZE separation of cytochrome C (1), carbonic anhydrase (2) and ribonclease A (3) in 10 mM HCl and Tris buffer at pH 3.50. The applied current was a constant 300 μA. The injected sample volume was 0.5 μl with protein quantities of 1 μg (1), 2.5 μg (2) and 6.5 μg (3). A capillary of 25 cm length was used. Additives are as indicated.

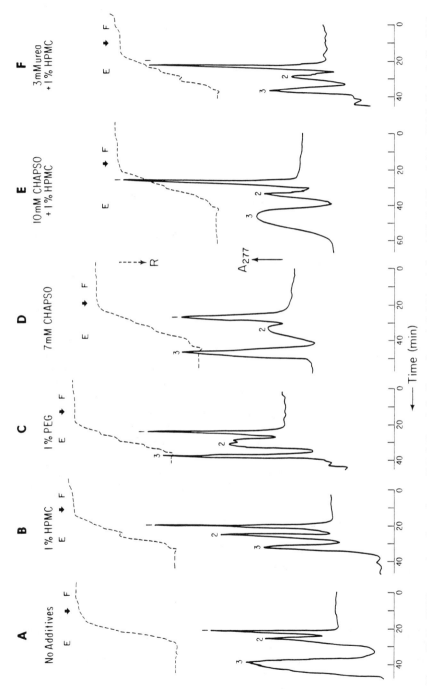

Figure 17-10. CIEF separation of cytochrome C (1, 0.18 mg/ml), ribonuclease A (2, 0.36 mg/ml) and Hb (3, 0.08 mg/ml) in pH 3.5–10 Ampholine (1 %) with 8 mM arginine. A 10 cm column was used together with 100 mM electrode solutions. Focusing (*F*) was performed at a constant voltage of 2000 V and elution (*E*) occurred at a constant current of 100 μA. Absorption measurements were made at 277 nm. Additives are as indicated.

observed in the absorption peaks but also illustrated in the enhanced conductivity signal. The presence of 4 % PEG (Fig. 17-8C) and 1 % PVA (Fig. 17-8D), as well as the combined use of 1 % HPMC with 10 mM CHAPSO or 3 M urea (data not shown), did not improve the result. The addition of 10 mM SDS was found to completely inhibit the separation of Hb and A1b (Fig. 17-8E). This is because the SDS protein complexes acquire almost the same zeta potential (and therefore equal mobilities) and separation can only be achieved in a sieving matrix. The plate heights measured for the examples presented are 0.2 to 0.3 cm, whereas 10 fold lower values were obtained with a 10 mM HCl/ammediol buffer at pH = 9.5. Examples of the cationic CZE separation of equine cytochrome C (1), bovine carbonic anhydrase (2) and bovine ribonuclease A (3) in 10 mM HCl and Tris buffer at pH 3.50 are shown in Fig. 17-9. Without the addition of a fluid stabilizing agent no complete separation of the three proteins is achieved with a column length of 25 cm (Fig. 17-9A), whereas the separation is markedly improved in presence of 1 % HPMC (Fig. 17-9B). Full separation was also obtained in systems containing 1 % HPMC and either 3M urea (Fig. 17-9C) or 10 mM CHAPSO (Fig. 17-9D). It is interesting to note that the cationic protein peaks are sharper than those in the anionic pherograms presented in Fig. 17-8. A possible explanation for this improvement could be reduced electro-osmosis at the lower pH.

Figure 17-10 depicts pherograms obtained for the CIEF separation of equine cytochrome C (1, 0.18 mg/ml), bovine ribonuclease A (2, 0.36 mg/ml) and canine Hb (3, 0.08 mg/ml) in pH 3.5-10 Ampholine (1 %) with 8 mM arginine. A 10 cm column was used together with conditions outlined elsewhere[34]. The use of 1 % HPMC was found to produce the best results for these proteins. This method is limited to amphoteric constituents which can be detected at wavelengths > 250 nm due to background absorption of the carrier ampholytes (see Fig. 17-5).

Capillary instruments featuring narrow bore Teflon tubes of 0.2 to 0.8 mm ID and on-line detection at the capillary end were originally constructed for ITP. A large number of ITP applications demonstrate the high resolution and speed obtained[8–12]. Such instruments can also be used for ZE and IEF. The performance in these modes, however, is characterized by a rather low efficiency, long analysis times and, for proteins, by the need of a viscosity enhancing agent as is shown with the examples given in this chapter. In our laboratory the Tachophor 2127 is used as a research tool for a number of endeavours such as, (i) the purity control of synthetic peptides[34], (ii) the investigation of buffer systems for preparative electrophoretic separations[64], and (iii) the validation of theory[65].

17.4   Electrical Field Flow Fractionation and Electropolarization Chromatography

The conceptual basis of EFFF in a ribbonlike channel of high aspect ratio is shown in Fig. 17-2B. The applied electric field forces the charged species towards one membrane which has to be impermeable to the solutes of interest. EFFF is an electrically induced, capillary-type separation method which is different from conventional CZE. The method demands low voltages and thereby avoids adverse thermal effects. In its nature as an elution technique EFFF constitutes a cascading process which leads to an enhancement of the number of resolvable components. EFFF separations are dependent on pH, ionic strength and temperature of the carrier electrolyte very much in the same way as in CZE. Electropolarization chromatography (EPC) as developed by Lightfoot et al.[50-52] is similar in concept to EFFF. The essence of the EPC procedure is to introduce a small mass of solutes containing macromolecules (proteins) into the buffer stream flowing through a hollow untrafiltration fibre in the presence of a transverse electric field. The fibre acts as a capillary (typical ID is 0.5 mm) with a semipermeable wall, which has to be impermeable to the solutes of interest but freely permeable to small electrolyte species necessary for the maintenance of the electric field inside the fibre lumen. Differential retardation is achieved by accumulation differences (differential polarization) in the slow moving fluid near the fibre wall, hence the relation to EFFF. EPC is characterized by total retention of proteins above a specified electric field strength, a phenomenon which is called electroretention[51,52].

The overall setup of these techniques is schematically presented in Fig. 17-11. The relation to an HPLC system in which the chromatographic column is replaced by the EFFF trough is obvious. An electrolyte pump providing the buffer flow and a flow through detector are mounted at the two ends of the separation capillary. A computer for the control of the pump and the power supply, as well as for data acquisition, data storage and data handling, permits versatile programming in time of the entire process. Detection of eluted solutes occurs by measuring of specific physical properties such as optical and electrochemical responses using flow through cells as in liquid chromatography and flow injection analysis.

In EFFF, developed by Giddings et al.[40-48], the separation channel consists of two parallel membranes which separate the flowing stream from the electrode compartments (Fig. 17-2B). Chamber thicknesses of 0.25 to 0.35 mm having channel widths of 15-25 mm were used with a length of about 50 cm. Reasonable protein separations could be achieved in flexible channels which were fabricated with regenerated dialyzer tubing having a molcular weight cut off of 12000[47] for which the theoretical predications could explain the principal trends. In a second approach a stiff channel assembly was constructed, representing a significant advance because of its capability of yielding predictable retention[48]. Flow rates of about 5 ml/h with electric fields of about 2 to 20 V/cm could be admitted. Analysis times of several hours were typical with this kind of equipment. Alternatively, Davis et al. designed an apparatus with an annular channel

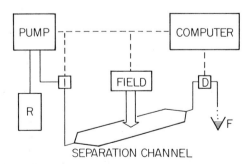

Figure 17-11. Schematic representation of the setup for FFF. R = buffer reservoir; I = sample inlet system; D = flow through detector; F = fraction collector.

utilizing concentric porous Vycor glass tubes of various diameters[49]. For this device, experimental retention ratios are reported to depart considerably from predictions based on theoretical EFFF. Interactions between solutes (macromolecules) and the channel walls in all approaches, as well as electrophoretic effects occurring in the proximity of the membrane walls, are disturbing factors responsible for many observed deviations from theoretical predictions. Efforts are currently being undertaken to develop electrical hyperlayer (focusing) FFF (EHFFF), a FFF method in which contact between most solutes and the channel walls is avoided[66–68]. Separation in EHFFF is based upon partitioning by the concomitant presence of a pH and an electrical potential gradient representing the same gradient combination as in IEF. Solute layers having different depths from the walls are thereby established.

In EPC the separation trough consists of an anisotropic ultrafiltration hollow fibre of circular cross section bathed in a circulating buffer solution and subjected to a transverse electric field. The fibres used by Reis et al.[50,51] were 50 cm long, had an ID of about 0.5 mm and had a nominal molecular weight cut off of 5000. The fibres were isolated from the driving electrodes by a dialysis bag. Carrier flow rates of about 12 μl/min were admitted together with field strengths of 5–10 V/cm in order to achieve separations without electroretention in about 2 hours. At these low voltages retardation was found to be quite consistent with the theoretical prediction. Higher field strengths, however, were shown to immobilize macromolecules in the column. The totally retained components at the hollow fibre's wall could be eluted only by decreasing the applied voltage for the case of reversible sorption. By programming in time of the electric field, separations according to differences in electroretention values are achieved. This is an analogy to the chromatofocusing principle[69]. For a great number of biopolymers however, electroretention has to be avoided because of sorption non-reversibility, causing a substantial loss of material.

17.5   Analytical Continuous Flow Electrophoresis

A small stream of sample is admitted into the laminar flow of the buffer except for conventional IEF where the sample can be evenly mixed with the carrier ampholytes prior to electrophoresis. For all techniques the liquid flow, and to a smaller extent also the applied electric field strength, define the transit time of the sample. Resolution is given in terms of differences in mobility or pI values respectively, sample quantity and composition, current applied, flow velocity, effectiveness of cooling, length of electrophoresis chamber and width of the flowing fluid film. If samples are recovered resolution is dependent on the density of outlet tubes of the separation chamber. For analytical purposes it usually suffices to monitor the solutes directly through a window at the end of the separation chamber with an optical scanning system. Good reproducibility and very short analysis times are achieved. Further assays can be performed on the fractions collected. Online detection also provides the means for feedback control and automation.

Two systems are currently available commercially, the analytical continuous flow system of Hannig[53] (ACE 710, Hirschmann, Unterhaching, FGR) and the CFE apparatus for analytical and preparative separations of Wagner et al.[54] (Elphor VaP 22, Bender & Hobein, Munich, FRG). The separation cells of these instruments are rectangular chambers in which laminar flow conditions are maintained by a small gap

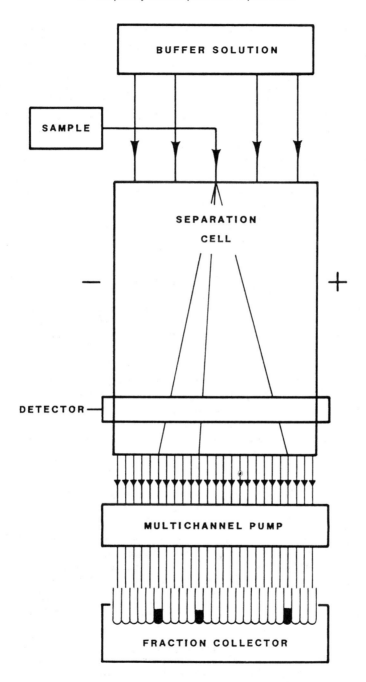

Figure 17-12. Schematic representation of the setup for CFE.

between two parallel plates. A plastic or glass front plate is held apart from the cooled (thermostated) back plate by a thin space of typically 0.5 mm thickness. A temperature controlled cooling system carries off the heat produced by the electrophoretic process. For analytical purposes separation chambers of 120 × 30 mm appear to be sufficient (ACE 710), whereas for the exploitation of the preparative capabilities of such a system chamber dimensions of 500 × 200 (Elphor VaP 22) are suggested. Typical running conditions include field strengths up to 300 V/cm. The sample transit times range from 20 to 3600 seconds. The essential parts of a CFE apparatus are depicted in Fig. 17-12. Separation in the Hirschmann machine is recorded by a video-camera with a Vidicon measuring system and the detection principle incorporated into the Bender & Hobein apparatus consists of a scanning light beam together with appropriate interference filters and photo sensors, allowing the monitoring of light scattering, absorbance or fluorescence.

17.6 Discussion and Future Prospects

Twenty years after the inception of conventional CE a considerable amount of progress has been made. The increased interest in this area during the last five years has led to the design of new capillary instruments using very narrow fused silica capillaries. The first commercial instruments based on that technology are now being introduced. At the same time, CFE has been improved and two instruments are currently being marketed. The FFF methods (i.e. EFFF, EPC and EHFFF), on the other hand, represent interesting alternatives which have scarcely been explored.

It is now apparent that CE can be applied successfully to any analytical problem currently solved by ion chromatography. In fact the application range of CE is considerably broader, since it can be used for the challenging analyses of peptides, proteins and nucleic acids. Quality control in production and research, as well as applications in clinical, pharmacological, food and environmental laboratories will greatly benefit from this technology.

17.7 Ackowledgements

Many thanks are due to those friends and colleages who have contributed to the work described and who have offered advice, criticism and discussion on the topic of capillary electrophoresis during the past 10 years. The generous donation of the Tachophor 2127 by its manufacturer, LKB, Bromma, Sweden is gratefully acknowledged. This work was financially supported by NASA grant NAGW-693.

17.8 References

1. W. Thormann, R. A. Mosher, M. Bier, In *Chemical Separations, Vol 1: Principles*; C. J. King, J. D. Navratil, eds: Litarvan Literature, Denver, CO, 1986, pp. 153–168.
2. S. Hjertén, K. Elenbring, F. Kilar, J. Liao, A. J. C. Chen, C. J. Siebert, M. Zhu, *J. Chromatogr. 403*, 47–61 (1987).
3. W. Thormann, D. Arn, E. Schumacher, *Electrophoresis 5*, 323–337 (1984).
4. S. W. Compton, R. G. Brownlee, *Biotechniques 6(5)*, 432–440 (1988).
5. J. W. Jorgenson, D. J. Rose, R. T. Kennedy, *Am. Lab.* April, 32–41 (1988).
6. B. P. Konstantinov, O. V. Oshurkova, *Sov. Phys. Tech. Phys. 11*, 693–704 (1966).

7. W. Preetz, *Naturwissenchaften 54*, 85-87 (1967).
8. F. M. Everaerts, J. L. Beckers, Th. P. E. M. Verheggan, *Isotachophoresis*, Elsevier, Amsterdam, 1976.
9. S. Hjalmarsson, A. Baldesten, *CRC Crit. Rev. Anal. Chem. 11*, 261-352 (1981).
10. D. Kaniansky, P. Havasi, *Trends Anal. Chem. 2*, 197-202 (1983).
11. P. Boček, *Top. Curr. Chem. 95*, 131-178 (1981).
12. J. Akiyama, T. J. Mizuno, *Chromatogr. 119*, 605 (1976).
13. F. E. P. Mikkers, F. M. Everaerts, Th. P. E. M. Verheggan, *J. Chromatogr. 169*, 1-20 (1979).
14. J. W. Jorgenson, K. D. Lukacs, *Anal. Chem. 53*, 1298-1302 (1981).
15. K. D. Lukacs, J. W. Jorgenson, *J. High Res. Chromatogr. Chromatogr, Commun. 8*, 407-411 (1985).
16. D. L. Rose, J. W. Jorgenson, *Anal. Chem. 60*, 642-648 (1988).
17. D. L. Rose, J. W. Jorgenson, *J. Chromatogr. 438*, 23-34 (1988).
18. K. Otsuka, S. Terabe, T. Andro, *J. Chromatogr. 348*, 39-47 (1985).
19. T. Tsuda, G. Nakagawa, M. Sato, K. Yagi, *J. Applied Biochem. 5*, 330-336 (1983).
20. T. Tsuda, K. Nomura, G. Nakagawa, *J. Chromatogr. 264*, 385-392 (1983).
21. A. S. Cohen, S. Terabe, J. A. Smith, B. L. Karger, *Annal. Chem. 59*, 1021-1027 (1987).
22. A. S. Cohen, B. L. Karger, *J. Chromatogr. 397*, 409-417 (1987).
23. H. H. Lauer, D. McManigill, *Anal. Chem. 58*, 166-170 (1986).
24. S. Fujiwara, S. Honda, *Anal. Chem. 58*, 1811-1814 (1986).
25. E. Gassmann, J. E. Kuo, R. N. Zare, *Science 230*, 813-814 (1985).
26. W. Thormann, J. P. Michaud, R. A. Mosher, In *Electrophoresis '86*, M. Dunn, ed.; VCH Verlagsgesellschaft, Weinheim, 1986, pp. 267-270.
27. F. M. Everaerts, W. M. L. Hoving-Keulemans, *Science Tools 17*, 25-28 (1970).
28. S. Hjertén, M. Zhu, *J. Chromatogr. 327*, 157-164 (1985).
29. S. Hjertén, *J. Chromatogr. 270*, 1-6 (1983).
30. S. Hjertén, *J. Chromatogr. 347*, 191-198 (1985).
31. A. Zhu, J. Liu, *Academia Sinica*, P. R. Beijing, China.
32. F. Foret, M. Deml, V. Kahle, P. Boček, *Electrophoresis 7*, 430-432 (1986).
33. M. Deml, F. Foret, P. Boček, *J. Chromatogr. 320*, 159-165 (1985).
34. M. A. Firestone, J. P. Michaud, R. Carter, W. J. Thormann, *J. Chromatogr. 407*, 363-368 (1987).
35. X. Huang, T. J. Pang, M. J. Gordon, R. N. Zare, *Anal. Chem. 59*, 2747-2749 (1987).
36. R. A. Wallingford, A. G. Ewing, *Anal. Chem. 59*, 1762-1766 (1987).
37. J. A. Olivares, N. T. Nguyen, C. R. Yonker, R. D. Smith, *Anal. Chem. 59*, 1232-1236 (1987).
38. W. Thormann, R. A. Mosher, M. Bier, In *Electrophoresis '84*, V. Neuhoff, ed.; Verlag Chemie, Weinheim, 1984, pp. 118-121.
39. W. Thormann, R. A. Mosher, M. Bier, *J. Chromatogr. 351*, 17-29 (1986).
40. R. A. Mosher, W. Thormann, M. Bier, *J. Chromatogr. 436*, 191-204 (1988).
41. W. Thormann, A. Tsai, J. Michaud, R. A. Mosher, M. Bier, *J. Chromatogr. 389*, 75-86 (1987).
42. S. Hjertén, M. Zhu, *J. Chromatogr. 346*, 265-270 (1985).
43. S. Hjertén, F. Kilar, J. Liao, M. Zhu, In *Electrophoresis '86*, M. Dunn, ed.; VCH Verlagsgesellschaft, Weinheim, 1986, pp. 451-461.
44. S. Hjertén, J. Liao, K. Yao, *J. Chromatogr. 387*, 127-138 (1987).
45. M. A. Firestone, W. Thormann, *J. Chromatogr. 436*, 309-315 (1988).
46. K. D. Caldwell, L. F. Kesner, M. N. Myers, J. C. Giddings, *Science 176*, 296-299 (1972).
47. L. F. Kesner, K. D. Caldwell, M. N. Myers, J. C. Giddings, *Anal. Chem. 48*, 1834-1839 (1976).
48. J. C. Giddings, G. Lin, M. N. Myers, *Sep. Sci. 11*, 553-568 (1976).
49. J. M. Davis, F. R. F. Fan, A. J. Bard, *Anal. Chem. 59*, 1339-1348 (1987).
50. J. F. G. Reis, E. N. Lightfoot, *AIChE J. 22*, 779-785 (1976).
51. A. S. Chiang, E. H. Kmiotek, S. M. Langan, P. T. Nobile, J. F. G. Reis, E. N. Lightfoot, *Sep. Sci. Technol. 14*, 453-474 (1979).
52. A. B. Shah, J. F. G. Reis, E. N. Lightfoot, R. E. Moore, *Sep. Sci. Technol. 14*, 475-497 (1979).
53. K. Hannig, *Electrophoresis 3*, 235-243 (1982).
54. H. Wagner, R. Kessler, *GIT Lab.-Med. 7*, 30-35 (1984).
55. P. A. David, P. J. Pellechia, D. L. Manning, M. P. Maskarinec, *ORNL/TM-9141 report*, 1984, (CA 101(14):12230ly).
56. S. Fujiwara, S. Honda, *Anal. Chem. 59*, 487-490 (1987).
57. A. Guttman, A. Paulus, A. S. Cohen, N. Grinberg, B. L. Karger, *J. Chromatogr. 448*, 41-53 (1988).
58. Y. Walbroehl, J. W. Jorgensen, *Anal. Chem. 58*, 479-481 (1986).

59. N. Catsimpoolas, *Electrophoresis 1*, 73–78 (1980).
60. S. Hjertén, *J. Chromatogr. 347*, 191–198 (1985).
61. S. Koboyashi, T. Shiogai, J. Akiyama, In *Analytical Isotachophoresis*, F. M. Everaerts, ed.; Elsevier, Amsterdam, 1981, pp. 47–53.
62. L. Arlinger, In *Electrophoresis, a Survey of Techniques and Applications, Part A: Techniques*, Z. Deyl, ed.; Elsevier, 1979, pp. 363–377.
63. W. Thormann, D. Arn, E. Schumacher, *Electrophoresis 6*, 10–18 (1985).
64. J. E. Sloan, R. A. Mosher, W. Thormann, M. A. Firestone, M. Bier, In *Protein Purification: Micro to Macro, UCLA Symposia on Molecular and Cellular Biology, New Series, Vol. 68*, R. Burgess, ed.; A. R. Liss, New York, NY, 1987, pp. 329–335.
65. W. Thormann, R. A. Mosher, In *Electrophoresis '88*, C. Schafer-Nielsen, ed., VCH Verlagsgesellschaft, Weinheim, 1988, pp. 121–140.
66. W. Thormann, M. A. Firestone, M. L. Dietz, T. Cecconie, R. A. Mosher, *J. Chromatogr. 461*, 95–101 (1989).
67. J. C. Giddings, *Sep. Sci. Technol. 18*, 765–773 (1983).
68. M. R. Schure, K. D. Caldwell, J. C. Giddings, *Anal. Chem. 58*, 1509–1516 (1986).
69. L. A. AE. Sluyterman, O. Elgersma, *J. Chromatogr. 150*, 17–30 (1978).

INDEX